HfO_2-Based Ferroelectric Materials

HfO_2-Based Ferroelectric Materials

Fabrication, Characterization and Device Applications

Edited by
Xubing Lu

WILEY VCH

Editor

Xubing Lu
South China Normal University
Guangzhou, Guangdong, China

Cover Design: Wiley
Cover Image: © FIRMAN LESTARI/Shutterstock

Typesetting: Lumina Datamatics, Madurai, India

Library of Congress Card No.: Applied for.

British Library Cataloguing-in-Publication Data A catalogue record for this book is available from the British Library.

Bibliographic information published by the Deutsche Nationalbibliothek The Deutsche Nationalbibliothek lists this publication in the Deutsche Nationalbibliografie; detailed bibliographic data are available on the Internet at <http://dnb.d-nb.de>.

Print ISBN: 9783527353187
ePDF ISBN: 9783527844883
ePub ISBN: 9783527844876
oBook ISBN: 9783527844869

Contents

Foreword

Research on ferroelectric materials has been continuously and extensively conducted since the discovery of ferroelectricity in Rochelle salt in the 1920s, particularly with regard to applications in nonvolatile information storage. Although the commercialization of ferroelectric memories, such as 1T1C-type FeRAMs based on traditional perovskite ferroelectric materials like $Pb(Zr,Ti)O_3$, has been achieved, their widespread industrial adoption has remained limited due to challenges such as incompatibility with standard CMOS processes and difficulties in large-scale integration.

In 2011, Böscke and his colleagues in Germany reported the discovery of ferroelectricity in silicon-doped HfO_2. Owing to its complete compatibility with CMOS processes and the persistence of ferroelectric properties at the nanoscale, HfO_2-based ferroelectric materials rapidly emerged as a major research focus in both academia and the semiconductor industry. As a researcher with extensive experience in the field of ferroelectric materials and ferroelectric memories, I firmly believe that the publication of this book, *HfO_2-Based Ferroelectric Materials: Fabrication, Characterization, and Device Applications*, will have a profound impact on future research and the industrial applications of HfO_2-based ferroelectric materials.

I am well acquainted with Professor Xubing Lu, who has an extensive background in the study of ferroelectric materials and their device applications. During his tenure at the Tokyo Institute of Technology and the National Institute of Advanced Industrial Science and Technology (AIST) in Japan, he conducted research on high-reliability ferroelectric memory devices and collaborated with Fujitsu Semiconductor, a leading manufacturer of ferroelectric memory. His deep understanding and insightful perspectives on ferroelectric materials and their device applications are truly impressive. Since returning to China, he and his research team have carried out a series of systematic and in-depth studies on the fabrication, characterization, and device applications of HfO_2-based ferroelectric materials, achieving remarkable research results in this field.

In this book, the authors begin with a systematic exploration of the origins of ferroelectricity in HfO_2, the methods for modulating its ferroelectric properties, and the underlying mechanisms governing these characteristics. They then provide an in-depth discussion of the fundamental device structures, operating principles, major recent advancements, existing challenges, and future development trends of

HfO_2-based non-volatile memories, including 1T1C and 1T architectures, as well as their applications in energy-efficient computing and neuromorphic devices. In addition, the final sections of the book address the use of HfO_2-based ferroelectric materials in high-performance energy storage and piezoelectric sensing applications.

The authors of the various chapters include distinguished scholars from leading universities in this research field, as well as researchers from renowned semiconductor companies involved in the development of ferroelectric memory devices. I believe that this book will not only provide a comprehensive framework of knowledge for students and faculty studying HfO_2-based ferroelectric materials and their device applications, but also serve as a valuable reference for industrial researchers engaged in engineering applications in the future.

I sincerely wish this book great success and hope it reaches a wide readership.

Hiroshi Ishiwara
Professor Emeritus, Tokyo Institute of Technology
(presently, Institute of Science Tokyo)
IEEE and MRS Fellow
Former President of Japan Society of Applied Physics

Preface

Ferroelectric materials represent a unique class of functional dielectrics characterized by the ability to exhibit two distinct polarization states at zero electric field, which can be switched by applying an electric field that exceeds the coercive field. Since the discovery of ferroelectricity in Rochelle salt in the early 1920s, various ferroelectric materials, notably $BaTiO_3$ and $Pb(Zr,Ti)O_3$, have been extensively studied for a wide range of device applications, particularly in nonvolatile memories and piezoelectric transducers. A typical example of such applications is the successful commercial production of ferroelectric random-access memories utilizing $Pb(Zr,Ti)O_3$ films. Despite the significant advancements made in perovskite ferroelectrics to date, many challenges remain for their application in high-performance nonvolatile memories. Traditional perovskite ferroelectric materials are often incompatible with standard CMOS processes and exhibit pronounced size effects on ferroelectricity, which hinder their large-scale commercial viability. In contrast, HfO_2-based ferroelectric materials, first reported by Böscke et al. in 2011, present a promising solution to these challenges, owing to their unique advantages over traditional perovskite ferroelectrics. These advantages include excellent compatibility with CMOS processes, low annealing temperatures, and favorable nanoscale ferroelectric properties. Within just over a decade, hafnium oxide ferroelectric materials have emerged as a research hotspot in both academia and industry, not only for high-performance nonvolatile memory applications but also for areas such as energy-efficient computing, neuromorphic devices, energy storage, and piezoelectric devices.

This book, comprising 14 chapters, offers a comprehensive review of the current research status and future outlook of HfO_2-based ferroelectric materials from various perspectives, including the origins of ferroelectricity, material fabrication and characterization, and potential device applications. The authors of each chapter include experienced researchers and renowned scholars from famous universities and well-known semiconductor companies in the field of ferroelectrics. I hope that this book will provide valuable insights and useful references to readers engaged in the study of HfO_2-based ferroelectric materials and devices.

This book is the result of more than 2 years of dedicated effort in writing, reviewing, and editing. I would like to extend my sincere gratitude to all the contributing authors for their outstanding work, the editors at Wiley for their diligence, and my colleagues and students from South China Normal University for their various forms of assistance. The chapter authors and I would also like to acknowledge the support from various funding agencies.

Xubing Lu
November 2025
Guangzhou, China

Acknowledgments

The editor of this book, Xubing Lu, thanks the support from the National Key R&D Program of China (Grant No. 2024YFA1208602) and the National Natural Science Foundation of China (Grant Nos. 52250281, 62174059).

1

Fundamentals of Ferroelectricity and Ferroelectric Materials

*Yan Zhang[1], Xubing Lu[2], and Jun-ming Liu[3]**

[1] *School of Optoelectronic Engineering, Henan Normal University, Xinxiang, Henan, China*
[2] *Guangdong Provincial Key Laboratory of Optical Information Materials and Technology and Institute for Advanced Materials, South China Academy of Advanced Optoelectronics, South China Normal University, Guangzhou, Guangdong, China*
[3] *Laboratory of Solid State Microstructures and Innovation Center of Advanced Microstructures, Nanjing University, Nanjing, Jiangsu, China*

*Corresponding author: liujm@nju.edu.cn

1.1 Ferroelectrics

Ferroelectrics are a special class of dielectrics whose ferroelectricity arises from the cooperative interaction of neighboring dipoles, exhibiting spontaneous polarization that can be reoriented under the influence of an external electric field. Ferroelectrics share many similarities with ferromagnets (magnetic materials with permanent magnetic moments) in terms of phase transitions and response to external fields. Historically, the term "ferroelectric" was coined by analogy with "ferromagnetic." Figure 1.1a schematically illustrates the unique properties and classification of polarized materials as ferroelectrics. Typically, the interaction between ferroelectric dipoles and the lattice structure directly influences the orientation and strength of the dipoles, i.e., the magnitude of spontaneous polarization. When the dipoles are arranged in a regular pattern within the ferroelectric material, those at the boundaries are unable to continue aligning or interacting, resulting in the accumulation of some uncompensated charges on the material's surface. Mechanical force and temperature are common external stimuli that alter the lattice structure of ferroelectric materials, thereby changing the surface charge distribution. The surface charge generated by an applied external stress is referred to as the piezoelectric effect, while the change in spontaneous polarization in response to temperature variations is known as the pyroelectric effect. Unlike the polarization response induced by stress, the polarization associated with the pyroelectric effect occurs spontaneously and is retained within the structure in the form of permanent dipoles. Ferroelectric polarization is similar to pyroelectric effect-associated

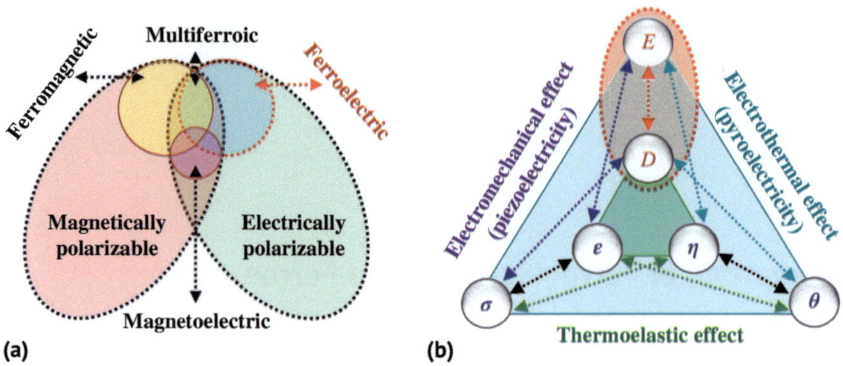

Figure 1.1 (a) Relationship between ferroelectric, magnetic, multiferroic, and magnetoelectric materials. (b) Diagram of the interrelationship between thermal, mechanical, and electrical properties of materials. Here, η denotes the entropy, and θ denotes the temperature.

polarization but differs in that it can be reversed by the application of an electric field. Figure 1.1b illustrates the reversible thermodynamic interactions among mechanical, thermal, and electric fields. The physical effects are associated with primary variables and their related ones, and a change in any one variable induces changes in the others. For piezoelectric materials, mechanical stress induces internal dielectric displacement, while applying an electric field results in material deformation. The interactions between electrical and mechanical variables define the converse and direct piezoelectric effects, as represented by the linear constitutive Equations (1.1) and (1.2). Here, σ_{ij}, D_i, ε_{ij}, and E_k represent mechanical stress, electric displacement, strain, and electric field, respectively, while e_{ijk}, C_{ijkl}, and x_{ij} denote the piezoelectric coefficients, elastic constants, and dielectric permittivity under constant strain, respectively:

$$\sigma_{ij} = C_{ijkl}\,\varepsilon_{kl} - e_{ijk}\,E_k \tag{1.1}$$

$$D_i = e_{ijk}\,\varepsilon_{jk} + x_{ij}\,E_j \tag{1.2}$$

Polarization switching is a fundamental feature of ferroelectric materials, where the domains can be reoriented to different polarization states under the application of an electric field or mechanical stress. The polarization behavior of ferroelectric materials is characterized by measuring their polarization response to an electric field (i.e., dielectric displacement), which reflects both the spontaneous polarization and its reorientation under the influence of the electric field. In uniaxial ferroelectrics, the polarization direction can be switched within a 180° range under the application of an electric field. In biaxial ferroelectrics, both the electric field and high stress can alter polarization, with the switching angle depending on the symmetry of the crystal. For ferroelectric materials, the mechanical strain and dielectric displacement responses can be divided into a linear component and a residual component (remanent polarization), with their relationships expressed by Equations (1.3) and (1.4):

$$\varepsilon_{ij} = \varepsilon_{ij}^{\mathrm{L}} + \varepsilon_{ij}^{\mathrm{R}} \tag{1.3}$$

$$D_i = D_i^{\mathrm{L}} + D_i^{\mathrm{R}} \tag{1.4}$$

$$\sigma_{ij} = C_{ijkl}(\varepsilon_{kl} - \varepsilon_{kl}^{R}) - e_{ijk}E_k \tag{1.5}$$

$$D_i = e_{ijk}\left(\varepsilon_{jk} - \varepsilon_{jk}^{R}\right) + x_{ij}E_j + P_i^{R} \tag{1.6}$$

Ferroelectric materials are typically described using linear constitutive equations for their piezoelectric properties, but this approach overlooks the inherent nonlinear characteristics of ferroelectrics. As shown in Equations (1.5) and (1.6), the polarization response of ferroelectric materials under an electric field exhibits pronounced nonlinear behavior, highlighting that ferroelectrics generally exhibit superior piezoelectric performance compared to non-ferroelectric materials. Specifically, the nonlinear effects of polarization switching are typically manifested as a characteristic butterfly-shaped hysteresis loop.

The emergence of spontaneous polarization in a crystal establishes a unique polar direction that is asymmetric with respect to all other directions in the crystal. The entire crystal exhibits polarity along this direction, with one end being positively polarized and the other negatively polarized. A unit cell of a ferroelectric crystal can be viewed as an equivalent dipole, characterized by a permanent dipole moment, with one end having a layer of positive bound charge and the other having a layer of negative bound charge. The electric field generated by the bound charge, which opposes polarization within the crystal, is known as the depolarization field. Under mechanical constraints, the strain induced by spontaneous polarization increases the strain energy, making a uniformly polarized state unstable. This instability drives the formation of multiple small regions, or domains, each containing electric dipoles aligned in the same direction, but with varying orientations across domains. The boundaries between these domains are called domain walls. The formation of domains minimizes the crystal's electrostatic and strain energies, while the presence of domain walls introduces additional domain wall energy. The stable configuration of electric domains is determined by the condition that minimizes the total free energy, which is the result of a balance between electrostatic, strain, and domain wall energies.

The hysteresis loop reflects the movement of domain walls and changes in the domain structure of ferroelectric materials, representing the evolution of regions with aligned spontaneous polarization, as shown in Figure 1.2. The characteristics

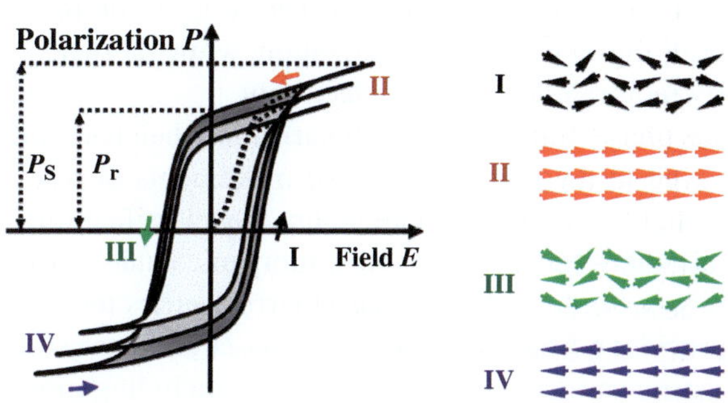

Figure 1.2 Arrangement of electric domains and corresponding polarization hysteresis loops, the arrowheads denote assigned electric domains.

of the hysteresis loop are typically described by parameters such as saturation polarization (P_s), remnant polarization (P_r), and coercive field (E_c). Under weak electric fields, ferroelectric polarization depends linearly on the electric field, where reversible domain wall motion dominates. The hysteresis loop becomes nonlinear when the applied electric field exceeds E_c, accompanied by irreversible nucleation of new domains, domain wall motion, and rapid reorientation and alignment of polarization. The contribution of induced polarization from the enhanced electric field further increases the total polarization until saturation is reached. The crystal retains a macroscopic polarized state even after the electric field is removed. Polarization weakens and reverses direction when the electric field is reversed, eventually tending toward saturation in the opposite direction. The hysteresis loop in ferroelectric crystals is typically observed within a specific temperature range. When the temperature exceeds a critical threshold, known as the Curie temperature (T_c), the material undergoes a phase transition to a paraelectric state, and spontaneous polarization vanishes.

1.2 Material Structure and Theory

1.2.1 Structural Evolution

The phase transition of ferroelectric material is a special structural one accompanied by atomic-scale symmetry breaking. The paraelectric phase exhibits higher symmetry, whereas the ferroelectric phase has lower symmetry during the transition. Therefore, ferroelectricity is a material property that exists only in crystals lacking an inversion center, placing specific constraints on the symmetry of the point group to which the ferroelectric crystal belongs. According to symmetry requirements, ferroelectric phases are confined to 10 polar point groups within the 21 non-centrosymmetric groups within the 32 crystallographic point groups. The polar point groups include $1(C_1)$, $2(C_2)$, $m(C_s)$, $mm2(C_{2v})$, $3(C_3)$, $3m(C_{3v})$, $4(C_4)$, $4mm(C_{4v})$, $6(C_6)$, and $6mm(C_{6v})$. Furthermore, due to the temperature-dependent nature of atomic configurations, ferroelectric crystals also exhibit pyroelectricity, and universally fall within the category of piezoelectric materials owing to inherent asymmetry-derived characteristics. To summarize, pyroelectricity is a common characteristic of all materials exhibiting spontaneous polarization, while piezoelectricity encompasses the symmetry requirements for ferroelectricity.

The microscopic origin of the phase transition and polarization behavior in ferroelectric materials can be traced to the specific localized movements of electrons, ions, or molecules. Accordingly, ferroelectrics can be broadly classified into two types: displacement type and order–disorder type. Most inorganic oxide ferroelectrics fall into the former category, while molecular-based ferroelectrics tend to belong to the latter. Figure 1.3 provides a brief overview of the research history on ferroelectric materials, encompassing various types of compounds, including inorganic oxides, metal–organic hybrid compounds, organic small molecules, and polymers. In 1920, Valasek[1,2] first reported the polarization hysteresis effect in Rochelle

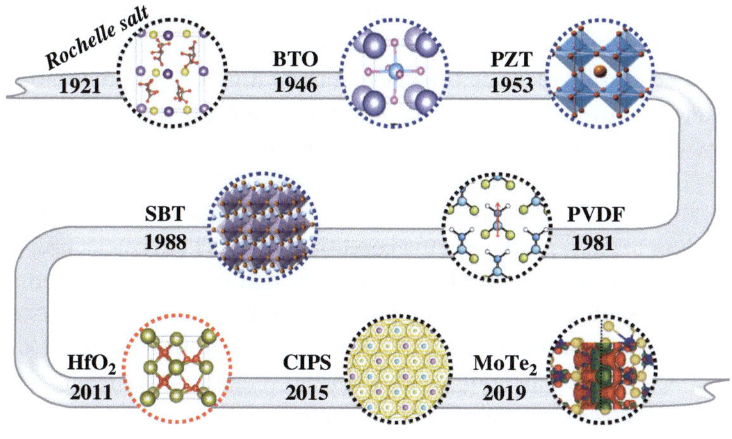

Figure 1.3 The development history of ferroelectrics, with the broken line presenting observations of some classical ferroelectric materials.

salt ($NaKC_4H_4O_6{\cdot}4H_2O$) and described the initial characteristic ferroelectric hysteresis loop relating charge to the applied electric field. The phase transition of Rochelle salt is particularly unique, featuring two transition temperatures. The ferroelectricity is believed to be associated with the movements of hydrogen atoms in hydroxyl groups and the oxygen atoms near sodium ions, characteristic of molecular-based ferroelectric compounds. Research on ferroelectric materials has been further advanced by the discovery of remarkable materials such as $BaTiO_3$ (BTO), $(Pb,Zr)TiO_3$ (PZT), KH_2PO_4, and various organic materials including polyvinylidene fluoride (PVDF).[3–5] Both BTO and PZT possess a typical non-centrosymmetric perovskite structure. As one of the most numerous classes of ferroelectrics, typical ferroelectric perovskites (general formula: ABO_3) also include $KNbO_3$, $KTaO_3$, $LiNbO_3$, $PbTiO_3$ (PTO), $PbMg_{1/3}Nb_{2/3}O_3$, $KNaNbO_3$, $SrBi_2Ta_2O_9$, and $BiFeO_3$, among others.[6,7] The electric field-driven displacement of cations (A^+ and B^+) relative to anions (O^-) in the strong polar bonds of ABO_3-type compounds induces spontaneous polarization, which is characteristic of displacement-type ferroelectrics. Currently, ABO_3 ferroelectrics, recognized as pioneering polycrystalline ceramic materials with ferroelectric properties, have found extensive applications across various fields.

The development of ferroelectric materials in the twenty-first century has taken a new direction, with nanometer-scale ultrathin ferroelectric films garnering significant attention, such as HfO_2, AlScN, and two-dimensional ferroelectrics. Notably, non-ferroelectric HfO_2 materials have been extensively utilized as a high-k gate dielectric layer in Si-based complementary metal–oxide–semiconductor (CMOS) technologies. Since the first report of ferroelectricity in doped HfO_2 with a fluorite crystal structure in 2011, extensive research has focused on novel ferroelectric HfO_2 (FE-HfO_2).[8,9] The FE-HfO_2 exhibits a distinct set of advantages over perovskite ferroelectrics, including robust ferroelectricity down to nanoscale, low-temperature annealing requirements, a high coercive electric field, and exceptional compatibility with contemporary CMOS fabrication processes. Currently, the fabrication techniques for FE-HfO_2 include atomic layer deposition (ALD), pulsed laser deposition, spin coating, and magnetron sputtering. ALD is highly valued for its ability to precisely control nanoscale thickness, ensure uniformity across large areas, achieve excellent three-dimensional conformality, and maintain stoichiometric accuracy, with these advantages made possible by the specific chemical reactions between the film or substrate surface and the precursor molecules. Commonly, ALD-fabricated FE-HfO_2 films exhibit a non-centrosymmetric orthorhombic phase (o-phase; $Pca2_1$) under specific conditions.[10,11] The ferroelectricity in FE-HfO_2 is believed to arise from the displacement of half of the oxygen atoms in the sublattice from the center of symmetry. This polarization, induced by lattice distortion, differs from the polar bond characteristics observed in perovskite ferroelectrics, which is crucial for understanding why FE-HfO_2 is more susceptible to the influence of oxygen vacancies. Accordingly, this chapter will delve into a detailed analysis of the relevant theoretical models and performance characteristics, with a particular focus on the distinctions between the two materials.

1.2.2 Physical Theory

The structural evolutions elucidate the essential symmetries required to support ferroelectricity, a process that can be further understood by examining the local atomic displacements in both the polar and nonpolar phases. Although both piezoelectric and ferroelectric crystals lack inversion symmetry, there is a significant difference in their electric dipole behaviors. In piezoelectric crystals, polarization is a transient phenomenon that responds to external stress and does not necessarily have a distinct polarization axis or a stable electric dipole. In contrast, the key feature of ferroelectric crystals is the presence of reorientable electric dipoles, which distinguishes them from other crystals. For example, as a piezoelectric material, the polarization switching behavior of AlN has certain limitations. When the direction of the applied electric field is opposite to the direction of the dipoles in the material, dielectric breakdown occurs before polarization switching can take place.[12] Currently, there are no definitive crystallographic criteria for identifying ferroelectric materials, but advances in ferroelectric material theory have provided crucial support for understanding ferroelectric structural phase transitions. The changes in the symmetry of ferroelectric systems reflect alterations in the degree of internal order. The Landau theory links the degree of order to symmetry changes, indicating that the order parameter representing spontaneous polarization is related to the changes in symmetry. With suitable modifications, the Ginzburg–Landau–Devonshire (GLD)[13,14] theory can also address certain first-order and second-order phase transitions. The core concept of GLD involves expanding the free energy as a power series in the order parameter (e.g., the polarization magnitude), determining the physical properties of the system from the minima of free energy, and relating the expansion coefficients to measurable parameters for comparison with experimental results.

The soft-mode theory explains the commonality of ferroelectric structural phase transitions from the lattice level, indicating that ferroelectric transitions are simply a special case of general structural phase transitions. Cochran[15] and Anderson,[16] within the framework of lattice dynamics, attributed the origin of ferroelectricity to the softening of a vibrational mode at the center of the Brillouin zone, a concept that is central to the soft-mode theory. When the frequency of the transverse optical (TO) mode softens to zero, atoms can no longer return to their original equilibrium positions, leading to a "freezing" of the vibrations. The freezing of the soft-mode eigenvector leads to a static displacement of atoms, giving rise to spontaneous polarization in the crystal. The frequency of lattice vibrations reflects contributions from both long-range dipole Coulomb interactions and short-range local forces. In the context of TO modes, the long-range dipole Coulomb forces and short-range local interactions cancel each other out. When their magnitudes are equal, the restoring force that drives atoms back to their equilibrium positions becomes zero, effectively freezing the atomic displacement from the equilibrium. In contrast, for longitudinal optical (LO) modes, these two forces work constructively and do not contribute to the ferroelectric phase transition. The soft-mode theory was originally applied to displacement systems, but it was later recognized that its fundamental concepts are also applicable to ordered–disordered systems.[17–19] The collective excitations that soften during the phase transition are pseudospin waves in ordered–disordered systems, which describe the distribution and motion of particles in a double-well potential rather than lattice vibrational modes. The excitations are primarily modeled by the transverse-field Ising model, capturing the dynamics of the particles within the potential.[19] The adjacent spins interact through nearest-neighbor coupling in the model, while the applied transverse field encourages spins to flip in a direction perpendicular to their orientation. The introduction of a transverse field complicates the phase transition characteristics of the system, which can describe the transition from an ordered to a disordered phase and provide support for the analysis of critical phenomena.

The soft-mode theory grounded in the adiabatic principle primarily focuses on the behavior of ions in ferroelectric materials. Electrons are considered in physics as an important medium that facilitates ion movement. The first-principles calculations from this perspective focus on describing the electronic interactions that control ion movement. By treating solids as multiparticle systems composed of electrons and atomic nuclei, ferroelectricity is explained at the level of both electrons and atoms based on the relationship between the system's total energy, electronic structure, and nuclear configuration. The local basis functions are typically used to capture the electronic states in the vicinity of atomic nuclei, while plane-wave expansions are more suitable for describing the delocalized electronic states in regions far from the atomic nuclei. The Linearized Augmented Plane Wave (LAPW) method was pioneered to represent wave functions as a linear combination of plane waves and local basis functions, enabling it to effectively describe electronic states in complex electronic environments and diverse crystal structures.[20] The hybrid approach allows the method to effectively handle boundary conditions of periodic structures while maintaining high precision. Furthermore, the theoretical foundation for first-principles calculations of ferroelectric materials includes the Hartree–Fock theory, which uses the wave function as the primary quantity to describe the state of many-particle systems, and density functional theory (DFT), which characterizes the ground-state physical properties of these systems through the electron density function. DFT offers the advantage of simulating electron–electron interactions through carefully designed exchange–correlation functionals, aiming to incorporate complex many-body effects into quantum systems.[21] By iteratively solving for and capturing the intricate correlations between electrons, DFT allows for the calculation of charge density distributions and performs self-consistent energy calculations. The refined functionals (Local Density and Generalized Gradient Approximations) possess multisystem compatibility and predictive power, providing a general framework for studying the electronic structure and properties of ferroelectric materials.

1.3 Phonon Mode

The advancement of microstructural theories has significantly enhanced our understanding of ferroelectric properties. For most known ferroelectrics, polarization is closely related to the non-centrosymmetric structure of crystals, primarily arising from the relative motion of charged ions or molecules within the lattice, including displacements of heavier ions/atoms and interactions with lattice vibrations (phonons). A one-dimensional ordered/disordered ferroelectric, based on rigid dipoles, correlates phonons involved in the polarization process with a ferroelectric. The polar phonon modes in displacive ferroelectrics are primarily optical in nature, whereas the acoustic modes, which are characterized by higher group velocities, play a pivotal role in the transport of heat and polarization. Taking the typical ferroelectric PTO as an example, the motion of titanium atoms can be particularly described by the TO phonon mode.[15,22,23] The displacement of the titanium atom from the central position within the unit cell causes the oxygen atoms in the surrounding octahedron to deviate slightly, which disrupts the symmetry of the charge distribution. As a result, a dipole moment is induced in the tetragonal phase due to the imbalance in the charge center within the unit cell. In the cubic phase, the titanium atom is symmetrically located at the center of the oxygen octahedron, maintaining a balanced charge distribution and preventing the formation of any net charge displacement or dipole moment. The atoms in the lattice undergo oscillations around their equilibrium positions at a frequency ω, and these vibrations are represented by phonon modes, which collectively form the phonon dispersion spectrum of the material. As depicted in Figure 1.4, the soft mode appears at higher frequencies above T_c, with titanium atoms potentially occupying distinct positions in neighboring unit cells. As the temperature decreases, the frequency of the soft mode at the Brillouin zone center decreases until it transitions into a low-frequency or zero-frequency mode at T_c. The wavelength of the mode becomes infinite, and the titanium atoms in adjacent unit cells move in the same direction, with matching polarization amplitude and vector.

A single unit cell is insufficient to generate the optical phonon response required for spontaneous polarization through the soft-mode mechanism. There is an

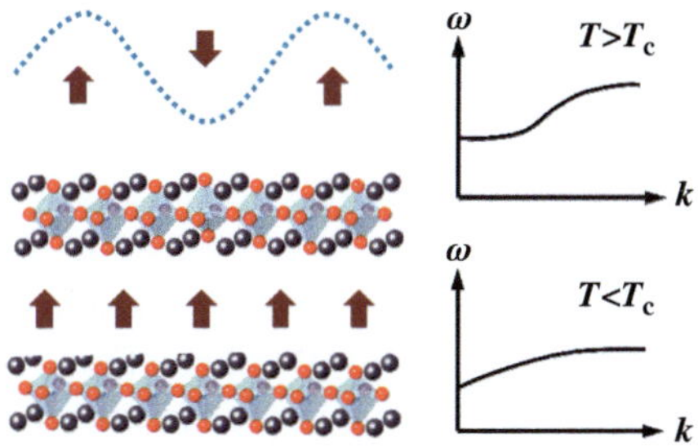

Figure 1.4 Schematic of the polarization direction, titanium ion position, and ferroelectric soft-mode phonon dispersion for a prototypical ferroelectric, PTO. *Source*: [22] / with permission of John Wiley & Sons.

inherent threshold in the number of unit cells required for stable ferroelectricity, as a collective response from multiple unit cells is necessary to induce the optical phonon mode that generates a ferroelectric dipole moment. The limitations of stable ferroelectricity in ABO_3 materials can be broadly classified into two categories: individual nanocrystals and semi-infinite films. A significant body of research has focused on epitaxial layers among the film categories. For example, ferroelectricity has been demonstrated in PTO films epitaxially grown on STO substrates, with the films exhibiting stable ferroelectric properties even at a thickness as thin as just three unit cells.[24] The formation of 180° domain walls in these films helps to stabilize the depolarization field, thereby maintaining ferroelectricity. In a similar vein, the minimum thickness at which ferroelectricity has been observed in BTO films varies across studies, with values ranging from seven unit cells down to four unit cells.[25–28] Due to the boundary effects introduced by the non-ferroelectric interface, the ferroelectric behavior in thin films is typically observed only when the thickness reaches at least three unit cells, which may represent the critical limit for ferroelectricity. When the size of nanocrystals surpasses that of ultrathin films, the ferroelectric phase becomes unstable. At the nanoscale, the relationship between ferroelectric domain stability and crystal size becomes more pronounced. As crystal size decreases, the enhanced surface energy significantly suppresses the domain wall formation and movement (size effects, discussed in Section 1.5.1).

The FE-HfO_2 demonstrates stable ferroelectricity at the nanoscale with switchability at the level of the irreducible unit cell, which seems to differ from the behavior observed in conventional ferroelectrics. The phase transition of FE-HfO_2 was predicted by first-principles calculations combined with group theory analysis, yet the optical soft mode driving this phase transition, specifically the critical dielectric relaxation in the dielectric spectrum, has not been observed in experiments to date.[29,30] The lack of a ferroelectric soft mode could also be due to the first-order character of the phase transition. The phonon modes related to FE-HfO_2 are illustrated schematically in Figure 1.5. First-principles calculations confirm that the transition from the tetragonal phase (t-phase; $P4_2/nmc$) to the o-phase crystal involves the condensation of four symmetric phonons (X_5^y, Y_5^z, Z_5^x, and Γ_{15}^z).[31] The zone-center phonon Γ_{15}^z leads to a displacement of all oxygen atoms along the z-axis, whereas the antipolar

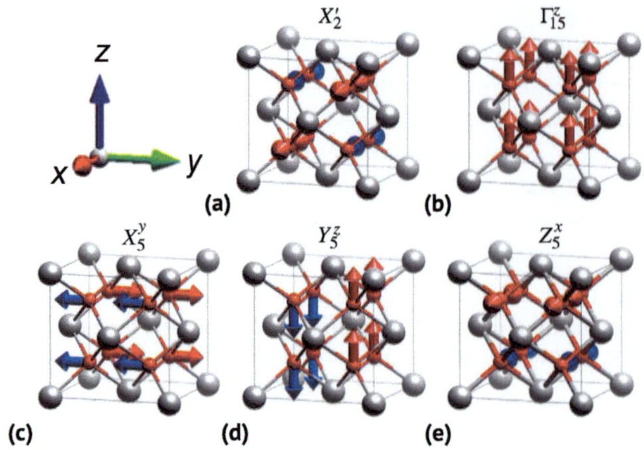

Figure 1.5 Oxygen displacements for the five phonon modes of FE-HfO_2 are illustrated as follows: (a) zone-boundary X_2' mode, (b) zone-center Γ_{15}^z mode, and (c)–(e) antipolar phonon modes X_5^y, Y_5^z, and Z_5^x, where the oxygen atoms in adjacent *yz*-planes move along the *x*-axis, in *xz*-planes move along the *z*-axis, and in *xy*-planes move along the *x*-axis in an antiparallel arrangement. As indicated in the literature, silver and red spheres represent Hf and oxygen atoms, respectively. *Source*: [31] / with permission of John Wiley & Sons.

phonon Y_5^z results in cell doubling, with oxygen atoms in neighboring *xz*-planes moving along the *z*-axis in opposite but parallel directions. Certain flat polar phonon modes as predicted based on the principle are expected to result in the extreme localization of electric dipoles within a width of half a unit cell.[32,33] The local dipoles are stable and can switch on a sub-nanometer scale in thin films, with minimal influence from domain walls and surface exposure. Additionally, it is hypothesized that the ferroelectricity of FE-HfO_2 arises from the nonlinear interaction between the stable Γ_{15}^z and X_5^y phonons and the primary instability of the X_2' phonon.[34–36] An analysis of the phase transition sequence and associated phonon modes in the literature suggests that the transition of FE-HfO_2 films from cubic to t-phase and subsequently to o-phase can be comprehensively characterized by eight distinct phonon modes. The interplay between polar and antiferroelectric modes induces destructive interference in atomic displacements for half of the atoms, while promoting constructive interference in the remaining half. This in-phase coupling within the cubic phase generates a characteristic alternating pattern of polar and nonpolar layers in the *Pca*21 phase, revealing the improper nature of ferroelectricity.[33,36]

FE-HfO_2 exhibits notable polymorphism, stemming from various distortions and structural deformations within the high-symmetry cubic fluorite structure, giving rise to multiple competing phases. Accurately identifying and understanding the symmetry of lattice modes in FE-HfO_2 is a critical step in pinpointing the origin of ferroelectricity and establishing an appropriate order parameter. Previous research has revealed that the transition from the cubic to the t-phase is induced by instabilities in the boundary phonon modes of the cubic structure.[37,38] The ferroelectric behavior observed in the polar phase of FE-HfO_2 can be interpreted by considering the high-symmetry, centrosymmetric t-phase, even though this tetragonal structure does not possess the polar instability necessary for ferroelectricity. First-principles calculations indicate that the stability of the polar o-phase arises from trilinear coupling between the polar, nonpolar, and antiferroelectric phonon modes, which not only induces polarity but also ensures the structural stability of the phase. Therefore, in the study of FE-HfO_2 and similar materials, the *Pbcn* (nonpolar) phase can serve as an important reference structure for understanding the transition from a nonpolar to a polar ferroelectric phase.[39] Meanwhile, the *Pbcm* (polar) phase provides a theoretical framework for exploring soft-mode-driven ferroelectric transitions, where the coupling between polar and nonpolar phonon modes is considered essential in the formation of ferroelectric o-phases.[37] For example, Aramberri and Íñiguez[37] proposed that FE-HfO_2 exhibits a uniaxial ferroelectric order, with the *Pbcm* phase serving as the reference state, and all low-energy polymorphs are connected via simple phonon distortions.

1.4 Thermodynamics and Kinetics

The generation, stability, and reversal of spontaneous polarization in ferroelectric materials can be comprehensively understood from both thermodynamic and kinetic perspectives. Thermodynamically, the polarization behavior of ferroelectric materials is governed by a free energy function, where different local minima represent distinct polarization states. The states represent different configurations of spontaneous polarization, with the material naturally favoring the lowest energy state, while the stability of these states is influenced by factors that can shift the system between various energy minima. Kinetics primarily investigates the reversal process of spontaneous polarization and the evolution of domain structures, which are mainly governed by processes such as domain nucleation, growth, and motion. The process is highly time-dependent, with dynamic responses driven by external stimuli or intrinsic material properties, understood in terms of transition rates and domain evolution. The two aspects often intersect, as key applications of ferroelectric materials typically involve ultrafast polarization reversal and dynamic domain evolution. The phase transition in ferroelectric materials can be comprehensively analyzed through both thermodynamics (energy minimization at equilibrium) and kinetics (transition rates and domain dynamics). Overall, thermodynamics provides the theoretical framework for the stability of ferroelectric polarization, while kinetics describes the free energy landscape for the spontaneous polarization switching and evolution during services,

$$G = G_0 + \frac{1}{2}A(T - T_c)P^2 + \frac{1}{4}BP^4 + \frac{1}{6}CP^6 \tag{1.7}$$

$$\frac{\partial G}{\partial P} = 0 \rightarrow E = AP + BP^3 + CP^5 \tag{1.8}$$

Existing literature has provided extensive documentation and explanations on the thermodynamic aspects of ferroelectric phase transitions. The GLD model for ferroelectric materials is a classical phenomenological theory that characterizes the ferroelectric phase transition by establishing a relationship between the macroscopic free energy and polarization. It employs an expanded form of free energy to describe the behavior of the system during transition. Free energy (G) is expressed as a polynomial expansion in terms polarization (P), as shown in Equation (1.7). Here, T represents temperature, while A, B, and C are coefficients that characterize the material's response behavior. The expression expands free energy as a sum of powers of polarization intensity, enabling the determination of system's physical structural characteristics from the free energy minimum, while also establishing a connection between the expansion coefficients and measurable parameters. The electric field that aligns with polarization is derived by optimizing the free energy in terms of polarization, as described in Equation (1.8). The free energy increases with polarization and the minimum free energy occurs at $P = 0$ when the temperature $T > T_c$, indicating that the material transitions to a paraelectric phase with the absence of spontaneous polarization. When the temperature $T < T_c$, the free energy adopts a double-well potential configuration, characterized by two symmetrical minima at finite polarization values, which correspond to the two directions of spontaneous polarization in ferroelectric materials. The first-order phase transition from the paraelectric to the ferroelectric state is characterized by a rapid shift in polarization, accompanied by notable changes in free energy and the release of latent heat. The phase transition (second-order transition) near T_c is continuous, with spontaneous polarization smoothly varying with temperature without the release of latent heat. Hysteresis phenomena are typically observed in second-order phase transitions, reflecting the nonlinear response characteristics of ferroelectric materials. This is well exemplified in BTO.[40–42] The first-order transition typically occurs during the transformation of BTO from the cubic paraelectric (Pm-$3m$) phase to the t-phase (P_4mm) at $T_c = 120\,°C$. The transition involves a sudden change in polarization accompanied by substantial variations in free energy and the release of latent heat. Free energy is no longer stable at zero polarization, prompting the system to adopt a nonzero polarization state. In contrast, the second-order transition features continuous changes in free energy–polarization (G–P) curve with temperature, as observed in the transition from the t-phase to the o-phase ($Amm2$) at lower temperatures, where the polarization direction gradually adjusts with structural changes.

The dynamic model of polarization characterizes the transient responses to changing external conditions by incorporating a free energy function, dissipation terms, and nonlinear effects. The Landau–Khalatnikov model serves as a classical framework for describing the dynamic behavior of ferroelectric polarization. It elucidates the time-dependent evolution of polarization under an applied electric field by capturing the variation of free energy with respect to polarization. The observed hysteresis and changes in polarization intensity validate the effectiveness of this model in elucidating the behavior of perovskite ferroelectrics.[43–45] It is worth noting that a thorough evaluation of the theory can be found in other references and is beyond the scope of this introduction. Additionally, researchers have investigated the growth processes of domains in perovskite ferroelectrics through a Nucleation and Growth model, emphasizing the impact of nucleation mechanisms and growth kinetics.[46,47] The emergence of the ferroelectric phase is driven by local supersaturation within the bulk material in homogeneous nucleation, where the phase transition occurs spontaneously throughout the volume. On the other hand, heterogeneous nucleation is facilitated by external factors such as crystal defects, impurities, or surface irregularities, which act as preferential sites for the initiation of domain formation. The surface and structural imperfections lower the energy barrier for nucleation, promoting the transition to the ferroelectric phase at specific locations. For instance, as the material temperature decreases and under the influence of external stress or electric fields, certain local regions within the PTO film may become oversaturated, leading to spontaneous formation of the ferroelectric phase. Conversely, defects or irregular surfaces in the PTO film lower the energy barrier for nucleation, thereby driving preferential nucleation in specific regions.

A retrospective review reveals that the existing thermodynamic theoretical framework has advanced the understanding of structural phase transitions in FE-HfO_2 films. The phase transition in fluorite-based FE-HfO_2 is generally considered to be of the first order, involving a transformation between the polar o-phase and nonpolar t-phase structures.[48–50] Figure 1.6a schematically illustrates the influence of various factors (doping, thermal budget, and film thickness) on the phase structure distribution of FE-HfO_2. As an example, the *G*–*P* and polarization–electric field curves of Zr-doped HfO_2 corresponding to first-order phase transitions at different temperatures are calculated and depicted using the GLD model, as shown in Figure 1.6b,c. In addition to T_c, three supplementary important indicative temperatures – T_0, T_1, and T_2 – can be identified. Here, T_0 is defined as the Curie–Weiss temperature, where the *U*–*P* curve undergoes a transition between double-well and triple-well configurations, influenced by factors such as Zr content and grain size. A "fracture" hysteresis phenomenon can be observed between T_c and T_1, which is the upper limit temperature for ferroelectricity. The two-step polarization switching process driven by an intermediate nonpolar phase occurs within a specific temperature range. As the temperature increases ($T_1 - T_2$), the metastable polar phase gradually becomes unstable under zero electric field. The ferroelectric o-phase that is induced displays a double hysteresis behavior within this temperature range when a sufficiently strong electric field is applied. Once the temperature exceeds

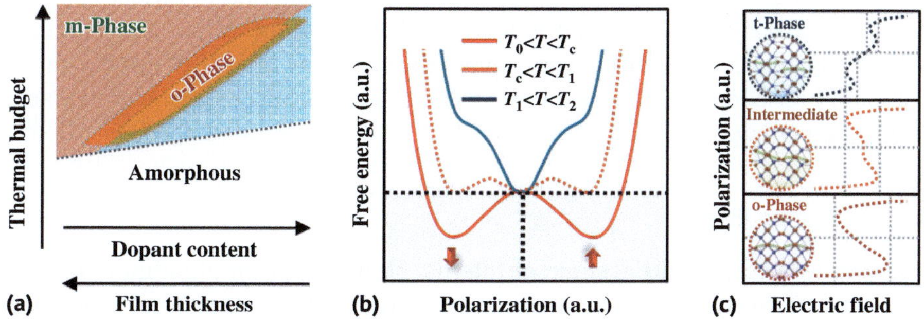

Figure 1.6 (a) Schematic diagram of the phase transition in FE-HfO_2; (b) G–P and (c) corresponding polarization–electric field diagrams based on the GLD model.

the field-induced ferroelectric transition temperature (T_2), the double hysteresis phenomenon disappears.

The common method to explain the formation and stability of metastable phases in nanoscale ceramic materials is by considering the effects of grain boundary and surface energies.[11,49] The principle extends to the stabilization of ferroelectric phases in FE-HfO_2 films, where the surface energy considerations play a pivotal role. The free energy calculations for various phase structures of HfO_2 and ZrO_2 suggest that the stability of the o-phase arises from the careful optimization of surface energy.[51,52] The stabilization of the o-phase relies on the interplay of surface energy, mechanical stress, and other factors.[53,54] Despite several proposed mechanisms for the thermodynamic stabilization of the ferroelectric phase, there are still discrepancies between experimental phase distributions and theoretical predictions based on current boundary energy models.[55–57] Additionally, recent phase transition mechanisms considering activation energy barriers suggest that the transition from the o/t-phase to the monoclinic phase (m-phase; $P21/c$) can be kinetically inhibited. Researchers have systematically investigated the formation mechanism of the ferroelectric phase within a comprehensive thermodynamic framework, integrating both kinetic factors and the Ostwald step rule to elucidate the crystalline phase transition patterns in FE-HfO_2.[58,59] According to this theoretical framework, initially amorphous or microcrystalline FE-HfO_2 films transition to the low-entropy t-phase during heating, driven by free energy reduction, with the t-phase remaining dominant over the m-phase due to a high energy barrier. During the subsequent cooling process, the relatively lower energy barrier facilitates the transition from the t-phase to the o-phase, thereby stabilizing the metastable o-phase rather than the thermodynamically favored m-phase. Moreover, reducing film thickness increases the surface area and enhances the contribution of surface energy to the total free energy, thereby stabilizing the t-phase during nucleation and growth. While further experimental validation is required, these kinetic insights provide valuable support for understanding the phase stability mechanisms in FE-HfO_2 films.

1.5 Functional Indicators

As well established, the key indicators for ferroelectric materials in various applications include scaling effects, imprint, wake-up, fatigue, and retention. These factors collectively impact the performances of both materials and integrated devices, making them critical considerations for applications such as ferroelectric memory and energy harvesting.

1.5.1 Scaling Effects

The scaling effects of biaxial strain or internal electric fields in ultrathin ferroelectrics can effectively stabilize ferroelectricity or enhance the functional properties in certain cases.[22,60] The effectiveness of the enhancement is closely tied to the material's crystal structure. Excessive strain or overly small dimensions can lead to performance inconsistencies or even degradation. From the perspective of ferroelectric phonons, fundamental size constraints exist for crystals that can sustain ferroelectricity (Section 1.3). The discretization of phonon modes resulting from small-scale scaling can impact the stability of ferroelectric phase transitions. On a macroscopic scale, this is reflected in weakened polarization, reduced dielectric constant, increased E_c, and suppressed transition temperatures. The scaling effects in ferroelectric crystals are not intrinsic but are primarily driven by surface and interface energies, strain, defects, and the heightened influence of extrinsic factors on smaller systems.[60–62] Both intrinsic and extrinsic factors exert an increasingly significant physical and chemical influence as the dimensionality of ferroelectrics decreases. In small-sized ferroelectrics, enhanced surface energy and the resulting competition between surface and bulk energies significantly suppress the formation and motion of domain walls. The depolarization field driven by charge separation within the material becomes more pronounced in smaller systems. The combined effects not only compromise the stability of ferroelectric domains but also degrade polarization performance. Additionally, local charge disturbances, arising from defects or their interactions with step edges, can influence the relative orientation of ferroelectric domains, further amplifying the depolarization field. In systems influenced by interfaces or defects, this enhanced depolarization field not only reorganizes domain structures but also reduces the overall polarization of the material.

Ferroelectricity remains even in displacement ferroelectrics of very small sizes, but as the critical thickness is approached, the polarization amplitude is suppressed. The reduction in ferroelectric response adversely affects the properties relevant to ferroelectric polarization, including dielectric and piezoelectric responses. Figure 1.7 illustrates the decrease in polarization or piezoelectric response of perovskite ferroelectric films with decreasing thickness, including epitaxial BTO, PZT, and BFO.[22] Similar thickness-dependent behavior is observed across all films, where the stable polarization response diminishes as the film thickness decreases, in agreement with previous computational predictions. In contrast to

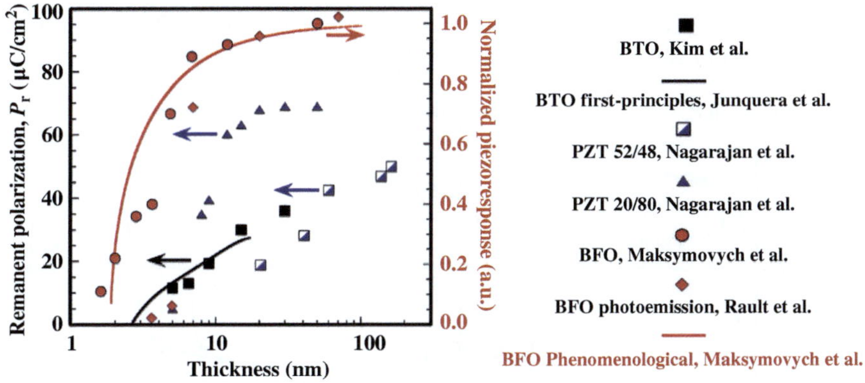

Figure 1.7 Thickness dependence of P_r or normalized piezoresponse in ABO_3 films. *Source*: [22] / with permission of John Wiley & Sons.

perovskite ferroelectrics, FE-HfO_2 has been experimentally demonstrated to maintain robust ferroelectricity even at the nanoscale (1–15 nm). From a crystallographic perspective (Section 1.2), the lattice mismatch between FE-HfO_2 films and substrate at the nanoscale induces lattice distortion, which may trigger phase transitions that promote ferroelectricity. Phonon interactions (Section 1.3) further reinforce this effect, as the strong coupling of alternating phonon modes in FE-HfO_2 generates positive feedback with the ferroelectric domains, stabilizing the material's polarization at the microscopic level. Thermodynamic and kinetic analyses (Section 1.4) reveal that surface and interface effects become more pronounced as the thickness of FE-HfO_2 approaches the nanoscale, favoring the stabilization of the tetragonal phase and facilitating the nucleation and growth of the ferroelectric o-phase. The discussions suggest the possibility that FE-HfO_2 films exhibit an inverse size effect, with the complex relationship between scaling effects and various influencing factors remaining an active area of investigation.

1.5.2 Imprint

The imprint phenomenon is quantitatively characterized by the horizontal shift of the *P*–*E* hysteresis loop as measured using the Sawyer–Tower circuit, indicating a change in the center of the negative and positive coercive voltages. The shift is attributed to one polarization state being more stable than the other in the ferroelectric bistability. As shown in Figure 1.8a, the polarization hysteresis and asymmetric electrical response lead to significant losses in the polarization state. Numerous efforts have been made to understand the origin of the imprint effect, with studies indicating that charge injection, built-in electric fields, defect dipoles, domain wall pinning, and surface/interface layers may all play a role in imprint formation.[63–66] In oxide ferroelectric thin films, common defects such as oxygen vacancies can interact with surrounding metal ions or defect sites to form dipoles. When subjected to an external electric field, these defect dipoles may undergo polarization reversal or flipping, which results in a persistent polarization offset. This phenomenon can be further exacerbated by the dynamics of defect migration and the local redistribution of charges, leading to a long-term shift in the film's overall polarization state. Oxygen vacancies in ferroelectric films can also interact

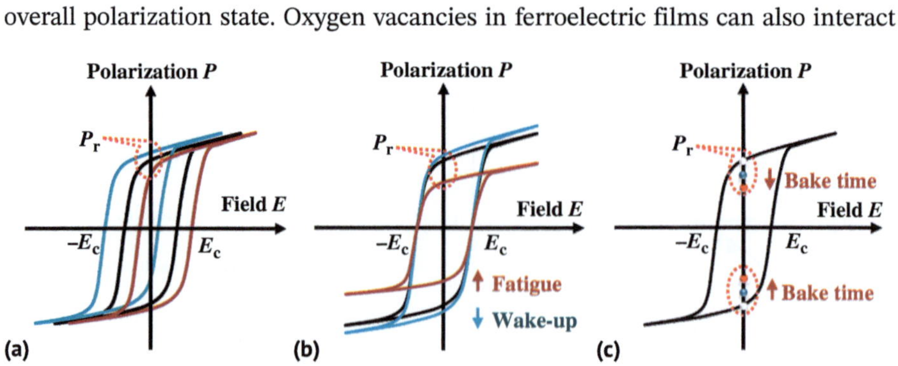

Figure 1.8 Schematic diagrams of the changes in polarization states of (a) imprint, (b) wake-up and fatigue, and (c) retention characteristics.

with other defect states, which generates additional energy levels that promote charge accumulation and release in localized regions, thereby exacerbating the imprint phenomenon. Additionally, electrode or interfacial layers may also introduce charge carriers (electrons or holes) through oxygen vacancies or defects. The injected charges can interact with the polarization state of the film, resulting in persistent charge accumulation that alters the local built-in electric field and causes a shift in polarization. On the other hand, theoretical calculations indicate that introducing a doped layer near the electrode region of the ferroelectric film may provide insight into the effects of nonuniform space charges and built-in electric fields.[67] Modeling that incorporates asymmetric electrode materials indicates that the asymmetry observed in the hysteresis loop may also stem from unshielded depolarization fields.[68,69] However, there remains a discrepancy between the experimentally observed offsets and modeling results, and a consensus on the exact mechanism of imprint has not yet been reached.

The imprint during prolonged use or high-frequency operation may lead to an increase in read error rates. Similar behaviors have been observed in both perovskite ferroelectrics and FE-HfO_2, and they can be mitigated by incorporating oxide electrodes (such as IrO_2 and RuO_3) and other dopants within the films.[70,71] This suggests that the imprints in different types of ferroelectrics may share a common physical origin. However, the thinner thickness (which allows the built-in electric field to dominate the shift of E_c), lower dielectric constant (which mitigates the interface effect), and defects (primarily neutral and positively charged oxygen vacancies) of FE-HfO_2 make its analysis comparatively more straightforward than that of perovskite ferroelectrics. Recent studies indicate that the imprint in FE-HfO_2 films primarily arises from a mechanism involving charge injection, the release of electrons from oxygen vacancies, changes in the built-in electric field, and shifts in E_c.[72] A scheme employing unipolar voltage pulses for polarization recovery has been proposed, offering a novel approach to mitigate the imprinting effect.

1.5.3 Wake-up and Fatigue

The electric field cycling behavior of ferroelectric materials typically consists of wake-up and fatigue phenomena, as shown in Figure 1.8b. Wake-up refers to the process in which the initially contracted electric hysteresis loop of the sample opens up during electric field cycling. The mechanisms underlying the wake-up effect, related to domain wall motion and defect behavior in perovskite ferroelectrics, have been systematically elucidated.[73,74] In the initial state, defects such as oxygen vacancies, domain wall defects, and charge traps hinder the reorientation of electric domains, resulting in limited polarization and a narrower hysteresis loop. During electric field cycling, defects migrate and redistribute, reducing the hindrance to domain wall movement and local electric field shielding, which in turn promotes polarization enhancement. The ferroelectric film begins to exhibit fatigue phenomena as the number of write cycles continues to increase.[73,75] As the external electric field increases during this stage, the polarization values in opposite directions gradually converge, leading to a reduction in P_r strength in both directions. Two defect-dominated models describe the origin of fatigue: (1) compared to the wake-up phase, sustained write cycles not only facilitate the diffusion of charged defects but also increase the generation of defects. These defects act as charge accumulation points and restrict domain wall movement, leading to a reduction in the external contributions to the measured response; (2) the formation and expansion of seed domains near the ferroelectric–electrode interface lock the polarization direction, preventing its reversal and ultimately leading to a reduction in external polarization. Additionally, for some ferroelectric materials, after a certain number of electric field sub-cycles, the transient current response may exhibit a split effect, where the displacement current peak splits into two peaks close to the sub-cycle amplitude. The split effect disappears at higher electric field cycling periods.[76]

The fatigue of FE-HfO_2 is characterized by approximately 10^8 cycles and accompanied by polarization wake-up, which still presents an order-of-magnitude gap compared to traditional perovskite ferroelectrics.[31,77–79] The limited fatigue characteristics are directly related to the narrow operational window between E_c and breakdown voltage in FE-HfO_2 films. While some reports indicate that lower oxygen supply may inhibit the formation of the m-phase and promote the nucleation and growth of the ferroelectric o-phase, the high concentration of oxygen vacancy defects in FE-HfO_2 films provides significant pinning sites for domain movement, accelerating breakdown failure. Employing high-thermal-stability metal electrodes, metal oxide electrodes, and dielectric capping/inter/seed layers can effectively suppress the formation of interface oxygen vacancies. Figure 1.9a exemplifies the fatigue behavior of VO_x-capped FE-HfO_2 films, with the application of bipolar triangular pulses enhancing the fatigue to over 10^9 cycles. The lower-energy V—O bond compared to the higher-energy Hf—O or Zr—O bonds is predicted to effectively suppress oxygen defects at the interface.[57] Furthermore, studies have investigated the correlation between structure and performance in FE-HfO_2 films during wake-up and fatigue from a microscopic viewpoint.[55,79] The polarization increase observed during wake-up is primarily due to the electric field-induced phase transition from the t-phase to the

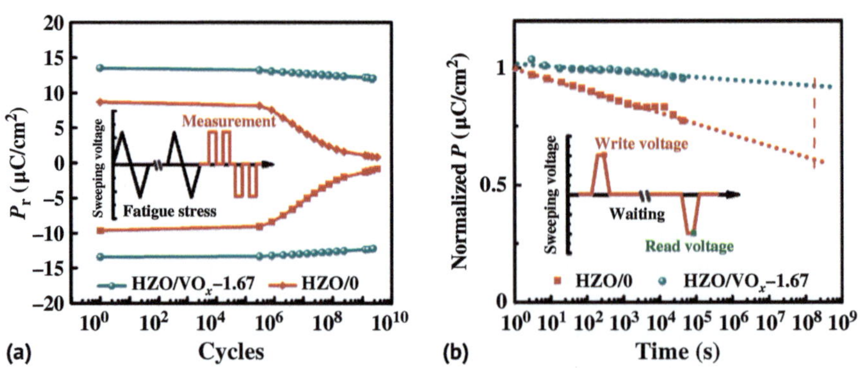

Figure 1.9 (a) Fatigue and (b) retention properties of the FE-HfO_2 films without or with the VO_x capping layer. Insets in the panels schematically show the applied pulses. *Source*: [57] / with permission of American Chemical Society.

ferroelectric o-phase. In contrast, the transition from the o-phase to the m-phase plays a crucial role in diminishing residual polarization during fatigue.

1.5.4 Retention

Retention is the ability to maintain a polarized state over time, as demonstrated in Figure 1.8c. An unstable polarization state can lead to the relaxation of switching domains back to their original configuration, causing the loss of information stored in ferroelectric polarization. This phenomenon is commonly attributed to the depolarization field generated by incomplete screening at the electrode interface, which is considered the primary factor contributing to retention loss.[80–82] The direction of the depolarization field is opposite to the direction of polarization within the film, effectively screening a portion of the polarized charges. The ideal electrode interface would result in the absence of a depolarization field for ferroelectric capacitors. The limited availability of free charges in the dielectric insulator layer means that most of the polarized charges in the ferroelectric films are not compensated promptly in metal–ferroelectric–insulator–metal structures, leading to a relatively large depolarization field. Furthermore, polarization retention failure also appears to be related to the flexoelectric effect induced by strain gradients.[83] It has been reported that the coupling of strain gradients with polarization generates a flexoelectric field within the film, which can also lead to retention failure. These mechanisms underscore the importance of maintaining ferroelectricity under complex operating conditions while highlighting the critical roles of interface and strain management in enhancing retention.

First-principles calculations provide a means to quantify the effect of electrical boundary conditions on ferroelectric materials. The simulation results indicate that as the thickness of the ferroelectric film decreases, the influence of the depolarization field becomes more pronounced.[60,84] In this case, the incomplete charge screening at the metal electrode of the perovskite ferroelectric leads to an increase in the depolarization field. However, E_c of FE-HfO_2 is approximately 1 MV/cm, much higher than that of perovskite ferroelectrics (E_c~10–100 kV/cm), and the defect concentration decreases during the formation of ultrathin films.[85,86] The high E_c and low density of defect states enable FE-HfO_2 to maintain a more stable polarization state. For instance, Figure 1.9b illustrates the retention performance of VO_x -capped FE-HfO_2 films. The inset shows the measurement of retained polarization through the application of writing and reading pulses, with adjustable waiting times between these pulses. The preset pulse and read pulse must have the same amplitude and opposite polarity, effectively preventing any one-sided cumulative polarization effects caused by differing pulse amplitudes during writing and reading.[57] The polarization loss of VO_x-capped FE-HfO_2 films is only 4.6% after 10^4 s with a projected retention of 93% after ~3×10^8 s (10 years), indicating significant and sustained polarization stability.

1.6 Summary

This chapter provides a brief overview of the fundamental properties of ferroelectric materials and conventionally utilized models/theories. It introduces the relevant background and research progress of ferroelectric materials from various aspects, including material origin, crystal configuration, phonon modes, thermodynamics, and kinetics. The discussion focuses on a comparative analysis between classical perovskite ferroelectrics and novel FE-HfO_2 materials, emphasizing the material and theoretical foundations that support ferroelectric characteristics and effects, such as crystal symmetry and structural phase transitions. While debates about certain ferroelectric mechanisms continue, it is evident that the fluorite-structured FE-HfO_2 offers indispensable perspectives and pathways for exploring ferroelectric physics. Additionally, this chapter outlines the fundamental functional metrics of ferroelectric materials, including common scaling effects, wake-up behavior, fatigue, and retention characteristics. It discusses the general trends in properties as crystal size decreases, the inverse scaling effects of FE-HfO_2 films, and the failure behaviors and mechanisms of films, providing valuable indicators and reference for the development of novel ferroelectric materials and devices.

Acknowledgments

This work is supported by the National Key R&D Program of China (2024YFA1208602), the Henan Province Natural Science Youth Funds (Grant No. 242300421713), and Key Program of Higher Education of Henan Province (Grant No. 24A510007). This work is also supported by the Natural Science Foundation of Guangdong Province of China (Grant No. 2024A1515013088).

References

1 Valasek, J. (1920). Piezoelectric and Allied Phenomena in Rochelle Salt. *Physical Review* 15: 537–538.

2 Valasek, J. (1921). Piezo-electric and allied phenomena in Rochelle salt. *Physical Review* 17: 475–481.

3 Busch, G. (1987). How I discovered the ferroelectric properties of KH_2PO_4. *Ferroelectrics* 71: 43–47.

4 Shirane, G. and Takeda, A. (1952). Phase transitions in solid solutions of $PbZrO_3$ and $PbTiO_3$ (I) small concentrations of $PbTiO_3$. *Journal of the Physical Society of Japan* 7: 5–11.

5 Merz, W. J. (1956). Switching time in ferroelectric $BaTiO_3$ and its dependence on crystal thickness. *Journal of Applied Physics* 27: 938–943.

6 Nuraje, N. and Su, K. (2013). Perovskite ferroelectric nanomaterials. *Nanoscale* 5: 8752–8780.

7 Fan, Z., Sun, K., and Wang, J. (2015). Perovskites for photovoltaics: A combined review of organic-inorganic halide perovskites and ferroelectric oxide perovskites. *Journal of Materials Chemistry A* 3: 18809–18828.

8 Böscke, T. S., Teichert, S., Bräuhaus, D. et al. (2011). Phase transitions in ferroelectric silicon doped hafnium oxide. *Applied Physics Letters* 9: 112904.

9 Müller, J., Böscke, T. S., Bräuhaus, D. et al. (2011). Ferroelectric $Zr_{0.5}Hf_{0.5}O_2$ thin films for nonvolatile memory applications. *Applied Physics Letters* 99: 112901.

10 Müller, J., Böscke, T. S., Schröder, U. et al. (2012). Ferroelectricity in simple binary ZrO_2 and HfO_2. *Nano Letters* 12: 4318–4323.

11 Park, M. H., Lee, Y. H., Kim, H. J. et al. (2015). Ferroelectricity and antiferroelectricity of doped thin HfO_2-based films. *Advanced Materials* 27: 1811–1831.

12 Hasegawa, K., Shimizu, T., Ohsawa, T. et al. (2023). Full polarization reversal at room temperature in unsubstituted AlN. *Applied Physics Letters* 123: 192903.

13 Ginzburg, V. L. (1949). On the theory of superconductivity. *Zhurnal Eksperimental'noi i Teoreticheskoi Fiziki* 20: 1064–1082.

14 Devonshire, A. F. (1954). Theory of ferroelectrics. *Advances in Physics* 3: 85–130.

15 Cochran, W. (1960). Theory of the lattice vibrations of ferroelectric crystals. *Advances in Physics* 9: 387–423.

16 Anderson, P. W. (1960). *Fiz. Dielektrikov*. Moscow: Acad. Nauk SSSR. pp. 258–266.

17 de Gennes, P. G. (1963). Polarization in a ferroelectric model. *Solid State Communications* 1: 132–137.

18 Onsager, L. (1944). Crystal statistics. I. A two-dimensional model with an order-disorder transition. *Physical Review* 65: 117–149.

19 Shirodkar, S. N. and Waghmare, U. V. (2014). A first-principles study of ferroelectricity in HfO_2. *Physical Review Letters* 112: 117602.

20 Andersen, O. K. (1975). Operator approach in the linearized augmented-plane-wave method: Efficient electronic-structure calculations including forces. *Physical Review B* 12: 3060–3083.

21 Kohn, W. and Sham, L. J. (1996). Density functional theory. In *Proceedings of the Conference of the Italian Physical Society*. Editrice Compositori. Vol. 49, pp. 561–572.

22 Ihlefeld, J. F., Harris, D. T., Keech, R. et al. (2016). Scaling effects in perovskite ferroelectrics: Fundamental limits and process-structure-property relations. *Journal of the American Ceramic Society* 99: 2537–2557.

23 Urfoot, J. C. and Taylor, G. W. (2023). *Polar Dielectrics and Their Applications*. Berkeley and Los Angeles, CA: University of California Press.

24 Tybell, T., Ahn, C. H., and Triscone, J. M. (1999). Ferroelectricity in thin perovskite films. *Applied Physics Letters* 75: 856–858.

25 Wen, Z., Li, C., Wu, D. et al. (2013). Ferroelectric-field-effect-enhanced electroresistance in metal/ferroelectric/semiconductor tunnel junctions. *Nature Materials* 12: 617–621.

26 Soni, R., Petraru, A., Meuffels, P. et al. (2014). Giant electrode effect on tunnelling electroresistance in ferroelectric tunnel junctions. *Nature Communication* 5: 5414.

27 Tenne, D. A., Turner, P., Schmidt, J. D. et al. (2009). Ferroelectricity in ultrathin $BaTiO_3$ films: Probing the size effect by ultraviolet Raman spectroscopy. *Physical Review Letters* 103: 177601.

28 Kim, D. J., Lu, H., Ryu, S. et al. (2012). Ferroelectric tunnel memristor. *Nano Letters* 12: 5697–5702.

29 Huan, T. D., Sharma, V., Rossetti, G. A. et al. (2014). Pathways towards ferroelectricity in hafnia. *Physical Review B* 90: 064111.

30 Kamba, S. (2021). Soft-mode spectroscopy of ferroelectrics and multiferroics: A review. *APL Materials* 9: 020704.

31 Banerjee, W., Kashir, A., and Kamba, S. (2022). Hafnium oxide (HfO_2) – a multifunctional oxide: a review on the prospect and challenges of hafnium oxide in resistive switching and ferroelectric memories. *Small* 18: 2107575.

32 Lee, H. J., Lee, M., Lee, K. et al. (2020). Scale-free ferroelectricity induced by flat phonon bands in HfO_2. *Science* 369: 1343–1347.

33 Kashir, A., Farahani, M. G., Lančok, J. et al. (2022). A grease for domain walls motion in HfO_2-based ferroelectrics. *Nanotechnology* 33: 155703.

34 Delodovici, F., Barone, P., and Picozzi, S. (2021). Trilinear-coupling-driven ferroelectricity in HfO_2. *Physical Review Materials* 5: 064405.

35 Fan, S., Singh, S., Xu, X. et al. (2022). Vibrational fingerprints of ferroelectric HfO_2. *npj Quantum Materials* 7: 32.

36 Zhu, T., Ma, L., Deng, S. et al. (2024). Progress in computational understanding of ferroelectric mechanisms in HfO_2. *npj Computational Materials* 10: 101.

37 Aramberri, H. and Íñiguez, J. (2023). Theoretical approach to ferroelectricity in hafnia and related materials. *Communications Materials* 4: 95.

38 Zhou, S., Zhang, J., and Rappe, A. M. (2022). Strain-induced antipolar phase in hafnia stabilizes robust thin-film ferroelectricity. *Science Advances* 8: eadd5953.

39 Raeliarijaona, A. and Cohen, R. E. (2023). Hafnia HfO_2 is a proper ferroelectric. *Physical Review B* 108: 094109.

40 Maeder, M. D., Damjanovic, D., and Setter, N. (2004). Lead free piezoelectric materials. *Journal of Electroceramics* 13: 385–392.

41 Kim, Y. J., Hyun, J. W., Kim, H. S. et al. (2009). Microstructural characterization and dielectric properties of barium titanate solid solutions with donor dopants. *Bulletin of the Korean Chemical Society* 30: 1267–1273.

42 Zhong, W., Vanderbilt, D., and Rabe, K. M. (1994). Phase transitions in $BaTiO_3$ from first principles. *Physical Review Letters* 73: 1861–1864.

43 Sivasubramanian, S., Widoni, A., and Srivastava, Y. (1998). Equivalent circuit and simulations for the Landau-Khalatnikov model of ferroelectric hysteresis. *Journal of Applied Physics* 50: 950–957.

44 Landau, L. D. (2013). On the anomalous absorption of sound near a second-order phase transition point. In: Ter Haar, D. (ed.). *Collected Papers of L.D. Landau*. Elsevier. pp. 626–629.

45 Onsager, L. and Machlup, S. (1953). Fluctuations and irreversible processes. *Physical Review* 91: 1505–1512.

46 Kittel, C. and McEuen, P. (2018). *Introduction to Solid State Physics*. Hoboken, NJ: John Wiley & Sons.

47 Gao, Q., Qi, J., Chen, K. et al. (2022). Halide perovskite crystallization processes and methods in nanocrystals, single crystals, and thin films. *Advanced Materials* 34: 2200720.

48 Kim, K. D., Lee, Y. H., Gwon, T. et al. (2017). Scale-up and optimization of HfO_2-ZrO_2 solid solution thin films for electrostatic supercapacitors. *Nano Energy* 39: 390–399.

49 Hoffmann, M., Schroeder, U., Künneth, C. et al. (2015). Ferroelectric phase transitions in nanoscale HfO_2 films enable giant pyroelectric energy conversion and highly efficient supercapacitors. *Nano Energy* 18: 154–164.

50 Xu, X., Huang, F. T., Qi, Y. et al. (2021). Kinetically stabilized ferroelectricity in bulk single-crystalline HfO_2:Y. *Nature Materials* 20: 826–832.

51 Materlik, R., Künneth, C., and Kersch, A. (2015). The origin of ferroelectricity in $Hf_{1-x}Zr_xO_2$: A computational investigation and a surface energy model. *Journal of Applied Physics* 117: 134109.

52 Batra, R., Tran, H. D., and Ramprasad, R. (2016). Stabilization of metastable phases in hafnia owing to surface energy effects. *Applied Physics Letters* 108: 172902.

53 Falkowski, M. and Kersch, A. (2021). Ab initio interphase characteristics in HfO_2 and ZrO_2 and nucleation of the polar phase. *Applied Physics Letters* 118: 032905.

54 Lee, I. Y. and Yu, J. (2023). Origin of morphotropic phase boundary in thin-film $Hf_{0.5}Zr_{0.5}O_2$ on the TiN electrode. *Journal of Applied Physics* 134: 074102.

55 Zhang, Y., Wang, D., Luo, C. L. et al. (2022). Controllable coercive field of ferroelectric HfO_2 films via UV-ozone surface modification. *IEEE Transactions on Electron Devices* 69: 3094–3099.

56 Zhang, Y., Wang, D., Wang, J. L. et al. (2023). Growth of the orthorhombic phase and inhibition of charge injection in ferroelectric HfO_2-based MFIS memory devices with a high-permittivity dielectric seed layer. *Science China Materials* 66: 219–232.

57 Zhang, Y., Fan, Z., Wang, D. et al. (2020). Enhanced ferroelectric properties and insulator–metal transition-induced shift of polarization-voltage hysteresis loop in VO_x-capped $Hf_{0.5}Zr_{0.5}O_2$ thin films. *ACS Applied Materials & Interfaces* 12: 40510–40517.

58 Park, M. H., Lee, Y. H., Kim, H. J. et al. (2017). Surface and grain boundary energy as the key enabler of ferroelectricity in nanoscale hafnia-zirconia: A comparison of model and experiment. *Nanoscale* 9: 9973–9986.

59 Park, M. H., Lee, Y. H., Mikolajick, T. et al. (2019). Thermodynamic and kinetic origins of ferroelectricity in fluorite structure oxides. *Advanced Electronic Materials* 5: 1800522.

60 Park, S. H., Kim, J. Y., Song, J. Y. et al. (2023). Overcoming size effects in ferroelectric thin films. *Advanced Physics Research* 2: 2200096.

61 Song, J., Zhuang, S., Martin, M. et al. (2021). Interfacial-strain-controlled ferroelectricity in self-assembled $BiFeO_3$ nanostructures. *Advanced Functional Materials* 31: 2102311.

62 Chen, W. J., Zheng, Y., Luo, X. et al. (2013). *Ab initio* study on the size effect of symmetric and asymmetric ferroelectric tunnel junctions: A comprehensive picture with regard to the details of electrode/ferroelectric interfaces. *Journal of Applied Physics* 114: 064105.

63 Zhou, Y., Chan, H. K., Lam, C. H. et al. (2005). Mechanisms of imprint effect on ferroelectric thin films. *Journal of Applied Physics* 98: 024111.

64 Warren, W. L., Tuttle, B. A., Dimos, D. et al. (1996). Imprint in ferroelectric capacitors. *Japanese Journal of Applied Physics* 35: 1521.

65 Abe, K., Komatsu, S., Yanase, N. et al. (1997). Asymmetric ferroelectricity and anomalous current conduction in heteroepitaxial $BaTiO_3$ thin films. *Japanese Journal of Applied Physics* 36: 5846.

66 Defrise, M. and De Mol, C. (1992). Super-resolution in confocal scanning microscopy: generalized inversion formulae. *Inverse Problems* 8: 175.

67 Wang, C. L., Zhang, L., Peng, Y. P. et al. (1998). Landau theory of asymmetric ferroelectric films under external electric fields. *Solid State Communications* 109: 213–216.

68 Baudry, L. (1999). Theoretical investigation of the influence of space charges on ferroelectric properties of $PbZrTiO_3$ thin film capacitor. *Journal of Applied Physics* 86: 1096–1105.

69 Hong, S., Colla, E. L., Kim, E. et al. (1999). High resolution study of domain nucleation and growth during polarization switching in $Pb(Zr,Ti)O_3$ ferroelectric thin film capacitors. *Journal of Applied Physics* 86: 607–613.

70 Hong, K., Yu, Y. S., and Lee, S. K. (1998). Effect of bottom electrode system on ferroelectric properties of sputter deposited PZT capacitors. *MRS Online Proceedings Library* 541: 107–112.

71 Alcala, R., Materano, M., Lomenzo, P. D. et al. (2023). The electrode-ferroelectric interface as the primary constraint on endurance and retention in HZO-based ferroelectric capacitors. *Advanced Functional Materials* 33: 2303261.

72 Yuan, P., Mao, G. Q., Cheng, Y. et al. (2022). Microscopic mechanism of imprint in hafnium oxide-based ferroelectrics. *Nano Research* 15: 3667–3674.

73 Mikolajick, T., Slesazeck, S., Mulaosmanovic, H. et al. (2021). Next generation ferroelectric materials for semiconductor process integration and their applications. *Journal of Applied Physics* 129: 100901.

74 Wu, J. and Wang, J. (2010). $BiFeO_3$ thin films of (111)-orientation deposited on $SrRuO_3$ buffered $Pt/TiO_2/SiO_2/Si(100)$ substrates. *Acta Materialia* 58: 1688–1697.

75 Ramesh, R., Gilchrist, H., Sands, T. et al. (1993). Ferroelectric La-Sr-Co-O/Pb-Zr-Ti-O/La-Sr-Co-O heterostructures on silicon via template growth. *Applied Physical Letters* 63: 3592–3594.

76 Schenk, T., Hoffmann, M., Ocker, J. et al. (2015). Complex internal bias fields in ferroelectric hafnium oxide. *ACS Applied Materials & Interfaces* 7 (36): 20224–20233.

77 Cao, R., Wang, Y., Zhao, S. et al. (2018). Effects of capping electrode on ferroelectric properties of $Hf_{0.5}Zr_{0.5}O_2$ thin films. *IEEE Electron Device Letters* 39: 1207–1210.

78 Cao, R., Song, B., Shang, D. et al. (2019). Improvement of endurance in HZO-based ferroelectric capacitor using Ru electrode. *IEEE Electron Device Letters* 40: 1744–1747.

79 Pešić, M., Fengler, F. P. G., Larcher, L. et al. (2016). Physical mechanisms behind the field-cycling behavior of HfO_2-based ferroelectric capacitors. *Advanced Functional Materials* 26: 4601–4612.

80 Pan, X. and Ma, T. P. (2011). Retention mechanism study of the ferroelectric field effect transistor. *Applied Physical Letters* 99: 013505.

81 Lomenzo, P. D., Slesazeck, S., Hoffmann, M. et al. (2019). Ferroelectric $Hf_{1-x}Zr_xO_2$ memories: Device reliability and depolarization fields. In: Shimeng Yu (ed.). *Proceedings of the 2019 19th Non-Volatile Memory Technology Symposium (NVMTS)*. IEEE. pp. 1–8.

82 Mo, F., Tagawa, Y., Jin, C. et al. (2020). Low-voltage operating ferroelectric FET with ultrathin IGZO channel for high-density memory application. *IEEE Journal of the Electron Devices Society* 8: 717–723.

83 Zou, M. J., Tang, Y. L., Zhu, Y. L. et al. (2020). Flexoelectricity-induced retention failure in ferroelectric films. *Acta Materialia* 196: 61–68.

84 Trieloff, M., Jessberger, E. K., Herrwerth, I. et al. (2003). Structure and thermal history of the H-chondrite parent asteroid revealed by thermochronometry. *Nature* 422: 502–506.

85 Müller, J., Böscke, T. S., Müller, S. et al. (2013). Ferroelectric hafnium oxide: A CMOS-compatible and highly scalable approach to future ferroelectric memories. In: Tahir Ghani (ed.). *Proceedings of the 2013 IEEE International Electron Devices Meeting*. IEEE. pp. 10.8.1–10.8.4.

86 Chen, H., Zhou, X., Tang, L. et al. (2022). HfO_2-based ferroelectrics: From enhancing performance, material design, to applications. *Applied Physics Reviews* 9: 011307.

2

Oxygen Vacancy-induced Ferroelectricity in HfO_2

*Chenxi Yu, Fei Liu, and Jinfeng Kang**

School of Integrated Circuits, Peking University, Beijing, China

*Corresponding author: Kangjf@pku.edu.cn

2.1 Introduction

The discovery of ferroelectric HfO_2[1] has attracted extensive research interest due to its robust ferroelectricity in nanoscale and excellent compatibility with complementary metal–oxide–semiconductor (CMOS). A robust ferroelectricity in HfO_2 thin films has been demonstrated by that in ultrathin (3 nm) films[2] and the extremely high endurance ($>10^{12}$),[3] while the conventional perovskite ferroelectrics suffer from deteriorated ferroelectricity when scaling down. Therefore, it is crucial to study the origin of ferroelectricity to understand the physics behind robust ferroelectricity in HfO_2 and to better exploit the HfO_2-based devices.

Generally, ferroelectricity in HfO_2 is attributed to the formation of a ferroelectric orthorhombic (o) $Pca2_1$ phase, the o-phase III (oIII).[4] A tetragonal (t) phase forms in the fabrication of HfO_2 thin film, and transforms into ferroelectric o-phase during annealing. Lee et al.[5] found that the flat-band phonon is the cause of robust ferroelectricity in HfO_2. Besides the ferroelectric $Pca2_1$ o-phase, other ferroelectric phases are also found, including the Hf-excess rhombohedral $R3m$ phase.[6] Recently, the relation between oxygen vacancy (V_O) and ferroelectricity has been attracting research interest. Yu et al.[7] demonstrated that the order–disorder transition of V_O can contribute to ferroelectricity through the formation of stable ferroelectric phases by order–disorder transition. The intrinsic V_O in HfO_2 thin films can induce localized dipoles through the symmetry-breaking mechanism, and the order–disorder transition of V_O can induce ferroelectric transition.

Although the o-phase theory is widely acknowledged, the ferroelectric o-phase is metastable under room temperature.[8] Researchers have made efforts to find the stabilization factors for the o-phase, including dopants, interface, strain, and vacancies. V_O is one of the factors that stabilize the ferroelectric o-phase.[9] Recent experiments demonstrate that V_O not only stabilizes ferroelectric phase, but is also closely related to ferroelectricity. For instance, the vacancy plays a crucial role in ferroelectric cycling.[10] It can migrate between electrodes and the ferroelectric layer during ferroelectric switching.[11] Therefore, it is necessary to study the relation between V_O and ferroelectricity.

Recent research has found that vacancies contribute to ferroelectricity besides the o-phase. Liu et al.[12] studied the role of V_O in ferroelectricity cycling behaviors. Based on first-principles calculations, they found that the vacancies in HfO_2 not only serve as electric traps, but also contribute to ferroelectricity by forming localized dipoles. Rushchanskii et al.[13] studied the ordering pattern of V_O in sub-stoichiometric $HfO_{2-\delta}$. Based on first-principles calculations, they predicted that the concentrated V_O can induce spontaneous polarization in nonpolar phases such as monoclinic (m) phase. These reports imply that the vacancy may contribute to ferroelectricity besides the widely acknowledged ferroelectric o-phase. Yu et al.[7] proposed an insight into the ferroelectricity from the principle of order–disorder transition. In order–disorder ferroelectrics, ferroelectricity results from the ordering transition of electric dipoles induced by the cations or vacancy defects. Yu et al.[7] made an assumption that the ferroelectricity in HfO_2 thin films is order–disorder ferroelectricity, in which the electric dipoles are induced by V_O. The phase structures and ferroelectric properties can be determined using first-principles calculations and systematic structure searching methods. Based on this principle of order–disorder transition, first-principles calculations were carried out. A stable ferroelectric phase (M_1) with ordered V_O was predicted to have a polarization of 27 μC/cm^2, and switch under an applied field with a barrier of 15 meV/atom by the order–disorder mechanism. Using combined transmission electron microscopy (TEM) techniques, the M_1 phase was observed in zirconium (Zr)-doped HfO_2 thin films.

2.2 Physical Model of Ferroelectricity Induced by Ordered Vacancy

The ferroelectricity is classified into two categories, displacive and order–disorder, based on the shape of one-body potential terms in the effective Hamiltonian.[14] The one-body potential of displacive ferroelectrics has one valley, while the one-body potential of order–disorder ferroelectrics has more than one valley. The vacancy-induced ferroelectricity in HfO_2 is of order–disorder type based on the classification. On the one hand, the vacancy-induced ferroelectricity can be classified into order–disorder type because the energy landscape has many valleys for each vacancy position. On the other hand, the displacive switching path, discussed in Section 2.4.3 and Figure 2.7, has small polarization and induces large strain which may not be favored in real samples.

Landau–Ginzburg–Devonshire (LGD) theory is the phenomenological theory of phase transition. Ferroelectricity in HfO_2 induced by the order–disorder transition of V_O can be formulated by the LGD theory. Free energy as a function of polarization

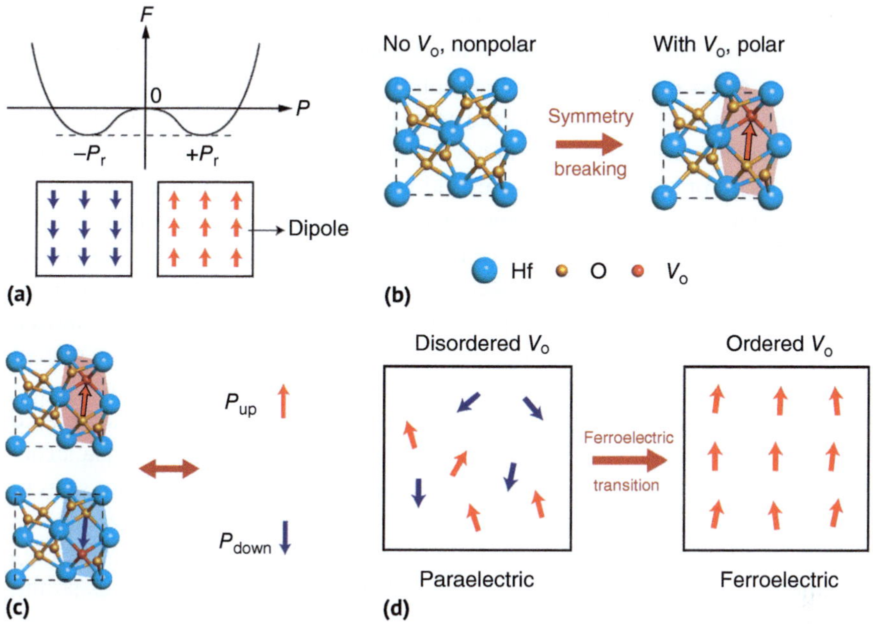

Figure 2.1 LGD model of ferroelectricity induced by ordered V_O. (a) Free energy *F* versus polarization *P* graph. The black square with arrows shows the direction of localized dipoles of ferroelectrics in the ferroelectric state. Red arrows are dipoles pointing upward and blue arrows are dipoles pointing downward. (b) Intrinsic O vacancies induce localized dipoles by symmetry breaking. This mechanism is not limited to the HfO_2 m-phase, which is used for demonstration. (c) Dipole direction relating to the position of V_O. (d) Ferroelectric transition from disordered paraelectric state to ordered ferroelectric state. *Source*: [7] / with permission of American Physical Society.

has two valleys corresponding to the states with positive polarization and negative polarization as shown in Figure 2.1a. For order–disorder ferroelectricity, the dipoles are maximally ordered in these two states, i.e., aligned to a certain direction. Based on first-principles calculations in Section 2.4.1, vacancies in HfO_2 can induce localized dipoles. The symmetry deduction principle of ferroelectricity states that non-ferroelectric phases have generally higher symmetry than the ferroelectric phases, and possible ferroelectric phases can be obtained by the symmetry reduction of high-symmetry non-ferroelectric phases.[4] In stoichiometric HfO_2, the nonpolar phases such as m- and t-phases have zero spontaneous polarization due to their nonpolar space groups. However, the formation of vacancies can break the symmetry of these nonpolar space groups, which induces localized dipoles as shown in Figure 2.1b. The direction of dipole is related to the position of vacancy as depicted in Figure 2.1c, which can be switched by the migration of oxygen (O) atoms to their adjacent positions. The HfO_2 phases with disordered dipoles induced by V_O are paraelectric, while those with ordered dipoles are ferroelectric. The ferroelectric transition of V_O-induced ferroelectricity in HfO_2 is related to the ordering transition of V_O. During ferroelectric transition, the disordered dipoles are aligned to a certain direction, accompanied by the rearrangement of vacancies from disordered state to ordered state, as shown in Figure 2.1d. The ferroelectric switching is related to the dynamic behaviors of V_O, including the displacement or migration of O atoms or ions (O^{2-}), generation and recombination of V_O similar to resistive switching. The dynamic behaviors of V_O also lead to the ferroelectric cycling behaviors in HfO_2-based ferroelectric capacitors. The wake-up effect may be caused by the redistribution of V_O. In pristine films, V_Os are randomly distributed, which causes a small ferroelectric signal during the first few cycles. During cycling, V_Os migrate under electric field and become ordered which causes the increase of spontaneous polarization. Fatigue happens when the broken central symmetry is resumed which leads to a decrease of polarization. The reappearance of central symmetry may be caused by the generation and concentration of more vacancies. Breakdown effect is related to the formation of a conduction filament, which leads to electric breakdown.

Based on this model, we can develop the corresponding theoretical methods to study ferroelectricity induced by V_O ordering transitions. First, the phase structures of HfO_2 with ordered V_O can be studied using a systematic structure searching method combined with first-principles software, which will be discussed in Section 2.3. To find the stable ferroelectric phases, we assume the ordered V_O arrangements in these phases, and include various impact factors such as strain, electrode, surface, dopants, V_O concentration, and other possible impact factors. Next, to find the most stable ferroelectric phases, we can calculate the free energy and select the phases with the least free energy. Furthermore, the dynamical stability of possible ferroelectric phases can be tested by phonon spectrum and molecular dynamics simulation. According to soft phonon theory of ferroelectricity, the ferroelectric phases have no negative frequencies in their phonon spectra. The ferroelectric phases are stable under room temperature if no phase transition takes place during a molecular dynamics simulation for a long simulation time under this temperature. Finally, the ferroelectric properties including spontaneous polarization and ferroelectric switching path can be calculated using first-principles methods, on the assumption that the ferroelectric transition is an order–disorder transition induced by V_O.

2.3 Phase Structure with Ordered Vacancies

V_Os are known to have a strong impact on phase structures of ferroelectric hafnia thin films. To study the ferroelectricity induced by ordered vacancies, we should first study the phase structures of HfO_2 with ordered vacancies, given the doping concentration, strain, V_O concentration, and other impact factors. For the HfO_2 thin films to show robust ferroelectricity, there must exist a stable ferroelectric phase with ordered vacancies, which can be switched under the application of external field. Crystal structure prediction (CSP) problem is the one of finding the most stable crystalline phase given only the chemical composition. CSP problem used to be an unsolved scientific problem due to the large dimension of energy landscape and the precise energy required, but nowadays can be solved by the USPEX software.[15,16] USPEX software uses an evolutionary algorithm to find the most stable structure, given only the chemical composition. External first-principles software is required to get the precise relative energy between different arrangements of atoms, for instance the widely used VASP code.[17,18]

Doping is an effective method to control the ferroelectric properties of HfO_2 thin films. Park et al.[19] found that with the Zr doping concentration increasing from 0% to 100%, the film undergoes change from ferroelectric to antiferroelectric. The different phase structures of Zr-doped HfO_2 films may account for the different ferroelectric orderings. Rushchanskii et al.[13] carried out a systematic structure searching on pure HfO_2 with V_O concentration of 12.5% using USPEX and VASP code, and Yu et al.[7] carried out structure searching on Zr-doped HfO_2 ($Hf_{0.5}Zr_{0.5}O_{2-\delta}$) with V_O concentrations of 6.25% and 12.5% (Figure 2.2). To study the effect of Zr doping, we set the V_O concentration to 12.5% and compare the different phase structures of HfO_2 and $Hf_{0.5}Zr_{0.5}O_2$, which are shown in Figure 2.2a,b, respectively.

The stoichiometric HfO_2 has five phases that are studied extensively in the studies on ferroelectricity, the m-phase ($P2_1/m$), o-phase I (oI, *Pbca*), oIII ($Pca2_1$), t-phase ($P4_2/nmc$), and cubic phase (c, $Fm\bar{3}m$). The energy ordering of these phases is m < oI < oIII < t < c. The formation of V_O in HfO_2 phase will generally break the symmetry. The space group of sub-stoichiometric phase is generally different from the corresponding stoichiometric phase, but we can group these sub-stoichiometric phases based on their similarity to the corresponding stoichiometric phases. In HfO_2 with 12.5% V_O, the M_1- and M_2-phases are m-phases with V_O; *O*- and *T*-phases are the o- and t-phases with V_O, respectively (Figure 2.2a). The *TT*- and *TM*-phases are t- and m-phases with V_O in tunnels. Compared to stoichiometric HfO_2, vacancies stabilize the t-phase. The energy ordering is m < t < o from Figure 2.2a. In $Hf_{0.5}Zr_{0.5}O_2$ with 12.5% V_O, the M_1-, O_1-, and T_1-phases are m-, o-, and t-phases with V_O, respectively (Figure 2.2b). The DT_1-phase is the t-phase with large distortion in [001] plane, with the space group $P2_1/m$. The structure data for these phases can be found in Tables A.1–A.13. In fact, the sub-stoichiometric phases can be constructed from stoichiometric ones by removing one O atom from the unit cell

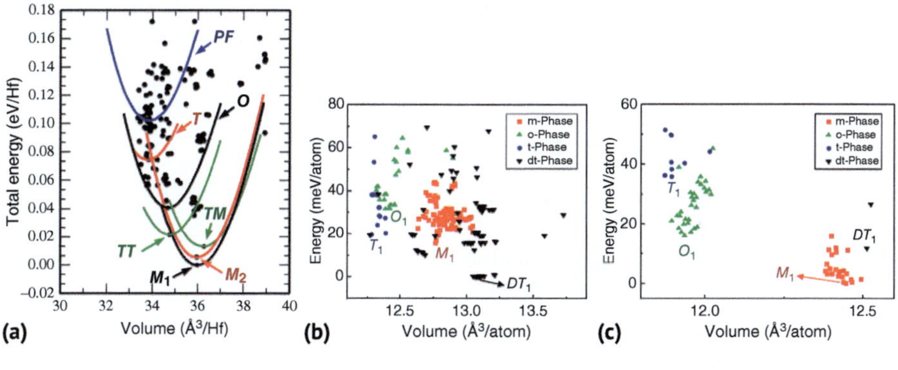

Figure 2.2 Phase structures with ordered V_O: (a) phase structure of HfO_2 with 12.5% V_O; (b) phase structure of $Hf_{0.5}Zr_{0.5}O_2$ with 12.5% V_O; and (c) phase structure of $Hf_{0.5}Zr_{0.5}O_2$ with 6.25% V_O. Panel (a): *Source*: [13] / with permission of American Physical Society. Panel (c): *Source*: [7] / with permission of American Physical Society.

with 12 atoms (the nonequivalent O atoms can be obtained by symmetry analysis), and doing structure relaxation using first-principles calculations. When constructing sub-stoichiometric phases from t-phase, one of the sub-stoichiometric phases (the DT_1-phase) will go through large lattice distortion such that the lattice angles deviate from 90° (the structure data is listed in Appendix). Though the formation of V_O stabilizes the t-phase in HfO_2, the energy of this phase is still higher than the m-phase. However, the DT_1-phase has the lowest energy in $Hf_{0.5}Zr_{0.5}O_2$ with 12.5% V_O, from which we can postulate that the ferroelectric ordering of Zr-doped HfO_2 may be relevant to the phase structure with ordered vacancies.

Next, we study the effect of V_O concentration. The phase structures of $Hf_{0.5}Zr_{0.5}O_2$ with lower V_O concentration of 6.25% are shown in Figure 2.2c. Compared to the case of 12.5% V_O, the M_1-phase has the lowest energy, not the DT_1-phase. This result is consistent with a previous report,[9] that the vacancies can stabilize non-m-phases. We will give a more detailed discussion in Section 2.5, where we will discuss the effect of V_O concentration on $Hf_{0.5}Zr_{0.5}O_{2-\delta}$ samples.

2.4 Polarization Induced by Vacancies

2.4.1 Local Dipoles Induced by Vacancies

The V_O in ferroelectric HfO_2 is widely recognized as one of the factors that can stabilize the ferroelectric $Pca2_1$ o-phase. Recently, Liu et al.[12] discovered that V_O in HfO_2 can induce localized dipoles, which indicates that V_O could also contribute to ferroelectricity. They constructed one vacancy in a 2×2×2m-supercell with 96 atoms (Figure 2.3) and calculated the spontaneous polarization using Berry phase method implemented in VASP code. In the original conference paper by Liu et al.[12] the spontaneous polarizations were 5.6 and 29.2 $\mu C/cm^2$ for different arrangements of V_O. However, we carried out Berry phase calculation on the structure in Figure 2.3, and found that the polarization is close to zero ($\approx 0.3\,\mu C/cm^2$) in spite of the arrangements of V_O. This result can be explained in the following two aspects. First, V_O concentration is not high enough to form macroscopic polarization. Second,

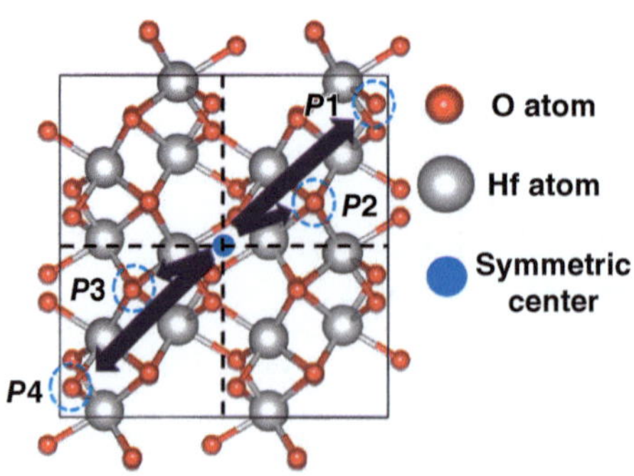

Figure 2.3 A 2×2×2 m-supercell with one vacancy. *P*1 and *P*4 as well as *P*2 and *P*3 are symmetric to the symmetric center. The formation of V_O will break the symmetry which induces spontaneous polarization. *Source*: [12] / with permission of IEEE.

polarization is not a well-defined quantity in systems with periodic boundary conditions according to the modern theory of polarization.[20] The quantity generally measured is the differential polarization relative to a reference state. A switching path is required to calculate polarization. In Section 2.4.3, we will point out that the switching of polarization induced by V_O is related to the dynamic behaviors of V_O, including V_O migration.

2.4.2 Polarization Formed by Vacancy Dipoles

From Sections 2.2 and 2.4.1, the concentrated distribution and ordered arrangements of V_O are crucial for the formation of macroscopic polarization. Figure 2.4 shows the polarization of M_1- and O-phases in HfO_2 with 12.5% V_O. These structures have ordered V_O arrangements due to the small size of unit cell and periodic boundary conditions. Polarization of O-phase is estimated to be 48 μC/cm^2 using Born effective charges. Though the m-phase in stoichiometric HfO_2 has no polarization due to the symmetry of $P2_1/m$ space group, V_O can induce a polarization of 12 μC/cm^2 in M_1-phase by the symmetry-breaking mechanism which is explained in Figure 2.3. The transition between M_1- and O-phases has a polarization of 47 μC/cm^2, though there may be large shear strain.

Zr and Hf are similar in their chemical properties. In Zr-doped HfO_2, nonpolar phases including the m-phase can also have polarization by similar symmetry-breaking mechanism. Yu et al.[7] calculated the polarization of M_1-phase in Figure 2.2c to be 27 μC/cm^2. Polarization and energy as a function of switching path are shown in Figure 2.5. One interesting feature of dipoles in HfO_2 is that they are separated by nonpolar regions (Figure 2.6). These nonpolar regions in $Pca2_1$ o-phases are named

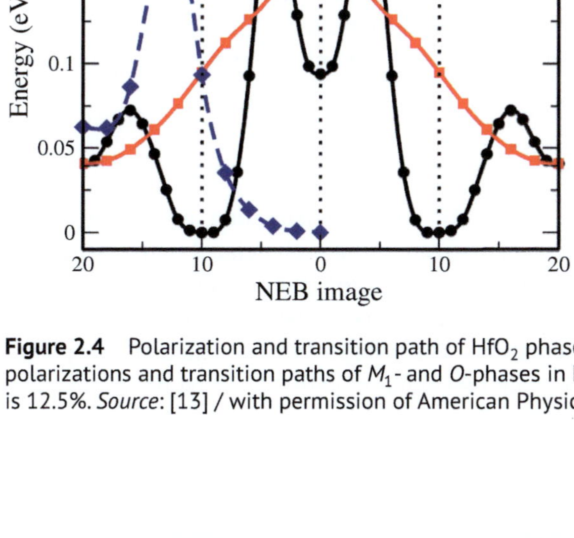

Figure 2.4 Polarization and transition path of HfO_2 phases with ordered V_O. The polarizations and transition paths of M_1- and O-phases in Figure 2.2a. The V_O concentration is 12.5%. *Source*: [13] / with permission of American Physical Society.

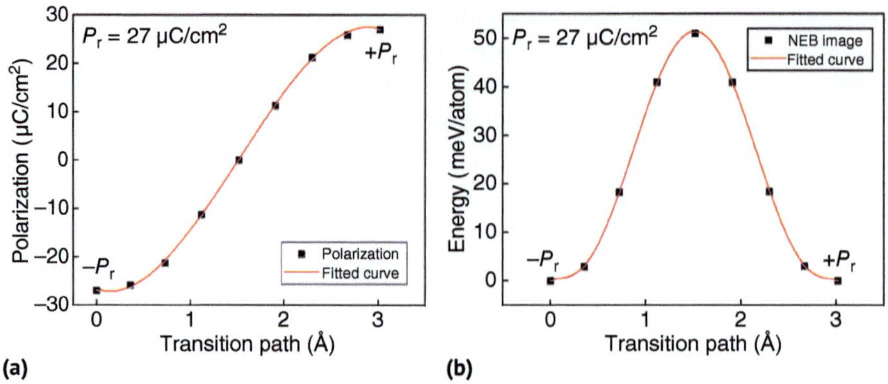

Figure 2.5 Polarization and transition path of $Hf_{0.5}Zr_{0.5}O_2$ M_1-phase with ordered V_O. (a) Polarization and (b) transition path of M_1-phase in Figure 2.2c. The V_O concentration is 6.25%. *Source*: [7] / with permission of American Physical Society.

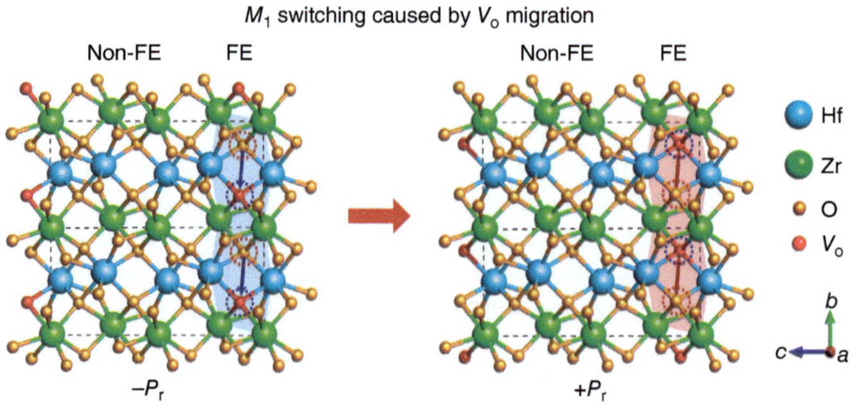

Figure 2.6 Structure and switching mechanism of $Hf_{0.5}Zr_{0.5}O_2$ M_1-phase. Polarization switching is caused by the migration of V_O in the M_1-phase to the adjacent O position highlighted using red dashed circle (or equivalently, the migration of O atom to the adjacent V_O). *Source*: [7] / with permission of American Physical Society.

"spacer" by Lee et al.[5] In conventional perovskite ferroelectrics, polarization distribution is continuous; while in HfO_2-based ferroelectrics, the dipoles may be discontinuously distributed because the existence of nonpolar regions leads to small domain wall energy.

2.4.3 Switching Path

The switching mechanism is crucial for the understanding of properties of ferroelectric materials. For instance, polarization is closely related to the switching path. According to the modern theory of polarization, the absolute value of polarization is not well defined. Only the change of polarization along a path can be calculated in first-principles calculations. For instance, we can calculate the polarization of *O*-phase in Figure 2.4 along two different paths. The first path is the one from negative polarization state of O-phase to the positive polarization state of O-phase, which gives a polarization of 48 μC/cm^2. The second path is the concatenation of three paths, the transition from O^- to M^-, the polarization reversal of M-phase (from M^- to M^+), and the transition from M^+ to O^+, which gives a polarization of 59 μC/cm^2. These two results do not lead to contradiction, for we can only calculate the difference of polarization along an arbitrary path given that the path does not have any metallic state. However, the transition barrier of these paths may be different. We propose that the path with lower barrier and higher differential polarization may be preferred in real samples.

We use the nudged elastic band (NEB) method to study the switching mechanism by first-principles calculations, taking the M_1-phase in Figure 2.2c as an example. We can construct a switching path of M_1-phase through DT_1-phase as the intermediate state, as shown in Figure 2.7. The mechanism of DT_1 switching along this path is similar to the switching of $Pca2_1$ o-phase,[4] which goes through $P4_2/nmc$ t-phase and is caused by the collective displacements of O atoms driven by electric field. The path has a barrier of 15.6 meV/atom and a polarization of 8.2 μC/cm^2. The spontaneous polarization is half of the differential polarization, 4.1 μC/cm^2. This path has some artifacts causing it not being preferred in real samples. First, the differences of lattice constants are large between M_1 and DT_1, which indicates there may be large strain during switching. Second, the polarization is small. The electric enthalpy functional is $H - \varepsilon \cdot P$, where H is the Hamiltonian, ε is the electric field, and P is the polarization. The electric field reduces the barrier during switching process, favoring ferroelectric switching. Therefore, the path with low polarization is not preferred under an external electric field.

We can also construct another switching path of M_1-phase inspired by the resistive switching of HfO_2 memristor. The switching path is shown in Figure 2.8.

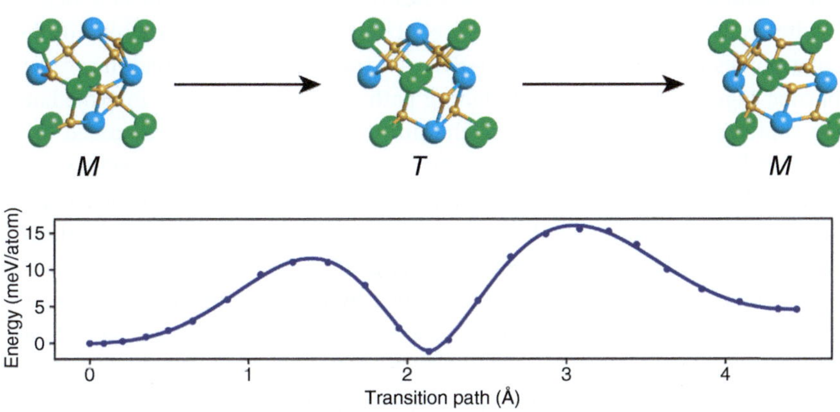

Figure 2.7 Switching path of $Hf_{0.5}Zr_{0.5}O_2$ M_1-phase through DT_1-phase. The upper figure shows the switching path and the bottom figure shows energy as a function of path. Hf, Zr, and O atoms are represented by blue, green, and yellow balls. The differential polarization is 8.2 μC/cm^2 and the barrier is 15.6 meV/atom. The barrier is not symmetric with respect to the direction of path, due to the asymmetry of the initial and final states which is caused by the nonequivalent positions of Hf and Zr atoms. The V_O concentration is 12.5%.

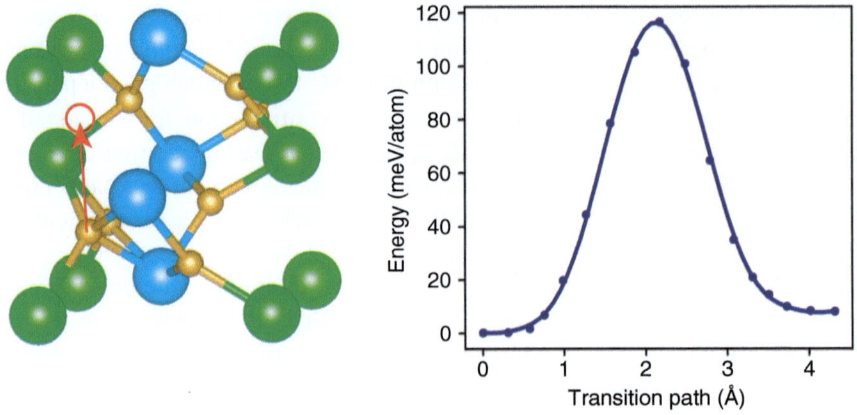

Figure 2.8 Switching path of $Hf_{0.5}Zr_{0.5}O_2$ M_1-phase by vacancy migration. The upper figure shows the switching path and the bottom figure shows energy as a function of path. Hf, Zr, and O atoms are represented by blue, green, and yellow balls. The red arrow denotes the migration direction of the O atom. The differential polarization is 112 μC/cm^2 and the barrier is 116.3 meV/atom. The barrier is not symmetric with respect to the direction of path, due to the same reason as in Figure 2.7. The V_O concentration is 12.5%.

The spontaneous polarization of M_1-phase is 56 μC/cm^2 and the barrier is 116.3 meV/atom. This path seems not preferred due to the high barrier. The high V_O concentration of 12.5% may account for the high barrier. The M_1-phase with 6.25% V_O has a polarization of 27 μC/cm^2 and a barrier of 51 meV/atom (Figure 2.5). The high V_O concentration may also lead to large leakage, which only exists near the interface with electrodes as shown in Section 2.5. The high barrier indicates that the polarization of O-deficient phases switches by another mechanism. Yu et al.[7] propose that the switching of the O-deficient phases is driven by the order–disorder transition of V_O, which has much lower barrier. Figure 2.9 shows the switching path through disordered intermediate states. To simulate the disordered states, the unit cell of M_1-phase is expanded to a 2×2×1 supercell containing four V_Os. Though a larger supercell can give a better simulation of the order–disorder transition, the size of the supercell is limited by the computational cost of first-principles calculations. The switching path has a barrier of 25 meV, and can be lowered to 15 meV if vacancies are charged.

In HfO_2, the large polarization is accompanied by the large displacements of O atoms. This fact can be seen by comparing the two paths in Figures 2.7 and 2.8, as well as the switching path of $Pca2_1$ o-phase.[4] The differential polarization ΔP along a path C can be formulated by the Born effective charge tensor Z_i of atom i as shown in Equation (2.1), where dr_i is the differential of the coordinates of atom i,

$$\Delta P = \int_C \sum_i Z_i \cdot \mathrm{d}r_i \tag{2.1}$$

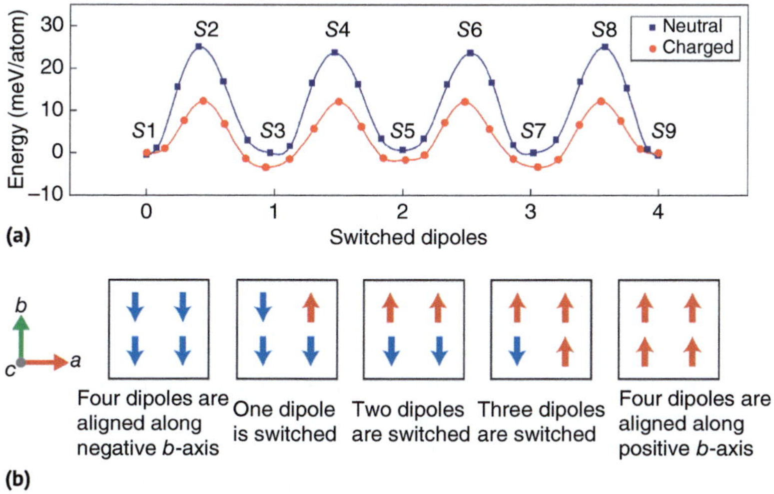

Figure 2.9 Switching path of $Hf_{0.5}Zr_{0.5}O_2$ M_1-phase by order–disorder transition. (a) The switching path of M_1-phase with a barrier of 25 meV for neutral vacancy and of 15 meV for charged vacancy. (b) Schematic showing the dipole configuration of states with different numbers of switched dipoles (*S*1, *S*3, *S*5, *S*7, and *S*9). All structures on the ferroelectric switching pathway are 2×2×1 supercells of M_1 with four V_Os containing 4×23=92 atoms. Initial and final structures are maximally ordered, with all dipoles aligned along the *a*-axis. Intermediate states are relatively disordered with partially switched dipoles.

If the Born effective charges are small, large displacements are required to induce large polarization. The Born effective charges in HfO_2 are 5.2 for Hf atom and −2.6 for O atom.[21] For comparison, the Born effective charge in $BaTiO_3$ is 9.96 (the effective charge of the soft local mode).[22] The large displacements in ferroelectric switching open the possibility for O atoms to go across the unit cell boundary and move toward one direction endlessly, for instance, the switching paths of $Pca2_1$ o-phase reported by Ma et al.[21] The SI-2 path and the inverse of SA path in the above work can form a loop, causing the polarization to go to infinity. The switching paths of m-phase with ordered vacancies also have this issue. We propose that this issue may be caused by the artifact of periodic conditions in calculations. In real samples, the migration of O atoms is restricted by interface.

2.5 Experimental Observations of Ordered Vacancies in Zr-doped Hafnia Samples

TEM techniques can be employed to characterize the phase structures of HfO_2 ferroelectric films. Yu et al.[7] fabricated TiN/$Hf_{0.5}Zr_{0.5}O_{2-\delta}$/TiN ferroelectric capacitor and employed combined TEM techniques to characterize the phase structure and vacancy ordering in the sample. The device fabrication flow is shown as follows.

First, 40nm TiN bottom electrode was formed on *p*-doped silicon (100) substrates by ion beam sputtering. Then, 15nm $Hf_{0.5}Zr_{0.5}O_{2-\delta}$ ferroelectric film was deposited by atomic layer deposition (ALD) at 280°C. Third, 30-nm TiN top electrode was sputtered and patterned into rectangles with different sizes. Finally, the films were annealed for 30s at 500°C in N_2 atmosphere to crystallize the films.

The $Hf_{0.5}Zr_{0.5}O_{2-\delta}$ films are ferroelectric with the hysteresis curve shown in Figure 2.10, measured by the positive-up–negative-down (PUND) method. These ferroelectric films were fabricated into cross-section samples by using dual-beam focused ion beam (FIB) technique. First, high-resolution transmission electron microscopy (HRTEM) was employed to select the grains with ordered V_O distribution. From the HRTEM image (Figure 2.11), the amount of virgin grains with double-modulated stripes can be found, which are possibly the target grains with ordered V_O arrangements. The preliminary phase calibration for the target grains showed that these grains have m-structure with a zone axis of ⟨021⟩. However, both the m-structure with $V_O(M_1)$ and the m-structure without $V_O(M_0)$ match the structure well, and further atomic-resolution STEM techniques are required to confirm the m-structure.

Next, high-angle annular dark-field (HAADF) imaging was used to acquire the atomic-resolution image of the TiN/$Hf_{0.5}Zr_{0.5}O_{2-\delta}$/TiN stack. Most grains with m-structure in the $Hf_{0.5}Zr_{0.5}O_{2-\delta}$ layer can be aligned toward the ⟨021⟩ zone axes. Note that the ⟨021⟩ zone axis is calibrated using the 1×1×2 supercell of m-phase which has 24atoms, which becomes the ⟨011⟩ zone axis if we use the unit cell for calibration. As a result, one of the grains with the double-modulated stacking is aligned to the ⟨021⟩ zone axes for HAADF imaging in Figure 2.12. The fast Fourier transform (FFT)

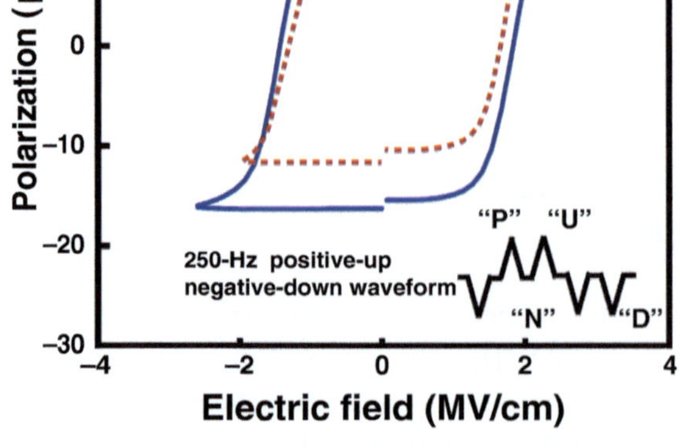

Figure 2.10 Ferroelectric hysteresis curve of TiN/$Hf_{0.5}Zr_{0.5}O_{2-\delta}$/TiN film. *Source*: [7] / with permission of American Physical Society.

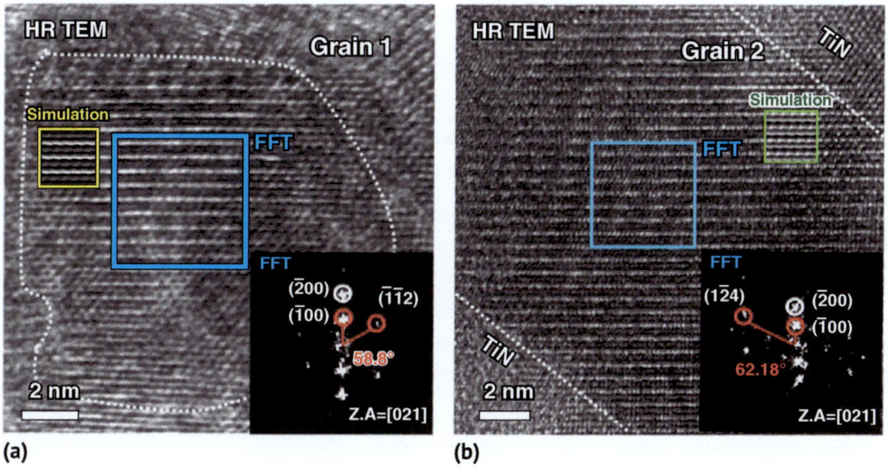

Figure 2.11 Preliminary phase calibration for the target grains with modulated stripes. (a), (b) HRTEM images of the selected grains (grains 1 and 2) in the virgin state with insets of FFT patterns, respectively. The calibration details are shown in Tables 2.1 and 2.2.

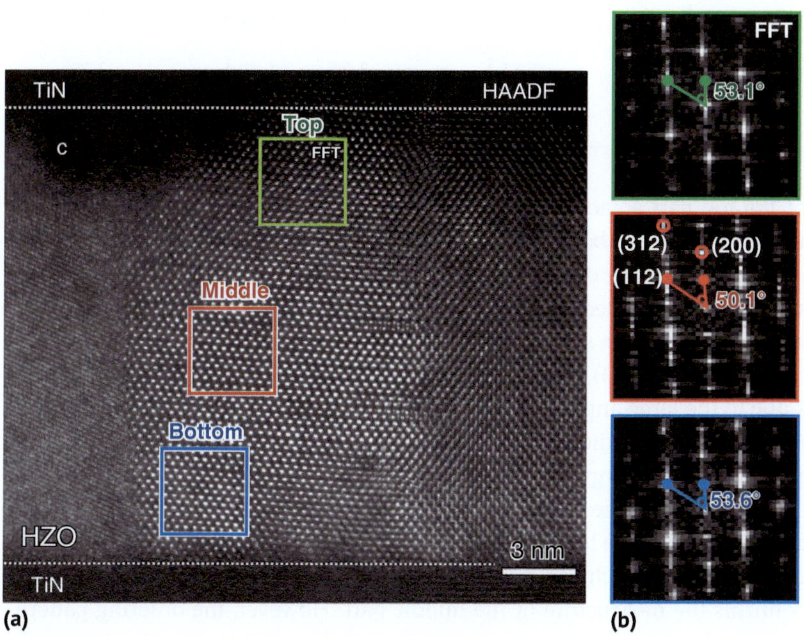

Figure 2.12 HAADF images of the TiN/$Hf_{0.5}Zr_{0.5}O_{2-\delta}$/TiN film. (a) HAADF image of the selected grain and (b) FFT patterns taken from the top, middle, and bottom regions of HAADF image in panel (a). Comparison between theoretical lattice constants and experimental measurements of the middle part is shown in Table 2.3 The structure in the middle part matches well with M_1-structure along the ⟨021⟩ direction. *Source*: [7] / with permission of American Physical Society.

Table 2.1 Calibrations for the phase structure in grain 1 of Figure 2.11a.

		M_0		M_1	
Direction	**Measurement**	**Calculation**	**Error**	**Calculation**	**Error**
$(\bar{1}00)$	5.14Å	5.09Å	1%	5.09Å	1%
$(\bar{1}1\bar{2})$	2.85Å	2.84Å	0.4%	2.84Å	0.4%
$(\bar{2}1\bar{2})$	1.97Å	1.99Å	1%	1.99Å	1%
Angle	58.8°	58.31°	0.5°	58.42°	0.4°

Table 2.2 Calibrations for the phase structure in grain 2 of Figure 2.11b.

		M_0		M_1	
Direction	**Measurement**	**Calculation**	**Error**	**Calculation**	**Error**
$(\bar{1}00)$	5.19Å	5.09Å	1.9%	5.09Å	1.9%
$(1\bar{2}4)$	1.62Å	1.68Å	3.5%	1.67Å	2.9%
$(\bar{1}\bar{2}4)$	1.87Å	1.81Å	3.2%	1.80Å	3.7%
Angle	62.18°	63.98°	1.8°	64.04°	1.9°

patterns of the top part adjacent to the top TiN electrode, the middle part, and the bottom part adjacent to the bottom TiN electrode are shown in Figure 2.12b, respectively. The phase structures of $Hf_{0.5}Zr_{0.5}O_{2-\delta}$ are calculated based on the fixed V_O concentration and zero applied strain (Figure 2.2), therefore, the middle part is focused upon. The middle part of the selected grain matches well with the M_1-structure along the ⟨021⟩ direction (Table 2.3), while the top and bottom parts are with totally different structures due to the large angle deviations. The non-m-structures of the top and bottom parts may be caused by the higher V_O concentration near the electrodes. In the following text, the electron energy loss spectroscopy (EELS) analysis will show that the V_O concentration is higher in the top and bottom parts. In Section 2.3, the first-principles calculations showed that the high V_O concentration will favor the non-m-phases such as T_1- and DT_1-phases, which can explain the phase calibration results. The HAADF imaging confirms the m-structure of the middle part. However, the ordering pattern of V_O still cannot be identified, due to the weak scattering of O atoms. The differentiated differential phase contrast (dDPC) imaging technique can directly identify the O atomic columns, which will be discussed later.

EELS analysis was conducted on the $Hf_{0.5}Zr_{0.5}O_{2-\delta}$ layer to study the V_O distribution in this layer, as shown in Figure 2.13. The O K energy loss near edge structure (ELNES) of HZO presents a characteristic doublet-peak (a and b). The formation of V_O reduces the intensity of peak a.[23,24] The V_O concentration is higher near the electrodes, which

Table 2.3 Calibrations for the phase structure in the middle part of Figure 2.12.

Direction	Measurement	Simulation of M_1	Error
(200)	2.5Å	2.6Å	3.8%
(112)	2.8Å	2.9Å	3.4%
(312)	1.6Å	1.5Å	6.7%
Angle	50.1°	49.7°	0.4°

Source: [7] / with permission of American Physical Society.

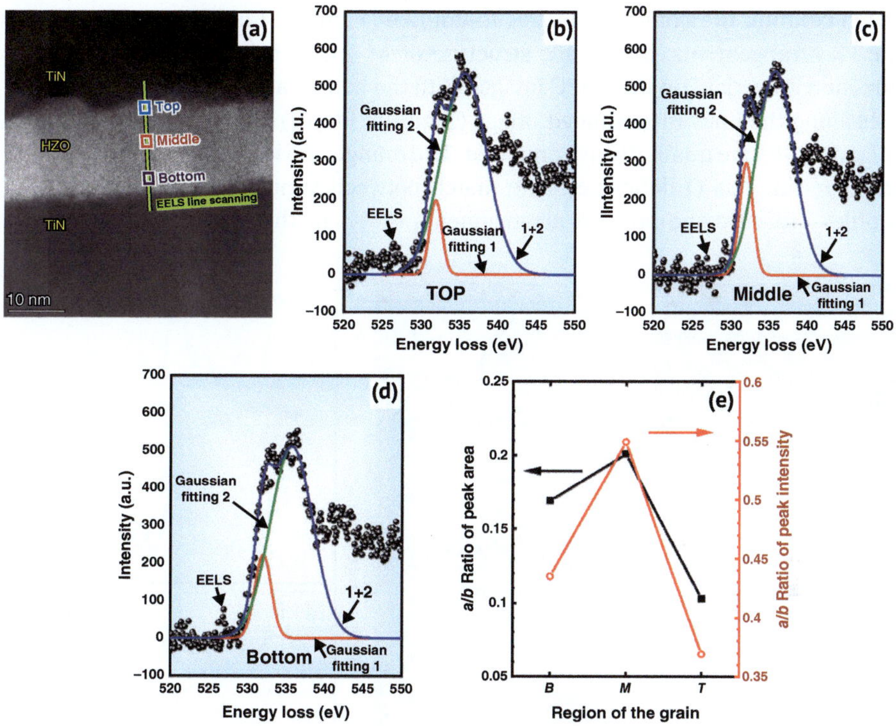

Figure 2.13 EELS analysis of $Hf_{0.5}Zr_{0.5}O_{2-\delta}$ layer. (a) HAADF image of the sample. The green line shows the direction of EELS line scanning. (b)–(d) O K EELS spectra of the top, middle, and bottom parts, averaged across TiN/$Hf_{0.5}Zr_{0.5}O_{2-\delta}$/TiN stack. (e) The measured *a*/*b* ratio values of the corresponding regions marked in panel (a). *Source*: [7] / with permission of American Physical Society.

can be seen from the intensity ratio of peak a to peak b and the peak areas of peaks a and b (Figure 2.13). The intensity ratio of peak a to peak b increases from 0.44 to 0.55, and then decreases to 0.37 in O K-edges when the line profile goes across the stack. The peak areas of peaks a and b also have the same trend.

The intensity of HAADF increases as the number of atoms and the atomic number increase in the atomic column due to the *Z*-contrast nature of HAADF imaging

(Z denotes the atomic number), therefore, the O atomic columns cannot be directly observed via HAADF due to their weak electron scattering compared to Hf and Zr atoms. Fortunately, the integrated differential phase contrast (iDPC) and dDPC imaging provides information of the local electrostatic potential field and charge density distribution, which can directly identify O atomic columns in $Hf_{0.5}Zr_{0.5}O_{2-\delta}$ thin films with high spatial resolution. The O atomic column shows a weaker contrast adjacent to Hf and Zr columns in both iDPC and dDPC images, however, in the $Hf_{0.5}Zr_{0.5}O_{2-\delta}$ sample, the dDPC has higher quality. The HAADF image and dDPC image of the $Hf_{0.5}Zr_{0.5}O_{2-\delta}$ sample are shown in Figure 2.14b and 2.14c. The simulated dDPC image of M_1-structure with ordered V_O shows that the O column along with the adjacent Hf or Zr column forms a tail-like contrast. If V_Os exist in the O column, the contrast changes to dot-like (Figure 2.14a[ii]) one. To observe the V_O arrangements, the atomic structure of M_1-phase oriented along the ⟨021⟩ direction is overlaid on the dDPC image. With the help of atomic structure, line profiles along the lines in simulated image (Figure 2.14a[ii]) and experimental image (Figure 2.14c) are qualitatively compared. The orange line is the line rich in O, while the blue line has O deficiency. The match between simulated and experimental profiles indicates the ordered arrangements of V_O in the $Hf_{0.5}Zr_{0.5}O_{2-\delta}$ sample

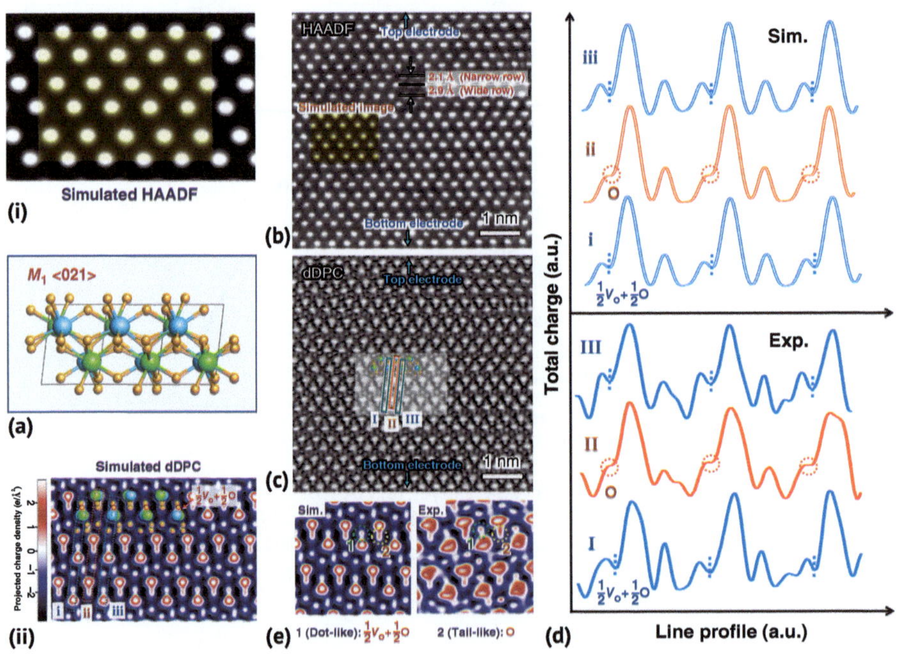

Figure 2.14 Atomic-resolution imaging of m-grain with ordered V_O. (a) Atomic model of M_1-structures along the ⟨021⟩ direction: (i) simulated HAADF image and (ii) simulated false-colored dDPC image. (b) HAADF image of the middle part in Figure 2.12. (c) The Wiener-filtered dDPC image (inverted contrast). (d) Line profiles taken along the lines i–iii in panel (a)(ii) and lines I–III in panel (c). (e) The simulated and experimental dDPC images show that the intensity changes from tail-like to dot-like at O-deficient sites. *Source*: [7] / with permission of American Physical Society.

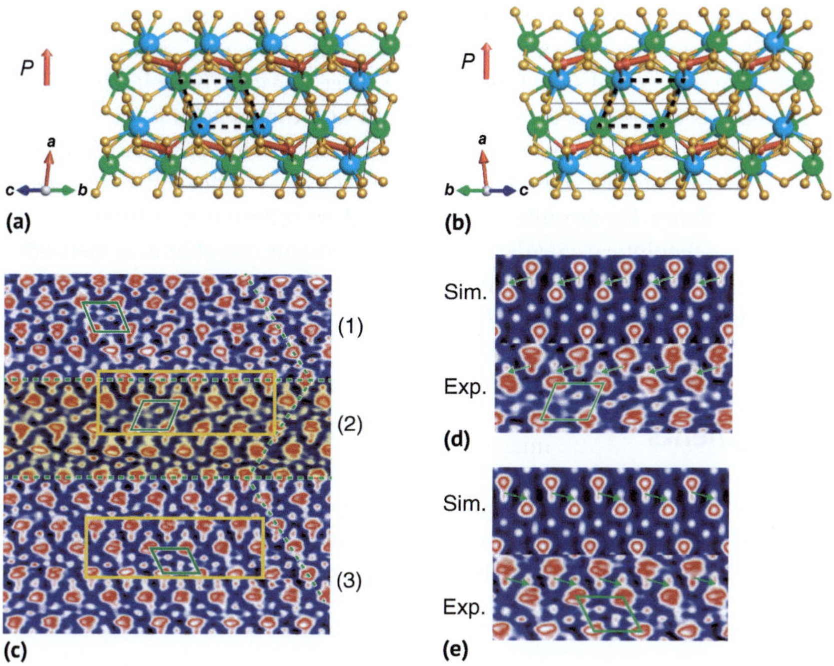

Figure 2.15 Polarization of selected grains identified from theoretical calculations. (a, b) M_1-structure oriented along the $\langle 0\bar{2}\bar{1} \rangle$ and $\langle 021 \rangle$ directions. Polarization is depicted using red arrow. The black dotted parallelogram is the half-cell which can be used to distinguish these two crystalline directions. (c) Part of the target grain with twin structure. (d) Comparison between simulated and experimental dDPC images of the selected regions (2) and (3) in panel (c). Polarization is depicted using light green arrow. *Source*: [7] / with permission of American Physical Society.

(Figure 2.14d). The sites of V_O can also be discerned from the enlarged charge density map in Figure 2.14e. In addition, the polarization of the selected grain is identified from the theoretical polarization of overlaid M_1-structure (Figure 2.15). The depicted arrows in Figure 2.15 are the projection of polarization, because the dDPC image is a projection onto the sample plane. The structures in Figure 2.15a,b compose the twin structure in Figure 2.15c, in which the twin boundary is depicted using dotted lines. The different V_O arrangements with different polarizations can be identified comparing the simulated and experimental dDPC images.

2.6 Summary

V_O is the dominant defect in fabricated ferroelectric HfO_2 thin films. V_O used to be attributed to one of the stabilization factors of ferroelectric o-phase, and plays a crucial role in ferroelectric cycling. Recently, V_Os are found to contribute to ferroelectricity through the order–disorder transition. In this chapter, the phase structures and ferroelectric properties including polarization and ferroelectric switching are discussed based on first-principles calculations. The phase structures of HfO_2 with various V_O concentrations and different dopants are given, as well as the spontaneous polarization and switching paths of these phases. The polarization of O-deficient phase is dependent on switching path and the switching mechanism of O-deficient phase is the order–disorder transition of V_O. In the last section, the experimental observations of ordered vacancies in $Hf_{0.5}Zr_{0.5}O_2$ thin films are given, which provide evidence for ferroelectricity induced by ordering transition of V_O. We hope that these theoretical calculations and experiments can offer new methods for the study on ferroelectric HfO_2, and the mechanism of vacancy-induced ferroelectricity provides new insights into the exploration on ferroelectric HfO_2-based devices.

Acknowledgments

This chapter was supported by the National Science Foundation of China (Grant Nos. T2293700 and T2293703) and the Programme of Introducing Talents of Discipline to Universities (Grant No. B18001).

References

1 Böscke, T. S., Müller, J., Bräuhaus, D. et al. (2011). Ferroelectricity in hafnium oxide thin films. *Applied Physics Letters* 99 (10): 102903. https://doi.org/10.1063/1.3634052.

2 Cheema, S. S., Kwon, D., Shanker, N. et al. (2020). Enhanced ferroelectricity in ultrathin films grown directly on silicon. *Nature* 580 (7804): 478–482. https://doi.org/10.1038/s41586-020-2208-x.

3 Luo, Q., Gong, T., Cheng, Y. et al. (2018). Hybrid 1T e-DRAM and e-NVM realized in one 10 nm node ferro FinFET device with charge trapping and domain switching effects. In: *Proceedings of the 2018 IEEE International Electron Devices Meeting (IEDM)*. IEEE. pp. 2.6.1–2.6.4. https://doi.org/10.1109/IEDM.2018.8614650.

4 Huan, T. D., Sharma, V., Rossetti, G. A. et al. (2014). Pathways towards ferroelectricity in hafnia. *Physical Review B* 90 (6): 064111. https://doi.org/10.1103/PhysRevB.90.064111.

5 Lee, H.-J., Lee, M., Lee, K. et al. (2020). Scale-free ferroelectricity induced by flat phonon bands in HfO_2. *Science* 369 (6509): 1343–1347. https://doi.org/10.1126/science.aba0067.

6 Wang, Y., Tao, L., Guzman, R. et al. (2023). A stable rhombohedral phase in ferroelectric $Hf(Zr)_{1+x}O_2$ capacitor with ultralow coercive field. *Science* 381 (6657): 558–563. https://doi.org/10.1126/science.adf6137.

7 Yu, C., Ma, H., Li, M. et al. (2024). Insights into the origin of robust ferroelectricity in HfO_2-based thin films from the order-disorder transition driven by vacancies. *Physical Review Applied* 22 (2): 024028. https://doi.org/10.1103/PhysRevApplied.22.024028.

8 Schroeder, U., Park, M. H., Mikolajick, T. et al. (2022). The fundamentals and applications of ferroelectric HfO_2. *Nature Reviews Materials* 7 (8): 653–669. https://doi.org/10.1038/s41578-022-00431-2.

9 Zhou, Y., Zhang, Y. K., Yang, Q. et al. (2019). The effects of oxygen vacancies on ferroelectric phase transition of HfO_2-based thin film from first-principle. *Computational Materials Science* 167: 143–150. https://doi.org/10.1016/j.commatsci.2019.05.041.

10 Chen, J., Jin, C., Yu, X. et al. (2022). Impact of oxygen vacancy on ferroelectric characteristics and its implication for wake-up and fatigue of HfO_2-based thin films. *IEEE Transactions on Electron Devices* 69 (9): 5297–5301. https://doi.org/10.1109/TED.2022.3190256.

11 Nukala, P., Ahmadi, M., Wei, Y. et al. (2021). Reversible oxygen migration and phase transitions in hafnia-based ferroelectric devices. *Science* 372 (6542): 630–635. https://doi.org/10.1126/science.abf3789.

12 Liu, C., Liu, F., Luo, Q. et al. (2018). Role of oxygen vacancies in electric field cycling behaviors of ferroelectric hafnium oxide. In: *Proceedings of the 2018 IEEE International Electron Devices Meeting (IEDM)*. IEEE. pp. 16.4.1–16.4.4. https://doi.org/10.1109/IEDM.2018.8614540.

13 Rushchanskii, K. Z., Blügel, S., and Ležaić, M. (2021). Ordering of oxygen vacancies and related ferroelectric properties in $HfO_{2-\delta}$. *Physical Review Letters* 127 (8): 087602. https://doi.org/10.1103/PhysRevLett.127.087602.

14 Blinc, R. and Žekš, B. (1972). Dynamics of order-disorder-type ferroelectrics and anti-ferroelectrics. *Advances in Physics* 21 (93): 693–757. https://doi.org/10.1080/00018737200101348.

15 Oganov, A. R., Lyakhov, A. O., and Valle, M. (2011). How evolutionary crystal structure prediction works – and why. *Accounts of Chemical Research* 44 (3): 227–237. https://doi.org/10.1021/ar1001318.

16 Lyakhov, A. O., Oganov, A. R., Stokes, H. T. et al. (2013). New developments in evolutionary structure prediction algorithm USPEX. *Computer Physics Communications* 184 (4): 1172–1182. https://doi.org/10.1016/j.cpc.2012.12.009.

17 Kresse, G. and Furthmüller, J. (1996). Efficient iterative schemes for ab initio total-energy calculations using a plane-wave basis set. *Physical Review B* 54 (16): 11169–11186. https://doi.org/10.1103/PhysRevB.54.11169.

18 Kresse, G. and Hafner, J. (1993). Ab initio molecular dynamics for liquid metals. *Physical Review B* 47 (1): 558–561. https://doi.org/10.1103/PhysRevB.47.558.

19 Park, M. H., Lee, Y. H., Kim, H. J. et al. (2015). Ferroelectricity and antiferroelectricity of doped thin HfO_2-based films. *Advanced Materials* 27 (11): 1811–1831. https://doi.org/10.1002/adma.201404531.

20 Resta, R. (1994). Macroscopic polarization in crystalline dielectrics: The geometric phase approach. *Reviews of Modern Physics* 66 (3): 899–915. https://doi.org/10.1103/RevModPhys.66.899.

21 Ma, L., Wu, J., Zhu, T. et al. (2023). Ultrahigh oxygen ion mobility in ferroelectric hafnia. *Physical Review Letters* 131 (25): 256801. https://doi.org/10.1103/PhysRevLett.131.256801.

22 Zhong, W., Vanderbilt, D., and Rabe, K. M. (1995). First-principles theory of ferroelectric phase transitions for perovskites: The case of $BaTiO_3$. *Physical Review B* 52: 6301–6312. https://doi.org/10.1103/PhysRevB.52.6301.
23 Wilk, G. D. and Muller, D. A. (2003). Correlation of annealing effects on local electronic structure and macroscopic electrical properties for HfO_2 deposited by atomic layer deposition. *Applied Physics Letters* 83 (19): 3984–3986. https://doi.org/10.1063/1.1626019.
24 Stemmer, S., Chen, Z. Q., Zhu, W. J. et al. (2003). Electron energy-loss spectroscopy study of thin film hafnium aluminates for novel gate dielectrics. *Journal of Microscopy* 210 (1): 74–79. https://doi.org/10.1046/j.1365-2818.2003.01175.x.

Appendix: Phase Structure Data

Structure data for M_1-, O_1-, DT_1-, and T_1-phases with different V_O concentrations are listed in Tables A.1–A.13. HZO is the acronym for $Hf_{0.5}Zr_{0.5}O_2$.

Table A.1 Indices for structure data.

Phase	Space group	HfO_2	HZO, 12.5% V_O	HZO, 6.25% V_O
M_1	$P2_1/c$	1.2	1.3	1.4
O_1	$Pca2_1$	1.5	1.6	1.7
DT_1	$P2_1/m$	1.8	1.9	1.10
T_1	$P4_2/nmc$	1.11	1.12	1.13

Table A.2 Structure data for $P2_1/c$.

Formula		Hf_4O_8	
a, b, c (Å)	5.14232	5.19515	5.32604
α, β, γ (deg)	90	99.6745	90
Atom		**Relative coordinates**	
Hf_1	0.724040985	0.457318991	0.292109013
Hf_2	0.275959015	0.957319021	0.207891002
Hf_3	0.275959015	0.542680979	0.707890987
Hf_4	0.724040985	0.042681001	0.792109013
O_1	0.551113009	0.742603004	0.022291999
O_2	0.448886991	0.242603004	0.477708012
O_3	0.448886991	0.257396996	0.977707982
O_4	0.551113009	0.757396996	0.522292018

Formula		Hf_4O_8	
O_5	0.932151020	0.330121994	0.652906001
O_6	0.067849003	0.830121994	0.847093999
O_7	0.067849003	0.669878006	0.347093999
O_8	0.932151020	0.169878006	0.152906001

Table A.3 Structure data for $P2_1/c$ with 12.5% V_O.

Formula		$Hf_2Zr_2O_7$	
a, b, c (Å)	5.16889	5.27315	5.29091
α, β, γ (deg)	91.2876	99.5843	90.0815
Atom		**Relative coordinates**	
Hf_1	0.719663316	0.451743295	0.295569980
Hf_2	0.729898810	0.050866183	0.788886615
Zr_1	0.279559564	0.542031116	0.712658113
Zr_2	0.273588778	0.948998114	0.211927580
O_1	0.452980943	0.238746844	0.481365451
O_2	0.457178084	0.267876129	0.981990534
O_3	0.556623726	0.755925242	0.523337756
O_4	0.923260773	0.343970608	0.663713484
O_5	0.073265966	0.847407990	0.830095491
O_6	0.063877848	0.655000633	0.328289421
O_7	0.918989191	0.154830846	0.159873575

Table A.4 Structure data for $P2_1/c$ with 6.25% V_O.

Formula		$Hf_4Zr_4O_{15}$	
a, b, c (Å)	5.17286	5.25179	10.66420
α, β, γ (deg)	89.6067	99.4018	90.2399
Atom		**Relative coordinates**	
Hf_1	0.274688000	0.952522000	0.102711000
Hf_2	0.724738000	0.038387000	0.395100000
Hf_3	0.278826000	0.956224000	0.603357000
Hf_4	0.721816000	0.039647000	0.896567000
Zr_1	0.720076000	0.451163000	0.143879000
Zr_2	0.272190000	0.542937000	0.354380000

(Continued)

Table A.4 (Continued)

Formula		$Hf_4Zr_4O_{15}$	
Zr_3	0.726337000	0.453266000	0.643995000
Zr_4	0.274998000	0.542599000	0.854485000
O_1	0.545957000	0.744761000	0.009014000
O_2	0.450395000	0.248315000	0.483274000
O_3	0.546332000	0.773546000	0.258731000
O_4	0.923727000	0.330757000	0.328507000
O_5	0.069716000	0.837178000	0.422276000
O_6	0.072850000	0.663879000	0.171763000
O_7	0.932224000	0.161006000	0.078310000
O_8	0.550161000	0.740956000	0.510582000
O_9	0.447511000	0.237961000	0.738464000
O_{10}	0.447574000	0.252314000	0.992688000
O_{11}	0.552394000	0.761496000	0.760337000
O_{12}	0.927073000	0.331487000	0.829354000
O_{13}	0.064260000	0.834487000	0.921616000
O_{14}	0.072862000	0.666235000	0.670581000
O_{15}	0.935486000	0.163130000	0.578242000

Table A.5 Structure data for $Pca2_1$.

Formula		Hf_4O_8	
a, b, c (Å)	5.04014	5.07445	5.26928
α, β, γ (deg)	90	90	90
Atom		**Relative coordinates**	
Hf_1	0.267435998	0.242569000	0.033794999
Hf_2	0.267435998	0.742569029	0.466205001
Hf_3	0.732563972	0.242569000	0.533794999
Hf_4	0.732563972	0.742569029	0.966205001
O_1	0.072143003	0.376846999	0.367911011
O_2	0.072143003	0.876847029	0.132089004
O_3	0.459796995	0.493584007	0.733287990
O_4	0.459796995	0.993583977	0.766712010
O_5	0.540202975	0.993583977	0.266712010
O_6	0.540202975	0.493584007	0.233288005
O_7	0.927856982	0.376846999	0.867910981
O_8	0.927856982	0.876847029	0.632089019

Table A.6 Structure data for $Pca2_1$ with 12.5% V_O.

Formula		$Hf_2Zr_2O_7$	
a, b, c (Å)	5.08444	5.04982	5.31225
α, β, γ (deg)	89.2709	89.2372	88.9638
Atom		**Relative coordinates**	
Hf_1	0.260596953	0.733937251	0.458118298
Hf_2	0.732315601	0.243865142	0.536954568
Zr_1	0.271279092	0.247903844	0.028637117
Zr_2	0.735246214	0.751533749	0.974189267
O_1	0.069353797	0.377977844	0.370448623
O_2	0.086065678	0.871261490	0.120383213
O_3	0.456438449	0.496149796	0.731307092
O_4	0.536275051	0.990773466	0.284287238
O_5	0.539277067	0.489776304	0.236042368
O_6	0.924857186	0.372816512	0.864393837
O_7	0.928497912	0.882420602	0.628526378

Table A.7 Structure data for $Pca2_1$ with 6.25% V_O.

Formula		$Hf_4Zr_4O_{15}$	
a, b, c (Å)	5.08939	10.15158	5.31489
α, β, γ (deg)	89.6319	89.6336	89.4950
Atom		**Relative coordinates**	
Hf_1	0.259979000	0.367250000	0.458601000
Hf_2	0.727117000	0.120807000	0.532703000
Hf_3	0.269069000	0.868114000	0.465534000
Hf_4	0.738310000	0.619807000	0.536125000
Zr_1	0.263278000	0.121756000	0.029235000
Zr_2	0.731812000	0.374670000	0.973671000
Zr_3	0.274504000	0.621532000	0.031023000
Zr_4	0.738412000	0.871058000	0.968419000
O_1	0.061738000	0.190431000	0.366875000
O_2	0.084518000	0.435442000	0.121684000
O_3	0.458233000	0.253909000	0.729893000
O_4	0.538363000	0.494132000	0.284436000
O_5	0.533567000	0.243529000	0.235369000

(Continued)

Table A.7 (Continued)

Formula		**$Hf_4Zr_4O_{15}$**	
O_6	0.920307000	0.183859000	0.862883000
O_7	0.932992000	0.441862000	0.632419000
O_8	0.073824000	0.688563000	0.371928000
O_9	0.077549000	0.936035000	0.132178000
O_{10}	0.458708000	0.737825000	0.729334000
O_{11}	0.460368000	0.994999000	0.762039000
O_{12}	0.538794000	0.994652000	0.270899000
O_{13}	0.543058000	0.744884000	0.235689000
O_{14}	0.928916000	0.688443000	0.867247000
O_{15}	0.932568000	0.939630000	0.626849000

Table A.8 Structure data for $P2_1/m$.

Formula		**Hf_2O_4**	
a, b, c (Å)	3.81659	3.42087	5.36778
α, β, γ (deg)	90	75.5926	90
Atom		**Relative coordinates**	
Hf_1	0.663708360	0	0.540772174
Hf_2	0.001404330	0.500000000	0.001861692
O_1	0.074749127	0	0.194701001
O_2	0.590339060	0.500000000	0.347968664
O_3	0.096499442	0	0.709469397
O_4	0.568548041	0.500000000	0.833227073

Table A.9 Structure data for $P2_1/m$ with 12.5% V_O.

Formula		**$Hf_2Zr_2O_7$**	
a, b, c (Å)	5.23979	5.24426	5.48020
α, β, γ (deg)	101.9923	101.9911	83.7476
Atom		**Relative coordinates**	
Hf_1	0.663708360	0	0.540772174
Hf_2	0.001404330	0.500000000	0.001861692
O_1	0.074749127	0	0.194701001
O_2	0.590339060	0.500000000	0.347968664

Formula		$Hf_2Zr_2O_7$	
O_3	0.096499442	0	0.709469397
O_4	0.568548041	0.500000000	0.833227073
Hf_1	0.548807990	0.049710165	0.472622595
Hf_2	0.952043963	0.954515220	0.019828273
Zr_1	0.054980592	0.545185740	0.477781109
Zr_2	0.460101783	0.456659562	0.018530881
O_1	0.758386046	0.763367588	0.658700635
O_2	0.246935190	0.240514149	0.182921892
O_3	0.735938891	0.746510849	0.171249378
O_4	0.733111575	0.234568653	0.827608898
O_5	0.233292962	0.730930460	0.836352962
O_6	0.251324499	0.779271830	0.321742107
O_7	0.775076509	0.248765784	0.317353269

Table A.10 Structure data for $P2_1/m$ with 6.25% V_O.

Formula		$Hf_4Zr_4O_{15}$	
a, b, c (Å)	5.83127	6.98094	8.23615
α, β, γ (deg)	115.0898	71.9667	90.0555
Atom		**Relative coordinates**	
Hf_1	0.463582898	0.035112016	0.811902915
Hf_2	0.526766018	0.982902167	0.203747390
Hf_3	0.031450370	0.232287027	0.702779456
Hf_4	0.963423496	0.788625366	0.311036321
Zr_1	0.962169227	0.285297742	0.309759727
Zr_2	0.531224759	0.484046839	0.204459026
Zr_3	0.460628969	0.542541026	0.810202574
Zr_4	0.026781111	0.734126623	0.702949699
O_1	0.824839576	0.938757037	0.606338219
O_2	0.589470711	0.799595622	0.333239101
O_3	0.162894484	0.081273712	0.406063193
O_4	0.401238181	0.225347670	0.687952465
O_5	0.098814017	0.546321054	0.819838465
O_6	0.103410270	0.037495342	0.820340416
O_7	0.901271802	0.970552186	0.179966133
O_8	0.906897220	0.474652036	0.183549982

(Continued)

Table A.10 (Continued)

Formula		$\mathbf{Hf_4Zr_4O_{15}}$	
O_9	0.163734488	0.589441649	0.408376792
O_{10}	0.656022585	0.837254983	0.904689919
O_{11}	0.827664713	0.430545691	0.606242578
O_{12}	0.330057230	0.183698939	0.107548268
O_{13}	0.674882195	0.330563540	0.909629396
O_{14}	0.330490197	0.687602868	0.107985023
O_{15}	0.585576483	0.297150864	0.334055942

Table A.11 Structure data for $P4_2/nmc$.

Formula		$\mathbf{Hf_2O_4}$	
a, b, c (Å)	3.59333	3.59333	5.22467
α, β, γ (deg)	90	90	90
Atom		**Relative coordinates**	
Hf_1	0.500000000	0.500000000	0.500000000
Hf_2	0	0	0
O_1	0	0.500000000	0.695308030
O_2	0.500000000	0	0.804691970
O_3	0.500000000	0	0.304692000
O_4	0	0.500000000	0.195308000

Table A.12 Structure data for $P4_2/nmc$ with 12.5% V_O.

Formula		$\mathbf{Hf_2Zr_2O_7}$	
a, b, c (Å)	5.08820	5.08820	5.22070
α, β, γ (deg)	90	90	91.2927
Atom		**Relative coordinates**	
Hf_1	0.507352891	0.992647109	0.504255738
Hf_2	0.992647109	0.507352891	0.504255738
Zr_1	0.005454522	0.005454522	0.002016079
Zr_2	0.494545478	0.494545478	0.002016079
O_1	0.250000000	0.250000000	0.676643672
O_2	0.750000000	0.750000000	0.700409009
O_3	0.750000000	0.750000000	0.208177719

Formula		$Hf_2Zr_2O_7$	
O_4	0.750000000	0.250000000	0.801460567
O_5	0.250000000	0.750000000	0.801460567
O_6	0.250000000	0.750000000	0.301998415
O_7	0.750000000	0.250000000	0.301998415

Table A.13 Structure data for $P4_2/nmc$ with 6.25% V_O.

Formula		$Hf_4Zr_4O_{15}$	
a, b, c (Å)	7.26380	6.28054	6.43148
α, β, γ (deg)	68.8691	91.0964	91.1055
Atom		**Relative coordinates**	
Hf_1	0.387451000	0.631945000	0.404968000
Hf_2	0.151904000	0.110834000	0.431274000
Hf_3	0.653902000	0.611611000	0.932206000
Hf_4	0.887626000	0.123646000	0.929272000
Zr_1	0.149862000	0.617705000	0.937629000
Zr_2	0.388893000	0.118344000	0.932832000
Zr_3	0.886547000	0.628032000	0.433539000
Zr_4	0.642080000	0.120048000	0.441195000
O_1	0.166135000	0.971658000	0.774458000
O_2	0.156670000	0.462276000	0.285186000
O_3	0.397823000	0.507422000	0.878302000
O_4	0.133285000	0.206040000	0.039657000
O_5	0.152133000	0.712099000	0.531789000
O_6	0.393145000	0.015675000	0.324857000
O_7	0.405799000	0.280259000	0.579105000
O_8	0.893849000	0.522203000	0.825862000
O_9	0.899503000	0.773166000	0.079139000
O_{10}	0.887357000	0.265024000	0.587817000
O_{11}	0.638649000	0.482975000	0.273123000
O_{12}	0.895417000	0.007330000	0.357205000
O_{13}	0.633901000	0.713755000	0.537712000
O_{14}	0.628702000	0.959239000	0.795006000
O_{15}	0.654164000	0.211857000	0.041883000

3

Origin and Multiple Regulations of Ferroelectric Properties in HfO_2-based Materials

Tian-Ling Ren, Houfang Liu, Dapeng Huang, and Yi Yang*

School of Integrated Circuits and Beijing National Research Center for Information Science and Technology, Tsinghua University, Beijing, China

*Corresponding author: rentl@tsinghua.edu.cn

Ferroelectricity, characterized by the spontaneous alignment of electric dipoles within certain dielectric materials and the reversibility of this polarization state under an external electric field, underpins critical functionalities in modern electronics. This nonvolatile, switchable polarization is particularly advantageous for applications such as nonvolatile memory, where it enables low-power data storage, as well as for sensors, actuators, and high-density capacitors leveraging coupled electromechanical or energy storage properties. For decades, the field has been dominated by perovskite oxides ($AB\ O_3$), including lead zirconate titanate (PZT), barium titanate ($BaTiO_3$), and sodium potassium niobate ($NaKNbO_3$). While these canonical ferroelectrics exhibit robust ferroelectric performance, their integration into nanoscale devices faces formidable challenges related to dimensional scaling, direct compatibility with silicon-based complementary metal–oxide–semiconductor (CMOS) platforms, and environmental/health concerns arising from the toxicity of constituent elements, notably lead in PZT.

Recently, the unexpected discovery of robust ferroelectricity in thin films of hafnium dioxide (HfO_2) and its doped derivatives has invigorated the field, offering a potentially transformative pathway for next-generation nanoelectronics. HfO_2, already a cornerstone high-k dielectric material in advanced CMOS transistors, exhibits ferroelectricity primarily in specific non-centrosymmetric orthorhombic crystal phases (e.g., $Pca2_1$), stabilized under nanoscale confinement, doping, or strain. This compatibility with silicon manufacturing, coupled with its lead-free nature and scalability, positions HfO_2-based ferroelectrics as exceptionally promising candidates for overcoming the limitations of traditional perovskites. This chapter critically examines the origins of ferroelectricity in hafnia systems, delineates their distinctive material characteristics compared to conventional ferroelectrics, and comprehensively reviews the diverse strategies employed to precisely control and optimize their ferroelectric behavior for targeted device applications.

3.1 The Origin and Characteristics of HfO_2-based Ferroelectricity

Hafnium dioxide (HfO_2), traditionally recognized for its high-*k* dielectric properties, has recently garnered considerable attention for its ferroelectric behavior, particularly when processed into thin films or nanometer-scale dimensions. Unlike most conventional ferroelectric materials that possess perovskite structures, HfO_2 crystallizes in a fluorite structure. This unique crystal structure, coupled with specific processing conditions, made the initial discovery of ferroelectricity in HfO_2 unexpected and has driven extensive research into its underlying mechanisms.

3.1.1 Crystal Structures and Non-centrosymmetric Ferroelectric Phase

The observation of ferroelectricity requires the presence of a non-centrosymmetric crystal structure, which allows for spontaneous polarization. In 2011, the groundbreaking discovery of ferroelectricity in Si-doped HfO_2 thin films challenged the previous understanding that the fluorite structure of HfO_2 would preclude such behavior due to its inherent centrosymmetric structure in stable bulk phases. Crystallographic considerations highlight the symmetries required to support ferroelectricity. Therefore, analyzing the space point groups can help identify the origin of ferroelectricity in HfO_2. Among the 32 crystallographic point groups, 21 exhibit a non-centrosymmetric structure. Only the point groups C_1, C_2, C_s, C_3, C_{3v}, C_{2v}, C_4, C_{4v}, C_6, and C_{6v} are considered polar point groups, which are likely to demonstrate spontaneous polarization. Huan et al.[1] identified several potential non-centrosymmetric phases that could exhibit ferroelectricity: orthorhombic $Pca2_1$ (o-III or PO), orthorhombic $Pmn2_1$ (o-IV), monoclinic *Pm*, monoclinic *Cc*, and triclinic *P*1. These polar phases can emerge through symmetry reduction from higher-symmetry, nonpolar phases, such as the cubic $Fm\bar{3}m$ phase (c-phase), as illustrated in the symmetry reduction flowchart in Figure 3.1a.

First-principles calculations comparing the energies of these five polar phases revealed that the *Pm*, *Cc*, and *P*1 phases have significantly higher energies compared to the $Pca2_1$ and $Pmn2_1$ phases under typical conditions. This suggests that $Pca2_1$ and $Pmn2_1$ phases are the most likely polar phases responsible for ferroelectricity in HfO_2 thin films. Although most experimental evidence points to orthorhombic $Pca2_1$ phase as the primary ferroelectric phase, the o-IV-phase also warrants consideration due to its similar free energy at certain temperature and pressure conditions. Notably, the orthogonal $Pca2_1$ and $Pmn2_1$ structures can be viewed as distortions of the nonpolar tetragonal $P4_2/nmc$ structure along the [110] and [100] directions, respectively.

The numerous phases of HfO_2 in the fluorite structure and the ease of phase interconversion pose significant challenges for the controlled synthesis and stabilization of the desired ferroelectric phases. In the bulk at room temperature and pressure, the most stable phase is the monoclinic $P2_1/c$ phase (m-phase). Upon heating, the

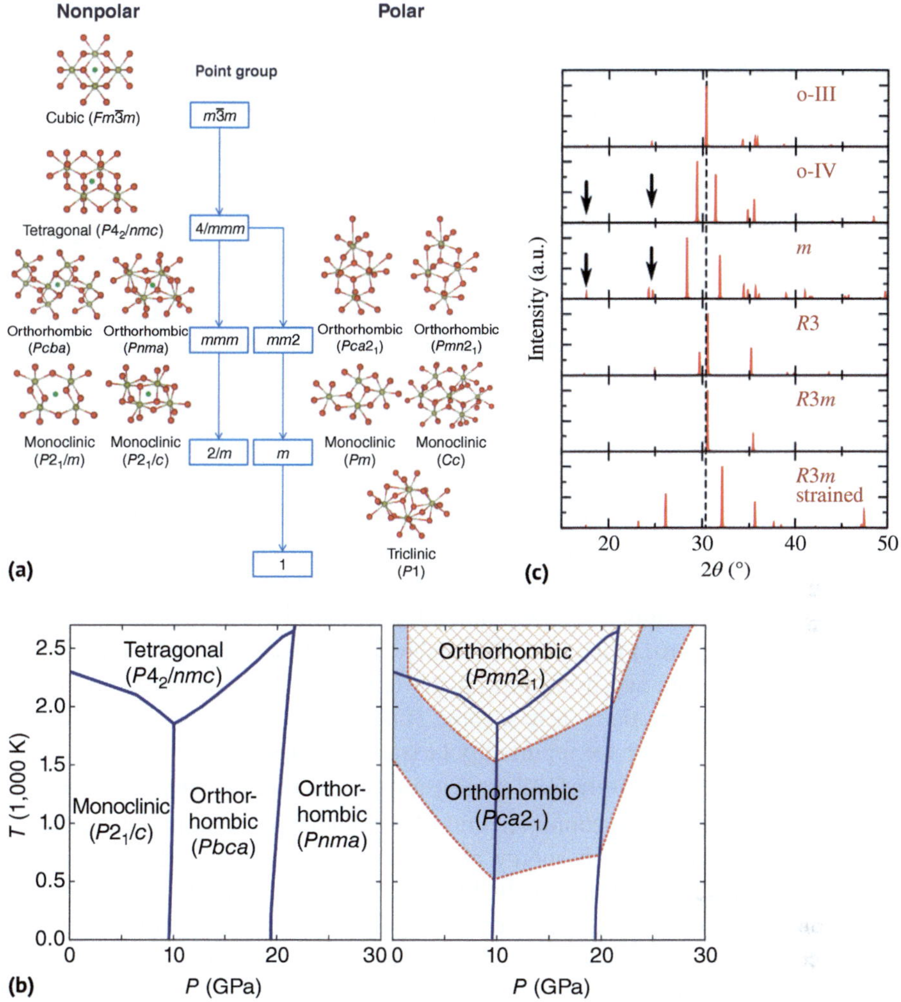

Figure 3.1 (a) Symmetry reduction flowchart of low-energy phases of hafnia, starting from the $m\bar{3}m$ cubic point symmetry. (b) Computed equilibrium phase diagram of hafnia and regimes in which $\Delta F(P,T)$, the free energy difference between $Pca2_1$ and $Pmn2_1$ phases and the equilibrium phases, is small, i.e., $\Delta F(P,T) < k_B T/5$. *Source*: [1] / with permission of American Physical Society. (c) Simulated XRD patterns for the various competing phases. *Source*: [7] / with permission of American Physical Society.

m-phase undergoes a first-order phase transition to the tetragonal $P4_2/nmc$ phase (t-phase) at approximately 2,000 K, and subsequently to the c-phase at temperatures above 2,870 K.[2] Under pressure, the m-phase can transform into the antipolar orthorhombic *Pbca* phase (o-I), and at even higher pressure, to the orthorhombic *Pnma* phase (o-II),[3] as shown in Figure 3.1b. All these bulk phases are nonpolar.

Given the nonpolar nature of the stable bulk phases, the observation of ferroelectricity in HfO_2 thin films was initially surprising. Following the discovery of the ferroelectric phase in 2011,[4] the non-centrosymmetric $Pca2_1$ (o-III) phase was primarily responsible for ferroelectricity. This understanding persisted until the more recent reports on potential contributions from rhombohedral (r) *R*3*m* and *R*3 phases.[5,6] However, the stabilization of these polar phases requires specific conditions, including doping, applied stress, and interface energy effects.

Simulated X-ray powder diffraction (XRD) patterns of the $Pca2_1$ (o-III), $Pmn2_1$ (o-IV), $P2_1/c$ (m), *R*3, and *R*3*m* phases (Figure 3.1c) show that the distinctive peaks of the polar phases closely resemble each other, making unambiguous phase identification based solely on the XRD results difficult, particularly in polycrystalline films.[7] Analysis using simulated electron diffraction patterns can provide additional insights,[8] revealing structural similarities, for instance, between the *R*3 and *R*3*m* phases and the distorted $Pca2_1$ structure, further contributing to experimental ambiguity.

Regarding the atomic origin of polarization, the non-centrosymmetric $Pca2_1$ phase contains both four-coordinated oxygen (O4) and three-coordinated oxygen (O3) atoms in addition to Hf atoms.[9] Displacement of O3 atoms primarily along the (001) direction breaks the inversion symmetry, leading to macroscopic spontaneous polarization along the (001) direction. The t-phase is typically considered a low-barrier precursor to the $Pca2_1$ phase at lower temperatures. In the $Pmn2_1$ phase, the inversion symmetry is broken through the displacement of the O3 atoms, mainly along the (−101) direction.[10] Notably, the orthogonal $Pca2_1$ and $Pmn2_1$ structures can be obtained by distorting the tetragonal $P4_2/nmc$ structure along its [110] and [100] directions, respectively. Theoretical calculations using density functional theory (DFT) have reported polarization values of approximately 54.43 and 60.81 $\mu C/cm^2$, respectively. DFT studies also suggest that in-plane stress can stabilize the $Pmn2_1$ phase. The rhombohedral *R*3*m* and *R*3 phases, proposed in 2018,[5] are generally considered high-energy phases. DFT calculations indicate that the unstressed *R*3*m* phase exhibits negligible spontaneous polarization,[8] effectively being nonpolar. Polarity in the *R*3*m* phase appears to be induced primarily under specific conditions, such as applied in-plane equibiaxial stress along the (111) orientation.[11] Based on comprehensive experimental observations (including XRD and HRTEM [high-resolution transmission electron microscopy]), theoretical calculations (DFT for energies and polarization), and thermodynamic stability analyses, the $Pca2_1$ phase remains widely considered as the primary non-centrosymmetric structure responsible for influencing ferroelectric polarization for the observed ferroelectricity in HfO_2-based thin films, although the potential contribution of other polar phases under specific conditions is an ongoing area of research.

3.1.2 Phonon-driven Ferroelectric Phase Transitions

While the ferroelectricity observed in HfO_2-based systems has been widely attributed to the stabilization of metastable polymorphs (e.g., orthorhombic $Pca2_1$ phase) through strategies like doping and defect engineering, the atomistic origins underlying their distinctive characteristics remain subjects of active investigation and debate within the ferroelectrics community. These unique characteristics include anomalously high coercive fields (typically 1–4.5 MV/cm), size-independent polarization retention down to sub-nanometer thickness, and exceptional domain structure. A significant theoretical advancement contributing to the understanding of HfO_2-based ferroelectrics is the prediction and observation of unusually flat polar phonon bands within the material. Polar phonons represent lattice vibrations that interact with the electric dipoles of the material, and their characteristics play a crucial role in the switching behavior of ferroelectrics. Understanding phase transitions in ferroelectrics requires considering the collective atomic displacements in both the nonpolar and polar phases. Since phonons are collective excitations involving multiple unit cells, the existence of ferroelectric soft modes (phonons with frequencies approaching zero at a phase transition) inherently requires a certain spatial scale to develop. In HfO_2, the presence of flat phonon bands is theorized to facilitate the localization of dipoles at a sub-nanometer scale, which has a significant influence on its ferroelectric properties.

To investigate the atomistic origin of the unique structural features of the ferroelectric o-phase of HfO_2 ($Pca2_1$), a sequence of symmetry-lowering steps was analyzed, starting from the cubic $Fm\bar{3}m$ structure of HfO_2. The c-phase is stable above 2,870 K.[12] As the temperature decreases, it transitions to the tetragonal $P4_2/nmc$ phase at 2,870 K, and further to the m-phase at 2,000 K. In contrast, the ferroelectric o-phase is stabilized in thin films at much lower temperature (room temperature).[13]

The phonon spectrum of the high-temperature c-phase (Figure 3.2a) exhibits a prominent phonon instability with $X X_2'$ symmetry at the Brillouin zone boundary $w = -i$ 228/cm. This instability involves antiparallel x-displacements of adjacent oxygen atoms within the yz-plane (Figure 3.2b). Condensation of an X_2' phonon drives the transition from cubic structure to tetragonal $P4_2/nmc$ phase, which results in zero net polarization. Among the four phonons involved in the transition from tetragonal to orthorhombic, two are particularly important: (i) the Γ_{15}^{Z} phonon, where all oxygen atoms in a certain sublattice move collectively along the z-axis (Figure 3.2c), capable of generating uniform polarization; and (ii) the cell-doubling antipolar Y_5^Z phonon, in which oxygen atoms in adjacent xz-planes move along the z-axis in an antiparallel fashion (Figure 3.2d), leading to A-type ordering with zero net polarization. Similar to the mechanism proposed for strained ZrO_2,[14] the ferroelectricity of HfO_2 is suggested to arise improperly due to the nonlinear interaction between stable Γ_{15}^{Z} and X_5 phonons with the primary instability of the X_2' phonon.

A distinctive feature of the ferroelectric order is the condensation of the polar Γ_{15}^{Z} and antipolar Y_5^Z phonons with exactly equal amplitudes, which results in dipolar partitioning into two types of alternating atomically thin layers (Figure 3.2e).[15] The first type consists of spacers, where relevant oxygen atoms exhibit no z-axis displacement, while the second type consists of the ferroelectric layers with parallel z-axis displacements of their oxygen atoms. Therefore, these spacer layers effectively act as inactive regions that can shield the elastic interactions between the ferroelectric active layers. Experimental validation for the existence of these layered

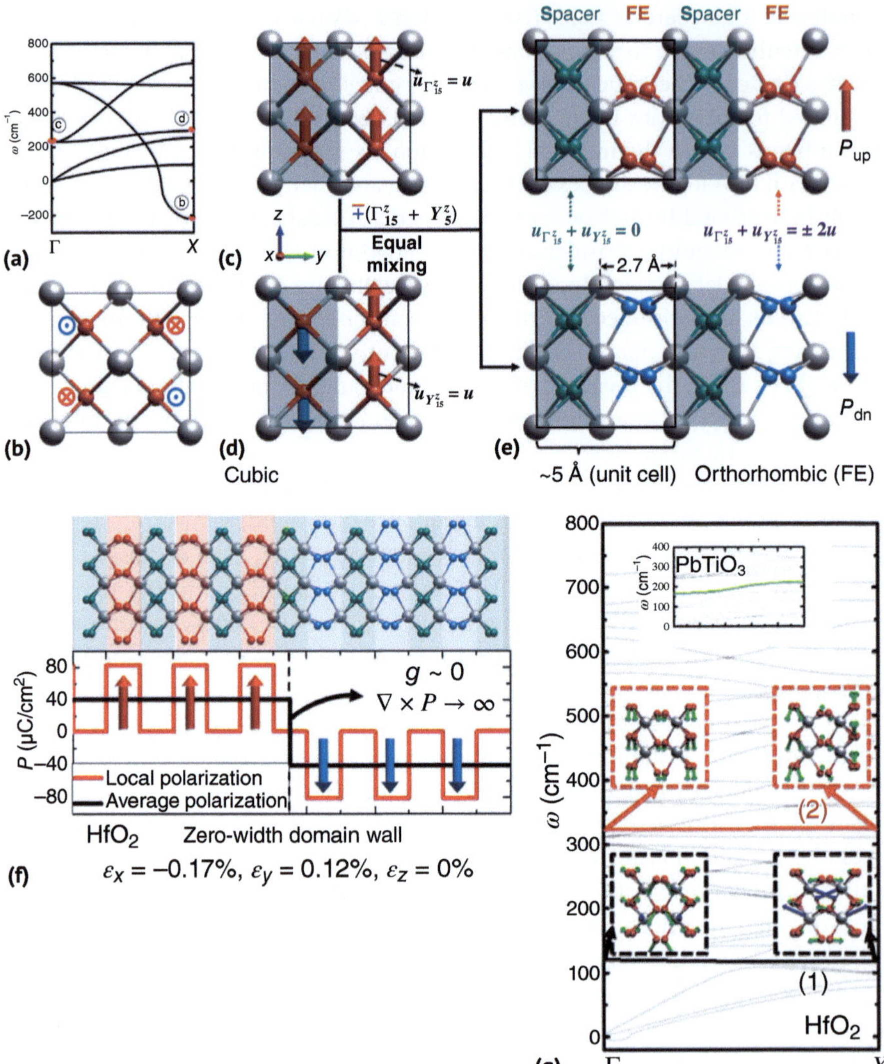

Figure 3.2 (a) Phonon dispersion of the c-phase. The red dots labeled b–d denote the primary instability of (b) X_2' mode, (c) Γ_{15}^{Z}, and (d) Y_5^Z, respectively, where the arrows denote u, the displacements of oxygen atoms. (e) Polar Γ_{15}^{Z} and antipolar Y_5^Z phonons condense in-phase with equal magnitude to generate an orthorhombic structure that consists of alternating spacer layers and ferroelectric layers with up (top) and down (bottom) polarizations, respectively. (f)Atomic structure of the domain wall and variation in local polarization along the direction perpendicular to the domain walls in HfO_2. Red and black lines correspond to the local polarization averaged over their half- and single-unit cells, respectively. (g) Flat polar bands are indicated with black and red lines, and eigenmodes at Γ and Y in the bands are depicted in the insets. *Source*: [9] / with permission of The American Association for the Advancement of Science.

structures has been provided by studies utilizing transmission electron microscopy (TEM), where they were referred to as minor and major layers, respectively.[16] The natural dipolar partitioning in orthorhombic HfO_2 has significant implications for its polarization domain structure, starkly contrasting with the behavior observed in perovskite ferroelectric $PbTiO_3$. In HfO_2, the local polarization in the spacer layer vanishes even in a uniformly polarized state, inherently leading to a domain wall of vanishing thickness between oppositely polarized (180°) domains (Figure 3.2f). Crucially, this domain wall is nearly strain-free (Figure 3.2f, ε_i, where $i = x, y$, and z), exhibiting minimal structural distortion and supporting unimpeded bulk polarization in its vicinity. One might expect a high energy cost for such an atomically sharp domain wall in HfO_2 due to the $g\left|\vec{\nabla} \times \vec{\mathbf{P}}\right|^2$ term. However, the near-zero values of g, evident in the flat band of polar phonons (Figure 3.2g), render such a sharp domain wall energetically feasible. The flatness of the polar bands results in a phonon velocity approaching zero. The Γ-point phonon within the lowest frequency flat band (Figure 3.2g, black line) involves atomic displacements related to all modes condensed during the cubic-to-orthorhombic transition, while the higher-frequency phonon (Figure 3.2g, red line) involves combinations of polar and antipolar modes.

The elastic interaction between ferroelectric domains is effectively screened by the spacer layer in HfO_2. This distinct domain wall structure contributes to the unusually large coercive field (E_c) for polarization switching in HfO_2, even comparable to the activation field (E_a). In contrast, the coercive field in conventional ferroelectrics is typically an order of magnitude smaller than the activation field (E_c~$E_a/10$),[17] as polarization switching occurs via nucleation and lateral growth of domains with reversed polarization. This collective behavior allows E_c to be reduced by a factor proportional to the domain wall width.[18] In HfO_2, however, domain walls are extremely thin, minimizing this reduction in E_c. Furthermore, with weakly interacting domain walls and the nearly zero group velocity (g~0) for the relevant polar modes, due to the flat bands, domain walls are predicted not to propagate efficiently. Instead, their motion may occur through hopping over significant energy barriers, suggesting that the experimentally observed sluggish motion of domain walls is an intrinsic characteristic.

The scale-free nature of polarization switching in HfO_2 was demonstrated by comparing the reversal of uniform and local polarizations. The energetics of switching local and uniform polarizations in HfO_2 are remarkably similar. Flipping the polarization of a single layer is almost energetically equivalent (per layer) to flipping the polarization of all layers. This contrasts sharply with materials like $PbTiO_3$, where the reversal of local polarization within a single unit cell-wide region is energetically highly unfavorable, if not forbidden.

Subsequently, theoretical investigations have explored the impact of doping, oxygen vacancies (V_Os), and stress on phonon flat bands.[19] Ai et al.[19] demonstrated that La-doped HfO_2 with oxygen vacancies exhibited notably flattened polar phonon bands compared to pure HfO_2. Increasing La doping concentration resulted in more phonon modes exhibiting flatness. The authors identified several key factors responsible for the emergence of these enhanced flat bands in the doped structures. First, the definition of a flat phonon band is not universally standardized. The frequency difference ($\Delta\omega$) between the Γ and Y wave vectors within the 0.5-THz range, which is often considered a flat phonon band, actually represents a softening of the phonon band, as observed in defect-engineered materials. Consequently, doping-induced disorder leads to phonon mode softening and the flattening of phonon bands. Second, local symmetry breaking caused by point defects can fundamentally alter lattice vibration modes compared to the pure phase. Notably, antipolar phonon bands may lose their antipolar character and even give rise to net polarization, effectively transforming into polar modes. Third, doping can reduce the effective spring-like interactions between ions around point defects, leading to softer lattice vibrations and thus flatter, more polarized phonon bands. This effect arises from the extended interaction times between Hf ions surrounding the point defects. From a thermodynamic perspective, after La doping, the energy difference between the o-III-phase and the m-phase is reduced. Overall, the flattening of phonon bands in La-doped MO_2 (M = Zr, Hf) can be attributed to local symmetry breaking due to point defects, the stability of the o-III-phase, and weaker interactions between Hf ions near the point defects.

The debate concerning whether hafnium-based ferroelectrics should be classified as "proper" or "improper" ferroelectrics has been a subject of intense discussion.[20–22] Proper and improper ferroelectrics differ significantly in the characteristics of their phonon modes, with the distinction primarily stemming from the origin of their ferroelectric behavior. Proper ferroelectricity typically arises from spontaneous crystal symmetry breaking in materials, leading to a non-centrosymmetric structure. The defining phonon mode in proper ferroelectrics is the soft-mode phonon, which generally involves a reduction in lattice symmetry, thereby generating and enhancing electrostatic dipole moments. The soft-mode phonon frequency decreases and approaches zero at the ferroelectric phase transition temperature, signifying the breakdown of lattice symmetry and the emergence of a stable polar phase. In contrast, improper ferroelectrics are usually driven by external factors such as stress or electric fields, where nonpolar primary order parameters interact with polar modes at the center of the Brillouin zone. This results in improper ferroelectricity being linked to a structural phase transition, such as a shift from a nonpolar to a polar phase, triggered by the instability of nonpolar modes (e.g., shear or tilt modes). This nonpolar distortion then induces net polarization via coupling with a polar mode, which itself would not drive a transition alone.

Theoretical models were developed, such as the Landau free energy model by Delodovici et al.[22] using *Ccce* as a reference, based on DFT calculations. Their model reveals a trilinear coupling between a central Γ_{3-} mode exhibiting polar instability and two stable boundary modes: Y_{2+} with nonpolar characteristics and Y_{4-} with antipolar characteristics, which drive the ferroelectric phase transition in HfO_2. Similarly, Qi et al.[21] proposed that the polarization reversal in HfO_2 requires not only the polar mode Γ_{15}^{Z} but also at least one antipolar mode at the Brillouin zone boundary. Zhou et al.[20] connected the high-symmetry cubic $Fm\bar{3}m$ phase to the ferroelectric $Pca2_1$ phase through a series of tetragonal, antipolar, and polar modes. The high-symmetry cubic $Fm\bar{3}m$ phase transitions into the $P4_2/nmc$ phase via the tetragonal mode. The structure then undergoes a further transformation into the antipolar *Pbcn* phase through the antipolar mode. By introducing polar modes into the *Pbcn* structure, the polar $Pca2_1$ phase is achieved. They suggest that the stretching strain along the tetragonal mode direction triggers the phase transition from $P4_2/nmc$ to *Pbcn*. The origin of the ferroelectric phase undergoes two transitions: initially, under tensile stress, the structure shifts through the antipolar mode into the antipolar *Pbcn* phase. The strong coupling between polar and antipolar modes helps stabilize the polarization against depolarization effects. These findings suggest that tensile strain and the coupling of polar and antipolar modes are key to the ferroelectricity in HfO_2, offering a previously unexplored mechanism for resisting depolarization that is not related to improper ferroelectricity. This framework provides an explanation for how hafnium-based ferroelectrics can sustain ferroelectric ordering even in extremely thin films.

Further investigations into the concept of a "parent structure" for HfO_2 have aimed to address the proper/improper classification debate. Raeliarijaona and Cohen[23] noted that hafnium-based ferroelectrics are ideal candidates for generating symmetric double potential wells through polar modes, a characteristic of ferroelectricity, observed in both the o-III-and r-phases. However, defining a universally accepted nonpolar parent structure for HfO_2 that accurately represents the minimal and maximal energy states during polarization switching under an electric field is complex. An appropriate parent structure is one that displays both minimal and maximal energy states during the polarization switching process under the influence of an external electric field. By selecting the antiferroelectric *Pbcn* phase as the parent structure, it can transition to the ferroelectric $Pca2_1$ phase via the Γ_2^- mode. Furthermore, it was discovered that using the $P4_2/nmc$ phase as the parent structure allows for a direct transformation to the $Pmn2_1$ phase through the $\Gamma_{5,X}^-$ mode. These findings provide strong evidence supporting the intrinsic ferroelectricity of HfO_2. Additionally, Aramberri and Íñiguez[24] further substantiated this by introducing an "o-ref"-phase, which confirmed that hafnium oxide is an intrinsic ferroelectric material. This o-ref-phase is an intermediate state between the upward and downward polarizations of the $Pca2_1$ phase, a phenomenon also observed in experimental studies of ZrO_2. The inclusion of the o-ref-phase helps explain not only the origin of the o-III-phase, caused by the imaginary frequency of the Γ_2^- mode, but also the m-phase resulting from shear softening induced by the Γ_4^+ mode.

Within the framework of a phonon theory model and advanced theoretical models (ranging from the Ginzburg–Landau model to effective Hamiltonian models and first-principles atomic potentials), the phase transition processes between ferroelectric phases, including polarization flipping and field-driven transformations between low-energy phases, are effectively described and understood. These models not only clarify the physical nature of ferroelectric phase transitions but also highlight the complexity of internal polarization dynamics within these materials.

Phonon-driven polarity phase transitions, in particular, underscore the crucial role of lattice dynamics in ferroelectric phase changes. During these transitions, specific phonon modes, often involving soft-mode phonons, are excited, leading to lattice instability, which plays a key role in the phase transition. This instability is frequently linked to temperature variations and can be modulated through external electric fields, stress, or chemical doping. Differentiating between proper and improper ferroelectricity is of significant practical importance for understanding and designing ferroelectric materials, especially considering their potential advantages in information storage and sensing applications.

3.1.3 Thermodynamic Perspective

From a thermodynamic viewpoint, the polar o-phase is not inherently stable but requires specific conditions, such as a particular electrode interface or pressure, to exist. The phase transition in fluorite ferroelectrics is considered first-order, involving both the polar o-and nonpolar t-phases.[25–30] A two-step polarization switching process, mediated by the intermediate nonpolar phase, can be observed between the Curie–Weiss temperature (T_0), the Curie temperature (T_c), and the ferroelectric limit temperature (T_1). Park et al.[29] explored an intriguing observation in $Hf_{0.4}Zr_{0.6}O_2$ thin films, a phenomenon typically seen within a very narrow temperature range in ferroelectric perovskites. In materials exhibiting first-order phase transitions, when there is no spatial distribution of T_0, the remanent polarization (P_r) disappears abruptly as the temperature exceeds T_c, and the stable phase shifts from a polar to a nonpolar state. However, experimental observations in ferroelectric fluorite thin films reveal a significantly different behavior in the change of P_r with temperature.[25,26,28] As depicted in Figure 3.3, P_r decreases with increasing temperature, but the disappearance of P_r is not as sudden as predicted by the Landau–Devonshire (LD) theory. This suggests that the T_0 distribution in ferroelectric fluorite thin films can be affected by spatial inhomogeneities due to various factors

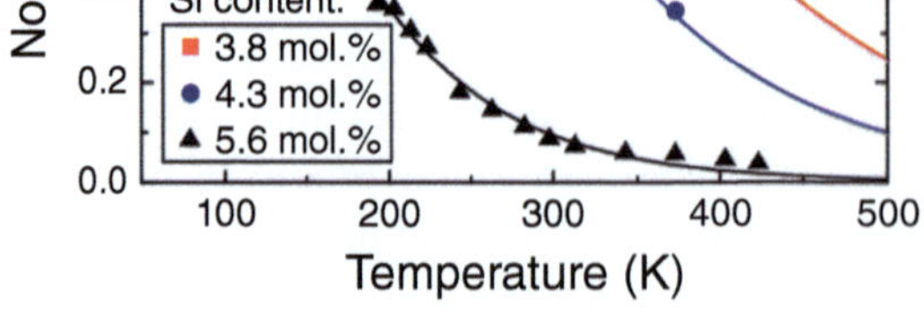

Figure 3.3 Normalized remanent polarization $P_r(T)$ values as a function of sample temperature for MIM capacitors with a 9-nm Si:HfO_2 layer having the Si contents of 3.4–5.6 mol.%. *Source*: [26] / with permission of Elsevier.

such as grain size, film thickness, strain, and impurity levels. As a result, the continuous change of P_r with temperature in ultrathin ferroelectric fluorite films can be better understood. According to the thermodynamic model, the fraction of the m-phase ($P2_1/c$) decreases as the annealing temperature increases. However, experimental observations show the opposite trend.[31,32] Therefore, relying solely on a simple thermodynamic model cannot fully explain the formation of the o-phase in HfO_2 ferroelectric films. To address these factors, a more comprehensive thermodynamic model for thin films is necessary, one that incorporates all relevant energy contributions. The Gibbs free energy of this model can be defined as:

$$G_\phi = U_\phi + U_{0,\phi} + PV_\phi - TS_\phi + \gamma_\phi A - V_\phi \mathbf{D} \cdot \mathbf{E} - V_\phi \sum_{ij} U_{ij,\phi}\sigma_{ij,\phi} \tag{3.1}$$

where subscript denotes the type of phase, bulk energy including the effects of doping, vacancies, and phonon bands, zero-point energy from vibrational modes, hydrostatic pressure, volume, temperature, entropy including vibrational and configurational entropy contributions (electronic entropy is neglected generally at room temperature because common phases are electrical insulators with wide bandgaps, around 6 eV), surface energy, associated surface or interface area, electric displacement field, electric field, strain tensor, and stress tensor. To investigate specific factors, unrelated energy contributions can be disregarded, as seen in the energy models presented in the literature.[15,33–36] The thermodynamic equilibrium phase is determined by minimizing the Gibbs free energy with respect to the phase type. These factors are discussed in the following subsections.

3.1.4 Coordination Number Perspective

The origin of hafnium-based ferroelectricity remains unclear, primarily due to the lack of comparable ferroelectric materials for reference. Additionally, the explanation of HfO_2's ferroelectric properties is not fully satisfactory when considered within the framework of soft-mode theory. HfO_2 exhibits a cubic fluorite phase at high temperatures, which possesses a soft mode at the Brillouin zone boundary, leading to a cubic-to-tetragonal phase transition. It is widely accepted that the t-phase, a high-temperature paraelectric phase, is associated with hafnium-based ferroelectricity. However, there is no soft mode at the center of the Brillouin zone in the t-phase. The Italian physicist group, led by Picozzi,[22] proposed that a three-phonon coupling mechanism could drive a paraelectric-to-ferroelectric phase transition in HfO_2. While this theory can explain how the t-phase transitions into an orthorhombic ferroelectric phase, it does not adequately address why the t-phase should transform into a ferroelectric phase. Moreover, this theory faces challenges in predicting new hafnium-based ferroelectric materials.

Recently, Chae et al. investigated the effect of tetravalent doping in fluorite-based ferroelectric oxides and proposed that the coordination number of 7 in the $Pca2_1$ phase is associated with oxygen displacement and the breaking of oxygen–metal bonds, which contribute to macroscopic polarization.[37] HfO_2, in both its monoclinic $P2_1/c$ ground state and stable ferroelectric orthorhombic $Pca2_1$ phase, exhibits a distinctive feature of cationic seven-coordination (7C). This 7C leads to two degrees of freedom for the anion, potentially enabling ferroelectric polarization reversal. Linking the ferroelectric phase with the spatial characteristics of seven-coordination is crucial for understanding the origin of hafnium-based ferroelectricity.

Certain AB_2 compounds, such as HfO_2, zirconium dioxide, and strontium iodide, are unique members that fall within a specific range of cationic and anionic radii ratios, lying between those favoring six-coordination (e.g., common rutile structures) and eight-coordination (8C; e.g., fluorite structures). This characteristic suggests that transitions between these coordination numbers are energetically accessible. Therefore, the transition of HfO_2 from 8C t-phase to lower-coordination phases, such as the 7C o-phase, can be fundamentally linked to this intermediate ionic radii ratio. For example, the ionic radius of Hf is not sufficiently large to ideally support eight adjacent oxygen ions in a stable configuration within the fluorite structure due to excessive repulsion between the anions. This helps explain why in the low-temperature phase of HfO_2, each Hf cation is primarily surrounded by only seven O anions (as indicated by the circled Hf cations in Figure 3.4),[38] unlike the eight O anions surrounding Ce4+ in fluorite CeO_2.[39] Similarly, the emergence of the $Pca2_1$ phase in HfO_2 may be driven by the need for the 8C-to-7C chemical configuration for Hf, as shown in Figure 3.4. The difference lies in the fact that, to form the $P2_1/c$ phase, the two oxygen anions within the ellipse move in opposite directions to achieve a symmetric structure, while in the $Pca2_1$ phase, the two oxygen anions move in the same direction, breaking the mirror symmetry along the c-axis. These two O anions become 3C configuration after distortion (denoted O_{III}), while the 4C O anions are labeled O_{IV} in the $Pca2_1$ phase. From $P4_2/nmc$ to $Pca2_1$, half of

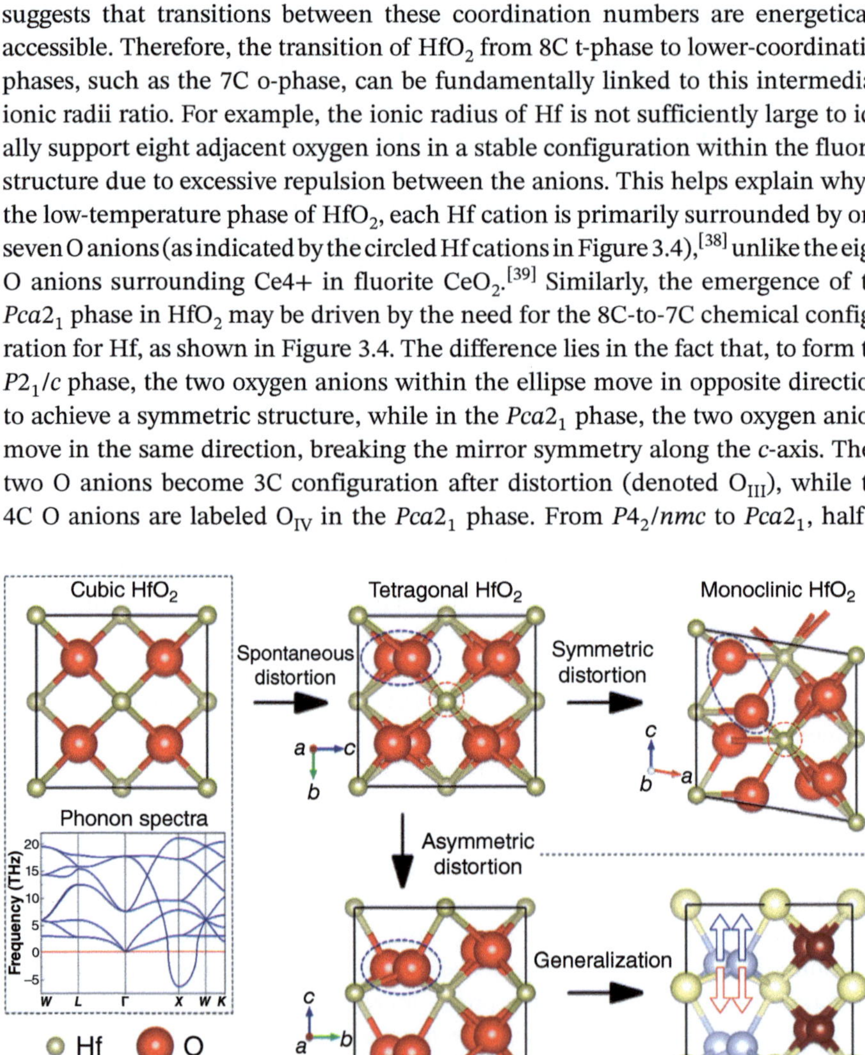

Figure 3.4 Crystal structures of cubic fluorite HfO_2, tetragonal $P4_2/nmc$ HfO_2, monoclinic $P2_1/c$ HfO_2, and orthorhombic $Pca2_1$ HfO_2, with their relations. *Source*: [38] / with permission of American Chemical Society.

the O_{IV} anions undergo slight movement along the *b*-axis, a distortion similar to that seen in the $P2_1/c$ phase. In the $Pca2_1$ phase of HfO_2, ferroelectric switching is primarily driven by the movement of O_{III} anions, while the O_{IV} anions remain relatively fixed. A potential advantage of the $Pca2_1$ distortion is that the unit cell does not need to expand as significantly as in the $P2_1/c$ case.[40]

Experimental evidence, such as the observation that nanoscale $Pca2_1$-HfO_2 tends to transition to the $P2_1/c$ phase (stable at low temperatures) upon heating,[31] presents an apparent paradox when viewed solely from the perspective of bulk thermodynamic equilibrium. However, this behavior can be effectively explained through chemical kinetics rather than thermodynamics, as the ferroelectric phase rapidly transforms into a thermodynamically stable phase with an increase in temperature. Such a kinetically limited phenomenon reduces the practical relevance of thermodynamic phase transition in understanding the origin of ferroelectric distortion in $Pca2_1$ thin films. Indeed, the ferroelectricity of hafnia is closely linked to energy minimization at the nanoscale, significantly influenced by surface/interface energy effects.[15,33] Therefore, adopting a chemical perspective that favors the 7C configuration for Hf in HfO_2 may offer a useful theoretical framework, as it does not rely on a thermodynamically stable phase and can accommodate surface energies, strains, and defects more naturally.

3.1.5 Summary

In summary, the origin of ferroelectricity in HfO_2 has garnered significant attention, leading to the identification of several interconnected factors that contribute to its unique properties and the stabilization of its ferroelectric phases. Understanding these aspects is crucial for the future development of HfO_2-based ferroelectrics. In terms of non-centrosymmetric ferroelectric phases, HfO_2 exhibits spontaneous polarization mainly in the orthorhombic $Pca2_1$ phase, where broken inversion symmetry enables the formation of spontaneous electric dipoles. From a microscopic perspective, phonon-driven ferroelectric phase transitions are critical, as flat polar phonon bands in HfO_2 lead to dipole localization at the sub-nanometer scale, influencing its switching behavior. From a thermodynamic perspective, a comprehensive model that accounts for various energy contributions, such as strain, surface energy, and electric fields, helps explain the stability of ferroelectric phases in thin films. The coordination number perspective is also crucial, with 7C around Hf cations in the orthorhombic $Pca2_1$ phase promoting oxygen displacement and bond breaking, facilitating macroscopic polarization. As research advances, achieving a deeper and more integrated understanding of these complex mechanisms – spanning from crystal structure and phonon behavior to thermodynamic stability and local bonding environments – will be pivotal. This fundamental knowledge is key to rationally designing stable and scalable HfO_2-based ferroelectric materials, ultimately unlocking their full potential for a wide range of applications in next-generation electronic devices, including ferroelectric memories, logic devices, and sensors.

3.2 The Effects of Doping and Defect on HfO_2-based Ferroelectricity

HfO_2, when intentionally doped or subjected to specific processing conditions that introduce defects, exhibits ferroelectric behavior, predominantly at nanoscale thicknesses. The ability to precisely tailor HfO_2's ferroelectric properties and stabilize metastable polar phases by the introduction of dopants or manipulation of intrinsic defects is a key factor driving its significant potential for integration into next-generation electronic devices. The influence of dopants and defects on fluorite-structured ferroelectrics like HfO_2 can be broadly categorized into two main aspects: (1) polymorphism of fluorite-structured ferroelectrics by influencing the relative free energy of various crystalline phases and (2) electrical properties by inducing charge trap sites that affect the domain switching dynamics, electrical conduction mechanisms, and reliability of the fabricated electronic devices. Nevertheless, the controllability in the crystallographic phase formation and their influence on the electronic properties remains a significant challenge in the field.

3.2.1 Doping Effects on HfO_2-based Ferroelectrics

Doping stands out as a highly effective method for engineering the properties of HfO_2, playing a pivotal role in enabling and optimizing its ferroelectric performance. The introduction of dopants significantly impacts the material's crystalline polymorphism by altering the relative free energies of various phases, and also modulates its electrical characteristics. Investigations into the influence of doping on the free energy landscape of HfO_2 phases predated the first experimental observation of ferroelectricity in Si-doped HfO_2.

Early theoretical work by Lowther et al.[41] in 1999 proposed that the polar orthorhombic $Pca2_1$ phase could become energetically favorable compared to the stable monoclinic $P2_1/c$ phase under high hydrostatic pressure (~4 GPa), although this lacked experimental confirmation at that time. Subsequently, Huan et al.[1] conducted a comprehensive re-examination of the free energy of various crystalline phases of HfO_2 across wide range of temperatures and pressures. They found that the free energy differences between the two polar o-phases (space groups: $Pca2_1$ and $Pmn2_1$) and the stable m-phase were relatively small, on the order of $k_B T/5$, where k_B is Boltzmann's constant and T is the temperature. They also suggested that the relatively small free energy barrier between these phases could be easily overcome through doping and strain.

The impact of doping becomes particularly significant during the annealing process, especially when a critical annealing temperature is reached. Doping effectively suppresses the phase transition rates leading to the formation of the equilibrium m-phase,[42] thereby kinetically trapping the material in a metastable polar phase. However, the precise atomic-level mechanism by which doping facilitates the formation of the o-III ($Pca2_1$) phase remains an active research area, partially because the o-III-phase is not predicted to be a thermodynamic equilibrium phase for HfO_2 in its bulk phase diagram.[43] Furthermore, no common doping element has been identified that can stabilize the o-III-phase as the ground state.[44] Nevertheless, doping, whether involving ionic (interstitial) or substitutional, can modify the crystal structure, spontaneous polarization, coercive field, and overall stability of HfO_2-based ferroelectrics.

3.2.1.1 Role of Isovalent Doping

Isovalent doping, where the dopant ion has the same valence state as the host ion it replaces, is a prominent strategy for inducing ferroelectricity in HfO_2. Si doping of HfO_2 to stabilize the polar o-III $Pca2_1$ phase is one of the most notable examples of isovalent doping. In 2011, Böscke et al.[13] observed the stabilization of the polar o-III $Pca2_1$ phase in 10-nm Si-doped atomic-layer-deposition (ALD) thin films, with doping concentrations starting from 2.6 f.u.%. At a silicon doping concentration of 4.3 f.u.%, the hysteresis loop begins to pinch, forming an antiferroelectric shape. Around 6 f.u.%, the ferroelectric behavior of Si-doped HfO_2 transforms into linear dielectric. At higher silicon concentrations, the t-phase is stabilized, which is used as a high-k material for dynamic RAM (DRAM) trench capacitors.[45] Since Si is isovalent to Hf, the dopant complex is assumed to be substitutional Si_{Hf}. In Kröger–Vink notation, this complex is represented as $SiO_2 \rightarrow Si_{Hf} + 2O_O$, with no lattice defects, as Si is isovalent to Hf. However, a substitution could be energetically more favorable if an oxygen is displaced, which can be controlled during fabrication through the external oxygen partial pressure. As a result, the complexes may not remain isovalent. A typical example of an aliovalent complex is $Si_{Hf}V_O$. The Kröger–Vink reaction for the aliovalent substitution of silicon is $SiO_2 \rightarrow Si_{Hf} + O_O + V_O^{\cdot\cdot} + 2e' + 1/2O_2$. In Kröger–Vink notation, positive charges are indicated by the "·"-symbol, and negative charges by the "′"-symbol. Aliovalent dopants are discussed in more detail in the following.

The formation energies of Si_{Hf}, $Si_{Hf}V_O$, and V_O defects in the o-III-phase have been calculated under two different oxygen partial pressures.[46] Interstitial silicon (Si_I) is excluded, as its formation energy is always higher. Under oxygen-rich conditions, Si_{Hf} is preferred over $Si_{Hf}V_O$. According to Materlik et al.[47], oxygen-poor conditions are more representative of the manufacturing environment, while oxygen-rich conditions are closer to those encountered during the device's lifetime. For oxygen-poor conditions, the oxygen chemical potential is derived from O_2, while for oxygen-rich conditions, it is derived from TiO_2 (with TiN commonly used as an electrode material for devices). The oxygen vacancies that form over the device's lifetime may react with the existing Si_{Hf} to form $Si_{Hf}V_O$. Additionally, electric fields applied during field cycling can promote the creation of oxygen vacancies. Once the relevant defect structures are identified, the implications for phase stability can be further explored. As shown in Figure 3.5a, at 3 and 6 f.u.%, the polar orthorhombic (also denoted by PO [o-III] $Pca2_1$) phase is favored by doping, but not enough to fully explain the observed stabilization without additional effects. The stabilization of the t-phase beyond 6 f.u.% aligns well with experimental observations. For a doping concentration of 12.5 f.u.%, the computational results suggest significant dopant–dopant interaction effects, necessitating more detailed calculations.

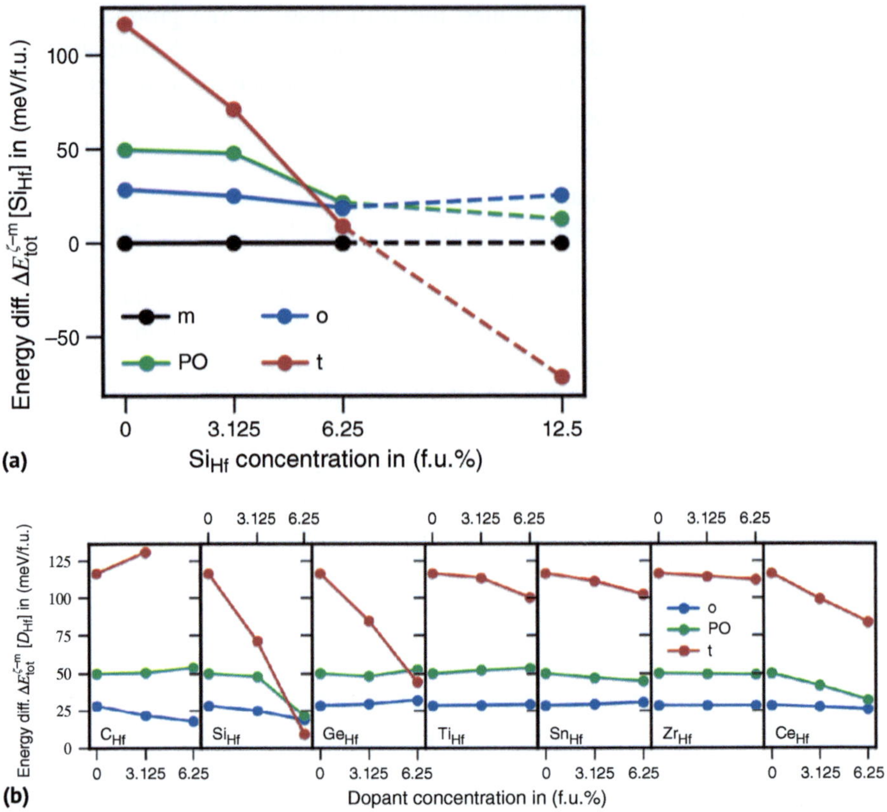

Figure 3.5 (a) Energy differences with respect to the m-phase for different Si_{Hf} concentrations. The atom positions and the lattice parameters of the calculations at 12.5 f.u.% doping concentration show a significant difference from the actual symmetry of the phase (dashed line). (b) Evolution of the energy differences for different dopants with respect to the m-phase with increasing doping concentration. *Source*: [46] / with permission of American Chemical Society.

Silicon can be considered a prototype system for studying the effects of isovalent Group IV dopants. Similarly, dopants such as C, Ge, and Sn from the carbon group, along with Ti and Zr, have been investigated. Additionally, Ce from the lanthanide series was studied due to its stable +4 oxidation state. Based on the analysis of silicon formation energies, the study can be focused on substitutional defects. Figure 3.5b illustrates the evolution of the energy differences of substitutional dopants relative to the m-phase, ranging from pure HfO_2 to a doping concentration of 6.25 f.u.%. While some dopants show a tendency to favor the PO (o-III) phase, none of the dopants investigated alone favor the PO (o-III) phase as the lowest-energy phase. By comparing the energy evolution with doping concentration for the o-phase across all seven subplots, it can be concluded that the o-phase (*Pbca*) is relatively insensitive to doping. In contrast, the PO (o-III) phase shows slight sensitivity to doping, particularly in the cases of Si and Ce doping. On the other hand, the t-phase demonstrates high sensitivity to doping, especially with Si, Ge, and Ce.

In general, it can be concluded that the energy sensitivity to doping increases with the symmetry of the crystal phases.

In 2020, Wu et al. used first-principles methods to demonstrate that Si doping increases the grain size of t-phase during the heating stage, suggesting that Si doping promotes the nucleation and growth of t-phase. During the annealing process, the authors found that the phase transition barrier from t-phase to the ferroelectric o-III-phase is significantly lower than the barrier to m-phase, making Si doping more effective in facilitating the transformation of t-phase into the ferroelectric o-III-phase. The high barrier to m-phase prevents t-phase from transitioning into it.[48] Regarding Zr, due to the similar atomic radii and the identical number of outermost electrons between Zr and Hf, these two elements are mutually soluble in any proportion. In other words, Zr can replace Hf throughout the entire compositional range.[49] Figure 3.6 presents the polarization–electric field (P–E) hysteresis loop of the HZO film.[50] When the Zr concentration is between 30 and 50 mol.%, the ferroelectric behavior of the HZO system stabilizes. When the Zr content exceeds the Hf content, the material exhibits antiferroelectric properties, and when the Hf content is reduced to zero, pure ZrO_2 demonstrates antiferroelectric characteristics.[51] At an ZrO_2 concentration of 50 mol.%, the P_r value reaches its peak, and at this concentration, the fraction of m-phase ($P2_1/c$) becomes zero. This result suggests that the centrally symmetric m-phase is suppressed as the doping concentration increases, and the reduced m-phase enhances the likelihood of forming the non-centrosymmetric $Pca2_1$ phase. In relation to the antiferroelectric behavior in ZrO_2, Reyes-Lillo et al.[14] calculated the free energies of various phases in both HfO_2 and ZrO_2, finding that the free energy difference between metastable and stable phases in ZrO_2 is much smaller than in HfO_2. The antiferroelectric properties of ZrO_2 arise from the phase transition between the field-induced t-and o-phases. The results demonstrate that in $Hf_{1-x}Zr_xO_2$ solid solutions, as the Zr concentration increases,

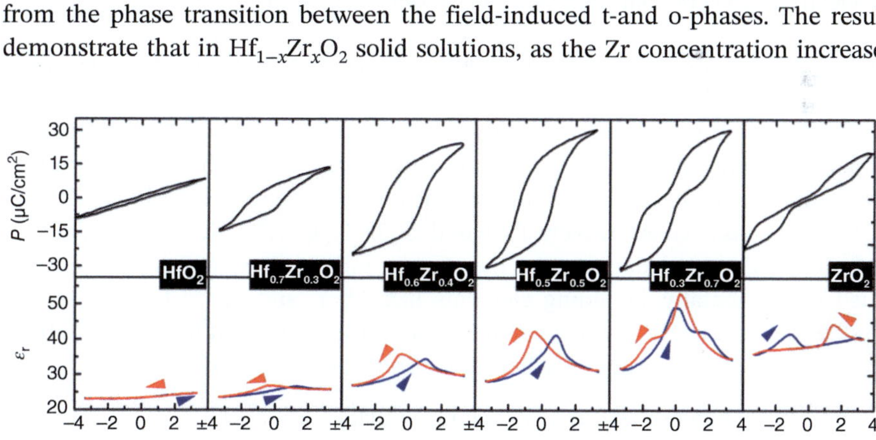

Figure 3.6 *P–V* hysteresis at 1 kHz and small-signal *C–V* hysteresis at 10 kHz (50-mV level) of 9-nm-thin HfO_2–ZrO_2-based metal–insulator–metal capacitors at room temperature. An evolution from paraelectric HfO_2 to ferroelectric HfO_2–ZrO_2 and then to an antiferroelectric-like behavior in ZrO_2 can be observed in *P–V* as well as *C–V* characteristics. *Source*: [50] / with permission of American Chemical Society.

the free energy difference between the orthorhombic $Pca2_1$ phase and the m-phase becomes smaller.[52–54]

Various mechanisms have been proposed to explain how doping influences ferroelectricity in HfO_2. One common explanation links the effect to the dopant's ionic radius. Dopants with smaller atomic radii can shorten the metal–oxygen bond length in HfO_2 thin films, enhance structural asymmetry, and induce ferroelectric or antiferroelectric properties. Conversely, larger dopants might increase bond lengths, also contributing to structural asymmetry. However, this explanation fails to account for ultrathin ferroelectric properties induced by the solid solution doping of Zr and Hf.[49] Another widely accepted mechanism is that doping can reduce the free energy difference between the o-and m-phases.[54] As the doping concentration increases, these modifications to the energy landscape led to distinct changes in the observed ferroelectric polarization behavior.

3.2.1.2 Role of Aliovalent Doping

Aliovalent dopants, having a different valence state than the host ion (Hf4+) they replace, introduce charge imbalances that require compensation. This need for charge compensation leads to a broader range of potential defect structures compared to isovalent doping,[55] making their effects more complex. The dopants Be, Mg, Ca, Sr, and Ba are divalent (II-valent) dopants in HfO_2. Given their II-valent nature, a compensation mechanism is necessary to balance two excess charges in HfO_2. The two primary complexes that compensate for these charges are as follows (where D represents the dopant):

- ionically compensated $D_{Hf}V_O$: $DO \rightarrow D''_{Hf} + O_O + V_{\ddot{O}}$;
- electronically compensated D_{Hf}: $\frac{1}{2}O_2 + DO \rightarrow D''_{Hf} + 2O_O + 2h^{\cdot}$.

In these notations, "′" denotes a negative charge and "·" represents a positive charge. An "ionically compensated" complex refers to charge compensation through the presence of other defects, such as oxygen vacancies. Conversely, "electronically compensated" indicates that charge compensation occurs through free electrons (or holes) within the material or at its surface. The role of these complexes in stabilizing the PO-phase has been extensively discussed for Sr-doped HfO_2 in Materlik et al.[47].

Trivalent (III-valent) dopants are the most common aliovalent dopants investigated experimentally, including elements like Al, Y, and La.[56] In the case of III-valent dopants in HfO_2, three distinct complexes are relevant for compensating the excess electronic charges[57] (where D represents the dopant):

- Ionically compensated $(2D_{Hf})V_O$: $D_2\,O_3 \rightarrow 2D'_{Hf} + 3O_O + V_{\ddot{O}}$;
- Electronically compensated D_{Hf}: $\frac{1}{2}O_2 + D_2O_3 \rightarrow 2D'_{Hf} + 4O_O + 2h^{\cdot}$;
- Mixed compensated $D_{Hf}V_O$: $D_2\,O_3 \rightarrow 2D'_{Hf} + 2O_O + 2V_{\ddot{O}} + 2e' + \frac{1}{2}O_2$.

The three complexes vary in the number of oxygen vacancies associated with them. Under high oxygen partial pressures, the electronically compensated defect D_{Hf} is favored, where the charge is balanced not by lattice defects but by electronic defects (holes), leading to a reduction in the Fermi level. In contrast, under low oxygen partial pressures, the mixed compensated defect $D_{Hf}V_O$ is preferred, with charge compensation achieved through both electrons and ions (e.g., oxygen vacancies), which raises the Fermi level. In ceramics, the ionically compensated defect $(2D_{Hf})V_O$ is typically the most stable, as charge compensation occurs solely through ions.

The following results were obtained when examining isovalent and II-valent dopants. For Y and La doping, all three defects tend to stabilize the PO-phase over the m-phase. However, the $D_{Hf}V_O$ and $(2D_{Hf})V_O$ defects also strongly favor the t-and c-phases, which complicates or even prevents the effective stabilization of the PO-phase. Experimentally, stabilization of the t-phase, or potentially the c-phase, over the PO-phase would manifest as an antiferroelectric-like hysteresis loop, which has not been observed for Y and La doping. Therefore, to align with experimental observations, the D_{Hf} defect is most likely responsible for the ferroelectric behavior in doped HfO_2, although the $D_{Hf}V_O$ defect might contribute within a narrow concentration range. This conclusion is supported by the fact that the D_{Hf} defect has the lowest formation energy among the three defect structures under oxygen-rich conditions. These conditions are likely provided by the introduction of an oxygen precursor at high temperatures during the ALD process or by rapid thermal annealing at even higher temperatures. The antiferroelectric behavior observed in Al-doped HfO_2 is consistent with the Al_{Hf} defect, which strongly favors the t-phase. Al doping does not significantly promote the PO-phase, prompting the investigation of additional interstitial and O-site defects for Al. However, none of these defects substantially enhance the PO-phase. The reason Al induces ferroelectric behavior within a very narrow concentration window, as seen experimentally, remains unclear. It is likely that the significant size difference between Al and Y/La plays a role, as the smaller size of Al could lead to dopant clustering, which has not yet been explored.

Theoretically, Batra et al.[44] conducted a comprehensive study on the impact of doping on the free energies of various HfO_2 phases using high-throughput calculations. Their research explored nearly 40 different doping scenarios at a concentration of 3.125 cat.%, revealing that no dopant was able to stabilize the polar $Pca2_1$ phase as the most stable phase. However, II-valent and III-valent dopants such as Ca, Sr, Ba, La, Y, and Gd, along with the synergistic effect of oxygen vacancies, were found to significantly reduce the free energy of polar o-phases ($Pca2_1$ and $Pmn2_1$). Additionally, their theoretical analysis indicated that dopants with larger ionic radii and lower electronegativity are particularly effective in stabilizing polar phases. This effect was attributed to the formation of stronger bonds between the dopant cations and the second-nearest oxygen atoms.

3.2.1.3 Role of Co Doping and Nonmetal Doping

While single cation doping is widely recognized as a crucial method for inducing ferroelectricity in HfO_2, more advanced strategies, such as co-doping (incorporating multiple dopant elements) and doping with nonmetal elements, have emerged as powerful tools to further tune and optimize HfO_2-based ferroelectric properties. Theoretical studies suggest that lanthanide metals (e.g., La, Gd), alkaline earth metals (e.g., Ca, Sr), and Y are the most promising dopants due to their larger ionic radii and lower electronegativity.[58] From another viewpoint, these dopants share a common characteristic of consisting of II-valent and III-valent elements, which inevitably introduce holes. However, previous research often overlooked the impact of carriers introduced by heterovalent dopants. The relationship between metallic and ferroelectric properties has been extensively studied in perovskite ferroelectrics, with various mechanisms proposed to explain how free carriers, either holes or electrons, affect polarization.[59–62] Typically, due to carrier electrostatic screening, ferroelectricity in barium titanate ($BaTiO_3$) vanishes above a critical carrier concentration.[61] Interestingly, some studies have found that small hole doping unexpectedly enhances polarization in $BaTiO_3$.[63,64] Recently, Cao et al.[65] proposed that hole doping plays a critical role in stabilizing polar phases in HfO_2, using virtual crystal approximation to suggest that even at high La doping concentrations, the o-III-phase can be stabilized. However, the remanent polarization in doped HfO_2 films generally shows an initial increase followed by a decrease within a small doping range.[66–68] Therefore, it is essential to explore the true roles of various dopants in stabilizing the o-III-phase, with a focus on their effects on the electronic structure.

Long and co-workers conducted a systematic investigation using first-principles calculations to explore the effects of various valence dopants (D^{2+}: Mg, Ca, and Sr; D^{3+}: Al, Y, La, and Gd; D^{4+}: Si, Ti, Zr, Sn, and Pb; D^{5+}: V, Nb, and Ta) on the crystal structure, electronic structure, phase stability, and the fraction of ferroelectric phases in HfO_2.[69] Their findings reveal that *p*-type dopants (D^{2+} and D^{3+}) that introduce holes generally enhance the stability of the o-III-phase and increase its fraction, while *n*-type dopants (D^{5+}) that introduce electrons exert an opposite effect. In comparison to single *p*-type doping, co-doping *p*-type elements with V not only maintains charge neutrality but also significantly boosts the o-III-phase fraction. Under specific co-doping compositions (such as Mg@V, Ca@V, Sr@V, and La@V), it is even possible to stabilize the ground-state o-III-phase. This beneficial co-doping effect is attributed to the formation of deep energy levels, or carrier recombination centers, through V doping, similar to the behavior of oxygen vacancies.

Complementary doping with lanthanides or rare earth metals that have smaller ionic radii than La^{3+}, is another co-doping strategy being explored. These smaller ions are expected to influence the lattice strain, grain size, and phase composition, consequently, the ferroelectric behavior of co-doped HZO. Previous studies have reported a reduction in m-phase formation accompanied by an increase in t-and c-phases in a binary zirconia system doped with III-valent ions such as Y^{3+}, Sr^{3+}, and Yb^{3+}.[70] Furthermore, it has also been demonstrated that very low doping concentrations (around 1–2%) can induce significant effects in complex oxides, including a reduction in dielectric loss and suppression of grain growth.[71,72] For example, Popovici et al.[73] reported a fatigue-resistant capacitor co-doped with La, Gd, and/or Y, achieving remnant polarization $\geq 15\ \mu C/cm^2$ after 1×10^{11} endurance cycles. Additionally, doping with Group VB metals significantly enhances the formation of oxygen vacancies in HfO_2 due to their pentavalent nature. The introduction of

Sc helps regulate the formation of oxygen defects by absorbing electrons ionized by VB elements, thus mitigating some structural distortions. The control of oxygen vacancies through co-doping plays a crucial role in improving the ferroelectric orthorhombic (o-III-phase) fraction in thin films. Simultaneously, Chen and co-workers refined the defect states in HfO_2 using the donor (Ta^{5+}) and acceptor (La^{3+}) co-doping strategy.[74] This approach resulted in a more organized and less defective microstructure, along with a reduced switching barrier. The co-doped HfO_2 samples exhibited a remarkable 83% increase in remanent polarization and a decrease in coercive fields. Most notably, the switching kinetics were significantly enhanced, resulting in the fastest switching process ever reported for HfO_2-based ferroelectric devices.

Carbon, nitrogen, and hydrogen are also key dopants that can influence the ferroelectric and electrical properties of fluorite-structured oxides. In the case of ferroelectric HfO_2 or $Hf_{1-x}Zr_xO_2$ thin films grown by the ALD method, residual C, N, and H impurities may stem from organic ligands of metal–organic precursors, the oxygen source, or the annealing process. Künneth and co-workers[75] investigated the effect of carbon concentration in HfO_2 on the free energy values of various crystalline phases. They suggested that C substitutes for Hf at concentrations of 3.125% and 6.25%, but observed a slight increase in the free energy difference between the ferroelectric o-and m-phases, indicating that C as a substitutional dopant does not effectively induce ferroelectricity in HfO_2. However, other forms of C doping, such as interstitial C, were not considered in their study. It's also worth noting that the 3.125% concentration is the minimum vacancy level that can be modeled in a DFT calculation with 96 atoms (2 × 2 × 2 unit cells). The role of C has also been explored in high-k gate oxides to determine whether C impurities can promote the formation of t-phase over the m-phase, given that the former has a dielectric constant almost twice as high as the latter. Jung et al.[76] examined the effects of C incorporation into Hf-substitutional, O-substitutional, and interstitial positions on the free energy difference between the t-and m-phases. Their findings showed that C incorporation into interstitial or oxygen-substitutional sites could further reduce the free energy of the t-phase compared to Hf substitution. In their study, the combined effect of C impurity and doubly charged V_Os was shown to significantly lower the free energy of the t-phase. Therefore, a more pronounced positive or negative impact from carbon in other configurations or combined with other defects cannot be ruled out based solely on limited theoretical studies like that of Künneth and co-workers.[75] The comprehensive role of carbon requires more in-depth investigation considering various defect sites and interactions.

In contrast to theoretical studies, several experimental investigations have demonstrated that C impurities significantly influence the ferroelectric properties of doped HfO_2. Kim et al.[77,78] explored the effect of deposition temperature during the ALD process on phase evolution and the associated ferroelectric performances of HfO_2 and $Hf_{0.5}Zr_{0.5}O_2$ films. Figure 3.7 illustrates the polarization versus electric field curves for $Hf_{0.5}Zr_{0.5}O_2$ (panels a–e) and HfO_2 (panels f–i) films, with deposition temperatures ranging from 280 to 200 °C. Additionally, Figure 3.7j–l show

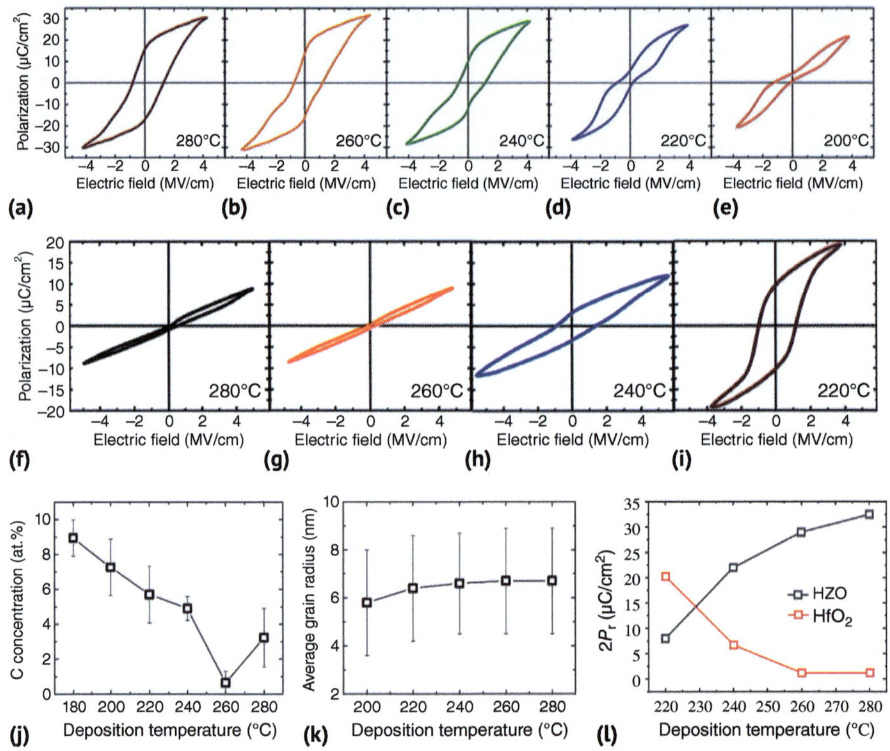

Figure 3.7 Polarization–electric field curves of TiN/$Hf_{0.5}Zr_{0.5}O_2$/TiN capacitors with $Hf_{0.5}Zr_{0.5}O_2$ film deposited at (a) 280, (b) 260, (c) 240, (d) 220, and (e) 200 °C, respectively. *Source*: [77] / with permission of Elsevier. Polarization–electric field curves of TiN/HfO_2/TiN capacitors with HfO_2 film deposited at (f) 280, (g) 260, (h) 240, and (i) 220 °C, respectively. *Source*: [78] / with permission of Royal Society of Chemistry. Changes in (j) C concentration analyzed using Auger electron spectrometry, (k) average grain radius analyzed using the watershed method implemented by Gwyddion software, and (l) $2P_r$ values as a function of deposition temperature. *Source*: [77] / with permission of Elsevier. (l) [78] / with permission of Royal Society of Chemistry.

the variations in C concentration, average grain size, and $2P_r$ values as functions of deposition temperature. As shown in Figure 3.7a–e, $Hf_{0.5}Zr_{0.5}O_2$ exhibited strong ferroelectric behavior with a remanent polarization greater than 16 μC/cm^2 and a high fraction of o-phase at a deposition temperature of 280 °C. However, at a deposition temperature of 200 °C, the films displayed antiferroelectric-like double hysteresis, which resulted from a field-induced phase transition between the t-and o-phases, or from depolarization fields. This observation suggests that at zero electric field, the dominant crystalline phases in $Hf_{0.5}Zr_{0.5}O_2$ films deposited at 280 and 200 °C are the o-and t-phases, respectively. Park et al.[31] studied the impact of forming gas (95% N_2 + 5% H_2) annealing on the ferroelectric properties of $Hf_{0.5}Zr_{0.5}O_2$ films with Pt and TiN top electrodes. Their findings confirmed that the incorporation of hydrogen did degrade the ferroelectric performance, but the effect was notably less severe compared to perovskite oxide (ABO_3) materials.

It is important to note that HfO_2 and ZrO_2 are widely used in commercial semiconductor chips (such as MOSFETs and DRAM capacitors), where hydrogen annealing is commonly employed. This process is particularly critical for metal/ferroelectric/semiconductor stacks, as it plays a key role in reducing the interfacial trap density at the ferroelectric–semiconductor interface.

Oh et al.[79] investigated the impact of high-pressure hydrogen annealing on the ferroelectric properties of W/Al:HfO_2/W and W/Al:HfO_2/Si capacitors, finding that the endurance of the W/Al:HfO_2/Si capacitor could be enhanced by reducing the interfacial trap density. The hydrogen in ferroelectric-doped HfO_2 can also influence the formation of internal fields and the electrical conduction mechanisms. Park et al.[80] studied the influence of different annealing atmospheres (N_2, O_2, and forming gas) on the chemical and ferroelectric properties of $Hf_{0.5}Zr_{0.5}O_2$ thin films, observing a shift in the *P–E* curves due to the formation of V_Os. Among the oxygen sites with three (O3) and four (O4) neighboring Hf ions, substituting O3 with nitrogen results in a more significant reduction in the free energy difference between the ferroelectric o-and m-phases (~20 meV per formula unit at 3.125% doping) compared to the substitution of O4 sites, which has a much smaller effect (less than 0.2 meV at 3.125% doping).[81] Zhou et al. reported that nitrogen doping had an even greater effect on reducing the free energy difference between the o-and m-phases than lanthanum doping at equivalent concentrations.[81] It is important to note that lanthanum is one of the dopants that can significantly lower the free energy of the ferroelectric o-phase,[31] which has been experimentally confirmed to result in large P_r (> 25 $\mu C/cm^2$). This suggests that nitrogen doping can be highly potent, potentially achieving significant ferroelectricity at relatively low concentrations compared to some cation dopants, although direct experimental confirmation and optimization are still areas of active research.

3.2.2 Defect Engineering in HfO_2-based Ferroelectrics

Beyond intentional doping with foreign elements, intrinsic point defects (such as oxygen vacancies and interstitials), extended defects (like grain boundaries), and interface-specific defects play critical roles in influencing the phase stability, polarization switching, and overall performance of HfO_2-based ferroelectric thin films and devices. Engineering or controlling these defects is therefore a vital aspect of optimizing ferroelectric behavior.

3.2.2.1 Oxygen Vacancies and Their Impact

Oxygen vacancies ($V_{\ddot{O}}$), being native point defects common in oxides, are particularly influential in HfO_2. Materano et al.[82] employed DFT calculations to investigate the impact of oxygen defects on the crystallization process and phase transitions between different crystalline phases in undoped HfO_2. Based on Ostwald's rule for nucleation, which suggests that initial crystallization occurs in the t-phase, their work examined how energy differences between the phases and the mean energy paths (MEPs, see Figure 3.8) varied depending on film stoichiometry. MEPs are defined as a series of structures connecting two (meta)stable phases, representing the path with the minimal increase in energy. The calculations were grounded in the phase dependence of

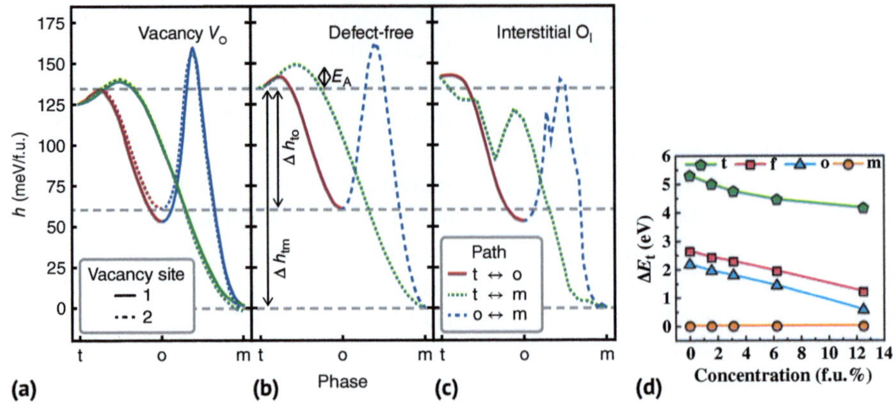

Figure 3.8 Energies of MEPs along the reaction coordinates between crystal phases at $T = 0$ for HfO_2, relative to the m-phase, (a) including a vacancy (3.125 f.u.%), (b) defect-free HfO_2, and (c) including an interstitial (3.125 f.u.%) within a 96-atom HfO_2 supercell. The transition rate assuming a nucleation-limited transformation depends on the values of Δh that are illustrated in (b) for the transitions between t-and o-phases (index to) and between t-and m-phases (index tm), respectively. *Source*: [81] / with permission of Elsevier. (d) Total energies (ΔE_t) of the t-, o-III (f)-, and o-phase HfO_2 relative to that of the m-phase as a function of oxygen vacancy concentrations. *Source*: [82] / with permission of American Chemical Society.

formation energy (h), which results in energy differences (Δh) between crystal phases as a function of defect concentration. This influences phase formation from the perspective of nucleation theory. The phase transition rate to the lower-energy phase, $R \sim \exp(-G^*/kT)$, depends on a thermodynamic barrier, $G^* = \alpha\gamma^3/|\Delta h|^2$, which is influenced by defect and doping concentrations.[83] Here, α is a geometry factor that defines the shape of the considered nuclei. In this model, only metastable equilibrium structures are relevant, characterized by the energy difference Δh and the crystalline interface energy γ.[84] Oxygen vacancies reduce the phase energy Δh for all phases relative to the energetically lowest m-phase (Figure 3.8d).[46,81,85] The formation of oxygen vacancies promotes the o-III-phase, but as vacancy concentration increases, the t-phase becomes more favorable.

From a kinetic perspective, the energy landscape of crystal structures plays a crucial role in phase formation. In this model of phase transitions, which occur without the creation of a phase boundary, the transition rate depends on the barrier height between the initial and final phases. The energy landscape for transitions between relevant phases and the transition states that define the barrier heights has been calculated with and without the inclusion of oxygen defects.[82,86] The results show that oxygen vacancies do not significantly impact the barrier height. However, the presence of oxygen interstitials lowers the barrier for the metastable t-phase, or may even eliminate this barrier entirely toward the m-phase. Consequently, the o-phase cannot be stabilized in the presence of oxygen interstitials, leading to the suppression of ferroelectricity.

3.2.2.2 Intrinsic Defects and Phase Stability

Beyond oxygen vacancies, other intrinsic defects such as interstitials and extended defects like grain boundaries also influence the phase stability of HfO_2. Liu and co-workers discovered that doping with interstitial Hf and Zr atoms causes lattice expansion and increases both in-plane and out-of-plane stresses. This stabilizes the r-phase and its ferroelectric properties, while reducing the stress and energy demands from the electrodes.[6] DFT calculations showed that the formation energy of the r-phase is lower than that of the o-III-and m-phases in Hf(Zr)-rich $Hf(Zr)O_2$ films when the Hf(Zr)-to-O ratio exceeds 1.079:2, favoring r-phase formation. These calculations also revealed that a 7% strained r-phase $Hf_{1.08}O_2$ exhibits a switching barrier of 7.6 meV/atom, significantly lower than the 20-meV/atom barrier in stoichiometric r-phase HfO_2. The control of interstitial atoms leads to remarkable ferroelectric properties, including a high remanent polarization of 22 $\mu C/cm^2$, a small saturation-polarization electric field (E_s) of 1.25 MV/cm, an ultralow coercive field of approximately 0.65 MV/cm, and a high breakdown electric field (E_{BD}) of 4.16 MV/cm.

Interstitial Hf dopants have also been investigated for their impact on the orthorhombic $Pca2_1$ phase. Zhu et al.[87] showed that interstitial Hf can reduce the energy difference between the polar orthorhombic $Pca2_1$ phase and the intermediate tetragonal $P4_2/nmc$ phase, leading to a decrease in both the switching barrier and coercive field. In contrast to the r-phase, which requires additional compressive strains for polarization, the $Pca2_1$ phase in $Hf_{1.03}O_2$, with a moderate doping level of 3%, already exhibits a low switching barrier and a high polarization. Large-scale molecular dynamics (MD) simulations at finite temperatures, using a DFT-derived machine learning force field, revealed that just 0.5% interstitial doping in $Pca2_1$-HfO_2 is sufficient to significantly reduce the switching field. Additionally, following a prepoling treatment in which interstitial dopants facilitate the formation of mobile domain walls, the switching field can drop below 1 MV/cm, aligning with experimental observations. Further DFT calculations on various interstitial dopants indicated an inverse relationship between dopant size and switching barrier, identifying several promising dopants that could further reduce the coercive field.

Grain boundaries, as two-dimensional defects, play a crucial role in shaping the electrical properties of ferroelectric HfO_2. The strong interactions between grain boundaries and point or line defects have been a key focus in materials science. It has been proposed that the energy of grain boundaries may influence polymorphism both thermodynamically and kinetically. These boundaries can serve as nucleation sites for ferroelectric phase transitions, thereby affecting the switching behavior of HfO_2 films. This topic was further explored in a progress report on ferroelectric phase formation in HfO_2-based ferroelectrics.[43] Additionally, precise control over deposition parameters such as temperature, pressure, and precursor composition can help manipulate or reduce the occurrence of intrinsic defects, leading to enhanced ferroelectric performance.

3.2.2.3 Interface Defects and Their Role

At the nanoscale, the significant contribution of surface or interface energy to the total free energy can stabilize phases that are typically metastable in bulk form. The interface between the ferroelectric layer and the electrodes plays a crucial role in determining the overall performance of the device. In addition to size effects, the ferroelectric properties are significantly influenced by the formation of often non-ferroelectric interfacial layers and the accumulation of defects at interfaces and interfacial grain boundaries. These interfacial layers typically arise from the oxidation of metal electrodes or semiconductor surfaces. Particularly pronounced effects are observed when the ferroelectric HfO_2 layer directly contacts an Si substrate, which forms part of the gate stack in ferroelectric field-effect transistors (FeFETs). The oxidation of the Si substrate has been a major challenge for both advanced MOSFETs with HfO_2 high-k gate dielectrics and FeFETs. It is important to note that HfO_2-based FeFETs suffer from the formation of low-k SiO_2 interfaces. Although the situation in ferroelectric HfO_2 is generally better than in zirconate titanate (PZT), it can still pose significant challenges for low-power device performance. When a gate bias is applied across the gate stack of an FeFET, a substantial portion of the electric field is applied across the SiO_2 interfacial layer, which increases the minimum voltage needed to reverse the polarization state. Lomenzo et al.[88] investigated the impact of an interfacial low-k non-ferroelectric layer by inserting an Al_2O_3 (k~9) layer of varying thicknesses between TiN and $Hf_{0.5}Zr_{0.5}O_2$ films. Their findings demonstrated that, for a given pulse height, the P_r value significantly decreases as the Al_2O_3 thickness increases, due to the greater voltage drop across the Al_2O_3 layer. The depolarization field can be calculated using Equation (3.2):

$$E_{dep} = P\left[\varepsilon_{FE}\,\varepsilon_0\left\{\left(\varepsilon_{IL}\,d_{FE}/\varepsilon_{FE}\,d_{IL} + 1\right)\right\}^{-1}\right] \tag{3.2}$$

Here, E_{dep}, P, ε_{FE}, ε_{IL}, ε_0, d_{FE}, and d_{IL} refer to the depolarization field, ferroelectric polarization, dielectric constant of the ferroelectric, dielectric constant of the interfacial layer, vacuum permittivity, thickness of the ferroelectric, and thickness of the interfacial layer, respectively. Using Equation (3.2), they also calculated the depolarization field and observed that it increased as the thickness of the Al_2O_3 layer grew from 0 to 1.5 nm. The increased depolarization field led to a deterioration in retention performance.

Another significant defect impacting device reliability is the non-ferroelectric secondary phase at the electrode–ferroelectric interfaces. This interface has been identified as a key factor that can degrade the ferroelectric properties in their pristine state, resulting in the so-called "wake-up" effect. The interfacial layer often has a much higher defect density, and charge trapping at this layer can become a critical issue. Ali et al.[89] investigated the retention properties of Si:HfO_2-based FeFETs with SiO_2 and SiON interfacial layers between Si:HfO_2 and Si. Based on the extrapolation from accelerated retention tests at 150 °C, the expected memory window values after 10 years were 0.15 and 0.5 V for the SiO_2 and SiON interfacial layers, respectively, compared to initial memory window values of 0.75 and 0.95 V. They concluded that improved charge trapping, rather than solely its dielectric constant difference, was the primary reason for the enhanced retention observed in FeFET with the SiON interfacial layer. Another issue is grain boundaries, where mobile oxygen vacancies ($V_{\ddot{O}}$) can accumulate due to attraction between the grain boundaries and defects. This accumulation results in limited endurance, as it leads to the formation of permanent conduction paths.

3.2.3 Summary

In conclusion, the intricate interplay of intentional doping and defect engineering is paramount to controlling and optimizing the ferroelectric properties of HfO_2-based materials. Doping, particularly with aliovalent elements like Sc, Ta, and La, can significantly influence the phase stability, reduce free energy differences between the o-and m-phases, and enhance remanent polarization. Aliovalent dopants introduce charge compensation mechanisms that enhance the ferroelectric characteristics by stabilizing the desired phase. On the other hand, defect engineering, particularly oxygen vacancies, further impacts phase transitions and ferroelectric behaviors. Oxygen vacancies can stabilize specific phases, such as the o-phase, and influence switching dynamics. The combination of doping and defect engineering offers a powerful strategy for optimizing HfO_2-based ferroelectric materials for practical applications. However, despite significant progress, achieving atomic-scale control over defect formation and distribution, and reliably predicting their complex interactions with dopants and their collective influence on phase transitions, electronic properties, and long-term device performance, remains a key challenge. Continued, in-depth exploration of these fundamental aspects is essential for overcoming current limitations and realizing the full potential of HfO_2-based ferroelectric technology in future electronic devices.

3.3 The Role of Surface/Interface/Grain Boundary Energy in HfO_2-based Ferroelectricity

3.3.1 The Role of Surface, Interface, and Grain Boundary Energies in Stabilizing Ferroelectric Phases

The emergence of ferroelectricity in HfO_2 is intrinsically linked to its crystallographic structure, specifically the stabilization of non-centrosymmetric phases, most importantly the orthorhombic $Pca2_1$ phase, amidst various competing polymorphs (monoclinic, tetragonal). While bulk HfO_2 is not ferroelectric under typical conditions, the unique environment of thin films, particularly at interfaces and grain boundaries, plays a critical role in stabilizing these metastable polar phases. This subsection examines the significant impact of surface, interface, and grain or phase boundary energies on the ferroelectric performance of HfO_2-based materials. A thorough understanding of how these energy contributions modify the thermodynamic and kinetic landscapes is essential for optimizing key device characteristics such as polarization switching speed, data retention, and fatigue resistance.

Real ferroelectric HfO_2 thin films are typically polycrystalline and are fabricated within complex stacks, often sandwiched between two electrodes. Therefore, the energy contribution from these interfaces is a key factor influencing the observed ferroelectricity. In general, the energy contribution from interfaces is characterized by the interface energy γ. The numerical effect of interface energy on the crystallographic phase formation in HfO_2 has been explored through theoretical modeling.[15,33,34] In these models, each crystal phase exhibits a distinct interface energy, and because of its influence on Gibbs free energy (G), the interface energy can alter the thermodynamic equilibrium phase, as described in Equation (3.3):

$$\mathcal{G}(p, A, E, \boldsymbol{\sigma}^d, T) = \min_{\xi} G^{\xi}(p, A, E, \boldsymbol{\sigma}^d, T) \tag{3.3}$$

where $\mathcal{G}$ is the total energy in thermodynamic equilibrium, p the hydrostatic pressure, A the surface or interface area, E the electric field, $\boldsymbol{\sigma}^d$ the deviatoric stress tensor, and T the temperature.

Calculations of free surface energies (relative to vacuum) for various HfO_2 phases and surface cleavage planes provide initial insights. Batra et al.[33] calculated the free surface energies (relative to vacuum) for various hafnia phases across different surface cleavage planes. As shown in Figure 3.9a, their findings indicate that the t-phase planes exhibit the lowest surface energies for all orientations examined, except for the (001) surface, where the o-IV-phase demonstrated the lowest surface energy. As a result, it is expected that the t-and o-IV-phases will be stable, rather than the bulk equilibrium m-phase, at small length scales where finite size effects outweigh the bulk energetics. To determine the length scales at which these phases can be stabilized, they developed a thermodynamic model for the total potential energy of a hafnia particle or thin film with finite dimensions, according to:

$$E^{\xi} = U_{\mathrm{B}}^{\xi} + \sum_k \gamma_k^{\xi} A_k^{\xi} \tag{3.4}$$

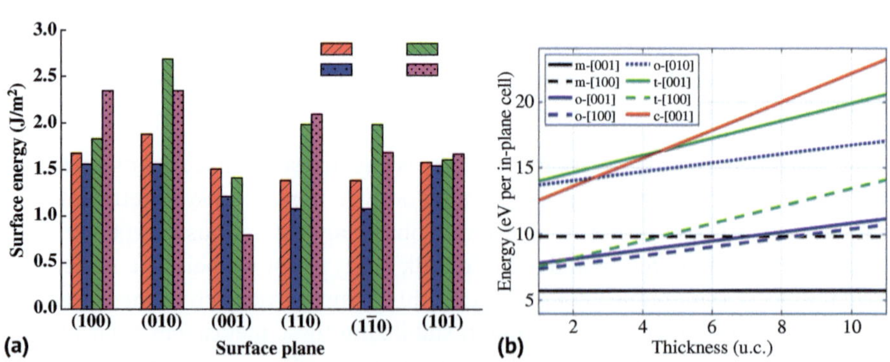

Figure 3.9 (a) DFT-computed surface energies of the m-, t-, o-III-, and o-IV-phases of HfO_2 corresponding to different surface orientations. *Source*: [33] / with permission of AIP Publishing LLC. (b) Thin film energies, computed via the energy model considering the interfacial energies and bulk energies, as a function of film thickness for stacks. The bulk energy of m-phase is set as the zero point of bulk energies. *Source*: [35] / with permission of Royal Society of Chemistry.

where U_B^ξ is the bulk energy of the phase ξ, A_k^ξ the area, and γ_k^ξ the surface energy of the surface k in the phase ξ. Following the thermodynamic rules for a particle with given dimensions, the phase with the least E^ξ will be the most stable in thermodynamic equilibrium. The free surface model outlined earlier highlights the significant role that surface energies play in the stabilization of crystal phases. It also suggests that the finite dimensions of fabricated HfO_2 thin films may be a key factor in the material's ferroelectric properties. This highlights the fundamental role surface energies play in stabilizing non-bulk phases and suggests that the finite dimensions of HfO_2 thin films are a key factor enabling ferroelectricity.

In real thin films, the interfaces with electrodes contribute additional interface energies that differ from those of free surfaces. To assess this, Dogan et al. created Ir/HfO_2/Ir stacks to calculate the interface energies between different HfO_2 phases and the Ir electrode material. They subsequently developed an energy model for films of varying thicknesses, incorporating both interfacial and bulk energies.[35] As shown in Figure 3.9b, their results indicate that the (001)-oriented m-phase grains exhibit the lowest energy configuration across all film thicknesses. However, the (001)-oriented o-III-phase, (100)-oriented o-III-phase, and (100)-oriented t-phase grains also remain competitive in ultrathin films. Notably, the energy of the (100)-oriented t-phase film surpasses that of the (001)-oriented o-III-phase film at a thickness of approximately two unit cells (around 1 nm), suggesting that the o-III-and m-phases dominate in thicker films (above two unit cells). Chen et al. developed a multiphase coexistence phase-field model, focusing on the interplay between these two phases.[90] In their model, the effects of grain boundaries were incorporated through a scale factor, τ, which adjusts the proportion of surface energy relative to the total energy in thin films. As τ increases, the proportion of surface energy rises, corresponding to smaller grain sizes. Due to the lower surface energy of the o-III-phase compared to the m-phase, the fraction of o-III-phase grains increases with rising τ and more pronounced surface effects. This finding supports the idea that grain boundary surface energies could foster the emergence of the new polar phase in HfO_2-based thin films.

Beyond the average effect modeled by factors like τ, the specific energy of internal phase boundaries between different crystallographic phases also impacts the kinetic transformations during processing. Künneth et al. developed an energy model that incorporates the internal interface energy between the t-phase and other phases, while also considering the experimental distribution of grain radii.[34] They discovered that the interface energy between the t-and m-phases is typically higher than that between the t-and o-III-phases. As a result, the t-and o-III-phase grains exhibit similar energy levels over a broader range of grain radii. Consequently, the kinetic transition from the t-phase to the o-III-phase is more favorable than the transition from the t-phase to the m-phase. A similar conclusion was reached by Wu et al. in their study on the formation of ferroelectric phases in Si-doped hafnia through nucleation.[48] Their DFT calculations showed that the t-phase is the thermodynamically stable phase when considering the effects of dopants and surface energy at high temperatures. They also found that the activation barrier for the phase transition from the t-phase to the m-phase is higher than that for transition to the o-III-phase. As the temperature decreases, the t-phase grains become metastable and transition to the o-III-phase. The transition from the t-phase to the o-III-phase was directly confirmed through *ab-initio* molecular dynamics (AIMD) simulations.

In addition to the surface effects previously mentioned, Lee et al. demonstrated that hydroxyl adsorption during the deposition process can significantly lower the surface energy of the (112)-oriented o-III-phase, promoting its appearance in ultra-thin films.[91]

3.3.2 Summary

In conclusion, the surface, interface, and grain boundary energies play a critical and multifaceted role in enabling and controlling ferroelectricity in HfO_2-based thin films. These energy contributions modify the thermodynamic energy landscape at finite dimensions, stabilizing the ferroelectric o-phase relative to the bulk stable m-phase, especially at small length scales and grain sizes. Furthermore, the specific energies of internal phase boundaries kinetically favor the transformation from processing-induced metastable phases (like tetragonal) to the ferroelectric o-phase. While significant progress has been made in modeling and understanding these energy effects, precisely controlling interfacial structure and chemistry, grain boundary networks, and internal phase boundaries during fabrication remains challenging. Future research should focus on leveraging this understanding to engineer these energy landscapes more effectively, for example, through advanced deposition techniques, tailored electrodes, and strategic co-doping that influences interface energies. Optimizing these energy contributions is crucial for enhancing key device performance metrics such as switching speed, data retention, and fatigue resistance, ultimately paving the way for robust HfO_2-based ferroelectric devices in practical applications.

3.4 External Field Driving Effects on HfO_2-based Ferroelectricity

The emergence of robust ferroelectric properties in HfO_2, particularly its ability to maintain a stable ferroelectric phase in nanoscale thin films, makes it a promising candidate for next-generation memory and integrated circuit technologies. Beyond the intrinsic material properties and defect engineering strategies, the behavior of HfO_2-based ferroelectrics is profoundly influenced by external factors present during processing and device operation. The term "external field driving effects" refers to the influence of various external stimuli, such as electric fields, mechanical stress or strain, and temperature fields (including gradients), on the formation, stability, and switching dynamics of the ferroelectric phase. A deeper understanding of these effects will enable the design of HfO_2-based ferroelectric devices with enhanced performance, including improved switching speed, retention, and resistance to fatigue.

3.4.1 Mechanical Field Effects on HfO_2 Ferroelectricity

Mechanical fields, primarily in the form of stress and strain, arise intrinsically from mismatches in lattice constants and thermal expansion coefficients between the HfO_2 film and the substrate, as well as from mechanical boundaries created by the capping electrode during thermal annealing. Considering the strain effect, it can be inferred that compressive strain reduces the energy difference between the polar phases (o-III-, o-IV-, and r-I-phases) and the equilibrium m-phase, thus promoting ferroelectricity due to the inverse relationship between the energies and volumes of the various phases of HfO_2.[92] This hypothesis has been supported by numerous first-principles computational studies.[11,15,35,36,86]

Studies investigating the effect of applied pressure, a form of isotropic compressive stress, have shown that it can induce phase transitions. Materlik et al.[15] found that isotropic compressive pressure could trigger the phase transition from the m-phase to the antipolar o-I-phase, aligning with the findings of Huan et al.[1] Their work also suggested that surfaces or interfaces could lower the pressure required for phase transformations.[15] However, strain arising from substrate lattice mismatch is typically anisotropic (often modeled by fixing the surface area and imposing zero stress in the film's normal direction). They noted that significant phase transitions are most likely to occur in spatial orientations of grains with a close match in lattice constants. Crucially, they found that the (001)-oriented o-III-phase, exhibiting measurable polarization in the normal direction, could not be stabilized by compressive film stress alone. However, when compressive stress and electric field effects were combined, the polar o-III-phase could be stabilized as the equilibrium phase. Batra et al. conducted a similar investigation into these combined effects and obtained consistent results, as illustrated in the computed phase diagram under the influence of both the electric field and in-plane stress (Figure 3.10).[36]

Dogan et al. also examined the energy variations of different phases under constrained in-plane grain areas. However, rather than identifying the factors that promote the o-III-phase as the equilibrium state, they focused on the kinetic stabilization of the polar o-III-phase during thermal annealing.[35] While they did not consider the influence of the electric field, they investigated the combined effects of in-plane stress, doping, and the out-of-plane confinement induced by the top electrode. Their findings revealed that out-of-plane confinement could inhibit the

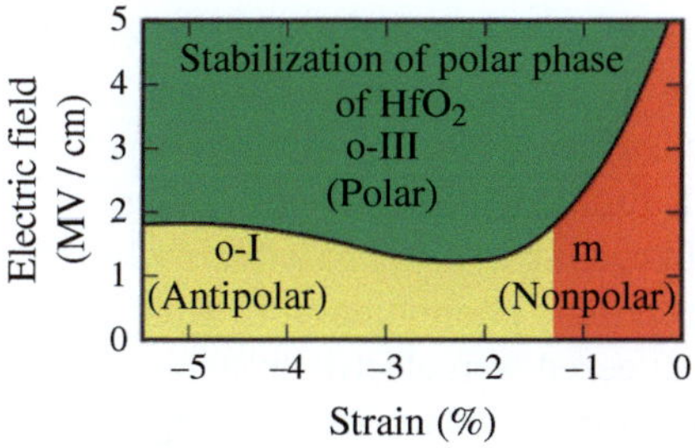

Figure 3.10 The computed phase diagram of hafnia under the influence of electric field and in-plane stresses. *Source*: [36] / with permission of American Chemical Society.

transition from the t-phase to the m-phase, which aligns with the capping effects observed in experiments.[13,50,93] Moreover, doping with appropriate concentrations (such as approximately 4% for Si and Al, and around 50% for Zr) could facilitate the transformation from the t-phase to the o-III-phase.

In addition to the in-plane strain effects on (100)-, (010)-, and (001)-oriented grains, the (111) in-plane strain has also received significant attention due to the prevalence of (111)-oriented grains in epitaxial HfO_2 thin films.[5,94] Qi et al. demonstrated that an in-plane shear strain could kinetically stabilize the polar o-IV-phase in (111)-oriented epitaxial films by inducing a transition from the t-phase.[7] Their simulated XRD and selected area electron diffraction (SAED) results for this o-IV-phase were found to be consistent with experimental observations.[5] Liu and Hanrahan examined the effects of various in-plane strains in different crystal orientations, finding that both the antipolar o-I-phase and the polar o-III-phase exhibit lower energy than the m-phase when subjected to the appropriate (111) in-plane strain. With the application of an electric field, the o-III-phase becomes the thermodynamically stable phase.[86] However, Zhang et al. argued that the m-phase remains the most stable phase under (111) in-plane strain, taking into account the variations in the in-plane areas of the m-phase for different <111> orientations (such as (111) and $(11\bar{1})$).[11] In contrast, Liu and Hanrahan reached a different conclusion.[86] Their focus was on the r-I-phase rather than the o-III-phase, and they found that (111) in-plane compressive stress could not stabilize the r-I-phase. However, it could enhance spontaneous polarization, with the r-I-phase potentially becoming the most stable phase under extreme film thickness conditions.

To investigate the physical origin of how strain influences specific phase transitions, Delodovici et al. examined the strain effect on symmetry-allowed distortions during the phase transition from t-phase to o-III-phase.[22] Using Landau theory based on symmetry distortion mode analysis, they developed a free energy landscape characterized by three critical modes. They then assessed the strain's impact on the stability of symmetry-allowed distortions and various components of the free energy landscape. Ultimately, they found that the strong trilinear coupling between the three modes plays a crucial role in stabilizing the polar o-III-phase, regardless of the specific strain effects. However, appropriate strain could soften the Y_{2+} mode, potentially rendering it unstable when tensile strain is applied along the [100] direction.

3.4.2 Electric Field Effects on HfO_2 Ferroelectricity

Considering the energy contribution of electric field, it is also effective to describe the impact of electric field by examining the forces on the ions induced by the applied electric field. Understanding how applied electric fields influence structural changes and polarization dynamics in HfO_2 is fundamental to device operation. A key phenomenon observed under electric field cycling in HfO_2-based ferroelectrics is the "wake-up" effect. This effect is a characteristic phenomenon in HfO_2-based ferroelectrics, marked by the appearance of squeezed ferroelectric hysteresis loops in the pristine state, along with an increase in remnant polarization during electric field cycling. Schenk et al.[95] determined that the activation energy for the wake-up process in Sr-doped HfO_2 is around 100 meV, which is an order of magnitude lower than that of conventional perovskite ferroelectrics, suggesting a more pronounced wake-up effect in FE-HfO_2. Zhou et al.[96] found that the wake-up process is more easily completed when a higher voltage and a lower-frequency cycling are applied, indicating domain depinning due to the reduction of defect concentration at the bottom electrode interface. The resistance changes observed in HfO_2 during repeated cycling further suggested oxygen vacancy movement, as indicated by valence change mechanism (VCM) measurements.[97] To better understand the mechanisms behind the domain and built-in field evolution during the wake-up process, first-order reversal curve (FORC) characterization was used for both the initial and woken-up FE-HfO_2. The results showed that the electric field and polarization density in FE-HfO_2 gradually became more uniform, which could contribute to the wake-up effect.[98] Additionally, *in-situ* XRD characterization of HZO revealed an increase in peak intensity near 30.5° corresponding to o-phase/t-phase with field cycling, suggesting that phase transformation occurred.[99]

Understanding the wake-up effect is crucial not only for enhancing the reliability of devices but also for shedding light on the underlying mechanisms of ferroelectric properties. Numerous studies and reviews have explored the wake-up effect in FE-HfO_2.[100–102] Lederer et al.[103] identified five distinct phenomena – classical, textured, non-split, sudden, and asymmetric wake-ups – that are commonly observed in almost all metal–ferroelectric–metal (MFM) and metal–ferroelectric–insulator–semiconductor (MFIS) capacitors. More recently, the structural transition from the *Pbca* phase to the $Pbc2_1$ phase and the charged domain wall (CDW) have been proposed as alternative explanations for the wake-up effect.[54] However, Xu et al.[104] fabricated bulk single-crystalline Y-doped HfO_2, which showed minimal wake-up, suggesting that preventing phase transformation and the redistribution of oxygen vacancies may effectively eliminate the wake-up effect.

Given the variability in wake-up behaviors exhibited by FE-HfO_2 produced through different fabrication techniques, the distinct wake-up schemes were categorized into two main groups and their possible mechanisms were discussed.

3.4.2.1 Antiferroelectric-type Wake-up

In this case, antiferroelectric-like behavior was observed in the fresh ferroelectric capacitor, typically exhibiting double current humps or a constricted *P–V* loop, as shown in Figure 3.11a. Upon electric field cycling, these features evolve, and the loop transforms into a well-developed ferroelectric hysteresis loop.[105–107] Qi and Rabe performed phase structure optimizations by incorporating electric field forces based on DFT calculations under finite electric fields.[108] They discovered that, combined with doping effects, the t-phase could transform into the polar o-III-or o-IV-phase under an electric field, exhibiting electrical hysteresis loops. Fan et al. also observed similar electric-field-induced nonpolar-to-polar phase transitions and hysteresis loops through AIMD simulations.[109] They found that in-plane strain could influence the hysteresis behavior. Under compressive in-plane strain conditions, the material exhibits antiferroelectric behavior driven by the out-of-plane electric field, while tensile in-plane strain results in the emergence of a ferroelectric hysteresis loop.

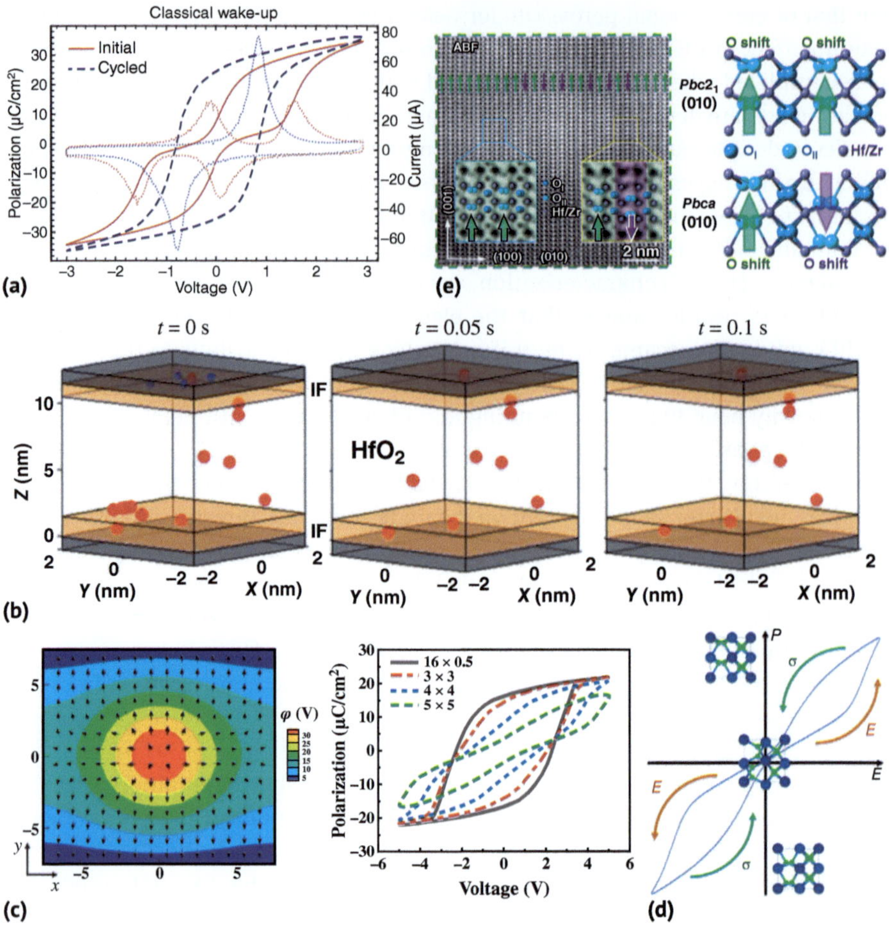

Figure 3.11 (a) Typical P–V and I–V curves for the initial and cycled ferroelectric capacitors. *Source*: [54] Springer Nature / CC BY 4.0. (b) Simulated vacancy diffusions of the MFM capacitors during wake-up. *Source*: [103] / with permission of John Wiley & Sons. (c) The impact of in-plane tensile stress and the consequential ferroelastic switching. *Source*: [110] / with permission of Elsevier. (d) The charged domain effect on the P–V hysteresis loops. (e) The observation of antiferroelectric *Pbca* phase and ferroelectric $Pbc2_1$ phase. *Source*: [111] / with permission of John Wiley & Sons.

Typically, this behavior depends on the number of electric field cycles rather than the operating time. The precise mechanism behind the antiferroelectric wake-up effect remains unclear. It has been suggested that phase transition, domain movement, and the built-in field may all contribute to this effect.[88,98,110] It has been established that a significant number of oxygen vacancies form at the interface, as many electrodes used to induce ferroelectricity in HfO_2 can readily scavenge oxygen. During the wake-up process, almost no new traps are created. As shown in Figure 3.11b, the redistribution of oxygen vacancies from the interface to the interior of the thin film leads to domain depinning and an increase in remnant polarization under field cycling, which aligns with the results from computational studies.[81,111] The reduction of the built-in field is facilitated by the diffusion of oxygen vacancies, leading to a more uniform field and a larger proportion of the ferroelectric phase. Additionally, Lomenzo et al.[88] proposed that the depolarization field created by the non-ferroelectric interface, the coexistence of ferroelectric and paraelectric phases, and limited charge screening could drive the phase transition in FE-HfO_2. The antiferroelectric wake-up effect can also be explained by the domain transition process, where CDWs and ferroelastic switching are considered potential mechanisms.[103,110] Recent studies have identified intrinsic tail-to-tail 90° domain structures and CDWs in the pristine-state HZO film.[110] It has been suggested that the distribution of CDWs throughout the film's thickness is the origin of the antiferroelectric wake-up effect, as described by the time-dependent Ginzburg–Landau model, shown in Figure 3.11c. The CDWs disappear with electric field cycling, which leads to an increase in remnant polarization. Regarding ferroelastic switching,[103] transmission Kikuchi diffraction results revealed an enhanced out-of-plane orientation of the polarization axis and a reduced ferroelastic axis in the out-of-plane orientation density, which supports ferroelastic switching during cycling, as illustrated in Figure 3.11d. This suggests that stable ferroelectric behavior can be achieved through stress relief during field cycling. Recently, the first experimental observation of the antiferroelectric phase of *Pbca* provided an additional explanation for the antiferroelectric wake-up effect, as shown in Figure 3.11e.[54] The 180° domain boundary structure of the $Pbc2_1$ phase was found to be identical to the crystal structure of the antiferroelectric phase. The energy of the antiferroelectric phase is slightly lower than that of the ferroelectric phase. A structural transition from the *Pbca* phase to the $Pbc2_1$ phase was observed under external electric field stimulation, as demonstrated by STEM-HAADF and STEM-ABF results, which provide synchronized imaging of oxygen atomic columns alongside heavier Hf and Zr atomic columns. Further analysis indicated that the field-induced phase transition is unlikely to occur due to the nearly constant spontaneous polarization and high Curie temperature, unless coupled effects from the redistribution of oxygen vacancies or other defects are involved. Additionally, studies on ferroelastic switching have suggested that, due to the high transition barrier, ferroelastic switching may be linked to the formation of new defects or the redistribution of the built-in field. Therefore, it is believed that the reductions in the built-in field, phase transition, and domain transitions are closely associated with the redistribution of existing oxygen vacancies at the metal–ferroelectric interface.

3.4.2.2 Polarization-enhanced-type Wake-up

Distinct from the "antiferroelectric-type wake-up" discussed previously, the term "polarization-enhanced wake-up" describes a scenario where the initial ferroelectric hysteresis loop in the pristine state does not exhibit antiferroelectric-like characteristics, yet the ferroelectric hysteresis loop still shows an increase in remnant polarization during electric field cycling.[102,112] Typically, the initial I–V curves for each bias direction lack distinct current peaks or show only a single peak. After several electric field cycles, a sudden widening or a saturated hysteresis loop may emerge.

The field-induced crystallization of HfO_2 in the o-phase is likely responsible for this wake-up phenomenon, as confirmed by grazing incidence XRD measurements that show the amorphous structure prior to wake-up.[103] Films produced by conventional methods may show such characteristics as annealing at low temperatures for short durations. Due to the significant changes in the dielectric constant of the ferroelectric films during cycling, it is believed that phase transformation between the m-, t-, and o-phases likely occurs. Structural analysis, including STEM-HAADF imaging, has also shown crystal structure evolution with polarization enhancement during the wake-up process.[111] It has been noted that the transition from m-to o-phase predominates in the bulk, while a reduction in the defect-rich t-phase primarily occurs at the metal–ferroelectric interface. Impedance spectroscopy measurements revealed an increase in resistance during field cycling, suggesting a reduction in defects. Due to the polycrystalline nature of the HfO_2 ferroelectric film, both mechanisms are typically present simultaneously in most cases, making it challenging to fully distinguish between their individual contributions.

3.4.3 Thermal Field Effects on HfO_2 Ferroelectricity

Temperature is a critical external field that profoundly influences the ferroelectric behavior of HfO_2. Thermal gradients or changes in temperature can significantly affect the ferroelectric behavior of HfO_2. The phase transition temperatures of the material, specifically the Curie temperature and the Néel temperature (T_N), are crucial in determining how the polarization responds to thermal stimuli. Following the discovery of the ferroelectric o-III-phase, various research groups have focused on studying the temperature-dependent phase transitions. Typically, as the temperature exceeds the Curie temperature, the ferroelectric phase transitions to a nonpolar paraelectric phase, causing the spontaneous polarization (P_s) to drop to zero. This loss of P_s can occur abruptly with thermal hysteresis (first-order transition) or gradually without hysteresis (second-order transition).[113] Additionally, according to the Curie–Weiss law, this phase transition is associated with a divergence in the dielectric constant.[113] However, phase transitions in fluorite-structured ferroelectrics differ significantly from those in conventional ferroelectrics, presenting several challenges for experimental study. One major difficulty lies in observing the reversible transition between the metastable polar o-III-phase and the nonpolar t-phase, all while preventing the formation of thermodynamically stable m-phase. Moreover, spatial inhomogeneities in thin films complicate the identification of the phase transition order.

Early work by Böscke et al. and Shimizu et al. highlighted temperature-dependent changes in the ferroelectric properties, including a reported Curie temperature of 460 °C based on temperature-dependent XRD patterns of an epitaxial Y-doped HfO_2 film.[93,114] Several subsequent studies covering both epitaxial[52] and polycrystalline films[25,26,115] have corroborated the presence of temperature-dependent phase transitions. Various o-III-to-t-phase transition temperatures have been reported depending on the film properties, doping, and orientation, such as in Y-doped HfO_2 films on YSZ substrates.[49]

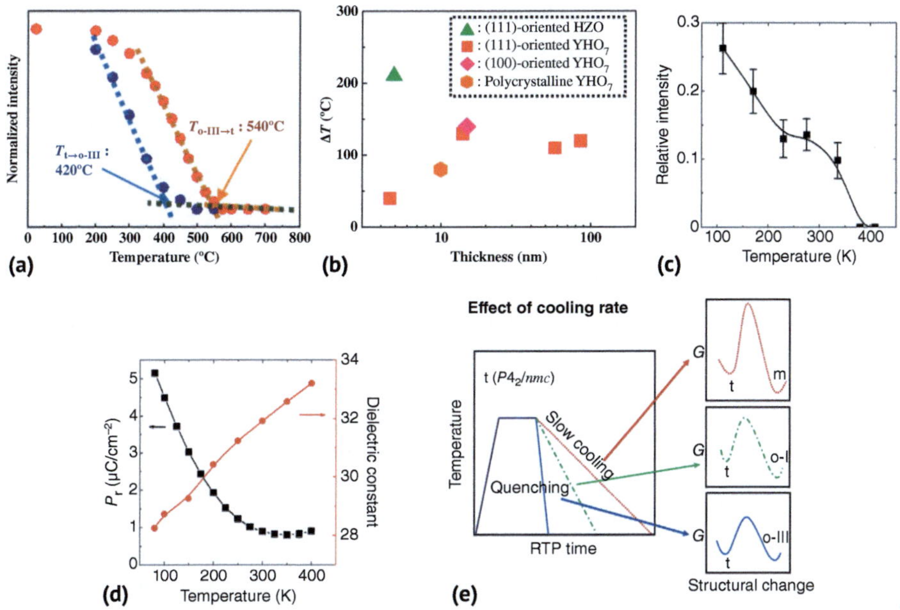

Figure 3.12 (a) Normalized intensity as a function of temperature of the (110) XRD peak from the o-phase (o-III) of an epitaxial 84-nm-thick Y-doped HfO_2 film. (b) Thermal hysteresis (ΔT) in epitaxial and polycrystalline 7% Y-doped HfO_2 (YHO_7) and $Hf_{0.5}Zr_{0.5}O_2$ films with various orientations obtained from XRD analysis with increasing and decreasing temperatures. (c) Changes in relative orthorhombic (110) peak intensity ($I_{o110}/(I_{o111} + I_{t101})$), where I_{xhkl} is the areal intensity of the (hkl) diffraction peak of the x-phase. (d) Corresponding changes in remanent polarization (P_r) and dielectric constant. (e) Schematic diagram for the secondary t-to-o-III-phase transition during the cooling step after RTP (left), which is strongly influenced by the cooling rate of RTP and by the energy landscape between various polymorphs (right). *Source*: [115] / with permission of Springer Nature.

Thermal hysteresis during both heating and cooling cycles has been widely observed in these temperature-dependent transitions, with values ranging from 40 to 150 °C for Y-doped HfO_2, and reaching up to 210 °C for $Hf_{0.5}Zr_{0.5}O_2$ (Figure 3.12a,b). As a result, a first-order martensitic phase transition accompanied by significant shear strain and volume expansion was identified for the o-III-to-t-phase transition, which was attributed to a gradual phase change. However, thermal hysteresis has not yet been confirmed in polycrystalline films produced by ALD. Furthermore, a divergence in the dielectric constant has not yet been observed in experimental studies of the phase transition between the o-III-and t-phases.[115] To address this, the temperature-dependent evolution of the crystallographic structure and the electrical properties of a 40-nm-thick 5.6% Si-doped HfO_2 film grown by ALD were characterized.[115] This film was selected to investigate the electrical property changes at low temperatures and the field-induced ferroelectricity at room temperature. A partial transition from the t-phase to the o-III-phase with decreasing temperature was observed (Figure 3.12c). The dielectric constant was measured across a broad temperature range extending beyond T_c, showing a linear increase without reaching a maximum (Figure 3.12d), although the remanent polarization grew as the temperature decreased.[115,116] One possible explanation for this could be the small grain size of the Si-doped HfO_2. In $Ba_{0.7}Sr_{0.3}TiO_3$ layers, the dielectric constant divergence disappears when the film thickness is reduced below 50 nm.[117] At T_c, soft phonon modes should become active in the infrared range for the dielectric maximum to manifest. For HfO_2 and ZrO_2, no infrared-active modes have been reported at 0 K,[118] although these modes may evolve with heating. In the case of the t-to-m-phase transition, mode softening is anticipated, and a martensitic first-order transition, potentially via an intermediate phase, is predicted.[43] Therefore, while temperature-dependent phase transitions are undoubtedly present and crucial for HfO_2 ferroelectricity, their complex nature, deviations from conventional models, and the precise role of specific phonon modes require further comprehensive research.

Beyond equilibrium or near-equilibrium phase transitions, thermal effects, particularly the cooling rate during post-deposition annealing, critically influence the final phase composition and ferroelectric properties. During the cooling phase following crystallization, the free energy of the t-phase increases at a significantly faster rate than those of the other phases due to its higher entropy. This rapid increase in free energy acts as the driving force for the transformation of t-phase into m-, o-III-, or o-I-phase. The kinetics of this transition, heavily influenced by the cooling rate, are paramount. Figure 3.12e schematically illustrates the temperature variations during crystallization at different cooling rates, along with the corresponding energy landscapes for phase transitions from t-phase to m-, o-III-, and o-I-phases.[119] Rapid quenching after high-temperature annealing is a common technique used to kinetically trap the material in the metastable ferroelectric phase.

Experimental studies on the kinetic effects of rapid quenching, such as the work by Lomenzo et al.,[88] have investigated residual polarization as a function of quenching rate and Y doping. Their findings indicate that the residual polarization of undoped HfO_2 is highly sensitive to the quenching rate, with a 50% reduction in residual polarization when the quenching rate drops from 10 to 4 °C/s. However, as the doping concentration increases, the sensitivity of polarization to the quenching rate diminishes. At a doping concentration of 1.7-mol.% Y, polarization becomes relatively insensitive to the quenching rate. At higher quenching rates, the residual polarization remains consistent, irrespective of the doping concentration. The variations in P_r values with decreasing quenching rate for different doping concentrations suggest that dopants may help stabilize oxygen vacancies, thus contributing to the stabilization of the ferroelectric phase. In the absence of dopants, prolonged annealing causes t-phase to relax into m-phase, resulting in a loss of ferroelectric properties. Therefore, in undoped or lightly doped HfO_2 thin films, a high quenching rate is essential for stabilizing the ferroelectric phase and preventing the formation of the thermodynamically stable m-phase. This observation aligns with the effects of mechanical confinement and activation barriers. As the cooling rate increases, mechanical confinement intensifies, reducing the likelihood of the transition from t-phase to the most stable m-phase via the tetragonal-to-monoclinic pathway.

Thermal effects can also influence the retention and fatigue behaviors of HfO_2-based devices. Elevated operating temperatures may accelerate the degradation of ferroelectric properties, leading to higher leakage currents and diminished switching performance. To address these challenges, research has concentrated on improving the temperature stability of HfO_2-based ferroelectrics by incorporating dopants and fine-tuning processing conditions. For example, doping HfO_2 with Zr or other elements has been shown to enhance the thermal stability of ferroelectric phase, making it more resilient to temperature-induced phase transitions. The thermal field-driven effects in HfO_2 are primarily due to the interaction between phonons (lattice vibrations) and spontaneous polarization. As temperature approaches T_c, increased thermal energy leads to larger-amplitude lattice vibrations, increasing domain wall mobility and fluctuations, which can contribute to polarization loss and degradation over time or repeated cycling.

3.4.4 Summary

In conclusion, HfO_2-based ferroelectric materials show remarkable potential, with external fields playing a pivotal role in controlling their ferroelectric properties. Mechanical field, primarily arising from strain induced by lattice mismatches and thermal expansion differences between the thin film and substrate, plays a critical role in modifying the energy landscape. This strain can effectively reduce the energy gap between polar and nonpolar phases, thereby enhancing phase stability. Electric field effects serve as the primary stimulus for driving polarization switching and inducing phase transitions, which are typically manifested by ferroelectric hysteresis loops. The nature and kinetics of these field-driven processes are notably influenced by mechanical fields, with strain potentially altering the transition pathways or favoring certain polar orientations (e.g., compressive strain potentially influencing antiferroelectric-like behavior, while tensile strain can favor ferroelectricity). Thermal fields, encompassing temperature changes and gradients, govern temperature-dependent phase transitions, including the transition to a paraelectric state above the Curie temperature.

Despite significant progress in elucidating the individual effects of mechanical, electric, and thermal fields, several key challenges remain. A limited understanding of the complex interplay and synergistic effects between these simultaneous external fields is a major hurdle. Furthermore, ensuring long-term polarization stability during prolonged operation under combined electrical bias, elevated temperatures, and mechanical stresses is critical for device reliability. Developing strategies to precisely manipulate multiple external fields concurrently and achieving robust and reliable ferroelectric behavior that is scalable for integration into practical applications are essential for future progress. These complexities represent significant avenues for future research aimed at fully harnessing the remarkable potential of HfO_2-based ferroelectric technology such as in high-performance computing and edge artificial intelligence.

3.5 Conclusion and Outlook

This chapter has provided a comprehensive examination of the origins and multiple regulation mechanisms of ferroelectric properties in HfO_2-based materials. The discovery of ferroelectricity in HfO_2 has ushered in a paradigm shift for post-Moore's Law electronics, driven primarily by its unprecedented compatibility with silicon-based complementary metal–oxide–semiconductor (CMOS) manufacturing. Unlike the perovskite materials that dominated the field for decades, the ferroelectricity in HfO_2 is an emergent nanoscale phenomenon, engineered through the precise stabilization of metastable, non-centrosymmetric crystal phases – predominantly the orthorhombic $Pca2_1$ phase. Current understanding reveals that this phase stability is governed by a complex and synergistic interplay of intrinsic and extrinsic factors, including fine-tuning of free energy via doping and defect management, nanoscale confinement at surface/interfaces, and the application of external mechanical stress and thermal field. Critically, the existence of unique flat phonon bands, which facilitate dipole localization, provides a fundamental explanation for its distinct properties, such as robust polarization stability at the sub-nanometer scale and anomalously high coercive fields (1–4.5 MV/cm).

While this foundational knowledge is extraordinary, the next frontier demands a shift from phenomenological control to predictive, atomic-scale design. Resolving fundamental debates, such as the intrinsic origin of ferroelectricity and the precise role of defects like oxygen vacancies, will be crucial. Achieving this requires a concerted effort combining advanced *in-situ* characterization techniques to probe atomic-scale switching dynamics with high-throughput computational screening and machine learning models. Such predictive frameworks will accelerate the discovery of novel dopant combinations and processing pathways, enabling the design of hafnium oxide-based materials with tailored ferroelectric and electrical properties for specific applications.

Translating these designer materials into reliable technology necessitates parallel innovations in manufacturing and integration. Deterministic phase engineering through advanced ALD and low-temperature processing routes is key to overcoming integration challenges and expanding the application landscape. This precision control is essential for tackling the critical reliability hurdles, including the wake-up effect, fatigue, and retention degradation – that currently limit commercial adoption. Mastering the intricate relationship between processing parameters, defect populations, and long-term device stability remains a paramount challenge.

The ultimate promise of HfO_2, however, extends far beyond conventional memory. Its unique properties provide the hardware foundation for new computing paradigms. FeFETs, in particular, enable the fusion of logic and memory, directly addressing the von Neumann bottleneck. By tuning the polarization switching dynamics, FeFETs can emulate the plasticity of neurons and synapses, making them a compelling platform for energy-efficient neuromorphic and in-memory computing architectures. Realizing this vision demands a holistic convergence of materials science, device physics, and manufacturing innovation. By continuing to unravel its fundamental complexities and engineer its multifaceted properties, hafnia-based ferroelectrics are positioned to define next-generation electronic, computing, and sensing technologies.

Acknowledgments

This work was supported in part by National Key R&D Program (2021YFC3002200 and 2020YFA0709800) of China, National Natural Science Foundation (62274101 and U20A20168) of China, in part by QYJS-2022-1600-B, in part by Beijing National Research Center Youth Innovation Foundation (BNR2024RC01002), National Postdoctoral Researcher Program (GZC20231216), and Research Fund from Tsinghua University Initiative Scientific Research Program.

References

1 Huan, T. D., Sharma, V., Rossetti, G. A., Jr. et al. (2014). Pathways towards ferroelectricity in hafnia. *Physical Review B* 90: 064111.

2 Ruh, R., Garrett, H. J., Domagala, R. F. et al. (1968). The system zirconia-hafnia. *Journal of the American Ceramic Society* 51: 23–28.

3 Kersch, A. and Falkowski, M. (2021). New low-energy crystal structures in ZrO_2 and HfO_2. *Physica Status Solidi (RRL) – Rapid Research Letters* 15: 2100074.

4 Böscke, T. S., Müller, J., Bräuhaus, D. et al. (2011). Ferroelectricity in hafnium oxide: CMOS compatible ferroelectric field effect transistors. In: *Proceedings of the 2011 International Electron Devices Meeting*. IEEE. pp. 24.5.1–24.5.4.

5 Wei, Y., Nukala, P., Salverda, M. et al. (2018). A rhombohedral ferroelectric phase in epitaxially strained $Hf_{0.5}Zr_{0.5}O_2$ thin films. *Nature Materials* 17: 1095–1100.

6 Wang, Y., Tao, L., Guzman, R. et al. (2023). A stable rhombohedral phase in ferroelectric $Hf(Zr)_{1+x}O_2$ capacitor with ultralow coercive field. *Science* 381: 558–563.

7 Qi, Y., Singh, S., Lau, C. et al. (2020). Stabilization of competing ferroelectric phases of HfO_2 under epitaxial strain. *Physical Review Letters* 125: 257603.

8 Zhu, T., Ma, L., Deng, S. et al. (2024). Progress in computational understanding of ferroelectric mechanisms in HfO_2. *npj Computational Materials* 10: 188.

9 Lee, H. J., Lee, M., Lee, K. et al. (2020). Scale-free ferroelectricity induced by flat phonon bands in HfO_2. *Science* 369: 1343–1347.

10 Lv, S., Cao, T., Wang, Z. et al. (2024). Physical origin of hafnium-based ferroelectricity. *Computational Materials Today* 4: 100010.

11 Zhang, Y., Yang, Q., Tao, L. et al. (2020). Effects of strain and film thickness on the stability of the rhombohedral phase of HfO_2. *Physical Review Applied* 14: 014068.

12 Terki, R., Bertrand, G., Aourag, H. et al. (2008). Cubic-to-tetragonal phase transition of HfO_2 from computational study. *Materials Letters* 62: 1484–1486.

13 Böscke, T. S., Müller, J., Bräuhaus, D. et al. (2011). Ferroelectricity in hafnium oxide thin films. *Applied Physics Letters* 99: 102903.

14 Reyes-Lillo, S. E., Garrity, K. F., and Rabe, K. M. (2014). Antiferroelectricity in thin-film ZrO_2 from first principles. *Physical Review B* 90: 140103.

15 Materlik, R., Künneth, C., and Kersch, A. (2015). The origin of ferroelectricity in $Hf_{1-x}Zr_xO_2$: A computational investigate-on and a surface energy model. *Journal of Applied Physics* 117: 134109.

16 Grimley, E. D., Schenk, T., Mikolajick, T. et al. (2018). Atomic structure of domain and interphase boundaries in ferroelectric HfO_2. *Advanced Materials Interfaces* 5: 1701258.

17 Wieder, H. H. (1957). Activation field and coercivity of ferroelectric barium titanate. *Journal of Applied Physics* 28: 367–369.

18 Choudhury, S., Li, Y., Odagawa, N. et al. (2008). The influence of 180° ferroelectric domain wall width on the threshold field for wall motion. *Journal of Applied Physics* 104: 084107.

19 Ai, P., Yan, F., Dong, W. et al. (2023). Point-defect-driven flattened polar phonon bands in fluorite ferroelectrics. *npj Computational Materials* 9: 119.

20 Zhou, S., Zhang, J., and Rappe, A. M. (2022). Strain-induced antipolar phase in hafnia stabilizes robust thin-film ferroelectricity. *Science Advances* 8: eadd5953.

21 Qi, Y., Singh, S., and Rabe, K. M. (2021). Polarization switching mechanism in HfO_2 from first-principles lattice mode analysis. arXiv preprint. arXiv:2108.12538 [cond-mat.mtrl-sci].

22 Delodovici, F., Barone, P., and Picozzi, S. (2021). Trilinear-coupling-driven ferroelectricity in HfO_2. *Physical Review Materials* 5: 064405.

23 Raeliarijaona, A. and Cohen, R. E. (2023). Hafnia HfO_2 is a proper ferroelectric. *Physical Review B* 108: 094109.

24 Aramberri, H. and Íñiguez, J. (2023). Theoretical approach to ferroelectricity in hafnia and related materials. *Communications Materials* 4: 95.

25 Park, M. H., Kim, H. J., Kim, Y. J. et al. (2015). Toward a multifunctional monolithic device based on pyroelectricity and the electrocaloric effect of thin antiferroelectric $Hf_xZr_{1-x}O_2$ films. *Nano Energy* 12: 131–140.

26 Hoffmann, M., Schroeder, U., Künneth, C. et al. (2015). Ferroelectric phase transitions in nanoscale HfO_2 films enable giant pyroelectric energy conversion and highly efficient supercapacitors. *Nano Energy* 18: 154–164.

27 Park, M. H., Kim, H. J., Kim, Y. J. et al. (2016). Giant negative electrocaloric effects of $Hf_{0.5}Zr_{0.5}O_2$ thin films. *Advanced Materials* 28: 7956–7961.

28 Park, M. H., Schenk, T., Hoffmann, M. et al. (2017). Effect of acceptor doping on phase transitions of HfO_2 thin films for energy-related applications. *Nano Energy* 36: 381–389.

29 Park, M. H., Kim, H. J., Lee, Y. H. et al. (2016). Two-step polarization switching mediated by a nonpolar intermediate phase in $Hf_{0.4}Zr_{0.6}O_2$ thin films. *Nanoscale* 8: 13898–13907.

30 Park, M. H., Lee, Y. H., Kim, H. J. et al. (2015). Ferroelectricity and antiferroelectricity of doped thin HfO_2-based films. *Advanced Materials* 27: 1811–1831.

31 Park, M. H., Kim, H. J., Kim, Y. J. et al. (2013). Evolution of phases and ferroelectric properties of thin $Hf_{0.5}Zr_{0.5}O_2$ films according to the thickness and annealing temperature. *Applied Physics Letters* 102: 242905.

32 Park, M. H., Lee, Y. H., Kim, H. J. et al. (2018). Understanding the formation of the metastable ferroelectric phase in hafnia–zirconia solid solution thin films. *Nanoscale* 10: 716–725.

33 Batra, R., Tran, H. D., and Ramprasad, R. (2016). Stabilization of metastable phases in hafnia owing to surface energy effects. *Applied Physics Letters* 108: 172902.

34 Künneth, C., Materlik, R., and Kersch, A. (2017). Modeling ferroelectric film properties and size effects from tetragonal interlayer in $Hf_{1-x}Zr_xO_2$ grains. *Journal of Applied Physics* 121: 205304.

35 Dogan, M., Gong, N., Ma, T. P. et al. (2019). Causes of ferroelectricity in HfO_2-based thin films: an *ab initio* perspective. *Physical Chemistry Chemical Physics* 21: 12150–12162.

36 Batra, R., Huan, T. D., Jones, J. L. et al. (2017). Factors favoring ferroelectricity in hafnia: A first-principles computational study. *The Journal of Physical Chemistry C* 121: 4139–4145.

37 Chae, K., Kummel, A. C., and Cho, K. (2022). Tetravalent doping in fluorite-based ferroelectric oxides for reduced voltage operations. *ACS Applied Materials & Interfaces* 14: 29007–29013.

38 Yuan, J. H., Mao, G. Q., Xue, K. H. et al. (2022). Ferroelectricity in HfO_2 from a coordination number perspective. *Chemistry of Materials* 35 (1): 94–103.

39 Skorodumova, N. V., Ahuja, R., Simak, S. I. et al. (2001). Electronic, bonding, and optical properties of CeO_2 and Ce_2O_3 from first principles. *Physical Review B* 64: 115108.

40 Kang, J., Lee, E. C., and Chang, K. J. (2003). First-principles study of the structural phase transformation of hafnia under pressure. *Physical Review B* 68: 054106.

41 Lowther, J. E., Dewhurst, J. K., Leger, J. M. et al. (1999). Relative stability of ZrO_2 and HfO_2 structural phases. *Physical Review B* 60 (21): 14485–14488.

42 Wen, Y., Wu, M., Cui, B. et al. (2024). Insights into the ferroelectric orthorhombic phase formation in doped HfO_2 thin films. *Journal of Applied Physics* 136: 124105.

43 Park, M. H., Lee, Y. H., Mikolajick, T. et al. (2019). Thermodynamic and kinetic origins of ferroelectricity in fluorite structure oxides. *Advanced Electronic Materials* 5: 1800522.

44 Batra, R., Huan, T. D., Rossetti, G. A., Jr. et al. (2017). Dopants promoting ferroelectricity in hafnia: Insights from a comprehensive chemical space exploration. *Chemistry of Materials* 29: 9102–9109.

45 Mueller, W., Aichmayr, G., Bergner, W. et al. (2005). Challenges for the DRAM cell scaling to 40nm. In: *Proceedings of the IEEE International Electron Devices Meeting, 2005: IEDM Technical Digest*. IEEE. pp. 4–339.

46 Künneth, C., Materlik, R., Falkowski, M. et al. (2018). Impact of four-valent doping on the crystallographic phase formation for ferroelectric HfO_2 from first-principles: Implications for ferroelectric memory and energy-related applications. *ACS Applied Nano Materials* 1: 254–264.

47 Materlik, R., Künneth, C., Mikolajick, T. et al. (2017). The impact of charge compensated and uncompensated strontium defects on the stabilization of the ferroelectric phase in HfO_2. *Applied Physics Letters* 111: 082902.

48 Wu, J., Mo, F., Saraya, T. et al. (2020). A first-principles study on ferroelectric phase formation of Si-doped HfO_2 through nucleation and phase transition in thermal process. *Applied Physics Letters* 117: 252904.

49 Mimura, T., Shimizu, T., Sakata, O. et al. (2021). Large thermal hysteresis of ferroelectric transition in HfO_2-based ferroelectric films. *Applied Physics Letters* 118: 112903.

50 Muller, J., Boscke, T. S., Schroder, U. et al. (2012). Ferroelectricity in simple binary ZrO_2 and HfO_2. *Nano Letters* 12: 4318–4323.

51 Müller, J., Böscke, T. S., Müller, S. et al. (2013). Ferroelectric hafnium oxide: A CMOS-compatible and highly scalable approach to future ferroelectric memories. In: *Proceedings of the 2013 IEEE International Electron Devices Meeting*. IEEE. pp. 10.8.1–10.8.4.

52 Sang, X., Grimley, E. D., Schenk, T. et al. (2015). On the structural origins of ferroelectricity in HfO_2 thin films. *Applied Physics Letters* 106: 162905.

53 Dou, X., Wei, W., Sang, P. et al. (2024). Polarization switching pathways of ferroelectric Zr-doped HfO_2 based on the first-principles calculation. *Applied Physics Letters* 124: 092901.

54 Cheng, Y., Gao, Z., Ye, K. H. et al. (2022). Reversible transition between the polar and antipolar phases and its implications for wake-up and fatigue in HfO_2-based ferroelectric thin film. *Nature communications* 13: 645.

55 Smyth, D. M. (2000). The effects of dopants on the properties of metal oxides. *Solid State Ionics* 129: 5–12.

56 Materlik, R., Künneth, C., Falkowski, M. et al. (2018). Al-, Y-, and La-doping effects favoring intrinsic and field induced ferroelectricity in HfO_2: A first principles study. *Journal of Applied Physics* 123: 164101.

57 Li, P., Chen, I. W., and Penner-Hahn, J. E. (1994). Effect of dopants on zirconia stabilization – an X-ray absorption study: III, charge-compensating dopants. *Journal of the American Ceramic Society* 77: 1289–1295.

58 Starsschich, S. and Boettger, U. (2017). An extensive study of the influence of dopants on the ferroelectric properties of HfO_2. *Journal of Materials Chemistry C* 5: 333–338.

59 Anderson, P. W. and Blount, E. I. (1965). Symmetry considerations on martensitic transformations: "ferroelectric" metals? *Physical Review Letters* 14: 217–219.

60 Kolodiazhnyi, T., Tachibana, M., Kawaji, H. et al. (2010). Persistence of ferroelectricity in $BaTiO_3$ through the insulator-metal transition. *Physical Review Letters* 104: 147602.

61 Wang, Y., Liu, X., Burton, J. D. et al. (2012). Ferroelectric instability under screened Coulomb interactions. *Physical Review Letters* 109: 247601.

62 Hickox-Young, D., Puggioni, D., and Rondinelli, J. M. (2020). Persistent polar distortions from covalent interactions in doped $BaTiO_3$. *Physical Review B* 102: 014108.

63 Li, G., He, C., Xiong, Y. et al. (2021). Coexistence of ferroelectricity and metallicity in M-doped $BaTiO_3$ (M = Al, V, Cr, Fe, Ni, and Nb): First-principles study. *Materials Today Communications* 27: 102394.

64 Michel, V. F., Esswein, T., and Spaldin, N. A. (2021). Interplay between ferroelectricity and metallicity in $BaTiO_3$. *Journal of Materials Chemistry C* 9: 8640–8649.

65 Cao, T., Ren, G., Shao, D. F. et al. (2023). Stabilizing polar phases in binary metal oxides by hole doping. *Physical Review Materials* 7: 044412.

66 Park, M. H., Schenk, T., Fancher, C. M. et al. (2017). A comprehensive study on the structural evolution of HfO_2 thin films doped with various dopants. *Journal of Materials Chemistry C* 5: 4677–4690.

67 Schroeder, U., Richter, C., Park, M. H. et al. (2018). Lanthanum-doped hafnium oxide: a robust ferroelectric material. *Inorganic Chemistry* 57: 2752–2765.

68 Lee, C. K., Cho, E., Lee, H. S. et al. (2008). First-principles study on doping and phase stability of HfO_2. *Physical Review B – Condensed Matter and Materials Physics* 78: 012102.

69 Li, G., Liu, Y., Yan, S. et al. (2024). Charge-compensated co-doping stabilizes robust hafnium oxide ferroelectricity. *Journal of Materials Chemistry C* 12: 6257–6266.

70 Jiang, S., Huang, X., He, Z. et al. (2018). Phase transformation and lattice parameter changes of non-trivalent rare earth-doped YSZ as a function of temperature. *Journal of Materials Engineering and Performance* 27: 2263–2270.

71 Xu, Z., Qiang, H., Chen, Y. et al. (2017). Microstructure and enhanced dielectric properties of yttrium and zirconium co-doped $CaCu_3Ti_4O_{12}$ ceramics. *Materials Chemistry and Physics* 191: 1–5.

72 Cao, H., Zheng, H., Fan, L. et al. (2020). Structural, morphological, dielectric and magnetic properties of Zn-Zr co-doped yttrium iron garnet. *International Journal of Applied Ceramic Technology* 17: 813–822.

73 Popovici, M. I., Walke, A. M., Bizindavyi, J. et al. (2022). High-endurance ferroelectric (La, Y) and (La, Gd) co-doped hafnium zirconate grown by atomic layer deposition. *ACS Applied Electronic Materials* 4: 1823–1831.

74 Zhou, C., Ma, L., Feng, Y. et al. (2024). Enhanced polarization switching characteristics of HfO_2 ultrathin films via acceptor-donor co-doping. *Nature Communications* 15: 2893.

75 Kersch, A., Materlik, R., Künneth, C. et al. (2018). Stabilization of ferroelectric properties in Hafnia with doping. In: Bastian, P., Kranzlmüller, D., Brüchle, H. et al. (eds.). *High Performance Computing in Science and Engineering*. Bayerische Akademie der Wissenschaften. pp. 78–79.

76 Jung, H. S., Jeon, S. H., Kim, H. K. et al. (2012). The impact of carbon concentration on the crystalline phase and dielectric constant of atomic layer deposited HfO_2 films on Ge substrate. *ECS Journal of Solid State Science and Technology* 1: N33.

77 Kim, K. D., Lee, Y. H., Gwon, T. et al. (2017). Scale-up and optimization of HfO_2-ZrO_2 solid solution thin films for the electrostatic supercapacitors. *Nano Energy* 39: 390–399.

78 Kim, K. D., Park, M. H., Kim, H. J. et al. (2016). Ferroelectricity in undoped-HfO_2 thin films induced by deposition temperature control during atomic layer deposition. *Journal of Materials Chemistry C* 4: 6864–6872.

79 Oh, S., Song, J., Yoo, I. K. et al. (2019). Improved endurance of HfO_2-based metal-ferroelectric-insulator-silicon structure by high-pressure hydrogen annealing. *IEEE Electron Device Letters* 40: 1092–1095.

80 Park, M. H., Kim, H. J., Kim, Y. J. et al. (2015). Study on the size effect in $Hf_{0.5}Zr_{0.5}O_2$ films thinner than 8 nm before and after wake-up field cycling. *Applied Physics Letters* 107: 192907.

81 Zhou, Y., Zhang, Y. K., Yang, Q. et al. (2019). The effects of oxygen vacancies on ferroelectric phase transition of HfO_2-based thin film from first-principle. *Computational Materials Science* 167: 143–150.

82 Materano, M., Mittmann, T., Lomenzo, P. D. et al. (2020). Influence of oxygen content on the structure and reliability of ferroelectric $Hf_xZr_{1-x}O_2$ layers. *ACS Applied Electronic Materials* 2: 3618–3626.

83 Park, M. H., Lee, Y. H., and Hwang, C. S. (2019). Understanding ferroelectric phase formation in doped HfO_2 thin films based on classical nucleation theory. *Nanoscale* 11: 19477–19487.

84 Falkowski, M. and Kersch, A. (2021). *Ab initio* interphase characteristics in HfO_2 and ZrO_2 and nucleation of the polar phase. *Applied Physics Letters* 118: 032905.

85 Hoffmann, M., Schroeder, U., Schenk, T. et al. (2015). Stabilizing the ferroelectric phase in doped hafnium oxide. *Journal of Applied Physics* 118: 072006.

86 Liu, S. and Hanrahan, B. M. (2019). Effects of growth orientations and epitaxial strains on phase stability of HfO_2 thin films. *Physical Review Materials* 3: 054404.

87 Zhu, T., Ma, L., Duan, X. et al. (2025). Origin of interstitial doping induced coercive field reduction in ferroelectric hafnia. *Physical Review Letters* 134: 056802.

88 Lomenzo, P. D., Richter, C., Mikolajick, T. et al. (2020). Depolarization as driving force in antiferroelectric hafnia and ferroelectric wake-up. *ACS Applied Electronic Materials* 2: 1583–1595.

89 Ali, T., Polakowski, P., Riedel, S. et al. (2018). High endurance ferroelectric hafnium oxide-based FeFET memory without retention penalty. *IEEE Transactions on Electron Devices* 65: 3769–3774.

90 Chen, Q., Zhang, Y., Liu, W. et al. (2021). Ferroelectric switching behavior of nanoscale $Hf_{0.5}Zr_{0.5}O_2$ grains. *International Journal of Mechanical Sciences* 212: 106828.

91 Lee, H., Choe, D. H., Jo, S. et al. (2021). Unveiling the origin of robust ferroelectricity in sub-2 nm hafnium zirconium oxide films. *ACS Applied Materials & Interfaces* 13: 36499–36506.

92 Schroeder, U., Hwang, C. S., and Funakubo, H. (eds.) (2019). *Ferroelectricity in Doped Hafnium Oxide: Materials, Properties and Devices*. Cambridge, MA: Woodhead Publishing.

93 Böscke, T. S., Teichert, S., Bräuhaus, D. et al. (2011). Phase transitions in ferroelectric silicon-doped hafnium oxide. *Applied Physics Letters* 99: 112904.

94 Nukala, P., Ahmadi, M., Wei, Y. et al. (2021). Reversible oxygen migration and phase transitions in hafnia-based ferroelectric devices. *Science* 372: 630–635.

95 Schenk, T., Schroeder, U., Pešić, M. et al. (2014). Electric field cycling behavior of ferroelectric hafnium oxide. *ACS Applied Materials & Interfaces* 6: 19744–19751.

96 Zhou, D., Xu, J., Li, Q. et al. (2013). Wake-up effects in Si-doped hafnium oxide ferroelectric thin films. *Applied Physics Letters* 103: 192904.

97 Starsschich, S., Menzel, S., and Böttger, U. (2016). Evidence for oxygen vacancies movement during wake-up in ferroelectric hafnium oxide. *Applied Physics Letters* 108: 032903.

98 Schenk, T., Hoffmann, M., Ocker, J. et al. (2015). Complex internal bias fields in ferroelectric hafnium oxide. *ACS Applied Materials & Interfaces* 7: 20224–20233.

99 Fields, S. S., Smith, S. W., Ryan, P. J. et al. (2020). Phase-exchange-driven wake-up and fatigue in ferroelectric hafnium zirconium oxide films. *ACS Applied Materials & Interfaces* 12: 26577–26585.

100 Chen, H., Zhou, X., Tang, L. et al. (2022). HfO_2-based ferroelectrics: From enhancing performance, material design, to applications. *Applied Physics Reviews* 9: 011307.

101 Chouprik, A., Negrov, D., Tsymbal, E. Y. et al. (2021). Defects in ferroelectric HfO_2. *Nanoscale* 13: 11635–11678.

102 Jiang, P., Luo, Q., Xu, X. et al. (2021). Wake-up effect in HfO_2-based ferroelectric films. *Advanced Electronic Materials* 7: 2000728.

103 Lederer, M., Olivo, R., Lehninger, D. et al. (2021). On the origin of wake-up and antiferroelectric-like behavior in ferroelectric hafnium oxide. *Physica Status Solidi (RRL) – Rapid Research Letters* 15: 2100086.

104 Xu, X., Huang, F. T., Qi, Y. et al. (2021). Kinetically stabilized ferroelectricity in bulk single-crystalline HfO_2:Y. *Nature Materials* 20: 826–832.

105 Lomenzo, P. D., Takmeel, Q., Moghaddam, S. et al. (2016). Annealing behavior of ferroelectric Si-doped HfO_2 thin films. *Thin Solid Films* 615: 139–144.

106 Kirbach, S., Lederer, M., Eßlinger, S. et al. (2021). Doping concentration dependent piezoelectric behavior of Si:HfO_2 thin-films. *Applied Physics Letters* 118: 012904.

107 Kozodaev, M. G., Chernikova, A. G., Korostylev, E. V. et al. (2019). Mitigating wakeup effect and improving endurance of ferroelectric HfO_2-ZrO_2 thin films by careful La-doping. *Journal of Applied Physics* 125: 034101.

108 Qi, Y. and Rabe, K. M. (2020). Phase competition in HfO_2 with applied electric field from first principles. *Physical Review B* 102: 214108.

109 Fan, P., Zhang, Y. K., Yang, Q. et al. (2019). Origin of the intrinsic ferroelectricity of HfO_2 from ab initio molecular dynamics. *The Journal of Physical Chemistry C* 123: 21743–21750.

110 Zhou, P., Zeng, B., Yang, W. et al. (2022). Intrinsic 90 charged domain wall and its effects on ferroelectric properties. *Acta Materialia* 232: 117920.

111 Pešić, M., Fengler, F. P. G., Larcher, L. et al. (2016). Physical mechanisms behind the field-cycling behavior of HfO_2-based ferroelectric capacitors. *Advanced Functional Materials* 26: 4601–4612.

112 Lederer, M., Abdulazhanov, S., Olivo, R. et al. (2021). Electric field-induced crystallization of ferroelectric hafnium zirconium oxide. *Scientific Reports* 11: 22266.

113 Lines, M. E. and Glass, A. M. (2001). *Principles and Applications of Ferroelectrics and Related Materials*. Oxford, UK: Clarendon Press.

114 Shimizu, T., Katayama, K., Kiguchi, T. et al. (2016). The demonstration of significant ferroelectricity in epitaxial Y-doped HfO_2 film. *Scientific Reports* 6: 32931.

115 Park, M. H., Chung, C. C., Schenk, T. et al. (2018). Origin of temperature-dependent ferroelectricity in Si-doped HfO_2. *Advanced Electronic Materials* 4: 1700489.

116 Materano, M., Lomenzo, P. D., Kersch, A. et al. (2021). Interplay between oxygen defects and dopants: effect on structure and performance of HfO_2-based ferroelectrics. *Inorganic Chemistry Frontiers* 8: 2650–2672.

117 Plonka, R., Dittmann, R., Pertsev, N. A. et al. (2005). Impact of the top-electrode material on the permittivity of single-crystalline $Ba_{0.7}Sr_{0.3}TiO_3$ thin films. *Applied Physics Letters* 86: 202908.

118 Sternik, M. and Parlinski, K. (2005). Lattice vibrations in cubic, tetragonal, and monoclinic phases of ZrO_2. *The Journal of Chemical Physics* 122: 064707.

119 Schroeder, U., Park, M. H., Mikolajick, T. et al. (2022). The fundamentals and applications of ferroelectric HfO_2. *Nature Reviews Materials* 7 (8): 653–669.

4

Design of HfO_2 Ferroelectric Materials with Superlattice-like Laminate Structure

Chunlai Luo[1], Ruiqiang Tao[2], and Wenwu Li[1]**

[1] *State Key Laboratory of Photovoltaic Science and Technology, Institute of Optoelectronics, College of Future Information Technology, Fudan University, Shanghai, China*
[2] *Institute for Advanced Materials, South China Academy of Advanced Optoelectronics, South China Normal University, Guangzhou, Guangdong, China*

*Corresponding authors: liwenwu@fudan.edu.cn, taoruiqiang@m.scnu.edu.cn

4.1 Introduction

Ferroelectric materials have garnered significant attention due to their exceptional properties, such as electrically switchable spontaneous polarization, fast switching speeds, and potential for low power consumption.[1–3] Among these materials, HfO_2-based ferroelectrics (FE-HfO_2) have shown great promise.[2,3] However, challenges such as polycrystalline-induced inhomogeneity, high coercive field, and limited reliability hinder their widespread application in advanced semiconductor technologies.[4,5]

This chapter explores the structure design, performance, and application of FE-HfO_2 films with superlattice-like laminate configurations. Superlattice-like laminate structures, which consist of alternating layers of different materials or compositions, greatly improve the ferroelectric properties of FE-HfO_2 films. These structures are classified into heterostructure and superlattice FE-HfO_2 laminate films – such as undoped HfO_2/ZrO_2-based films and $Hf_{0.5}Zr_{0.5}O_2$ (HZO)/Al_2O_3-based films – and have been demonstrated to be effective in solving issues related to switching dynamics, leakage current, and reliability. Additionally, this chapter examines the potential of superlattice laminate films, particularly those based on HfO_2/ZrO_2 combinations, in improving the reliability and switching speed of ferroelectric films. These films open up new avenues for the development of high-performance devices, including low-power transistors, high-density memory, and energy storage applications.

The following section will delve into the design considerations for FE-HfO_2 superlattice-like laminate films, the effects of laminate structures on their properties, and the application of FE-HfO_2 heterostructure and superlattice films in various device configurations. By understanding the design principles and applications of these advanced materials, this chapter aims to contribute to the development of more efficient and reliable ferroelectric devices for next-generation technologies.

4.2 Concept and Design of FE-HfO_2 Superlattice-like Laminate Films

4.2.1 Design and Classification of FE-HfO_2 Superlattice-like Laminate Films

This section focuses on FE-HfO_2 superlattice-like laminate films, categorized into heterostructure and superlattice configurations. It provides an in-depth discussion of their key classifications, structural designs, and material compositions, while also highlighting the advantages of these ferroelectric laminate films, with an emphasis on their unique features and the benefits derived from their specific configurations.

Heterostructure laminate films are structures formed by the combination of two or more different materials (including various phases of the same material). A defining feature of heterojunctions is that the layers of materials are not necessarily arranged in a strict periodic manner; in other words, the layers may not follow a fixed repetitive pattern (Figure 4.1a). This nonperiodic arrangement leads to significant interface effects, which can modulate the electrical and physical properties of the materials.

In contrast, superlattice laminate films are structures in which two or more different materials alternate periodically, with highly precise interlayer (IL) thicknesses (on the nanometer scale; Figure 4.1b). The thickness of each material layer is carefully controlled during fabrication, and the layers are arranged periodically in space. The key feature of superlattices is the periodic repetition of interfaces between layers, resulting in an ordered structure. By adjusting the thickness, material type, or number of layers, the properties of the superlattice can be finely tuned.

The material composition of FE-HfO_2 laminate films varies, typically including ferroelectric, antiferroelectric, dielectric thin films, and even metallic materials (Figure 4.1c). For FE-HfO_2 laminate films, the most commonly reported material systems are $Hf_xZr_{1-x}O_2$, HfO_2, ZrO_2, and Al_2O_3.

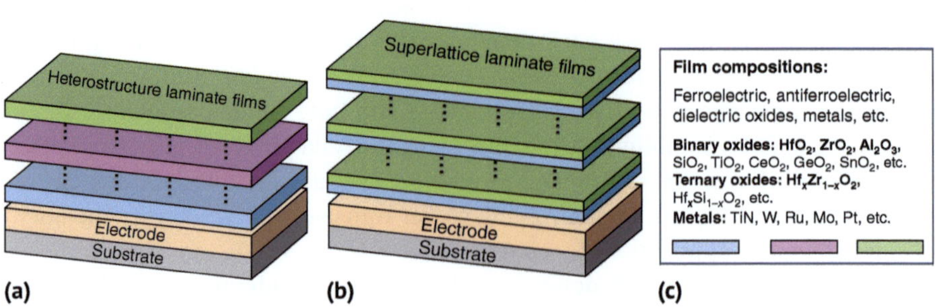

Figure 4.1 The classifications, structural designs, and film compositions of FE-HfO_2 laminate films. Schematics of FE-HfO_2 films with (a) heterostructure and (b) superlattice laminate structures. (c) Film composition of the FE-HfO_2 laminate films.

4.2.2 Impact of Superlattice-like Laminate Structures on FE-HfO_2 Films

Superlattice-like laminate structures can give rise to unique physical effects that are not achievable in single-solid-solution ferroelectric films, as already reported in laminated perovskite ferroelectrics.[6] Superlattice laminate structure can induce novel physical properties by tuning the delicate balance among spin, charge, orbital, and lattice degrees of freedom. This is achieved through methods such as strain engineering, interfacial effects, and quantum confinement, enabling control over material properties and the emergence of exotic phenomena. This approach of tuning physical properties based on the design of artificial microstructures provides important insights for the research of FE-HfO_2 films.

In FE-HfO_2 films, the physical origin of ferroelectricity is closely related to the metastable orthorhombic $Pca2_1$ phase. Unlike the nonpolar monoclinic $P2_1/c$ phase, the orthorhombic phase exhibits thermodynamic instability below the Curie temperature. To stabilize the ferroelectric orthorhombic phase, numerous theoretical and experimental methods have been proposed,[7,8] including the introduction of oxygen vacancies, dopants, stresses, and surface and interface effects. These factors contribute thermodynamically or kinetically to the stabilization of the polar orthorhombic phase.

In this context, it can be anticipated that the design of various heterostructure and superlattice laminates will inevitably affect the ferroelectric properties of FE-HfO_2 films. For example, compared to single-solid-solution FE-HfO_2 films, laminate structures with different film compositions can modify the oxygen vacancy content, strain, and stress within the films, thus influencing the formation of the polar orthorhombic phase. Moreover, the laminate structure introduces more interfacial effects, which may affect the polarization response and the dielectric constant of FE-HfO_2 films. Additionally, the stacking of different functional nanolayers in the heterostructure or superlattice laminate films can improve film quality with reduced defects and inhibit the formation of leakage current channels in FE-HfO_2, thereby improving the breakdown resistance. The novel physical properties and performance impacts that FE-HfO_2 laminate films remain to be explored. The following section will summarize, consolidate, and discuss the literature reported on FE-HfO_2 laminate films to date.

4.3 FE-HfO_2 Heterostructure Laminate Films

The majority of the reported FE-HfO_2 heterostructure laminate films are developed based on HZO, HfO_2, ZrO_2, and Al_2O_3. The performance of FE-HfO_2 heterostructure laminate films can be affected by many parameters such as nanolayer thickness, number of layers, and stacking configurations. Accordingly, in this section, different laminate structures will be categorized and discussed to illustrate the advantages and drawbacks that various stack designs bring to FE-HfO_2 films.

4.3.1 Heterostructure Films with Enhanced Ferroelectricity

HfO_2 and ZrO_2 films, with their simple composition and ease of fabrication through atomic-layer deposition (ALD), are widely used in the design of heterostructure laminate films. In 2016, Lu et al.[9] first fabricated ZrO_2/HfO_2 bilayer films (with thicknesses ranging from 7.5 to 18 nm) on a $Pt/Ti/SiO_2/Si$ substrate. By varying the thickness and stacking sequence of the HfO_2 and ZrO_2 layers, they achieved excellent ferroelectric properties (remnant polarization $[P_r] \sim 15\ \mu C/cm^2$) with an ZrO_2(6-nm)/HfO_2(6-nm) structure (Figure 4.2). The authors believe that the excellent ferroelectricity of the ZrO_2/HfO_2 bilayer stems from two factors: (1) effective confinement of the HfO_2 layer by the ZrO_2 layer and Si substrate to promote the ferroelectric orthorhombic phase; and (2) reducing the thickness and bulk characteristics of ZrO_2 and HfO_2 helps suppress the paraelectric monoclinic phase.

In 2017, Weeks et al.[10] investigated the ferroelectricity of HfO_2/ZrO_2 laminate films, focusing on the effects of nanolayer number, deposition temperature, and electric field cycling. As the nanolayer's thickness decreased, high-temperature deposition led to increased lattice distortion and enhanced remnant polarization. This was attributed to the crystallization at high deposition temperature, which promotes the formation of the orthorhombic phase during annealing. At 285 °C, the (1-nm HfO_2/1-nm ZrO_2) × 4 laminate structure achieved the highest P_r value ($50\ \mu C/cm^2$), outperforming solid-solution HZO films.

In 2019, Park et al.[11] systematically investigated the formation of the polar orthorhombic phase in bilayer and trilayer HfO_2/ZrO_2 stacks, examining the effects of the bottom and top nanolayers on the ferroelectric properties. They found that the $HfO_2/ZrO_2/HfO_2$ (HZH) laminate samples exhibited good ferroelectricity and could endure up to 10^{10} switching cycles with a P_r of $12\ \mu C/cm^2$. It is concluded that the development of the polar orthorhombic phase is governed by the kinetic factors

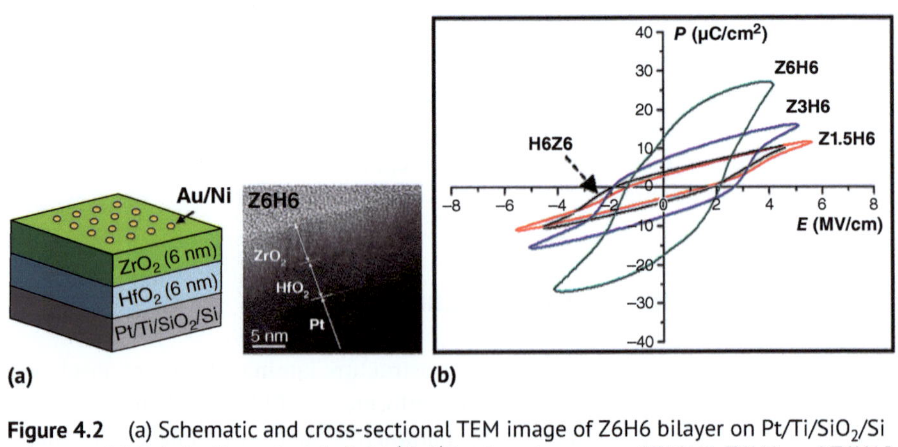

Figure 4.2 (a) Schematic and cross-sectional TEM image of Z6H6 bilayer on $Pt/Ti/SiO_2/Si$ substrate. (b) Polarization–electric field (*P*–*E*) hysteresis curves of Z12H6, Z6H6, and Z6H12 bilayers. *Source*: [9] / with permission of Elsevier.

of the amorphous or partially crystalline films during post-deposition annealing. Furthermore, the study suggests that the formation of the orthorhombic phase is governed by the types of initial layer grown on the TiN bottom electrode. And the phase transitions during the crystallization of the laminate films also play a critical role in the evolution of the orthorhombic phase.[11]

Since then, an increasing number of designs based on HfO_2 or ZrO_2 heterostructure laminate films have been reported.[12–16] Chen et al.[13] and Lee et al.[14], respectively, examined the effects of sublayer thickness on the ferroelectric properties of HfO_2/ZrO_2 and HfO_2/Al_2O_3 laminate films, finding that thinner sublayers promote the formation of the orthorhombic phase. Compared to HZO films, HfO_2/ZrO_2 heterostructure laminate films also show lower leakage currents.[13,17] They attributed the differences in P_r value and leakage current to variations in grain size, highlighting that reduced roughness, smaller grain sizes, and sufficient crystallization are crucial for ferroelectric properties.

In FE-HfO_2 films, there is a reverse size effect,[18] where ferroelectricity enhances as the film thickness decreases, even to the level of a half-unit-cell width (~3 Å).[18–20] This reverse size effect may originate from a phase stability inversion dominated by surface/interface energy. Specifically, the laminate structures reduce the grain size of the films, which introduces more surface and interface effects for FE-HfO_2 and promotes the formation of intermediate phases (e.g., tetragonal $P4_2/nmc$ phase) during the annealing and cooling processes.[21–23] These intermediate tetragonal phases are energetically more easily converted into ferroelectric polar orthorhombic phases,[24] making it the thermodynamically stable phase. Accordingly, the design of laminate structures can effectively inhibit the continuous growth of solid-solution FE-HfO_2 films. This is attributed to the surface/interface effect, where the laminate structure splits the FE-HfO_2 films into multiple thinner nanolayers to boost ferroelectricity.

4.3.2 Heterostructure Films with Low Leakage Current

In 2014, Kim et al.[25] inserted a 1-nm-thick Al_2O_3 IL at the mid-thickness of an HZO film, which alleviated the degradation of ferroelectric properties with increasing film thickness. This allows HZO/Al_2O_3/HZO films to demonstrate superior ferroelectric properties (i.e., polarization value) compared to pure HZO films at the same film thickness of 40 nm (Figure 4.3a,b). Studies show that the Al_2O_3 intermediate layer effectively prevents the continuous growth of the HZO film, while reducing grain size helps suppress the formation of the non-polar monoclinic phase. Furthermore, the Al_2O_3 IL blocks the grain boundaries in HZO, preventing the formation of leakage current paths (Figure 4.3c).[26] This not only significantly lowers the leakage current, but also improves the breakdown endurance of the HZO film (Figure 4.3d).[26–30] In addition, Chen et al.[31] reported that in the Al_2O_3/HZO stack, the interface near the bottom electrode can be effectively modified by the Al_2O_3 film, creating a higher barrier at the TiN–Al_2O_3 interface. This prevents charge injection from the bottom electrode, thereby further improving the reliability of the HZO film.

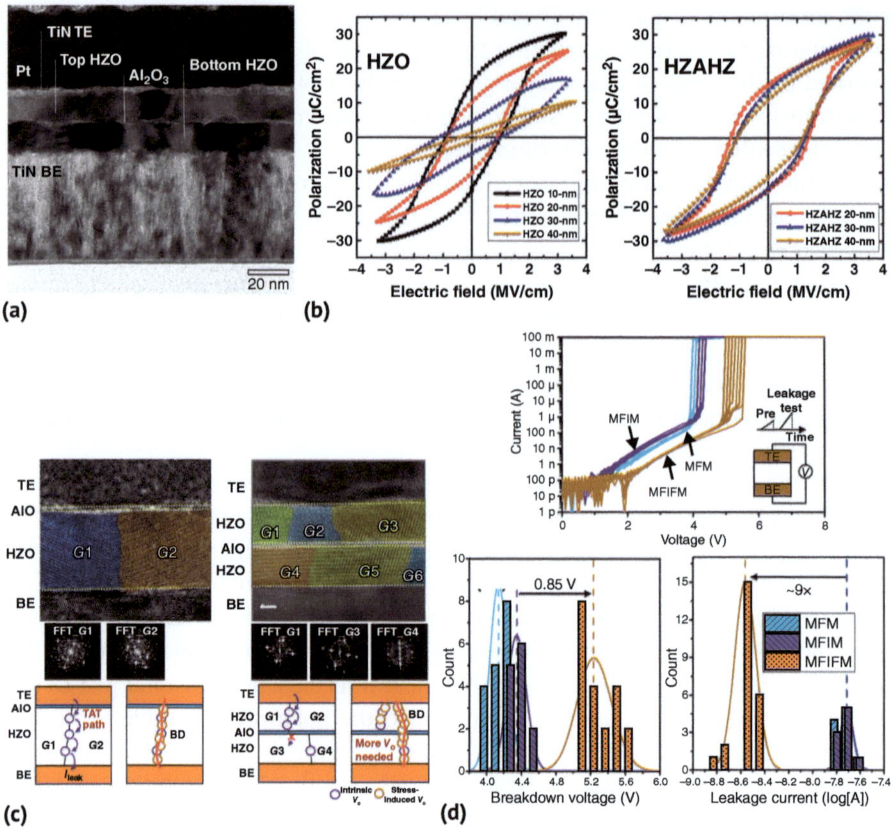

Figure 4.3 (a) Cross-sectional TEM image of the Pt/TiN/HZO/Al_2O_3/HZO/TiN structure. (b) *P–E* curves of HZO and HZO/Al_2O_3/HZO films with various thicknesses. *Source:* [25] / with permission of AIP Publishing. (c) HRTEM images, FFT patterns, and leakage and breakdown processes of HZO and HZO/Al_2O_3/HZO films. (d) Leakage and DC breakdown of HZO, HZO/Al_2O_3, and HZO/Al_2O_3/HZO laminate films. *Source:* [26] / with permission of IEEE.

The hafnia–zirconia–alumina (HZO/Al_2O_3) structures can also improve the P_r of the film by altering the concentration of oxygen vacancies within the laminate layers.[32,33] Specifically, Shin et al.[32] incorporated ultrathin Al_2O_3 or Y_2O_3 ILs (one–two ALD cycles) into HZO, which induces compressive stress for HZO and promotes the growth of polar orthorhombic phase grains. They claimed that the Al^{3+} and Y^{3+} ions diffuse into Hf^{4+} and Zr^{4+} sites, which creates charge-neutral oxygen vacancies and forms an oxygen-deficient phase, thus reducing the monoclinic phase. An Al_2O_3/HZO/Al_2O_3/HZO capacitor was designed and fabricated by Li et al.[33] demonstrating the significant effect of Al_2O_3 IL on oxygen vacancy behaviors at the dielectric–ferroelectric interface and under electric fields, which leads to enhanced ferroelectricity ($2P_r$~36.2 μC/cm^2).

Although HZO/Al_2O_3 laminate films offer advantages such as low leakage current, high breakdown resistance, and enhanced ferroelectricity, existing studies highlight their poor fatigue characteristics.[26,27,29,31,33,34] Zeng et al.[34] investigated the fatigue mechanism of HZO/Al_2O_3/HZO films under field cycling, and observed domain wall pinning in the fatigued films without field-induced phase transitions. Domain wall pinning[34] is believed to be the primary cause of the fatigue effect, which is ascribed to the increased trap charges at the Al_2O_3–HZO interface, analyzed by pulse measurement.

4.3.3 Heterostructure Films with Negative Capacitance Effect

The fluorite-structured gate oxide of the FE-HfO_2 laminate films with negative capacitance (NC) effect can surpass the state-of-the-art equivalent oxide thickness (EOT) scaling based on high-k (dielectric constant) HfO_2.[35,36] The NC effect in HZO films has been proposed,[37] but it remains unclear whether this effect is intrinsic or extrinsic.

Hoffmann et al.[38] studied symmetric Al_2O_3/HZO/Al_2O_3 and asymmetric HZO/Al_2O_3 heterostructure laminate structures to explore the relationship among depolarization field, interface charge, domain growth, and NC (Figure 4.4a,b). They observed a significant negative charge density at the HZO–Al_2O_3 interface, approximately $-10\,\mu C/cm^2$, which helps stabilize the negative polarization state in the asymmetric HZO/Al_2O_3 heterostructure.[38] In contrast, for the symmetric Al_2O_3/HZO/Al_2O_3 heterostructure, no preference for polarization direction was observed, as the electrostatic effects of the interface charges canceled each other out on average. The symmetric Al_2O_3/HZO/Al_2O_3 samples exhibited approximately equal upper and lower regions. In small-signal capacitance–voltage ($C - V$) measurements, a higher domain wall density resulted in an increased capacitance compared to the asymmetric samples. By utilizing pulse charge–voltage measurements to characterize charge injection in the device, the study extracted the double "S"-shaped $P - E_f$ characteristics and the three-well energy landscape for the symmetric Al_2O_3/HZO/Al_2O_3 samples (Figure 4.4c–e).[38] Additionally, the measured coercive field remained constant regardless of the thicknesses of the Al_2O_3 and HZO layers, indicating that it is governed by the intrinsic properties of the material rather than external factors. This supports the conclusion that the NC effect observed in ferroelectric HZO is intrinsic, rather than arising from domain structures or domain wall motion.

Hoffmann et al.[39] also demonstrated the NC effect and a reduction in EOT by constructing ultrathin HZH laminate films. This study revealed a combination of orthorhombic/tetragonal phases with partial in-plane polarization, which leads to energy landscape flattening and NC stabilization. The findings show that the achievable EOT is primarily influenced by the polarization axis angle, phase fraction, and the period of the ferroelectric domains. This work emphasizes the critical role of heterostructure laminate design in optimizing NC and minimizing EOT. Furthermore, it affirms that traditional approaches to reducing EOT through engineering ILs are also effective for NC gate stacks.

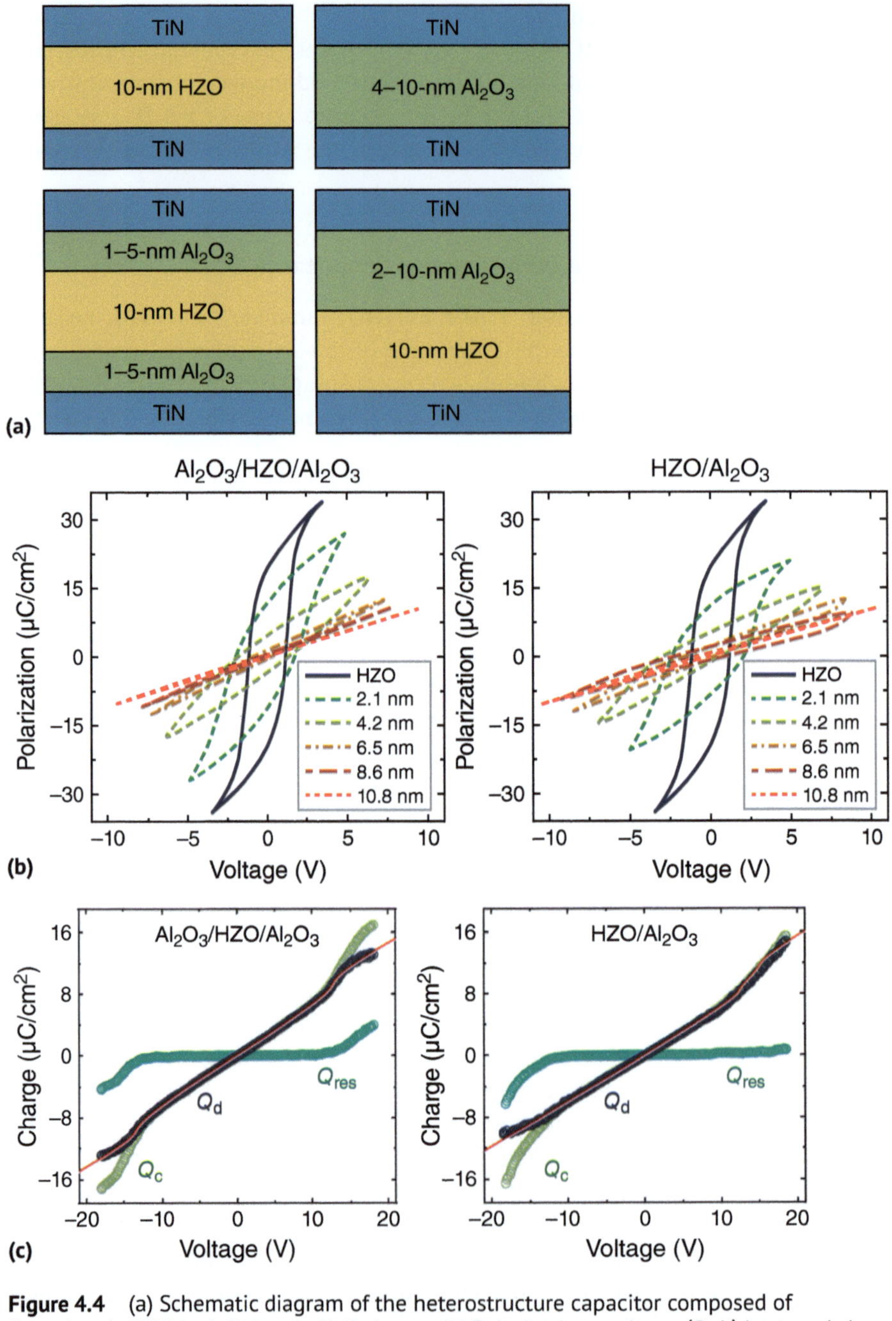

Figure 4.4 (a) Schematic diagram of the heterostructure capacitor composed of ferroelectric HZO and dielectric Al_2O_3 layers. (b) Polarization–voltage (*P*–*V*) hysteresis loops for symmetric and asymmetric HZO/Al_2O_3 ferroelectric/dielectric samples with various total Al_2O_3 thicknesses. (c) Pulsed charge–voltage measurement results for symmetric and asymmetric laminate films with identical 10.8-nm total Al_2O_3 thickness. The red line shows the fitting of the data with a simple Landau model.

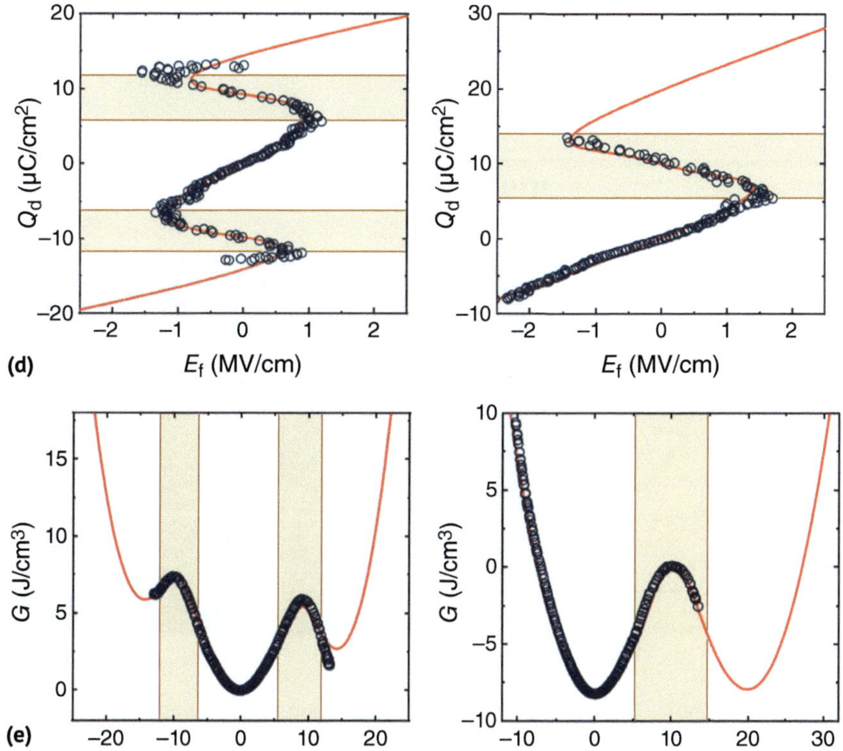

Figure 4.4 (Contd.) (d) The $Q_d - E_f$ curves extracted from panel (c) for symmetric and asymmetric laminate films. By integrating panel (d) with respect to Q_d, the free energy G can be obtained and is shown in panel (e). *Source*: [38] / with permission of John Wiley & Sons.

4.3.4 Heterostructure Films with High Dielectric Constant and Low Coercive Electric Field

The dielectric constant can be significantly engineered in HfO_2–ZrO_2 ferroelectric laminate films. Kashir et al.[40,41] achieved a high k of ~60 by fabricating ferroelectric/antiferroelectric HZO/ZrO_2 (HZZ) laminate films with optimized thicknesses and compositions (Figure 4.5). Liu et al.[42] successfully achieved a high dielectric constant of 52.7 and a large $2P_r$ value of up to 72.3 μC/cm^2 in HZZ (6/9-nm) bilayer laminate films by precisely optimizing the thickness ratio of HZO and ZrO_2, without requiring any wake-up operations. Lo et al.[43] fabricated ultrathin metal–insulator–metal capacitors and compared the performance of bilayer HZO consisting of 5-nm-thick antiferroelectric HZO (Zr = 75%) and 5-nm-thick ferroelectric HZO (Zr = 40%). The bilayer HZO laminate film demonstrated a combination of antiferroelectric and ferroelectric properties, particularly offering a high dielectric constant (k~58) while maintaining the characteristics of both phases. Zhao et al.[44] engineered ferroelectric/antiferroelectric phases in HZO by controlling oxygen vacancies and optimizing the interface energy. Their experimental and theoretical findings reveal

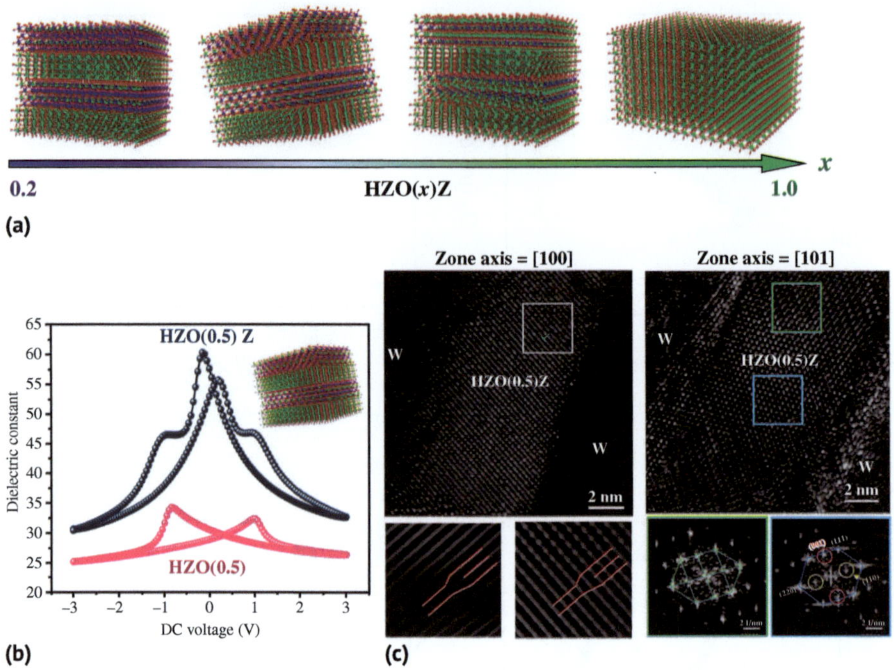

Figure 4.5 (a) Schematics of HZZ nanolaminates with different film compositions. Purple: Hf; green: Zr; and red: O. (b) Dielectric constant (k)–voltage curves of HZO and HZZ nanolaminate films. (c) HRTEM cross-sectional images from the HZZ nanolaminate films along the [100] and [010] zone axes, respectively. *Source*: [40] / with permission of American Chemical Society.

a decrease in oxygen vacancy concentration (V_O) and an increase in the ZrO_2–HfO_2 interface area, which promote the formation of the ferroelectric orthorhombic phase. Conversely, an increase in V_O and a reduction in the ZrO_2–HfO_2 interface area (due to alloy formation) favor the formation of antiferroelectric tetragonal phase, which resulted in enhanced remanent polarization ($2P_r = 51\ \mu C/cm^2$) and high dielectric constants (k~50).

The simultaneously enhanced dielectric constant and ferroelectric polarization of HZZ nanolaminate films can be explained through a morphotropic phase boundary (MPB).[40,41,45,46] MPB is a region in the compositional phase diagram that contains different ferroelectric orderings and separates areas of distinct crystal symmetries by adjusting the composition.[45] When different structural phases are nearly degenerate near the MPB, external electric fields cause polarization rotation rather than a change in its magnitude. As a result of the coexistence of multiple phases with minimal energy barriers between them, dielectric materials exhibit their highest piezoelectric and dielectric responses near the MPB. In MPB, composition serves as the intrinsic driving factor for transition, providing an alternative method to enhance the relative dielectric constant.

The reported HZZ and ferroelectric/antiferroelectric HZO laminate films can also lower the coercive electric field (E_c).[40,41,43,44,47,48] Kashir et al.[47] developed

HZZ nanolaminate films followed by annealing at 700 °C. The results showed a 40% reduction in E_c compared to solid-solution HZO. *C–V* measurements revealed a significant increase in dielectric constant near E_c, indicating improved domain wall mobility (Figure 4.6a–c). The tetragonal domain walls between polar orthogonal regions, identified by HRTEM, lower the polarization reversal barrier. The tetragonal regions act as lubricants, which facilitate smoother domain wall movement (Figure 4.6d,e). Jang et al.[49] reported low-E_c and fast-switching-speed HZZ laminate films, in which the switching barrier is also reduced by the tetragonal-phase spacers between ferroelectric regions.[50] Park et al.[48] fabricated a bilayer laminate film consisting of ferroelectric and antiferroelectric $Hf_xZr_{1-x}O_2$ with varying compositions.[51] High dielectric constant, low E_c, and high P_r have all been demonstrated. They attributed the larger dielectric constant difference between the ferroelectric and antiferroelectric layers in the bilayer stack to the increased voltage drop across the ferroelectric layer. As the voltage drop across the ferroelectric layer increased, the E_c decreased, influencing the dielectric response of the films.

Other HZO-based heterostructure laminate films, including HZO/$La_{0.67}Sr_{0.33}MnO_3$ (LSMO),[52] HZO/TiO_2,[53] HZO/W,[54] HZO/Mo,[54] and HZO/Ti,[55] have also been reported with improved ferroelectric performance, which is primarily affected by the IL stresses between heterostructure laminate films. In 2022, Liu et al.[52] constructed HZO/LSMO heterostructure laminate films and systematically investigated how symmetry engineering and anisotropic strain enhance the ferroelectric properties. By growing the heterostructure on a [110]-oriented perovskite substrate, considerable anisotropic strain was applied to the LSMO bottom electrode. An anisotropically strained LSMO layer serves as a structural template, which effectively tunes the structural symmetry and polar/nonpolar phase ratio of the HZO layer. As a result, the anisotropic tensile strain stabilizes the ferroelectric rhombohedral and orthorhombic phases,[52] enhancing the P_r up to 22 μC/cm^2. In contrast, the anisotropic compressive strain promotes the formation of the nonpolar tetragonal phase, resulting in a reduced P_r of 8 μC/cm^2.

The HZO/TiO_2/SiO_2/HZO,[53] HZO/W/Mo/HZO,[54] and HZO/Ti/HZO[55] heterostructure laminate films were fabricated to control the grain size and the applied tensile stress of HZO, thereby promoting the formation of the orthorhombic phase. The insertion of TiO_2 and Ti ILs in HZO films results in improved 2P_r values of 30.2 and 57 μC/cm^2, respectively. Grazing incidence X-ray diffraction (GIXRD) and stress measurements confirm the tensile stress caused by the mismatch of the coefficients of thermal expansion (CTEs) between the HZO film and IL.

The research reported so far indicates that stress and strain regulations are key factors in inducing and stabilizing the polar orthorhombic phase in FE-HfO_2 films.[7] For conventional solid-solution FE-HfO_2 capacitors, the stability of the orthorhombic phase partially depends on the mismatch in CTEs between the top/bottom electrodes, which generates controllable tensile stress during the cooling step in the rapid thermal annealing process.[56,57] Compared to single-solid-solution FE-HfO_2 films, the laminate structure designs offer more diverse dimensions for

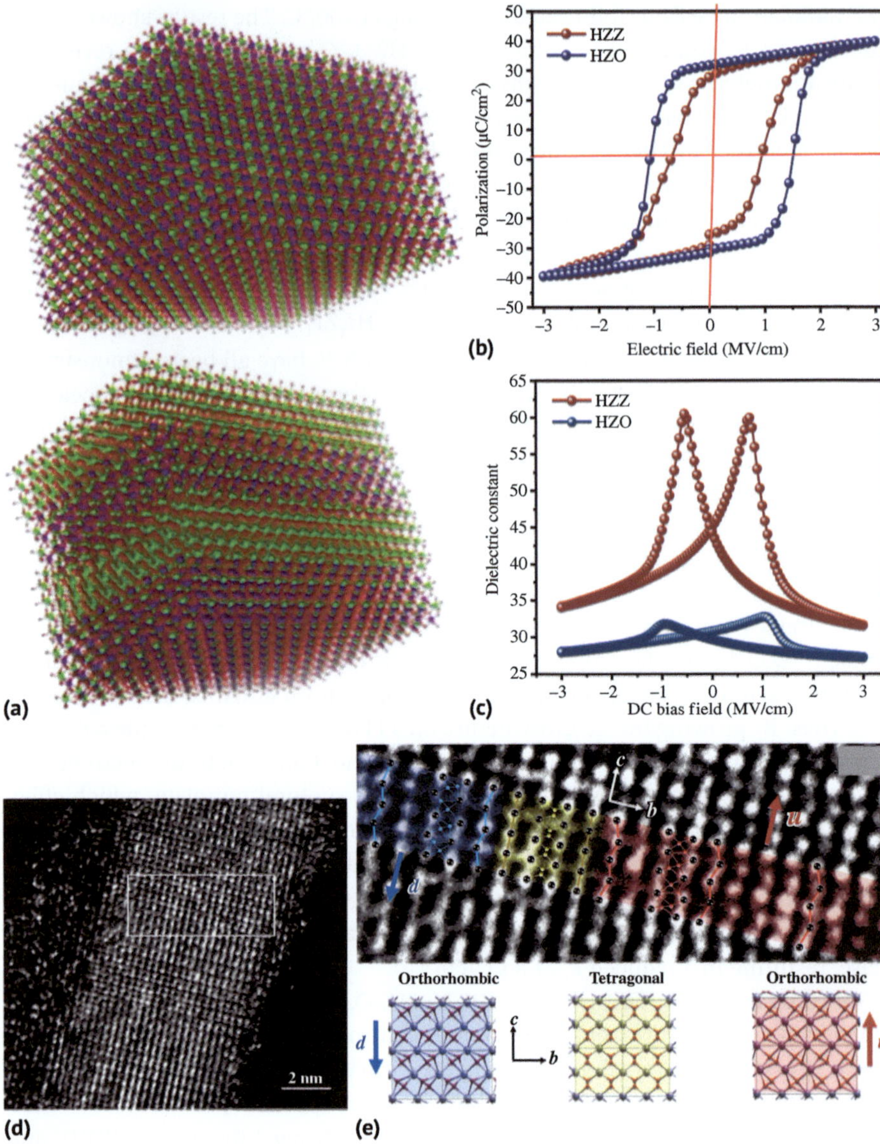

Figure 4.6 (a) Schematics of HZO and HZZ nanolaminate films. (b) *P*–*E* hysteresis loops and (c) dielectric constant (*k*)–DC bias field curves of HZO and HZZ nanolaminate films. (d) Cross-sectional HRTEM image of the HZZ nanolaminate films corresponding to the bottom of panel (a). (e) Tetragonal phase-like spacers (yellow shaded areas) between the domains with opposite directions (red and blue shaded regions). The black circles indicate the Hf and Zr atoms, while the red, blue, and yellow circles are candidates of oxygen atoms. *Source*: [47] / with permission of IOP Publishing.

stress regulation, including stresses at the electrode–film interface and induced by additional nanoscale IL.[53]

Specifically, the stress regulation mechanisms include: (1) thermal mismatch stress at the electrode–ferroelectric interface, (2) stress from differences in CTEs between two heterogeneous ILs, and (3) local strain fields caused by lattice mismatch at the interfaces.[58] Therefore, by precisely controlling parameters such as IL thickness and the number of stacking cycles, the spatial gradient distribution of stress can be effectively optimized.[11] Cui et al.[59] demonstrated that the laminate structure enhances ferroelectric excitation by introducing local stresses in HfO_2/ZrO_2 laminate films, which promote faster alignment of the polarization during annealing, effectively accelerating the polarization saturation process. As a result, the required annealing time for the laminate structure is shortened, leading to exceptional electrical performance.

4.4 FE-HfO_2 Superlattice Laminate Films

The research progress of FE-HfO_2 heterostructure laminate films highlights their numerous advantages, including enhanced ferroelectricity, reduced leakage current, improved breakdown field strength, NC effects, higher dielectric constants, and lower coercive fields. This section focuses on FE-HfO_2 superlattice structures, which consist of much thinner periodic ferroelectric nanolayers when compared to heterostructure laminate films. These thinner nanolayers result in a more uniform crystalline phase composition that extends throughout the entire film. Since FE-HfO_2 superlattices possess most of the advantages of heterostructure laminate films, these overlapping benefits will not be reiterated here. Instead, this section will delve into the latest advancements in FE-HfO_2 superlattice laminate films, offering a detailed examination of their distinct advantages, beyond those associated with heterostructures.

4.4.1 Superlattice Films with High Reliability

The reliability of FE-HfO_2 films, particularly their fatigue resistance, is crucial for applications in information storage and computing. Recent studies have shown that constructing superlattice laminate films can significantly enhance their fatigue performance. In particular, Peng et al.[60] fabricated HfO_2/ZrO_2 superlattice capacitors, demonstrating the superior endurance and fatigue recovery compared to conventional solid-solution HZO-based capacitors (Figure 4.7a,b). The HfO_2/ZrO_2 superlattice capacitors achieved endurance exceeding 10^{12} switching cycles, and polarization degradation could be fully recovered after approximately 30 s of power-off. In contrast, HZO devices showed only partial recovery under high electric field cycling. This degradation in HZO is ascribed to charge trapping and detrapping processes at the metal–ferroelectric interface during low electric field cycling.[60,61] In

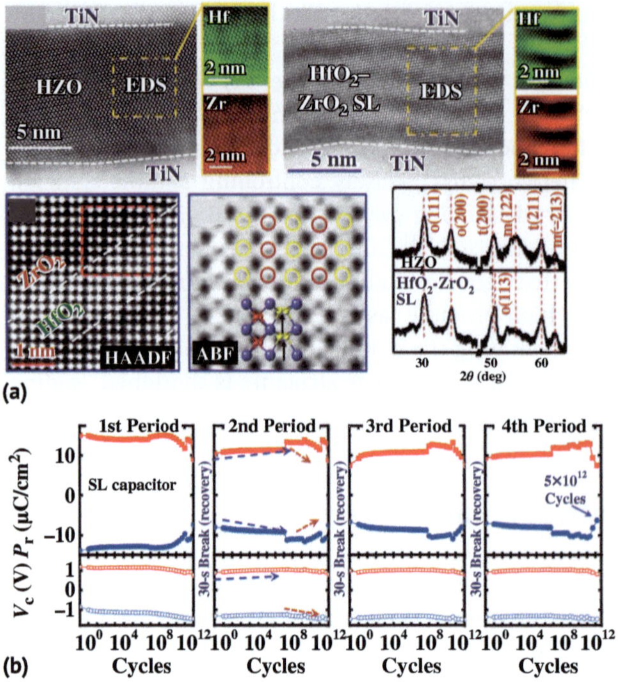

Figure 4.7 (a) HRTEM images, STEM HAADF images, and GIXRD curves of HZO and HfO_2/ZrO_2 superlattice laminate films. (b) The endurance performance of HfO_2/ZrO_2 superlattice capacitors. *Source*: [60] / with permission of IEEE.

comparison, the HfO_2/ZrO_2 superlattices exhibit a lower defect density, which more effectively suppresses these trapping processes. Meanwhile, Liang et al.[62] achieved 10^{11} cycles of fatigue endurance without P_r degradation after 600 °C annealing process (Figure 4.7c,d), owing to reduced charged defect accumulation and suppressed charge capture/release at the electrode–ferroelectric interface of ZrO_2/HfO_2 superlattice films.[63] It is worth noting that the difference between HfO_2/ZrO_2 and ZrO_2/HfO_2 superlattices lies in the stacking order of the HfO_2 and ZrO_2 layers.

Subsequently, more and more research groups have reported the fatigue resistance characteristics based on HfO_2/ZrO_2 superlattices.[64–68] For example, Cui et al.[65] reported on Back-End-of-Line (BEOL)-compatible HfO_2/ZrO_2 superlattice capacitors, which were treated at a low temperature of 400 °C. These capacitors exhibit a maximum $2P_r$ value of 32 μC/cm^2 at 2.5 MV/cm and demonstrate endurance of up to 2×10^9 cycles. The authors suggested that the periodic superlattice structure disrupts the uniform distribution of oxygen vacancies, leading to a reduced concentration of oxygen vacancies in the superlattice film compared to that in HZO.[65,66] Additionally, the different oxygen vacancy migration energy barriers in HfO_2 and ZrO_2 within the superlattice prevent the diffusion of oxygen vacancies,[69] thereby blocking the leakage paths. Zhao et al.[68] successfully fabricated HfO_2/ZrO_2 superlattices with a well-aligned *c*-axis of the orthorhombic phase along the deposition direction on smooth

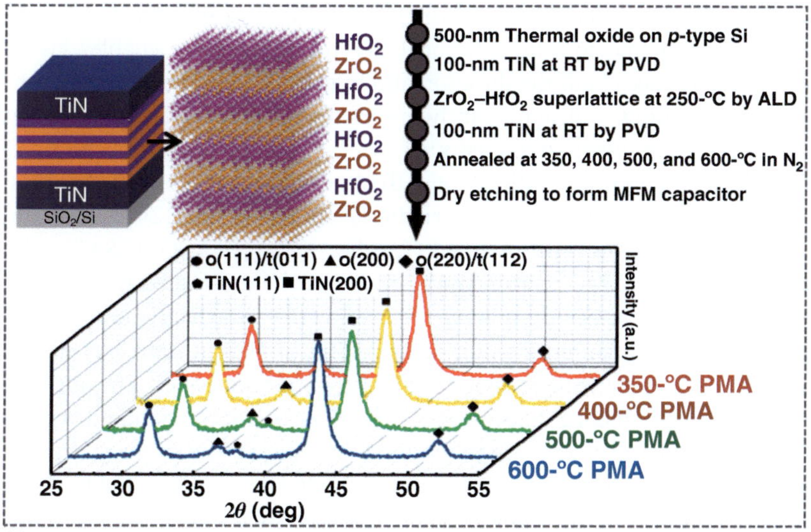

(c)

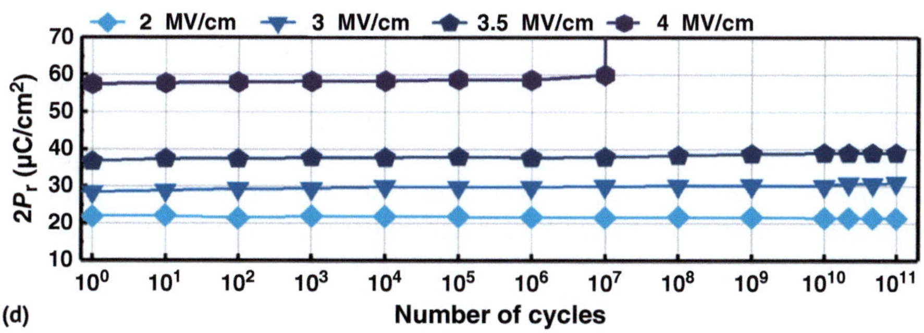

(d)

Figure 4.7 (Contd.) (c) Schematic diagram and GIXRD results of ZrO_2/HfO_2 superlattice capacitor. (d) The endurance characteristics of ZrO_2/HfO_2 superlattice capacitors with 600-°C annealing tested at different electric fields. *Source*: [62] / with permission of IEEE.

and amorphous TiN electrode substrate (with a roughness of 0.3 nm). The superlattice structure exhibited a record-high $2P_r$ value of 56 $\mu C/cm^2$ and demonstrated endurance exceeding 4×10^{12} cycles. They proposed that amorphous TiN is more suitable for forming the orthorhombic phase than the crystalline one, due to the larger lattice mismatch between cubic TiN and ZrO_2. Additionally, the smooth TiN surface creates a uniform electric field that minimizes the formation of oxygen vacancies $\left(V_O{}^{2+}\right)$, which reduces their concentration and prevents polarization pinning,[68] thereby mitigating fatigue characteristics.

Other HZO-based superlattice films, such as HZZ[70] and HZO/LSMO,[71] have also been reported. Jia et al.[70] developed an HZZ superlattice structure that effectively suppressed charge injection during polarization switching. The structure shows endurance beyond 10^{12} cycles while maintaining a high $2P_r$ value of 40.2 $\mu C/cm^2$. The improved performance was attributed to the reduction of high-energy charge injection during early domain nucleation, which is a major cause of defect formation and breakdown. Compared to solid-solution HZO, the superlattice enables faster domain switching, reduces both the duration and intensity of charge injection during polarization switching, and significantly lowers the depolarization field, which further decreases charge injection intensity.[70]

An HZO/LSMO superlattice with a total thickness greater than 50 nm was reported by Jiao et al.[71] consisting of periodic layers of 10-nm HZO and 1 nm LSMO, epitaxially grown on [110]-oriented $(La_{0.3}Sr_{0.7})(Al_{0.65}Ta_{0.35})O_3$ (LSAT) substrates. The 1-nm-thick LSMO spacer layers interrupt the grain growth of HZO, which prevents the superlattice from being affected by the paraelectric monoclinic phase, thus resulting in a large remanent polarization of 19.3 $\mu C/cm^2$ and maintaining over 90% of its polarization after 10^9 bipolar switching cycles. The authors attributed the structural distortion in thick superlattices to a strain gradient resulting from strain relaxation.[71] This strain gradient, considered the primary driving factor, induces a rhombohedral distortion through the flexoelectric field effect. The resulting distorted structure is more stable than the metastable orthorhombic ferroelectric phase, thereby enhancing resistance to ferroelectric fatigue. Moreover, this distortion enables the superlattice to retain a pure ferroelectric orthorhombic phase even at thicknesses exceeding 50 nm. To provide a more direct comparison of the fatigue resistance characteristics of the reported HfO_2/ZrO_2-based superlattices, the research advancements to date are summarized in Table 4.1.

Recent studies, aiming at enhancing the fatigue resistance of FE-HfO_2 superlattice films, have primarily concentrated on the charge injection and domain wall pinning mechanisms. However, it is important to note that the polar orthorhombic phase, responsible for the ferroelectricity of FE-HfO_2 films, is metastable.[7,8] Phase transitions during the fatigue process could also lead to the degradation of the ferroelectric properties.[72–74] Therefore, stabilizing the ferroelectric orthorhombic

Table 4.1 Ferroelectric properties of HfO_2/ZrO_2-based superlattice films as reported in the literature.

Structure	t_{SL} (nm)^a	E_c (MV/cm)	$2P_r$ ($\mu C/cm^2$)^b	Endurance	Ref.
HfO_2/ZrO_2	~10	~1.4	~30@2 MV/cm	10^{12}	[60]
ZrO_2/HfO_2	~10	~1	~20@2 MV/cm	10^{11}	[62]
HfO_2/ZrO_2	~10	~1	~20@2.75 MV/cm	10^7	[64]
HfO_2/ZrO_2	~8	~1.2	~30@2.5 MV/cm	10^9	[65]
HZH	~12	~1.4	~20@2.5 MV/cm	10^{11}	[66]
ZrO_2/HfO_2	~6	~1	~30@2 MV/cm	10^{10}	[67]
ZrO_2/HfO_2	~10	~1.5	~56@3 MV/cm	4×10^{12}	[68]
HZO/LSMO	~50	~2.5	~19@3 MV/cm	10^9	[71]
HZZ	~10	~1	~40@3 MV/cm	10^{12}	[70]

$^a t_{SL}$ represents the total thickness of the HfO_2/ZrO_2-based superlattices; $^b 2P_r$ value refers to the initial $2P_r$ value for the fatigue test.

phase is critical for improving both the ferroelectricity and fatigue properties of FE-HfO_2; however, there remains a paucity of studies exploring this aspect.

Superlattice design offers numerous advantages and holds significant potential for stabilizing the metastable orthorhombic phase. For instance, Mukherjee et al.[75] demonstrated through theoretical calculations for designed superlattice laminate FE-HfO_2 films with varying compositions and stacking orientations that it is possible to create fully polar, fully antipolar, or mixed heterostructures. These structures exhibit improved thermodynamic stability of the polar orthorhombic phase compared to that in bulk HfO_2. Their findings also suggest that combining HfO_2 with other oxides, such as ZrO_2, SiO_2, CeO_2, and TiO_2 – none of which possess a monoclinic ground state – tends to push the superlattice away from the nonpolar phase, thereby promoting the stability of ferroelectric structures and minimizing both elastic and electrostatic penalties.[75]

Additionally, the superlattice design of FE-HfO_2 films can effectively suppress the wake-up effect.[67,76] This phenomenon is thought to result from the redistribution of charged defects, such as oxygen vacancies, along with the associated enhancement of ferroelectric phase transitions after repeated electric field cycles. On the one hand, the superlattice structure facilitates the crystallization of the ferroelectric phase and reduces defects both in the bulk of ferroelectric films and at the ferroelectric–electrode interfaces.[62] On the other hand, the alternating interfaces in the superlattices can disrupt rapid vacancy diffusion and charge conduction pathways,[67,69] such as defects in grain boundaries, thereby mitigating the wake-up effect in FE-HfO_2 films.

4.4.2 Superlattice Films with Ultrafast Switching

It is reported that the HfO_2/ZrO_2 superlattice films can significantly enhance the switching speed of ferroelectric films. Li et al.[66] designed the HZH and $ZrO_2/HfO_2/ZrO_2$ (ZHZ) superlattices to investigate their switching dynamics (Figure 4.8a,b), using the Inhomogeneous Field Mechanism (IFM) model to consider the nonuniform field distribution within the superlattice films.[77] The HZH device achieved a higher remanent polarization, while the ZHZ structure showed weak wake-up effect, frequency stability, and faster switching speed. The activation field (E_a) reflects the difficulty of the polarization switching process: the larger the E_a, the more energy is required for switching. ZHZ capacitors exhibit a lower E_a than HZH, resulting in faster switching speeds (Figure 4.8c–e).

The switching dynamics of HfO_2/ZrO_2 superlattice with varying nanolayer thicknesses were investigated by Ke et al.[78] The metal–ferroelectric–metal (MFM) capacitors are fabricated with alternating HfO_2/ZrO_2 layers in ALD cycles of n ($n = 1$, 8, 15, 25, and 37). This study found that as the nanolayer thickness increases, the switching speed decreases, with HZO samples (HfO_2 and ZrO_2 layers in ALD cycles of $n = 1$) showing the fastest switching due to fewer domain pinning sites. Using the IFM model, it was shown that appropriate layer thickness enhances remanent polarization and switching speed, with significant performance improvements observed

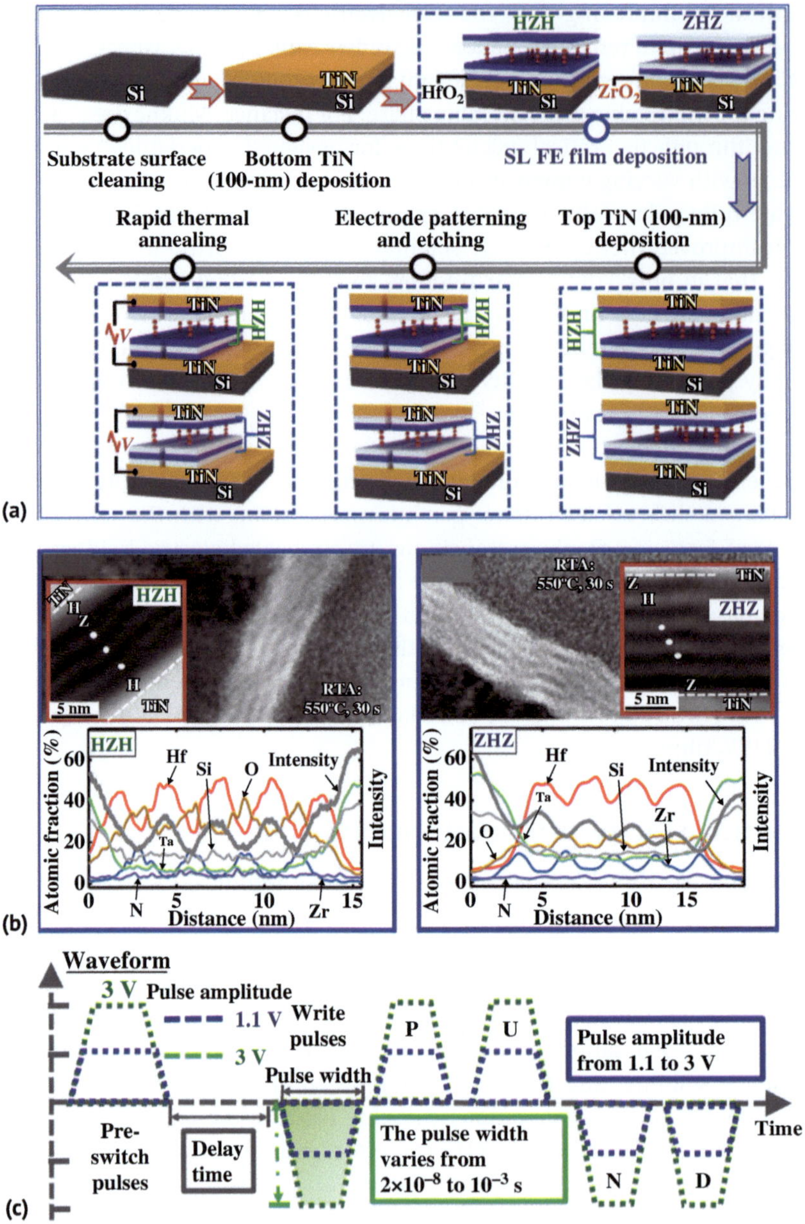

Figure 4.8 (a) Key process steps of fabricating HZH and ZHZ capacitors. (b) HRTEM images of MFM stacks with HZH and ZHZ. The line-scanned energy-dispersive X-ray spectroscopy (EDS) results for the distributions of Hf and Zr atoms in HZH and ZHZ are also shown. (c) Pulse sequence for the polarization switching (P_{sw}) measurement.

at 8 and 15 cycles.[78] These enhancements were attributed to a higher proportion of polar orthorhombic phases, disrupted grain boundaries, and facilitated domain nucleation. Meanwhile, Kang et al.[79] systematically studied the ferroelectric

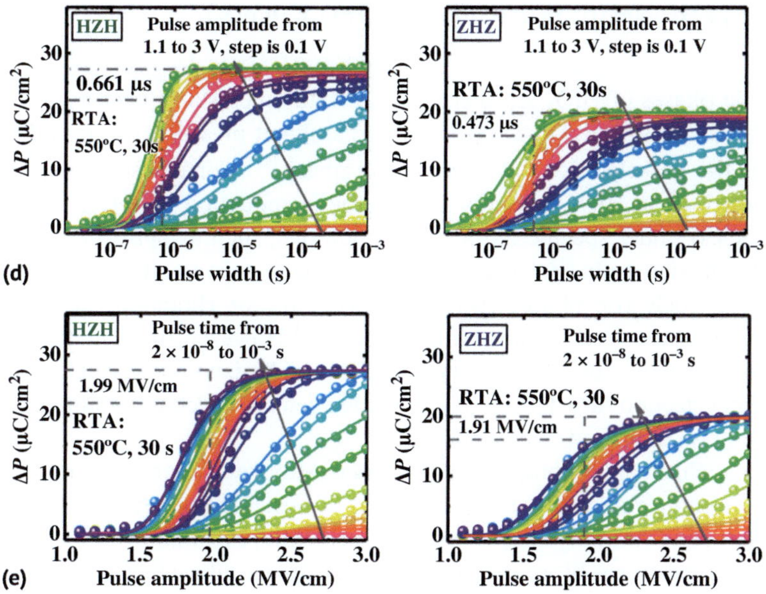

Figure 4.8 (Contd.) (d) P_{sw} versus pulse width of HZH and ZHZ superlattice capacitors at different pulse amplitudes. (e) P_{sw} versus pulse amplitude of HZH and ZHZ devices. The dots are experimental data and the solid curves are simulated results based on the IFM model. *Source*: [66] / with permission of IEEE.

properties and switching dynamics of HfO_2/ZrO_2 superlattice films with different ALD cycle numbers (5, 10, and 15) for HfO_2 and ZrO_2, and compared them with HZO films. The results showed that the HfO_2/ZrO_2 superlattice with HfO_2 and ZrO_2 layers of five ALD cycles exhibited superior ferroelectric performance, such as stronger fatigue stability and higher switching speed, due to the effective suppression of defect diffusion during the *P–V* cycling.[79] These findings highlight the potential of HfO_2/ZrO_2 superlattice structures for future high-speed ferroelectric memory applications.

In addition, Jang et al.[49] designed a 10-nm HZZ nanolayer superlattice structure. The composition of the nanolayers was optimized to maximize the effect of topological domain walls in reducing the polarization switching barrier, which can be confirmed by the reduced E_c and saturation electric field. As a result, the polarization switching speed of the HZZ film was more than twice that of solid-solution HZO. It was supposed that the tetragonal-like interphases between polarization regions play a key role in lowering the switching barrier and accelerating domain propagation.[49] Hao et al.[80] applied a compositional gradient strategy to $Hf_{1-x}Zr_xO_2$, which significantly boosts the polarization switching speed. The gradient HZO films demonstrated over twice the switching speed of traditional uniform films without deteriorating the ferroelectricity and durability. The switching mechanism shifted from nucleation-limited to domain wall growth, enhancing both polarization switching speed and the durability.[80]

4.4.3 Creating Ferroelectricity in Monoclinic $(HfO_2)_1/(CeO_2)_1$ Superlattices

The ferroelectric phases of HfO_2 are thermodynamically metastable over a broad temperature range (e.g., from 0 to 1,500 °C) and at ambient pressure, whereas the stable phase is the paraelectric monoclinic $P2_1/c$ phase.[81] Current strategies to realize ferroelectricity in HfO_2 primarily involve the introduction of defects (e.g., dopants, vacancies) under specific fabrication conditions, such as optimized film thickness and epitaxial strain.[7,82] The role of crystal defects is to either stabilize the metastable ferroelectric phases (being orthorhombic or rhombohedral)[7,82] or induce ferroelectricity in the monoclinic phase.[83,84] However, these materials often suffer from imperfections such as crystal defects and mixed non-ferroelectric phases. Such imperfections are detrimental to the ferroelectric cycling stabilities in FE-HfO_2, causing wake-up and fatigue effects.

In this context, Zhao et al.[85] proposed a theoretical strategy for achieving robust ferroelectricity in monoclinic HfO_2-based materials by designing epitaxial $(HfO_2)_1/(CeO_2)_1$ superlattices. Symmetry analysis and first-principles calculations show that Hf and Ce ions can be precisely ordered in the superlattice to induce ferroelectricity in the thermodynamically stable monoclinic phase.[85] Excellent ferroelectric properties of a polarization strength greater than 25 $\mu C/cm^2$ were induced with a low polarization switching energy barrier of around 2.5 meV/atom.[85] In contrast to previously reported FE-HfO_2 films, the proposed ferroelectricity arises from the stable monoclinic $P2_1/c$ phase of HfO_2 and naturally transitions to a ferroelectric state through nanostructure designing. However, while the theoretically predicted monoclinic-phase ferroelectricity is promising, its experimental realization still requires further investigation.

4.5 Device Applications of FE-HfO_2 Superlattice-like Laminate Films

FE-HfO_2 superlattice-like laminate films demonstrate significant potential for diverse device applications due to their exceptional properties, such as high stability, excellent dielectric characteristics, and robust ferroelectric behavior. These films are primarily employed in low-power NC transistors, leveraging their distinctive ferroelectric properties to decrease power consumption and improve electrical performance. These laminate films are also potentially used in multi-level memory systems, where their ability to store multiple bits of information per cell increases data density and retrieval speed. Furthermore, these films are employed in energy storage devices, providing efficient charge retention and rapid discharge capabilities, as well as in nanoelectromechanical system (NEMS) resonators, where their outstanding mechanical and electrical properties enable precise sensing and actuation.

4.5.1 Low-Power NC Transistors

To maintain gate control over the channel and to lower the operating voltage as a lateral dimension in advanced transistors' scale-down, an increased gate capacitance is essential. In 2021, Li et al.[35] were the first to report a significant enhancement in gate capacitance (C_{gg}) in a 1.8-nm HZH heterostructure gate stack on *n*-MOSFETs. Gate stacks combining this heterostructure laminate with an unscavenged SiO_2 IL (~8 Å), show a combined EOT of 7.5 Å on both bulk and SOI (silicon-on-insulator) *n*-MOSFETs. The implementation of a diminished EOT HZH gate stack has enabled the attainment of an exceptionally high intrinsic transconductance of 1.5 mS/μm in 90-nm SOI *n*-MOSFETs. The 2-Å improvement in EOT and the achievement of sub-8-Å EOT with the HZH stack are due to the NC effect found within the HZH laminate, which arises from the mixed ferroic order and the associated phase competition.[35] Meanwhile, Cheema et al.[86] from the same group reported the integration of HfO_2/ZrO_2 heterostructures in silicon transistors. It was stabilized with mixed ferroelectric and antiferroelectric orders, and scaled down to approximately 20 Å, which meets the requirements of gate oxide thickness for high-performance transistors (Figure 4.9). In this study, the effective oxide thickness in metal–oxide–semiconductor capacitors is comparable to an effective SiO_2 thickness of approximately 6.5 Å and demonstrates no penalty in mobility for transistors. Subsequently, they conducted a comprehensive investigation into the physical origins of NC in HZH gate stacks by integrating pulse voltage and impedance measurements with compact model and 2D phase-field simulations. Their findings indicate that the achievable EOT is significantly influenced by the pole axis angle, phase fraction, and ferroelectric domain period.[39]

Zou et al.[87] fabricated NC transistors by integrating a nanoscale gate stack of 2-nm Al_2O_3 and a 6-nm 4×[HfO_2 (10-Å)–ZrO_2 (5-Å)] superlattice as the gate dielectric onto a 2D MoS_2 semiconductor, achieving a negative-capacitance field-effect transistor (NCFET) with a channel length of 2 μm. The devices exhibited impressive performance, including an ultralow subthreshold swing (SS) of 28 mV/dec, a negligible hysteresis of 20 mV, and a high on/off current ratio of 10^7. These results are attributed to the enhanced NC effect of the laminate films, superior interface quality between the laminate and Al_2O_3, and a low defect density within the laminate layer.

4.5.2 High-density Multilevel Memory

The development of FE-HfO_2 superlattice-like laminate films has potential in high-performance multilevel memory, which is crucial for enhancing data storage density and faster read and write speeds in modern storage technology. In 2019, Ali et al.[88] demonstrated 1–3-bit/cell ferroelectric field-effect transistor (FeFET) operation using optimized $Si_xHf_{1-x}O_2$ (HSO) and HZO laminate layers with Al_2O_3 ILs, achieving a memory window (MW) of up to 3.5 V, a switching speed of 300 ns, retention of 10 years, and endurance of 10^4 cycles. These HSO/HZO-laminated

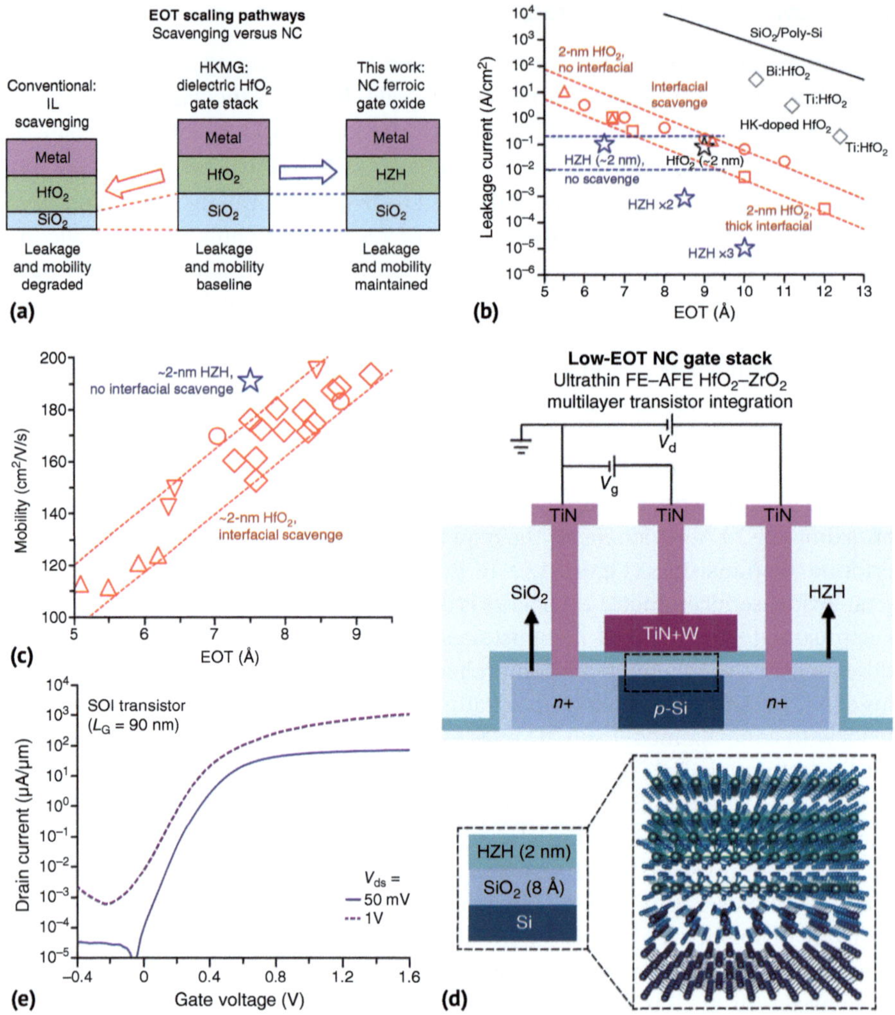

Figure 4.9 Device performance benefits from using ultrathin mixed-ferroic HfO_2/ZrO_2 gate stacks. (a) EOT scaling pathways. (b) Leakage versus EOT scaling of the HZH multilayer gate stacks (blue) benchmarked against the reported high-*k* metal gate (HKMG) literature. (c) Raw mobility versus EOT for long-channel transistors integrating the 2-nm HZH (blue) versus industry-reported long-channel transistors integrating standard 2-nm HfO_2 (red), reported with a charge density of 10^{13} cm^{-2}. (d) Schematic transistor device layout, which integrates the 2-nm HZH gate stack. (e) Current–voltage ($I - V$) transfer characteristics ($I_d - V_{gs}$),

FeFETs exhibit thickness-independent ferroelectric operation, enabling an optimized MW for reliable multilevel cell (MLC) storage (Figure 4.10a–c).[88] Subsequent systematic studies of the hybrid ferroelectric–antiferroelectric HSO/HZO laminates revealed key ferroelectric properties, including coercive field characteristics, highlighting their potential applications in low-power FeFET and FeRAM devices.[89,90] Ali et al.[91] further developed metal–ferroelectric–metal–insulator–semiconductor

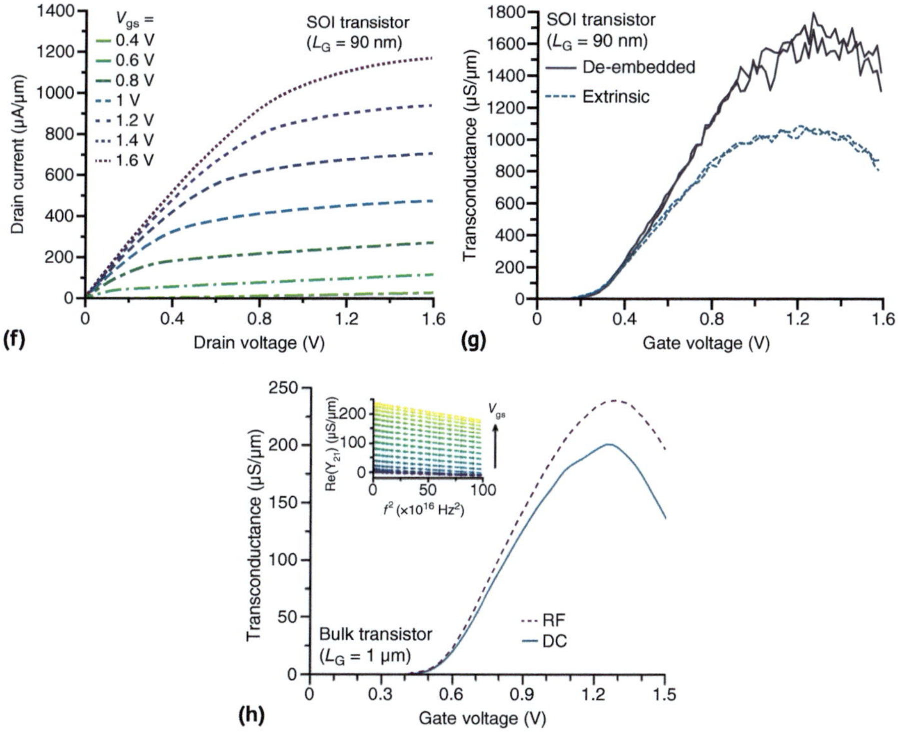

Figure 4.9 (Contd.) (f) output characteristics ($I_d - V_{ds}$), and (g) transconductance ($g_m - V_{gs}$) for short-channel ($L_G = 90$ nm) SOI transistors. V_{gs}, Gate-to-source voltage; V_{ds}, drain-to-source voltage. (h) Transconductance versus gate voltage for long-channel bulk transistors ($L_G = 1$ μm) measured via DC (derivative of $I_d - V_{gs}$) and radio frequency (RF) measurements at $V_{ds} = 1$ V. *Source*: [86] / with permission of Springer Nature.

(MFMIS) HSO/HZO-laminated-based FeFETs for low-voltage operation, demonstrating reduced variability and enhanced reliability. Lower-temperature annealing leads to a larger MW than that observed in MFMIS structures. Additionally, it reduces variability when compared to metal–ferroelectric–insulator–semiconductor (MFIS) structures, both subjected to the same high-temperature oxide (HSO) laminate of 2×10 nm. The endurance has been increased to 10^5 cycles due to a reduced electric field and optimal ferroelectric annealing (Figure 4.10d,e).[91] Lee et al.[92] developed a planar silicon-based FeFET that features a substantial storage window of up to 4.79 V, along with 10 years of data retention and excellent resistance to interference. The incorporation of a 2.4-Å-thick Al_2O_3 IL in a 20-nm-thick HZO film enhances both the coercive field and effective polarization, significantly improving its threshold voltage shift.[92] These studies highlight that the layered FE-HfO_2 structures enhance both the storage window and reliability, while also enabling multibit write capability of FeFETs, making them promising for high-density storage and neuromorphic applications.[93–95]

Peng et al.[96] demonstrated that HfO_2/ZrO_2 superlattice FeFETs show superior endurance, fatigue recovery, and retention properties compared to conventional

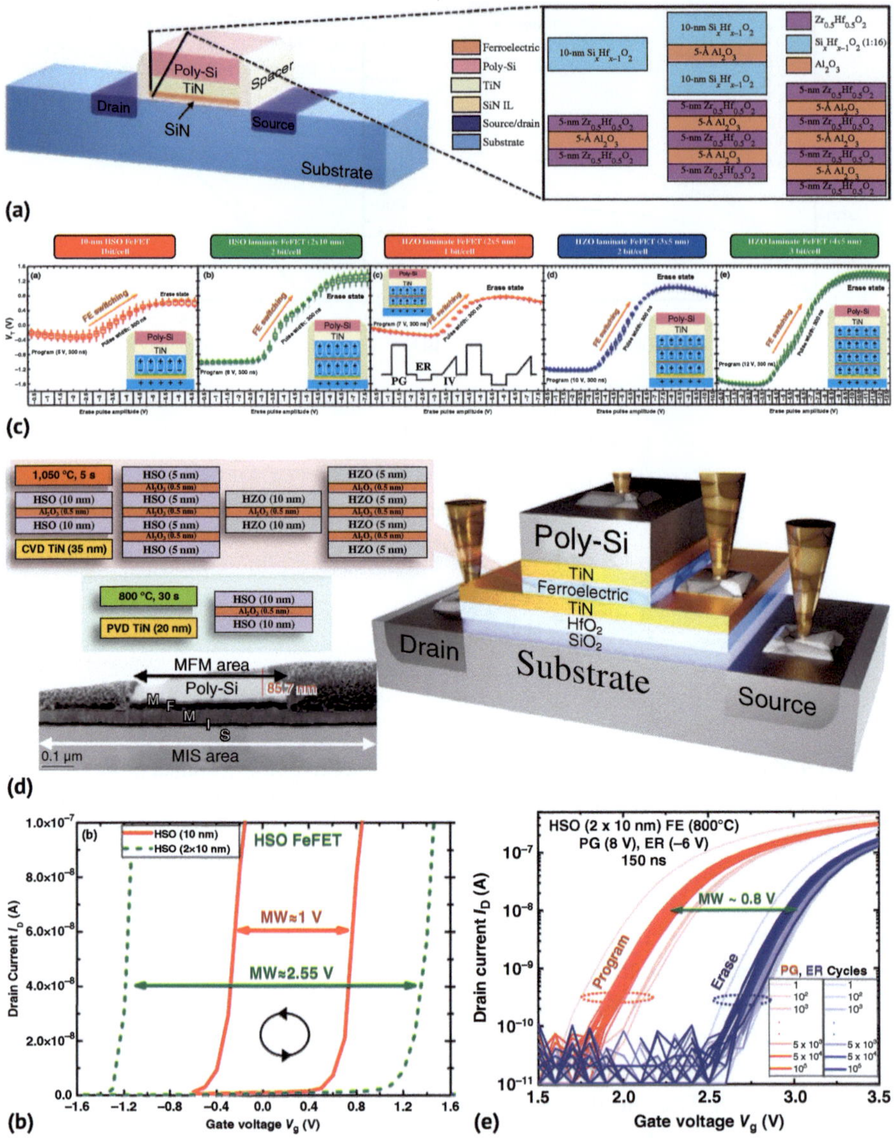

Figure 4.10 (a) Schematic of laminate-based FeFET with the inset of HSO and HZO stacks. (b) Transfer characteristics $(I_d - V_g)$ of the laminate FeFET based on HSO stacks. (c) The cross-wafer statistical erase V_T shift, extracted using a constant current criterion (10 nA) for incremental erase amplitude at 300 ns across different stack structures (inset), shows that ferroelectric switching has a higher dynamic range as the laminate thickness increases. This enhancement allows for a greater variety of bit/cell storage in the FeFET. *Source*: [88] / with permission of IEEE. (d) MFMIS schematic featuring HSO or HZO ferroelectric laminates (2 × 10 nm, 4 × 5 nm). (e) The $I_d - V_g$ curves of HSO (2 × 10 nm) MFMIS FeFET endurance (10^5 cycles). *Source*: [91] / with permission of IEEE.

HZO devices. The superlattice FeFETs show reduced interface trap generation during program/erase (P/E) cycling due to their larger EOT, which lowers the electric field across the SiO_2 interfacial layer. This effect mitigates MW degradation, allowing the superlattice FeFET to maintain an MW above 500 mV after 10^{10} P/E cycles.[96] Liao et al.[97] developed FeFETs based on HfO_2/ZrO_2 superlattices for multibit nonvolatile memory. These FeFETs, optimized using ALD and rapid thermal annealing processes, have a ferroelectric layer with a physical thickness of just 5 nm, demonstrating uniform ferroelectric domains. The devices achieved ultralow bit error rates (7.5×10^{-16}) at a P/E voltage of 4 V, with endurance of up to 10^9 cycles and data retention stability exceeding 10^4 s.[97] This work emphasizes that the HfO_2/ZrO_2 superlattice has the potential to minimize device-to-device variability in three-dimensional (3D) FeFETs, enabling high-density, high-endurance, nonvolatile memory technologies.

4.5.3 Energy Storage

Dielectric electrostatic capacitors are highly sought after for high-power energy storage applications due to their ultrafast charge–discharge capabilities.[98,99] Leveraging FE-HfO_2 laminate films, Cheema et al.[100] demonstrated the unprecedented electrostatic energy storage density (ESD) and power density in HfO_2/ZrO_2 superlattice micro-capacitors integrated on silicon through a three-pronged strategy (Figure 4.11). First, ALD antiferroelectric HZO/Al_2O_3 superlattices were engineered near a field-driven ferroelectric phase transition to enhance charge storage through the NC effect, achieving a volumetric ESD of 115 J/cm^3. Second, the use of antiferroelectric superlattice engineering overcomes traditional thickness limitations for energy storage. Finally, the incorporation of these superlattices into 3D capacitors demonstrates record energy storage (80 mJ/cm^2) and power density (300 kW/cm^2).

The superior energy storage performance of FE-HfO_2 superlattices can be attributed to their unique structure and material properties[100]: (1) HZO/Al_2O_3 superlattices optimize internal electric field distribution, thereby enhancing charge response and improving charge storage capacity; (2) the NC effect in superlattices enables higher charge storage at lower electric fields, increasing energy density; (3) compared to continuous FE-HfO_2 layers, HZO/Al_2O_3 superlattices exhibit higher breakdown strength, improving film stability and reliability; and (4) the superlattice can achieve high permittivity during the antiferroelectric-to-ferroelectric phase transition, further enhancing energy storage performance. The successful demonstration of FE-HfO_2 superlattices with ultrahigh energy and power density breaks the traditional capacity–speed trade-off in energy storage. Additionally, the use of ultrahigh-density, ultrafast-charging thin films in a compatible process enables the development of on-chip micro-capacitors, enhancing energy storage and power delivery in electronic microsystems. These advantages make FE-HfO_2 superlattices highly promising for next-generation energy storage applications.

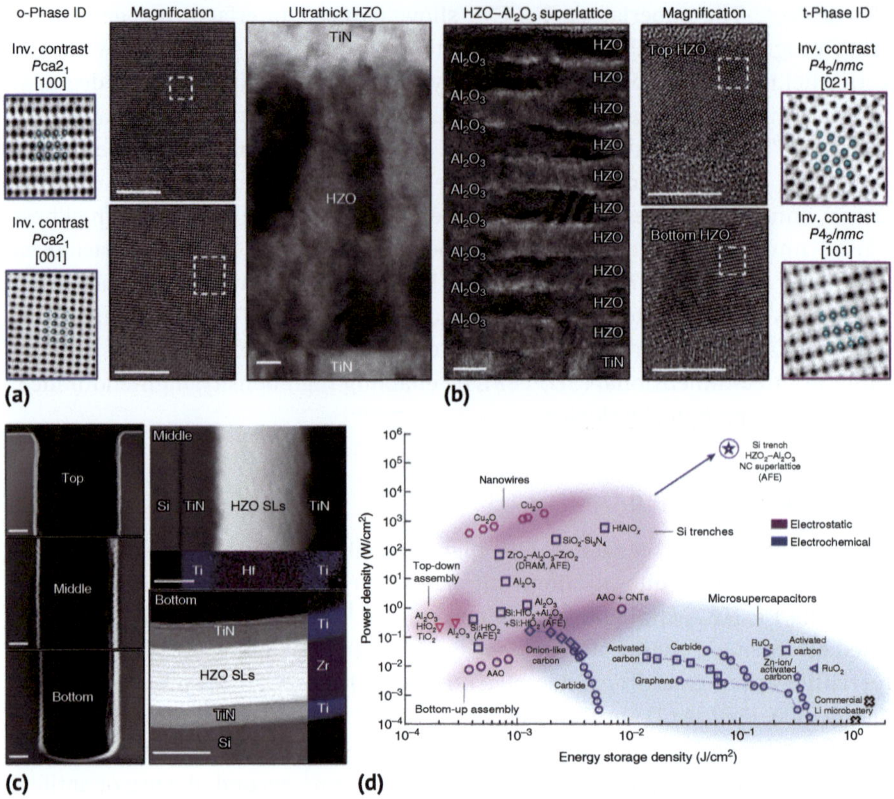

Figure 4.11 Cross-sectional HRTEM images of approximately 100-nm-thick (a) HZO film (continuous HZO × 10) and (b) HZO/Al_2O_3 superlattice film (superlattice HZO × 10). (c) TEM elemental mapping of the approximately 100-nm-thick HZO/Al_2O_3 superlattice conformally filling a 3D trench capacitor (about 100 μm depth) on the Si substrate. (d) Power density versus ESD for the highest reported micro-capacitors. *Source*: [100] / with permission of Springer Nature.

4.5.4 NEMS Resonators

NEMS resonators are key components for frequency control applications in electronic systems. These resonators rely on strong electromechanical coupling and high-quality factors, particularly in bulk acoustic wave (BAW) modes, which are achieved through superlattice structures combined with pulsed-poling-induced ferroelastic reorientation. Tharpe et al.[101] developed a NEMS resonator based on HZO/Al_2O_3 superlattices, which exhibits impressive performance characteristics: a frequency range of 0.4–17.3 GHz, a frequency–quality product of up to 4.04×10^{12} Hz, and an electromechanical coupling factor of 2.5% (Figure 4.12). Furthermore, these resonators achieve 37-dB on/off isolation using direct current (DC) bias depolarization, highlighting their switching capability.

The HZO/Al_2O_3 superlattice transducers offer several advantages that make them ideal for high-performance NEMS resonators.[101] This is due to their ability to maintain significant piezoelectric coupling even when the resonator is released

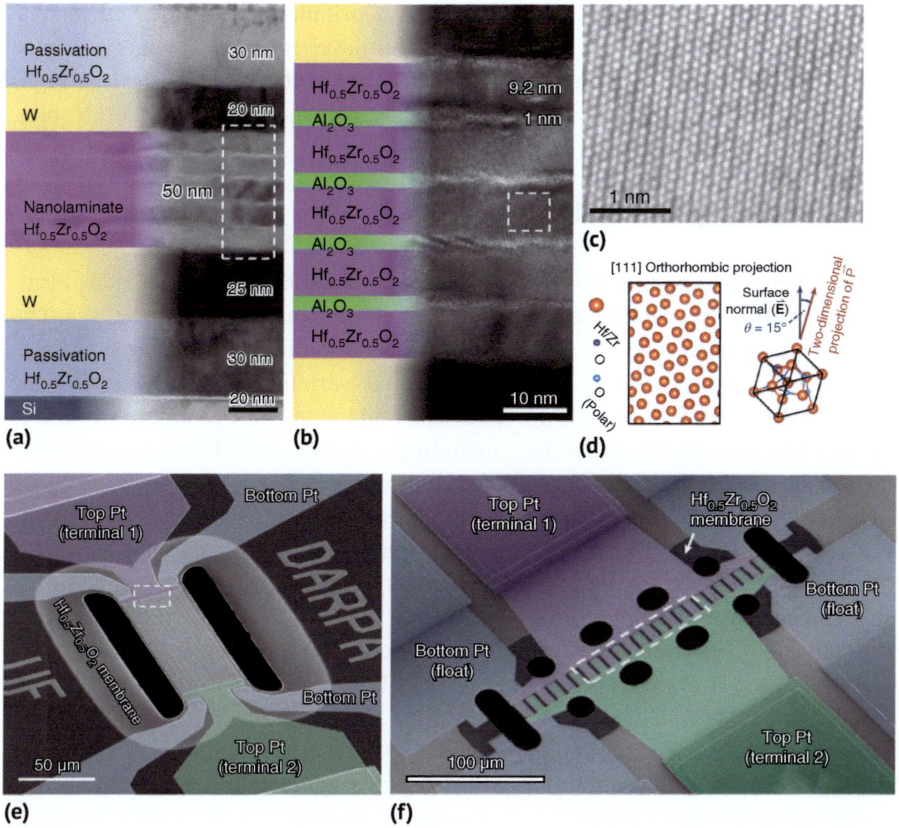

Figure 4.12 (a) HRTEM image of the superlattice HZO/Al_2O_3 NEMS resonator, with the high-level layering annotated. (b) High-magnification HRTEM image of individual HZO/Al_2O_3 superlattice layers, taken from the dashed box in panel (a). (c) STEM HAADF image of the third HZO layer from the dashed box in panel (b). (d) Atomic mapping of Hf/Zr atom within a [111]-projected orthorhombic structure, reconstructed from the dashed box in panel (c). (e) SEM image of a released 1,000-cycle-pulsed-poled HZO/Al_2O_3 lateral bulk acoustic wave (L-BAW) resonator with 2-μm electrode-finger pitch (λ) and wide tethers for tensile stress preservation. (f) SEM image of a released 1,000-cycle-pulsed-poled HZO/Al_2O_3 thickness bulk acoustic wave (T-BAW) resonator with 6 μm pitch (λ) and wide tethers for tensile stress preservation. *Source*: [101] / with permission of Springer Nature.

from its host substrate, allowing for the relaxation of stress boundary conditions. The mismatch in the CTEs of HZO and Al_2O_3, along with the large elastic constant of Al_2O_3, induces tensile stress within the transducer,[101,102] which helps sustain the orthorhombic phase after cavity release. Additionally, these superlattice transducers have a higher mechanical *Q*-factor than pure HZO,[102] contributing to improved performance. These advantages make the HZO/Al_2O_3 superlattice transducers crucial for the development of switchless, CMOS-compatible, monolithic frequency control systems, which are essential for multiband operation in the super-high-frequency (SHF) spectrum.

FE-HfO_2 superlattice-like laminate films have primarily been reported for use in low-power NC transistors, high-density multilevel memories, energy storage devices, and NEMS resonators. The successful application of these devices highlights the critical role FE-HfO_2 laminate films play in advancing modern electronics.

4.6 Summary and Outlook

This chapter has explored the concept, design, and application of FE-HfO_2 films, focusing particularly on heterostructure and superlattice laminate structures. The design and classification of FE-HfO_2 superlattice-like laminate films were discussed, highlighting how structures and compositional variations critically determine material performance. The analysis specifically addresses how laminate structures enhance key ferroelectric properties, including polarization switching speed, leakage current reduction, and operational reliability.

The development of FE-HfO_2 heterostructure laminate films was reviewed, including the design of undoped HfO_2/ZrO_2-based films with enhanced ferroelectricity, HZO/Al_2O_3 stacks with low leakage currents, and HZZ films that achieve high dielectric constants and low coercive electric fields. The NC effect in certain laminate films and its potential for improving device performance were also discussed. The potential of HfO_2/ZrO_2-based superlattice films for high reliability and ultrafast switching was explored in Section 4.4 on FE-HfO_2 superlattice films, along with the emergent ferroelectricity in monoclinic $(HfO_2)_1/(CeO_2)_1$ superlattices. These superlattice structures open up new opportunities for high-performance memory and energy storage devices. Finally, the diverse device applications of FE-HfO_2 laminate films were examined, including their use in low-power NC transistors, high-density multilevel memory, energy storage systems, and NEMS resonators.

Looking ahead, the development of FE-HfO_2 superlattice-like laminate films holds great promise for the next generation of ferroelectric devices. Future research should focus on further optimizing the design and synthesis of heterostructure and superlattice structures, exploring new material combinations, and enhancing the scalability of these materials for industrial applications. Additionally, there is significant potential for exploring their integration into more advanced device configurations, such as 3D memory arrays and energy-efficient circuits. As the understanding of these materials deepens, it is expected that FE-HfO_2 films will play a critical role in future electronics, particularly for nonvolatile memory, neuromorphic computing hardware, and high-power-density energy storage technologies.

Acknowledgments

The authors acknowledge the financial support from the National Key R&D Program of China (2024YFA1208602) and the Natural Science Foundation of Guangdong Province (2025A1515012545).

References

1 Fernandez, A., Acharya, M., Lee, H. G. et al. (2022). Thin-film ferroelectrics. *Advanced Materials* 34: 2108841. https://doi.org/10.1002/adma.202108841.

2 Silva, J. P. B., Alcala, R., Avci, U. E. et al. (2023). Roadmap on ferroelectric hafnia- and zirconia-based materials and devices. *APL Materials* 11: 089201. https://doi.org/10.1063/5.0148068.

3 Park, J. Y., Choe, D. H., Lee, D. H. et al. (2023). Revival of ferroelectric memories based on emerging fluorite-structured ferroelectrics. *Advanced Materials* 35: 2204904. https://doi.org/10.1002/adma.202204904.

4 Banerjee, W., Kashir, A., and Kamba, S. (2022). Hafnium oxide (HfO_2) – A multifunctional oxide: A review on the prospect and challenges of hafnium oxide in resistive switching and ferroelectric memories. *Small* 18: 2107575. https://doi.org/10.1002/smll.202107575.

5 Zagni, N., Puglisi, F. M., Pavan, P. et al. (2023). Reliability of HfO_2-based ferroelectric FETs: A critical review of current and future challenges. *Proceedings of the IEEE* 111: 158–184. https://doi.org/10.1109/jproc.2023.3234607.

6 Ramesh, R. and Schlom, D. G. (2019). Creating emergent phenomena in oxide superlattices. *Nature Reviews Materials* 4: 257–268. https://doi.org/10.1038/s41578-019-0095-2.

7 Schroeder, U., Park, M. H., Mikolajick, T. et al. (2022). The fundamentals and applications of ferroelectric HfO_2. *Nature Reviews Materials* 7: 653–669. https://doi.org/10.1038/s41578-022-00431-2.

8 Chen, H., Zhou, X., Tang, L. et al. (2022). HfO_2-based ferroelectrics: From enhancing performance, material design, to applications. *Applied Physics Reviews* 9: 011307. https://doi.org/10.1063/5.0066607.

9 Lu, Y. W., Shieh, J., and Tsai, F. Y. (2016). Induction of ferroelectricity in nanoscale ZrO_2/HfO_2 bilayer thin films on $Pt/Ti/SiO_2/Si$ substrates. *Acta Materialia* 115: 68–75. https://doi.org/10.1016/j.actamat.2016.05.029.

10 Weeks, S. L., Pal, A., Narasimhan, V. K. et al. (2017). Engineering of ferroelectric HfO_2–ZrO_2 nanolaminates. *ACS Applied Materials & Interfaces* 9: 13440–13447. https://doi.org/10.1021/acsami.7b00776.

11 Park, M. H., Kim, H. J., Lee, G. et al. (2019). A comprehensive study on the mechanism of ferroelectric phase formation in hafnia-zirconia nanolaminates and superlattices. *Applied Physics Reviews* 6: 041403. https://doi.org/10.1063/1.5118737.

12 Migita, S., Ota, H., Asanuma, S. et al. (2021). Accelerated ferroelectric phase transformation in HfO_2/ZrO_2 nanolaminates. *Applied Physics Express* 14: 051006. https://doi.org/10.35848/1882-0786/abfa75.

13 Chen, Y., Wang, L., Liu, L. et al. (2021). Thickness-dependent ferroelectric properties of HfO_2/ZrO_2 nanolaminates using atomic layer deposition. *Journal of Materials Science* 56: 6064–6072. https://doi.org/10.1007/s10853-020-05680-6.

14 Lee, J., Eom, D., Lee, C. et al. (2022). Sublayer thickness dependence of nanolaminated HfO_2-Al_2O_3 films for ferroelectric phase stabilization. *Applied Physics Letters* 120: 222902. https://doi.org/10.1063/5.0092125.

15 Yan, F., Liao, J., Cao, K. et al. (2023). Achieving excellent ferroelectric and dielectric performance of $HfO_2/ZrO_2/HfO_2$ thin films under low operating voltage. *Journal of Alloys and Compounds* 968: 172267. https://doi.org/10.1016/j.jallcom.2023.172267.

16 Mandal, B., Philippe, A. M., Valle, N. et al. (2025). Ferroelectric HfO_2–ZrO_2 multilayers with reduced wake-up. *ACS Omega* 10: 13141–13147. https://doi.org/10.1021/acsomega.4c10603.

17 Kim, Y., Min, K. K., Yu, J. et al. (2022). Lamination method for improved polarization-leakage current relation in HfO_2-based metal/ferroelectric/insulator/semiconductor structure. *Semiconductor Science and Technology* 37: 045001. https://doi.org/10.1088/1361-6641/ac4edd.

18 Cheema, S. S., Kwon, D., Shanker, N. et al. (2020). Enhanced ferroelectricity in ultrathin films grown directly on silicon. *Nature* 580: 478–482. https://doi.org/10.1038/s41586-020-2208-x.

19 Lee, H.-J., Lee, M., Lee, K. et al. (2020). Scale-free ferroelectricity induced by flat phonon bands in HfO_2. *Science* 369: 1343–1347.

20 Cheema, S. S., Shanker, N., Hsu, S. L. et al. (2022). Emergent ferroelectricity in subnanometer binary oxide films on silicon. *Science* 376: 648–652.

21 Batra, R., Tran, H. D., and Ramprasad, R. (2016). Stabilization of metastable phases in hafnia owing to surface energy effects. *Applied Physics Letters* 108: 172902. https://doi.org/10.1063/1.4947490.

22 Wu, J., Mo, F., Saraya, T. et al. (2020). A first-principles study on ferroelectric phase formation of Si-doped HfO_2 through nucleation and phase transition in thermal process. *Applied Physics Letters* 117: 252904. https://doi.org/10.1063/5.0035139.

23 Künneth, C., Materlik, R., and Kersch, A. (2017). Modeling ferroelectric film properties and size effects from tetragonal interlayer in $Hf_{1-x}Zr_xO_2$ grains. *Journal of Applied Physics* 121: 205304. https://doi.org/10.1063/1.4983811.

24 Park, M. H., Lee, Y. H., Mikolajick, T. et al. (2018). Thermodynamic and kinetic origins of ferroelectricity in fluorite structure oxides. *Advanced Electronic Materials* 5: 1800522. https://doi.org/10.1002/aelm.201800522.

25 Kim, H. J., Park, M. H., Kim, Y. J. et al. (2014). Grain size engineering for ferroelectric $Hf_{0.5}Zr_{0.5}O_2$ films by an insertion of Al_2O_3 interlayer. *Applied Physics Letters* 105: 192903. https://doi.org/10.1063/1.4902072.

26 Xu, Y., Yang, Y., Zhao, S. et al. (2022). Robust breakdown reliability and improved endurance in $Hf_{0.5}Zr_{0.5}O_2$ ferroelectric using grain boundary interruption. *IEEE Transactions on Electron Devices* 69: 430–433. https://doi.org/10.1109/ted.2021.3126283.

27 Chen, H.-y., Chen, Y.-H., Liang, Q.-J. et al. (2023). Effect of Al_2O_3 insertion layer on ferroelectricity in HfO_2/ZrO_2 nanolaminates. *Transactions of Nonferrous Metals Society of China* 33: 3113–3121. https://doi.org/10.1016/s1003-6326(23)66321-6.

28 Go, S., Noh, H., Lee, D. K. et al. (2024). Impact of Zr precursor on electrical characteristics of HfZrO laminated structure. *IEEE Electron Device Letters* 45: 2001–2004. https://doi.org/10.1109/led.2024.3435547.

29 Wang, D., Lu, Z., Wang, J. et al. (2024). Improvement of leakage and fatigue properties of $Hf_{0.5}Zr_{0.5}O_2$ thin film by embedding ultra-thin Al_2O_3 interlayer. *Journal of Materials Science* 60: 328–338. https://doi.org/10.1007/s10853-024-10527-5.

30 Si, M., Lyu, X., and Ye, P. D. (2019). Ferroelectric polarization switching of hafnium zirconium oxide in a ferroelectric/dielectric stack. *ACS Applied Electronic Materials* 1: 745–751. https://doi.org/10.1021/acsaelm.9b00092.

31 Chen, H., Tang, L., Liu, L. et al. (2021). Significant improvement of ferroelectricity and reliability in $Hf_{0.5}Zr_{0.5}O_2$ films by inserting an ultrathin Al_2O_3 buffer layer. *Applied Surface Science* 542: 148737. https://doi.org/10.1016/j.apsusc.2020.148737.

32 Shin, J., Seo, H., Ye, K. H. et al. (2024). Understanding phase evolution of ferroelectric $Hf_{0.5}Zr_{0.5}O_2$ thin films with Al_2O_3 and Y_2O_3 inserted layers. *Journal of Materials Chemistry C* 12: 5035–5046. https://doi.org/10.1039/d4tc00061g.

33 Li, Y.-C., Li, X. X., Wu, M. K. et al. (2022). Dual $Al_2O_3/Hf_{0.5}Zr_{0.5}O_2$ stack thin films for improved ferroelectricity and reliability. *IEEE Electron Device Letters* 43: 1235–1238. https://doi.org/10.1109/led.2022.3185246.

34 Zeng, B., Xie, S., Zhang, S. et al. (2024). Polarization fatigue mechanism of laminated hafnium zirconium oxide ferroelectric thin films. *Acta Materialia* 272: 119920. https://doi.org/10.1016/j.actamat.2024.119920.

35 Li, W., Wang, L. C., Cheema, S. S. et al. (2021). Demonstration of low EOT gate stack and record transconductance on Lg = 90 nm nFETs using 1.8 nm ferroic HfO_2-ZrO_2 superlattice. In: *Proceedings of the 2021 IEEE International Electron Devices Meeting (IEDM)*. https://doi.org/10.21203/rs.3.rs-413053/v1.

36 Hoffmann, M. and Salahuddin, S. (2021). Ferroelectric gate oxides for negative capacitance transistors. *MRS Bulletin* 46: 930–937. https://doi.org/10.1557/s43577-021-00208-y.

37 Hoffmann, M., Fengler, F. P., Herzig, M. et al. (2019). Unveiling the double-well energy landscape in a ferroelectric layer. *Nature* 565: 464–467. https://doi.org/10.1038/s41586-018-0854-z.

38 Hoffmann, M., Gui, M., Slesazeck, S. et al. (2021). Intrinsic nature of negative capacitance in multidomain $Hf_{0.5}Zr_{0.5}O_2$-based ferroelectric/dielectric heterostructures. *Advanced Functional Materials* 32: 2108494. https://doi.org/10.1002/adfm.202108494.

39 Hoffmann, M., Cheema, S. S., Shanker, N. et al. (2022). Quantitative study of EOT lowering in negative capacitance HfO2-ZrO2 superlattice gate stacks. In: *Proceedings of the 2022 International Electron Devices Meeting (IEDM)*. IEEE. pp. 13.12.11–13.12.14.

40 Kashir, A., Farahani, M. G., Kamba, S. et al. (2021). $Hf_{1-x}Zr_xO_2/ZrO_2$ nanolaminate thin films as a high-κ dielectric. *ACS Applied Electronic Materials* 3: 5632–5640. https://doi.org/10.1021/acsaelm.1c01105.

41 Kashir, A., Farahani, M. G., and Hwang, H. (2021). Towards an ideal high-κ HfO_2-ZrO_2-based dielectric. *Nanoscale* 13: 13631–13640. https://doi.org/10.1039/d1nr02272e.

42 Liu, L., Jiang, C., Yuan, X. et al. (2025). Large ferroelectric polarization and high dielectric constant in HfO_2-based thin films via $Hf_{0.5}Zr_{0.5}O_2/ZrO_2$ nanobilayer engineering. *Journal of Materiomics* 11: 100922. https://doi.org/10.1016/j.jmat.2024.07.007.

43 Lo, C., Chen, C. K., Chang, C. F. et al. (2022). High endurance and low fatigue effect of bilayer stacked antiferroelectric/ferroelectric $Hf_xZr_{1-x}O_2$. *IEEE Electron Device Letters* 43: 224–227. https://doi.org/10.1109/led.2021.3134178.

44 Zhao, Z., Chen, Y. R., Wang, J. F. et al. (2022). Engineering $Hf_{0.5}Zr_{0.5}O_2$ ferroelectric/anti-ferroelectric phases with oxygen vacancy and interface energy achieving high remanent polarization and dielectric constants. *IEEE Electron Device Letters* 43: 553–556. https://doi.org/10.1109/led.2022.3149309.

45 Guo, R., Cross, L. E., Park, S. E. et al. (2000). Origin of the high piezoelectric response in $PbZr_{1-x}Ti_xO_3$. *Physical Review Letters* 84: 5423–5426.

46 Liu, Y., Yang, J., Zhang, H. et al. (2024). Enhanced ferroelectricity and reliability in sub-6 nm ferroelectric $Hf_{0.5}Zr_{0.5}O_2/ZrO_2/Hf_{0.5}Zr_{0.5}O_2$ stack film compatible with BEOL process. *ACS Applied Electronic Materials* 6: 8507–8512. https://doi.org/10.1021/acsaelm.4c01745.

47 Kashir, A., Farahani, M. G., Lančok, J. et al. (2022). A grease for domain walls motion in HfO_2-based ferroelectrics. *Nanotechnology* 33: 155703. https://doi.org/10.1088/1361-6528/ac4679.

48 Park, G., Nguyen, A. H., Nguyen, M. C. et al. (2024). Tailoring of ferroelectric coercive field and polarization with ferroelectric and antiferroelectric $Hf_xZr_{1-x}O_2$ bilayer structure. *IEEE Electron Device Letters* 45: 1997–2000. https://doi.org/10.1109/led.2024.3435381.

49 Jang, H., Kashir, A., Oh, S. et al. (2022). Improvement of endurance and switching speed in $Hf_{1-x}Zr_xO_2$ thin films using a nanolaminate structure. *Nanotechnology* 33: 395205. https://doi.org/10.1088/1361-6528/ac79bb.

50 Choe, D.-H., Kim, S., Moon, T. et al. (2021). Unexpectedly low barrier of ferroelectric switching in HfO_2 via topological domain walls. *Materials Today* 50: 8–15. https://doi.org/10.1016/j.mattod.2021.07.022.

51 Yang, Y., Xu, W., Huang, D. et al. (2025). A comprehensive study on the sandwich stacking structures of antiferroelectric/ferroelectric doped hafnium oxide. *Applied Physics Letters* 126: 023504. https://doi.org/10.1063/5.0238806.

52 Liu, K., Jin, F., Zhang, X. et al. (2022). Anisotropic strain-mediated symmetry engineering and enhancement of ferroelectricity in $Hf_{0.5}Zr_{0.5}O_2/La_{0.67}Sr_{0.33}MnO_3$ heterostructures. *Advanced Functional Materials* 33: 2209925. https://doi.org/10.1002/adfm.202209925.

53 Joh, H., Jung, T., and Jeon, S. (2021). Stress engineering as a strategy to achieve high ferroelectricity in thick hafnia using interlayer. *IEEE Transactions on Electron Devices* 68: 2538–2542. https://doi.org/10.1109/ted.2021.3068246.

54 Nguyen, A.-D., Song, S. U., Nguyen, A. H. T. et al. (2024). Metal interlayer insertion for grain engineering in ferroelectric $Hf_{0.5}Zr_{0.5}O_2$. *IEEE Transactions on Electron Devices* 71: 6647–6651. https://doi.org/10.1109/ted.2024.3447614.

55 Tai, L., Wei, W., Sang, P. et al. (2023). Significant reliability improvement by inducing dual atomic-thin titanium intercalation layers in $Hf_{0.5}Zr_{0.5}O_2$ films. *IEEE Electron Device Letters* 44: 753–756. https://doi.org/10.1109/led.2023.3263294.

56 Shiraishi, T., Katayama, K., Yokouchi, T. et al. (2016). Impact of mechanical stress on ferroelectricity in $(Hf_{0.5}Zr_{0.5})O_2$ thin films. *Applied Physics Letters* 108: 262904. https://doi.org/10.1063/1.4954942.

57 Cao, R., Wang, Y., Zhao, S. et al. (2018). Effects of capping electrode on ferroelectric properties of $Hf_{0.5}Zr_{0.5}O_2$ thin films. *IEEE Electron Device Letters* 39: 1207–1210. https://doi.org/10.1109/led.2018.2846570.

58 Wang, Y., Zhu, C., Sun, H. et al. (2024). Phase transition kinetics and sublayer optimization of HfO_2/ZrO_2 superlattice ferroelectric thin films. *Applied Physics Letters* 125: 083503. https://doi.org/10.1063/5.0221607.

59 Cui, B., Ye, S., Wang, X. et al. (2024). Unveiling the role of local stress in enhancing ferroelectric properties and endurance of HfO_2/ZrO_2 superlattice structures. *IEEE Electron Device Letters* 46: 107–110. https://doi.org/10.1109/led.2024.3496720.

60 Peng, Y., Xiao, W., Liu, Y. et al. (2022). HfO_2-ZrO_2 superlattice ferroelectric capacitor with improved endurance performance and higher fatigue recovery capability. *IEEE Electron Device Letters* 43: 216–219. https://doi.org/10.1109/led.2021.3135961.

61 Gong, T., Tao, L., Li, J. et al. (2021). $10^5\times$ Endurance improvement of FE-HZO by an innovative rejuvenation method for 1z node NV-DRAM applications. In: *Proceedings of the 2021 Symposium on VLSI Technology*. IEEE.

62 Liang, Y.-K., Li, W. L., Wang, Y. J. et al. (2022). ZrO_2-HfO_2 superlattice ferroelectric capacitors with optimized annealing to achieve extremely high polarization stability. *IEEE Electron Device Letters* 43: 1451–1454. https://doi.org/10.1109/led.2022.3193383.

63 Pešić, M., Fengler, F. P. G., Larcher, L. et al. (2016). Physical mechanisms behind the field-cycling behavior of HfO_2-based ferroelectric capacitors. *Advanced Functional Materials* 26: 4601–4612. https://doi.org/10.1002/adfm.201600590.

64 Lehninger, D., Prabhu, A., Sünbül, A. et al. (2023). Ferroelectric [HfO_2/ZrO_2] superlattices with enhanced polarization, tailored coercive field, and improved high temperature reliability. *Advanced Physics Research* 2: 2200108. https://doi.org/10.1002/apxr.202200108.

65 Cui, B., Li, D., Gong, Z. et al. (2023). Back-End-of-Line compatible HfO_2/ZrO_2 superlattice ferroelectric capacitor with high endurance and remnant polarization. *IEEE Electron Device Letters* 44: 1011–1014. https://doi.org/10.1109/led.2023.3265516.

66 Li, K., Peng, Y., Xiao, W. et al. (2023). A comparative study on the polarization, reliability, and switching dynamics of HfO_2-ZrO_2-HfO_2 and ZrO_2-HfO_2-ZrO_2 superlattice ferroelectric films. *IEEE Transactions on Electron Devices* 70: 1802–1807. https://doi.org/10.1109/ted.2023.3248538.

67 Liang, Y.-K., Liu, Z., Cai, Z. et al. (2024). Demonstration of wake-up free 6 nm ultrathin ZrO_2-HfO_2 superlattice ferroelectric capacitors with high endurance against fatigue. *IEEE Electron Device Letters* 45: 1468–1471. https://doi.org/10.1109/led.2024.3410378.

68 Zhao, Z., Chen, Y.-R., Chen, Y.-W. et al. (2023). Towards epitaxial ferroelectric HZO on n^+-Si/Ge substrates achieving record $2Pr = 84\ \mu C/cm^2$ and endurance > 1E11. In: *Proceedings of the 2024 IEEE Symposium on VLSI Technology and Circuits*. IEEE. pp. 1–2. https://doi.org/10.23919/VLSITechnologyandCir57934.2023.10185233.

69 Gong, Z., Chen, J., Peng, Y. et al. (2022). Physical origin of the endurance improvement for HfO_2-ZrO_2 superlattice ferroelectric film. *Applied Physics Letters* 121: 242901. https://doi.org/10.1063/5.0127136.

70 Jia, S., Liao, J., Yang, Q. et al. (2025). Developing HZO-based superlattices to enhance fatigue-resistance by charge injection suppression. *Advanced Functional Materials* 35: 2501470. https://doi.org/10.1002/adfm.202501470.

71 Jiao, P., Liu, Z., Chen, H. et al. (2024). Greatly enhanced fatigue resistance in $Hf_{0.5}Zr_{0.5}O_2/La_{0.7}Sr_{0.3}MnO_3$ superlattices. *Physical Review Applied* 21: 054024. https://doi.org/10.1103/PhysRevApplied.21.054024.

72 Gong, T., Li, J., Yu, H. et al. (2021). Observation and characterization of recoverable fatigue process under low-electric field (<1.8 MV/cm) in HfZrO ferroelectric film. *IEEE Electron Device Letters* 42: 1288–1290. https://doi.org/10.1109/led.2021.3094831.

73 Zheng, Y., Zhong, C., Zheng, Y. et al. (2021). In-situ atomic visualization of structural transformation in $Hf_{0.5}Zr_{0.5}O_2$ ferroelectric thin film: From nonpolar tetragonal phase to polar orthorhombic phase. In: *Proceedings of the 2021 Symposium on VLSI Technology*.

74 Zheng, Y., Zheng, Y., Gao, Z. et al. (2021). Atomic-scale characterization of defects generation during fatigue in ferroelectric $Hf_{0.5}Zr_{0.5}O_2$ films: vacancy generation and lattice dislocation. In: *Proceedings of the 2021 IEEE International Electron Devices Meeting (IEDM)*. IEEE. pp. 33.35.31–33.35.34. https://doi.org/10.1109/IEDM19574.2021.9720565.

75 Mukherjee, B., Fedorova, N. S., and Íñiguez-González, J. (2024). First-principles predictions of HfO_2-based ferroelectric superlattices. *npj Computational Materials* 10: 153. https://doi.org/10.1038/s41524-024-01344-0.

76 Bai, N., Xue, K. H., Huang, J. et al. (2022). Designing wake-up free ferroelectric capacitors based on the HfO_2/ZrO_2 superlattice structure. *Advanced Electronic Materials* 9: 2200737. https://doi.org/10.1002/aelm.202200737.

77 Peng, Y., Wang, Z., Xiao, W. et al. (2022). Effect of thickness scaling on the switching dynamics of ferroelectric HfO_2–ZrO_2 capacitors. *Ceramics International* 48: 28489–28495. https://doi.org/10.1016/j.ceramint.2022.06.163.

78 Ke, X., Chai, J., Shao, X. et al. (2024). Switching dynamics of HfO_2-ZrO_2 nanolaminates with different laminate thicknesses. *IEEE Transactions on Electron Devices* 71: 3651–3658. https://doi.org/10.1109/ted.2024.3393935.

79 Kang, M., Peng, Y., Xiao, W. et al. (2024). HfO_2–ZrO_2 ferroelectric capacitors with superlattice structure: improving fatigue stability, fatigue recovery, and switching speed. *ACS Applied Materials & Interfaces* 16: 2954–2963. https://doi.org/10.1021/acsami.3c15732.

80 Hao, P., Zheng, S., Zeng, B. et al. (2023). Highly enhanced polarization switching speed in HfO_2-based ferroelectric thin films via a composition gradient strategy. *Advanced Functional Materials* 33: 2301746. https://doi.org/10.1002/adfm.202301746.

81 Ohtaka, O., Fukui, H., Kunisada, T. et al. (2004). Phase relations and volume changes of hafnia under high pressure and high temperature. *Journal of the American Ceramic Society* 84: 1369–1373. https://doi.org/10.1111/j.1151-2916.2001.tb00843.x.

82 Zhu, T., Ma, L., Deng, S. et al. (2024). Progress in computational understanding of ferroelectric mechanisms in HfO_2. *npj Computational Materials* 10: 188. https://doi.org/10.1038/s41524-024-01352-0.

83 Rushchanskii, K. Z., Blügel, S., and Ležaić, M. (2021). Ordering of oxygen vacancies and related ferroelectric properties in $HfO_{2-\delta}$. *Physical Review Letters* 127: 087602. https://doi.org/10.1103/PhysRevLett.127.087602.

84 Liu, C., Liu, F., Luo. Q. et al. (2018). Role of oxygen vacancies in electric field cycling behaviors of ferroelectric hafnium oxide. In: *Proceedings of the 2018 IEEE International Electron Devices Meeting*. IEEE.

85 Zhao, H. J., Fu, Y., Yu, L. et al. (2024). Creating ferroelectricity in monoclinic $(HfO_2)_1/(CeO_2)_1$ superlattices. *Physical Review Letters* 132: 256801. https://doi.org/10.1103/PhysRevLett.132.256801.

86 Cheema, S. S., Shanker, N., Wang, L. C. et al. (2022). Ultrathin ferroic HfO_2–ZrO_2 superlattice gate stack for advanced transistors. *Nature* 604: 65–71. https://doi.org/10.1038/s41586-022-04425-6.

87 Zou, J., Zou, X., Xu, J. et al. (2024). HfO_2–ZrO_2 superlattice ferroelectric films as gates for 2D MoS_2-based negative-capacitance transistors with enhanced subthreshold and endurance characteristics. *ACS Applied Nano Materials* 7: 26996–27004. https://doi.org/10.1021/acsanm.4c04974.

88 Ali, T., Polakowski, P., Kühnel, K. et al. (2019). A multilevel FeFET memory device based on laminated HSO and HZO ferroelectric layers for high-density storage. In: *Proceedings of the 2019 IEEE International Electron Devices Meeting (IEDM)*.

89 Ali, T., Lehninger, D., Lederer, M. et al. (2022). Tuning hybrid ferroelectric and antiferroelectric stacks for low power FeFET and FeRAM applications by using laminated HSO and HZO films. *Advanced Electronic Materials* 8: 2100837. https://doi.org/10.1002/aelm.202100837.

90 Ali, T., Polakowski, P., Riedel, S. et al. (2018). Silicon doped hafnium oxide (HSO) and hafnium zirconium oxide (HZO) based FeFET: A material relation to device physics. *Applied Physics Letters* 112: 222903. https://doi.org/10.1063/1.5029324.

91 Ali, T., Mertens. K., Olivo, R. et al. (2022). Impact of stack structure control and ferroelectric material optimization in novel laminate HSO and HZO MFMIS FeFET. In: *Proceedings of the 2022 International Symposium on VLSI Technology, Systems and Applications (VLSI-TSA)*. IEEE. pp. 1–2. https://doi.org/10.1109/VLSI-TSA54299.2022.9771003.

92 Lee, H. J., Nam, S., Lee, Y. et al. (2024). Laminated ferroelectric FET with large memory window and high reliability. *IEEE Transactions on Electron Devices* 71: 2411–2416. https://doi.org/10.1109/ted.2024.3371945.

93 Aabrar, K. A., Kirtania, S. G., Liang, F. X. et al. (2022). BEOL-compatible superlattice FEFET analog synapse with improved linearity and symmetry of weight update. *IEEE Transactions on Electron Devices* 69: 2094–2100. https://doi.org/10.1109/ted.2022.3142239.

94 Bégon-Lours, L., Slesazeck, S., Falcone, D. F. et al. (2024). Back-End-of-Line integration of synaptic weights using HfO_2/ZrO_2 nanolaminates. *Advanced Electronic Materials* 10: 230649. https://doi.org/10.1002/aelm.202300649.

95 Park, K., Chung, C., Ku, B. et al. (2025). $Hf_{0.5}Zr_{0.5}O_2/HfO_2/Hf_{0.5}Zr_{0.5}O_2$ laminated thin films and CF_4 plasma passivation for improved memory and synaptic characteristics of ferroelectric field-effect transistors. *Nanoscale* 17: 5689–5699. https://doi.org/10.1039/d4nr04592k.

96 Peng, Y., Xiao, W., Liu, F. et al. (2023). HfO_2–ZrO_2 superlattice ferroelectric field-effect transistor with improved endurance and fatigue recovery performance. *IEEE Transactions on Electron Devices* 70: 3979–3982. https://doi.org/10.1109/ted.2023.3279063.

97 Liao, C.-Y., Lou, Z.-F., Lin, C.-Y. et al. (2022). Superlattice HfO2-ZrO2 based ferro-stack HfZrO2 FeFETs: homogeneous-domain merits ultra-low error, low programming voltage 4 V and robust endurance 109 cycles for multibit NVM. In: *Proceedings of the 2022 International Electron Devices Meeting (IEDM)*. IEEE. pp. 36.36.31–36.36.34. https://doi.org/10.1109/IEDM45625.2022.10019369.

98 Hao, X. (2013). A review on the dielectric materials for high energy-storage application. *Journal of Advanced Dielectrics* 03: 1330001. https://doi.org/10.1142/s2010135x13300016.

99 Yang, K., Lee, E. B., Lee, D. H. et al. (2022). Energy conversion and storage using artificially induced antiferroelectricity in HfO_2/ZrO_2 nanolaminates. *Composites Part B: Engineering* 236: 109824. https://doi.org/10.1016/j.compositesb.2022.109824.

100 Cheema, S. S., Shanker, N., Hsu, S. L. et al. (2024). Giant energy storage and power density negative capacitance superlattices. *Nature* 629: 803–809. https://doi.org/10.1038/s41586-024-07365-5.

101 Tharpe, T., Hershkovitz, E., Hakim, F. et al. (2023). Nanoelectromechanical resonators for gigahertz frequency control based on hafnia–zirconia–alumina superlattices. *Nature Electronics* 6: 599–609. https://doi.org/10.1038/s41928-023-00999-9.

102 Zheng, X.-Q., Tharpe, T., Enamul Hoque Yousuf, S. M. et al. (2022). High quality factors in superlattice ferroelectric $Hf_{0.5}Zr_{0.5}O_2$ nanoelectromechanical resonators. *ACS Applied Materials & Interfaces* 14: 36807–36814. https://doi.org/10.1021/acsami.2c08414.

5

High-energy-efficiency Computing Applications for HfO_2-based Ferroelectric Materials

Xiao Yu[1,2], Peiyuan Du[1,2], Kechao Tang[3], Zhiyuan Fu[4], Jin Luo[3], Lin Chen[5], Tianyu Wang[6], Xueqing Li[7], Chengji Jin[1,2], Jiajia Chen[1,2], Huan Liu[1,2], Yan Liu[2], and Genquan Han[1,2]*

[1] *Hangzhou Institute of Technology, Xidian University, Hangzhou, Zhejiang, China*
[2] *Faculty of Integrated Circuits, Xidian University, Xi'an, Shaanxi, China*
[3] *School of Integrated Circuits, Peking University, Beijing, China*
[4] *School of Integrated Circuits, Southeast University, Nanjing, Jiangsu, China*
[5] *State Key Laboratory of Integrated Chips and Systems, School of Microelectronics, Fudan University, Shanghai, China*
[6] *School of Integrated Circuits, Shandong University, Jinan, Shandong, China*
[7] *Department of Electronic Engineering, Tsinghua University, Beijing, China*

*Corresponding author: gqhan@xidian.edu.cn

5.1 FeRAM Endurance and Performance Optimization

Since the discovery of ferroelectricity in hafnium oxide-based compounds, a wide variety of memory devices have been proposed to exploit their inherent nonvolatile properties. Among the most prominent HfO_2-based technologies are ferroelectric field-effect transistors (FeFETs), ferroelectric random-access memory (FeRAM), and ferroelectric tunnel junctions (FTJs). In particular, FeRAM benefits from its architectural resemblance to conventional dynamic random-access memory (DRAM), offering a straightforward integration path due to its compatibility with well-established fabrication techniques.[1–5]

A typical FeRAM cell adopts a one-transistor one-capacitor (1T1C) structure, comprising a single ferroelectric capacitor (FeCAP) connected in series with an access transistor, as shown in Figure 5.1a. Here, the word line (WL) biases the gate of the transistor to enable access control, whereas the bit line (BL) and the plate line (PL) coordinate the operations of reading and writing. Data storage relies on remanent polarization (P_r) within the FeCAP, where binary states are encoded through reversible polarization orientations. During readout, the capacitor is driven to saturation, and the distinction between logic states "1" and "0" is determined by analyzing

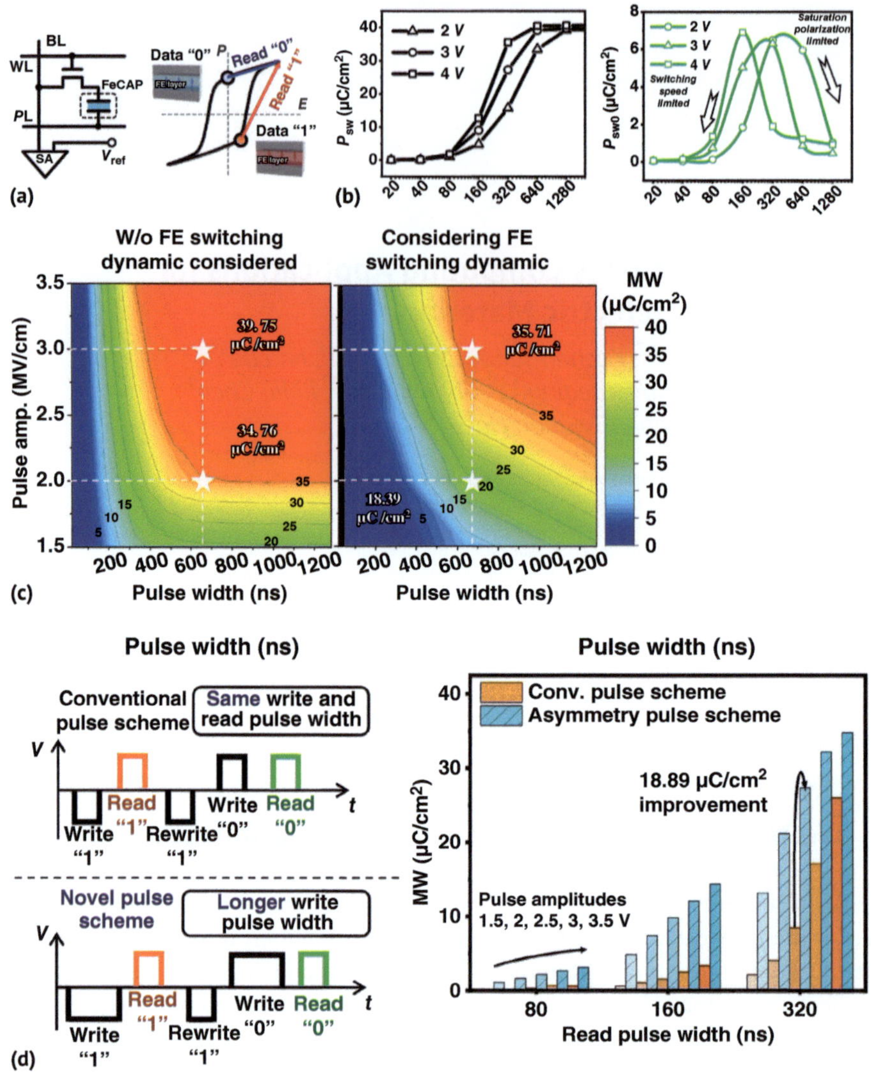

Figure 5.1 (a) Schematic of 1T1C FeRAM structure and its readout mechanism; (b) ferroelectric switching behavior during operations for different data states, where read "0" varies monotonically with pulse width; (c) comparison of FeRAM MW considering and ignoring ferroelectric switching dynamics; and (d) asymmetry operation scheme for FeRAM and its impact on enhancing MW. *Source*: [6] / with permission of IEEE.

the differential voltage on BL, which reflects the amount of switched charge in the selected memory cell (see Figure 5.1a).

5.1.1 Endurance and Performance Engineering of FeRAM

Although FeRAM exhibits advantages such as low-power operation and rapid switching speed, it still encounters critical limitations in terms of endurance and scalability. One major constraint arises from the multigrain microstructure and the inherently high coercive field required to switch polarization in HfO_2-based ferroelectric layers. As a result, FeCAPs built on HfO_2 typically suffer from dielectric breakdown after approximately 10^{10} switching cycles under full-polarization conditions, falling significantly short of the endurance thresholds demanded by DRAM-grade applications.[7–9]

To mitigate this endurance bottleneck, alternative operating methodologies have been introduced, particularly sub-loop switching combined with periodic reactivation, also referred to as "wake-up" pulses. These techniques reduce the applied electric field during cycling and have proven effective in prolonging device endurance to around 10^{14}cycles under scaled voltage conditions.[10–12] Nonetheless, operating at lower voltages presents its own set of challenges. Diminished switchable polarization and altered switching kinetics can give rise to reliability concerns, particularly during readout operations.

One prominent issue is the degradation of the memory window (MW), attributed to dynamic switching behavior that adversely impacts signal contrast between binary states.[13] This phenomenon, observed during repeated cycling, leads to a significant reduction in MW as illustrated in Figure 5.1b,c.[6] In response, asymmetric drive schemes have been proposed, wherein pulse amplitude and timing are selectively adjusted to stabilize switching and improve readout fidelity. The deployment of such techniques is essential for ensuring that FeRAM can fulfill the stringent reliability and performance requirements of next-generation memory systems.

5.1.2 High-density FeRAM Integration

Achieving higher integration density has become a central goal in the ongoing development of FeRAM technologies. For conventional 1T1C FeRAM architectures, methods such as trench capacitor formation and vertical stacking have been demonstrated as effective routes to enhance storage density.[14] To push density scaling even further, alternative memory topologies featuring fewer selector devices, such as one-transistor *n*-capacitor (1T*n*C) arrays and selector-less arrays, have been demonstrated to further improve scalability, as illustrated in Figure 5.2a.[15,16]

In the 1T*n*C scheme, a single access transistor regulates multiple FeCAPs that are connected in parallel. Each bit cell in such an array typically employs one WL, one BL, and multiple PLs to individually address the associated capacitors. This shared-selector approach enables more compact layouts, effectively improving cell density. However, applying standard switching pulses can inadvertently activate nontarget capacitors within the same bit cell, potentially corrupting data. To alleviate this, voltage partitioning schemes such as V/2 or V/3 biasing have been introduced.[18] Despite these efforts, the inherent multigrain microstructure and nonuniform switching behavior of hafnia-based ferroelectrics continue to pose significant challenges, especially in complex access scenarios involving large arrays.

To overcome these limitations, a new generation of ultralow-disturbance, selector-less FeRAM arrays has been developed. By engineering grain size and optimizing interfacial properties, high-uniformity FeCAPs have been fabricated to suppress switching variability (Figure 5.2b). Furthermore, design–technology co-optimization (DTCO) has facilitated the development of "disturb-recovery" access mechanisms, which help restore disturbed cells and maintain data integrity

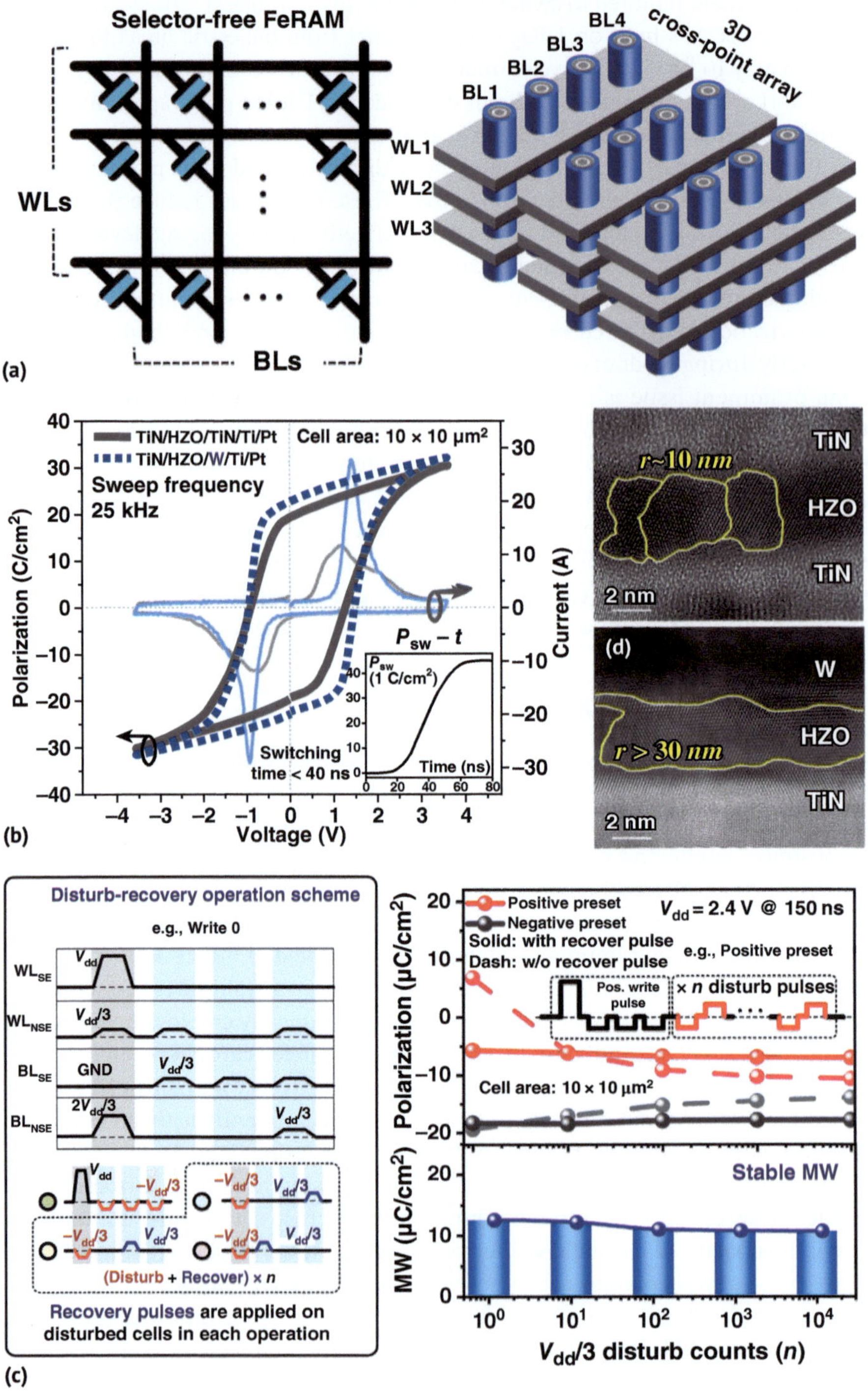

Figure 5.2 (a) Three-dimensional selector-less FeRAM array structure for high-density integration; (b) high-uniformity FeRAM cell with enlarged grain size and immunity to electrical disturbance; (c) "disturb-recovery" operation method for selector-less FeRAM arrays, enabling reliable memory performance;

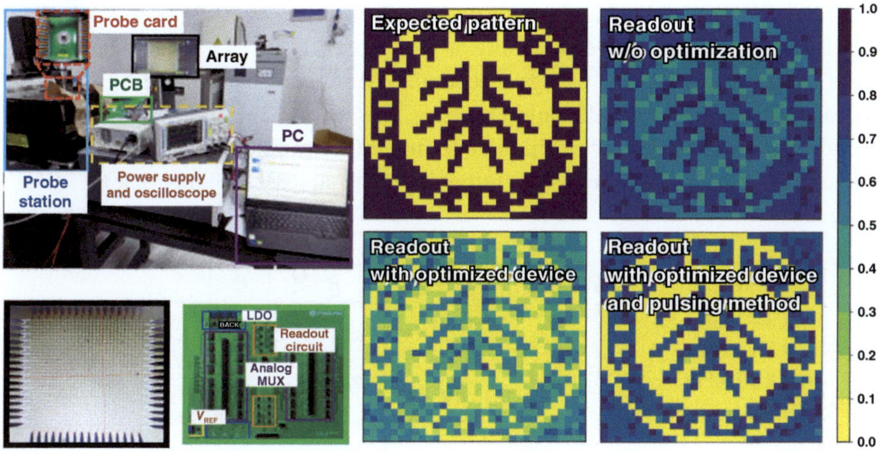

Figure 5.2 (Cont'd) and (d) demonstration of stable read operations in a 1-kbit selector-less FeRAM with and without optimization. *Source*: [17] / with permission of IEEE.

across repeated operations. With these innovations, reliable operation has been demonstrated in highly compact selector-less arrays with memory densities reaching up to $4F^2/N$ (Figure 5.2c,d).[17,19] These advancements mark a significant step toward scalable, high-performance FeRAM suitable for future memory-intensive applications.

5.1.3 FeCAP-based Energy-efficient Computing

FeCAPs have recently attracted significant attention for their promising role in energy-efficient computing paradigms, such as in-memory search, computing-in-memory (CIM), and processing-in-sensor (PIS) applications. These architectures aim to minimize data movement and conversion overhead, which is the key contributor to energy inefficiency in conventional computing. However, standard CMOS-based implementations often suffer from high circuit complexity, increased area overhead, and static power dissipation, mainly due to the use of redundant logic cells for computation.

To overcome these issues, FeCAP-based content-addressable-memory (CAM) arrays have been developed, exploiting the distinctive nonmonotonic small-signal capacitance–voltage (*C*–*V*) behavior of FeCAPs, as shown in Figure 5.3a.[20] By encoding weights directly into the remanent polarization states and applying carefully tailored sub-loop read pulses, individual FeCAP devices can perform XNOR-based search operations without requiring complementary branches. This compact design enables ultralow energy consumption during search operations and offers high endurance.

Additionally, these same *C*–*V* properties have been exploited to implement CIM functionality within FeCAP arrays, achieving multiply–accumulate (MAC) operations with high linearity. This architecture has achieved inference accuracy exceeding 96% in few-shot learning applications (Figure 5.3b). Building on this foundation, a resonant adiabatic capacitive CIM (RA C-CIM) framework has been proposed and experimentally validated using three-dimensionally stacked FeCAP arrays.[21,22] This design leverages the capacitive switching of HfO_2-based FeCAPs to enable both non-volatile write operations and energy-recoverable MAC computation. Through resonant energy recycling and adiabatic charging, the RA C-CIM achieves substantial reductions in dynamic power dissipation compared to traditional pulse-driven CIM architectures, while simultaneously minimizing thermal losses (Figure 5.3c).

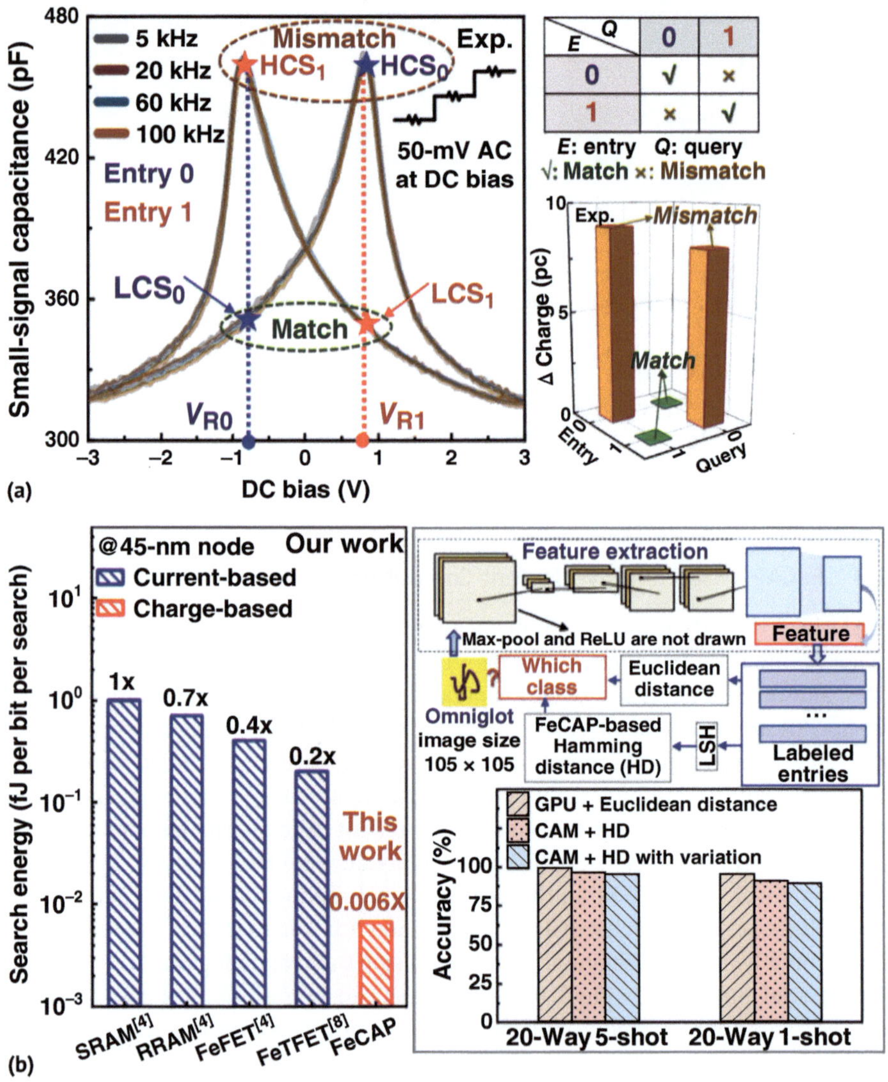

Figure 5.3 (a) Small-signal C–V characteristics of FeCAP exhibiting XNOR operation within a single device, and high reliability. (b) FeCAP array-based high-efficiency comparison achieving CMOS-comparable accuracy.

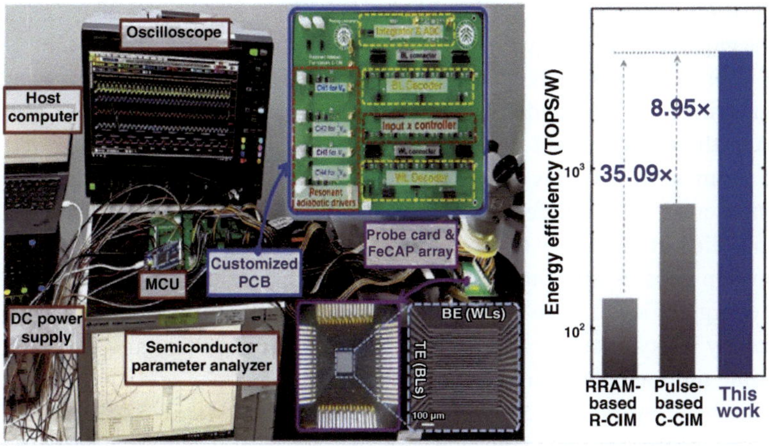

(c)

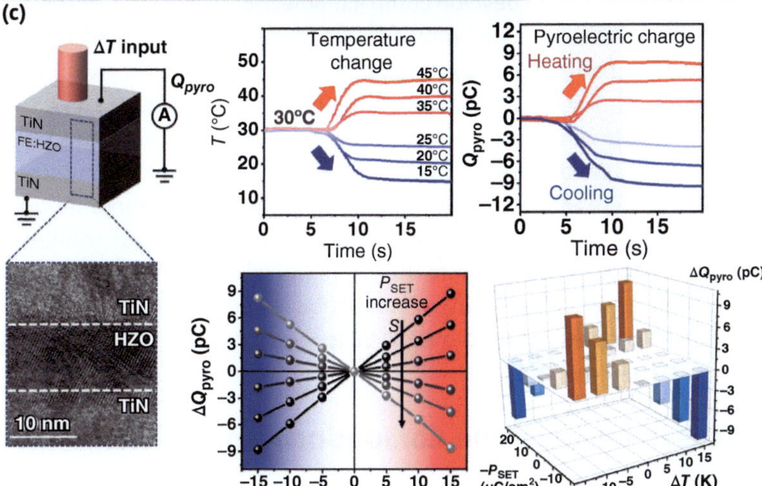

(d)

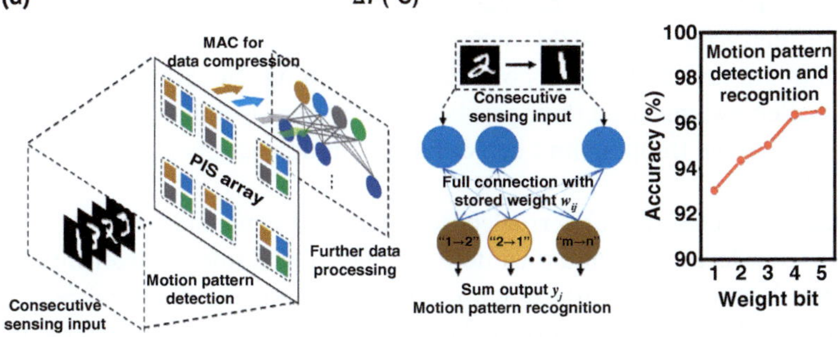

(e)

Figure 5.3 (Cont'd) (c) Experimental demonstration of RA C-CIM utilizing a capacitive FeCAP array. (d) Pyroelectric FeCAP for *in-situ* motion sensing and in-sensor multiplication. (e) FeCAP-based PIS array with MAC operations and *in-situ* motion recognition under ultralow power and high accuracy. *Source*: [20] / with permission of IEEE.

Collectively, these innovations highlight the transformative potential of FeCAPs in bridging the gap between energy-efficient hardware platforms and artificial intelligence (AI) inference workloads, particularly for edge computing systems that operate under stringent power and area constraints.

In addition to their role in CIM architectures, FeCAP arrays have demonstrated considerable promise in ultralow-power PIS applications, leveraging the diverse physical phenomena inherent to ferroelectric materials. PIS systems integrate sensing and computation at the device level, thereby significantly reducing data transfer overhead between sensors and processors. While this approach holds substantial potential for energy-efficient edge computing, traditional PIS implementations are often hindered by bulky circuitry and high energy demands, particularly for tasks such as real-time motion detection.

Recent studies have leveraged the pyroelectric effect and multidomain polarization dynamics of HfO_2-based FeCAPs to develop compact, power-efficient PIS devices.[23] These novel FeCAP-PIS units can directly transduce temperature variations (ΔT) into signed MAC outputs by modulating charge release through polarization switching, enabling true in-sensor motion detection. Such functionality enables four-quadrant MAC operations with nearly zero static power consumption, as shown in Figure 5.3d. Experimental evaluations have confirmed excellent linearity (>99%) across multilevel signed weight states and demonstrated 99% recognition accuracy in infrared motion sensing tasks, surpassing the performance of conventional CMOS-based solutions (Figure 5.3e). These advancements illustrate the feasibility of embedding intelligent processing within sensing modules, laying a robust foundation for highly energy-efficient, edge-native AI systems powered by FeCAP technology.

5.2 FeFET-based Memory and Computing Applications

FeFETs based on HfO_2 have been considered as promising contenders for next-generation nonvolatile memory (NVM) owing to their excellent scalability, ultralow power consumption, and high-speed polarization switching. Nonetheless, one of the primary challenges impeding their practical deployment is limited endurance, largely driven by charge trapping and defect formation during repeated program/erase (P/E) cycles.[24]

In the widely adopted metal–ferroelectric–interlayer–semiconductor (MFIS) stack, both the ferroelectric film and the interlayer play pivotal roles in determining charge transport behavior, particularly with regard to trap generation and retention characteristics. As a result, optimizing the gate stack composition is essential to extend FeFET device lifetime.

For HfO_2-based ferroelectric layers, engineering a lower coercive field (E_c) and reducing spontaneous polarization (P_s) are effective strategies to suppress the internal electric field across the interlayer (E_{IL}), thereby minimizing stress-induced degradation. Simultaneously, selecting interlayer materials with a high dielectric constant (κ) and robust thermal stability is essential to suppress the formation of low-κ interfacial oxides and to maintain favorable electrical characteristics at the semiconductor interface. Additionally, achieving a low interface trap density (N_{it}) at the ferroelectric–interlayer boundary, along with process compatibility, is vital for ensuring reliable and scalable FeFET integration. These material and interface considerations are illustrated in Figure 5.4a.

5.2.1 High-endurance Si-channel-based FeFETs

A co-engineered gate stack integrating $Hf_{0.95}Al_{0.05}O_2$ (HAO) as the ferroelectric layer and Al_2O_3 as the interfacial has been successfully demonstrated (Figure 5.4b).[25] Compared to conventional HZO, HAO shows smaller spontaneous polarization P_s and coercive field E_c, which help alleviate electrical stress within the gate stack. Meanwhile, Al_2O_3 serves as a superior interlayer material, offering a higher dielectric constant and better thermal stability than SiO_2, thereby minimizing the formation of parasitic low-κ oxides at the silicon interface.

This combination of materials delivers improved thermal compatibility and enhanced interface quality, resulting in a significant boost in FeFET endurance. Specifically, the HAO+Al_2O_3 stack achieves cycling stability beyond 5×10^9 cycles, an improvement of approximately three orders of magnitude over standard HZO-based FeFETs (Figure 5.4c). To validate this enhancement, double quasi-static capacitance–voltage (DQSCV) measurements were employed, revealing a substantial reduction in trapped charge densities for both electrons (N_e) and holes (N_h). These findings confirm a lower N_{it} at the ferroelectric–interlayer boundary (Figure 5.4d), underscoring the effectiveness of the co-optimization strategy in promoting device reliability.

FeFET devices can also operate in multilevel cell (MLC) mode by exploiting the partial polarization of ferroelectric domains, thereby increasing storage density per unit area. However, implementing MLC functionality in FeFETs introduces substantial challenges, such as device-to-device (D2D) variability and endurance degradation.[28] To mitigate these problems, a robust MLC FeFET structure was developed through a comprehensive co-optimization strategy that simultaneously refines the ferroelectric gate stack and the programming pulse conditions (Figure 5.4e).[26]

To enhance device uniformity, a thin Al_2O_3 layer was inserted between two HZO ferroelectric films. This interlayer engineering promotes grain size reduction, which effectively suppresses D2D variations. Additionally, programming pulses with higher amplitudes and shorter durations were adopted to minimize random domain nucleation events, leading to more consistent switching behavior. As a result, the average standard deviation (σ_{avg}) of the threshold voltage (V_{th}) distributions was reduced by 36% compared to conventional HZO-only FeFETs, as shown in Figure 5.4f,g. From a reliability standpoint, the inserted Al_2O_3 layer also acts as a barrier to defect migration and helps mitigate the formation of leakage paths, preventing early breakdown. Furthermore, the use of shorter pulses helps limit charge trapping and suppresses the generation of new traps during cycling. Collectively, these optimizations enable endurance to exceed 10^9 cycles (Figure 5.4h).

Beyond their role as memory, FeFETs are increasingly being utilized in CIM frameworks, where partial polarization switching is harnessed for analog weight representation.[29] However, analog CIM operations demand high linearity and long

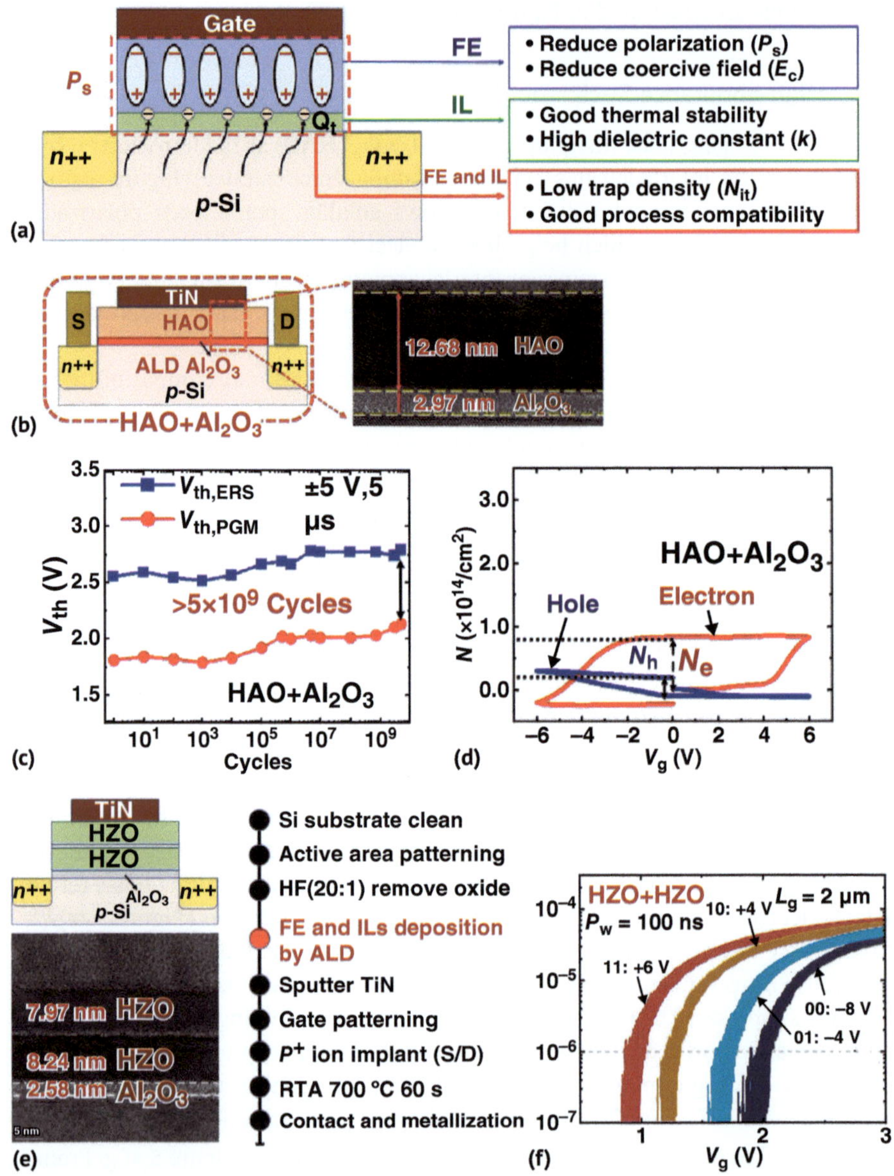

Figure 5.4 High-endurance FeFETs achieved through co-optimization of ferroelectric and interlayer: (a) design principle; (b) schematic and HRTEM image of HAO+Al_2O_3 FeFET; (c) endurance; and (d) trapped charge density. *Source*: [25] / with permission of IEEE. Multilevel FeFETs with enhanced endurance and uniformity and by gate stack and writing pulse optimization: (e) schematic, HRTEM image, and process flow of HZO+HZO FeFET; (f) $I_d - V_g$ characteristics.

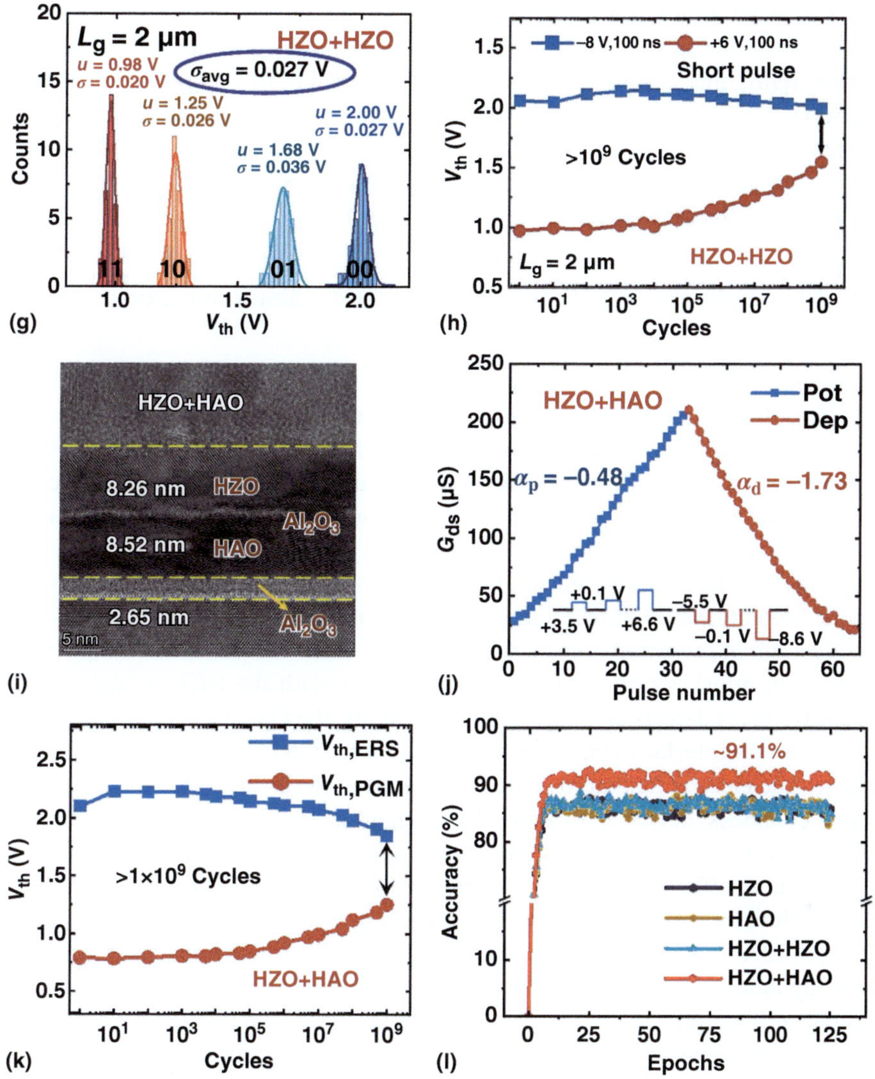

Figure 5.4 (Cont'd) (g) V_{th} distribution; and (h) endurance data. *Source*: [26] / with permission of IEEE. Hybrid-layer FeFETs with improved linearity and endurance: (i) HRTEM image; (j) conductance versus pulse counts; (k) endurance; and (l) simulated training accuracy of HZO+HAO FeFET. *Source*: [27] / with permission of IEEE.

endurance from the weight elements, in which conventional FeFETs still require improvement.[30] To this end, a novel FeFET structure incorporating a hybrid ferroelectric layer (HZO + HAO) and an Al_2O_3 interlayer has been proposed (Figure 5.4i).[27] The mixed HZO–HAO composition leads to a broader E_c distribution, which enables smoother, more linear weight updates due to gradual domain switching. As demonstrated in Figure 5.4j, the device exhibits improved linearity in weight updates, with reduced nonlinearity coefficients ($\alpha_p = -0.48$ and $\alpha_d = -1.73$).

The enhanced endurance of this hybrid FeFET originates from the reduced P_s and E_c of HAO, along with defect suppression and high-κ characteristics of Al_2O_3. Experimentally, this design achieves endurance beyond 10^9 cycles (Figure 5.4k). Simulation studies on MNIST digit classification further validate the architecture's performance, showing an inference accuracy of 91.1%, which is superior to other FeFET-based implementations, thereby confirming its potential for accurate and reliable CIM applications (Figure 5.4l).

5.2.2 High-endurance FeFETs Incorporating Oxide–Semiconductor Channels

HfO_2-based FeFETs can also employ oxide–semiconductor (OS) channels such as In_2O_3, ZnO, InGaZnO (IGZO), and IWO.[31–34] Compared to their silicon-channel counterparts, FeFETs with OS channels offer two key advantages. First, these OSs can be deposited at significantly lower temperatures than polycrystalline silicon, making the devices highly compatible with back-end-of-line (BEOL) integration, particularly within monolithic three-dimensional architectures.[35] Such compatibility paves the way for advanced heterogeneous integration opportunities in future systems. Second, the inherent oxide nature of these channel materials allows for the realization of interlayer-free device structures. Unlike Si-based FeFETs, where a native SiO_x interlayer often forms at the interface with the HfO_2 ferroelectrics, which introduces defects and high interfacial fields, the OS-based approach naturally suppresses interfacial oxide formation. Eliminating the interlayer effectively decreases the electric field across the ferroelectric-interfacial region (E_{IL}), which in turn enhances device endurance by mitigating stress-induced degradation mechanisms. This structural simplification not only improves reliability but also facilitates more efficient fabrication pathways for next-generation FeFET devices.

High-endurance FeFETs can also be realized by integrating OS channels with antiferroelectric (AFE) gate dielectrics.[36] AFE behavior in hafnia-based systems is typically induced by increasing the ZrO_2 content in HZO films beyond a certain threshold, as illustrated in Figure 5.5a,b. In AFE materials, the characteristic double-hysteresis loop allows one polarization branch to be exploited for memory operation, effectively reducing E_c. This reduction in E_c helps lower the interlayer electric field E_{IL}, which in turn mitigates degradation mechanisms related to high-field stress.

However, in conventional Si-based FeFETs, the use of AFE layers can make the polarization states more susceptible to depolarization effects due to the presence of an interfacial oxide. In contrast, OS-based FeFETs benefit from an interlayer-free structure, which significantly reduces depolarization fields and enhances retention stability (Figure 5.5c). As a result, AFE-based OS FeFETs exhibit improved endurance, reaching up to 10^9cycles (Figure 5.5d), while retaining nonvolatile performance exceeding 10years. Moreover, the combination of a low E_c and elimination of an interlayer minimizes the programming voltage drop across the gate stack, ensuring reliable operation even at a supply voltage of 2V (Figure 5.5e). These attributes make AFE-OS FeFETs particularly attractive for high-speed, low-power memory applications (Figure 5.5f).

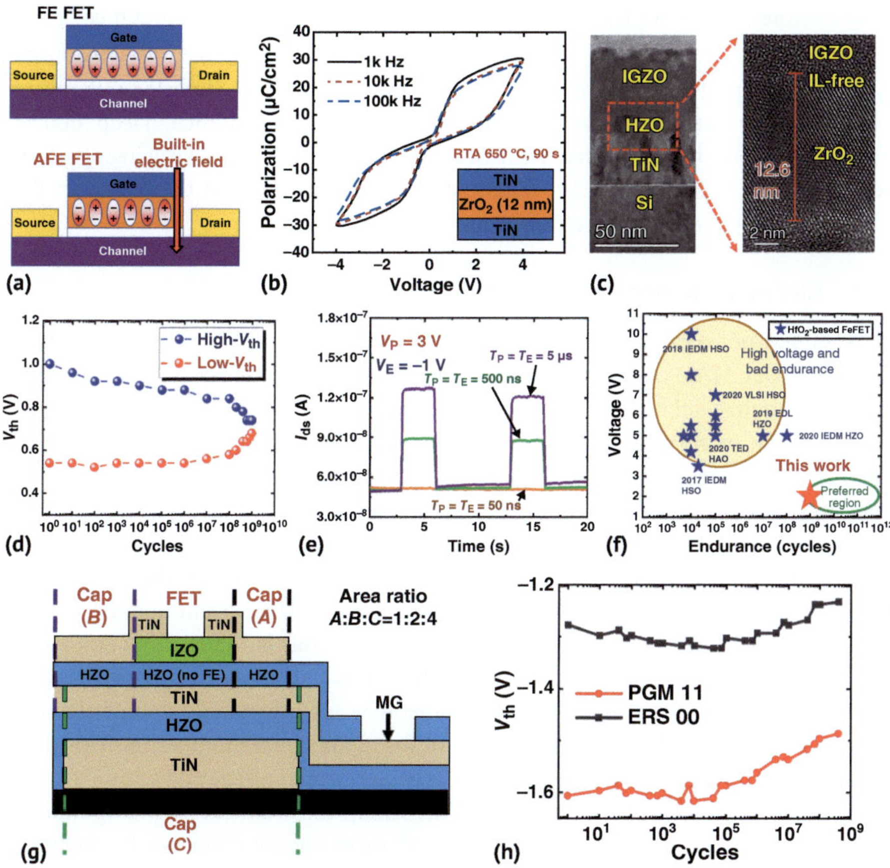

Figure 5.5 High-endurance FeFETs based on OSs and AFE materials: (a) device schematics; (b) P–V curves of the ZrO_2 AFE; (c) TEM image of the interlayer-free structure; (d) V_{th} evolution with cycling (up to endurance); (e) I_{ds} under 2 V operation; and (f) endurance-operation voltage benchmarking. *Source*: [36] / with permission of IEEE. Multibit FeFETs with 1T3C configuration: (g) device schematic illustrating triple-capacitor storage and (h) endurance characteristics. *Source*: [37] / with permission of IEEE.

To address the trade-off between endurance and memory density in embedded nonvolatile memory (eNVM) use cases, a novel one-transistor three-capacitor (1T3C) FeFET structure has been introduced.[37] This architecture employs three FeCAPs of varying areas, each contributing distinct polarization states. Unlike the traditional MLC FeFETs that rely on partial polarization in a single ferroelectric layer, this approach circumvents the need for a large-E_c material. Consequently, it reduces programming voltage and improves endurance. The 1T3C FeFET structure enables dual-bit data storage within a single memory cell and demonstrates enhanced cycling stability, with endurance up to 4×10^8 cycles (Figure 5.5g,h).

In neuromorphic computing, Lu et al. developed a symmetric nonlinear (SN) model to describe the conductance modulation of FeFETs.[38] Using a metal–ferroelectric–metal–interlayer–semiconductor (MFMIS) configuration with an

IGZO channel, these FeFETs can achieve up to seven discrete conductance states and a dynamic range exceeding $10^3(G_{max}/G_{min} > 10^3)$. Compared to traditional asymmetric nonlinear (AN) models, the SN model exhibits superior symmetry in conductance tuning, thereby enhancing learning performance. Deep neural networks (DNNs) implemented with SN-modeled FeFETs achieve a recognition accuracy of approximately 94.1%, demonstrating the potential of these devices for building high-density, energy-efficient neuromorphic systems.

5.2.3 CIM Applications Based on FeFET

Conventional nonvolatile-based computing-in-memory (nvCIM) architectures typically carry out MAC operations within the analog voltage domain. However, this approach requires relatively high supply voltages (V_{DD}) to support a wide signal dynamic range, along with complex data conversion circuits such as analog-to-digital converters (ADCs) and DACs, which contribute to substantial power consumption. To overcome these limitations in voltage scalability and energy efficiency, a novel time-domain (TD) nvCIM architecture based on FeFETs has been proposed and experimentally demonstrated (Figure 5.6a–c).[39] This TD design exploits the cumulative delay produced by cascaded inverter chains to perform MAC accumulation in TD, enabling a more energy-efficient computation model that is inherently compatible with digital circuit design flows. Thanks to the intrinsic three-terminal structure of the FeFET, the same device can serve simultaneously as a nonvolatile storage element for synaptic weight and a dynamic modulator for input signal modulation. This dual functionality allows local MAC operations to be performed efficiently within a compact 2T-1FeFET ferroelectric delay unit. Additionally, the architecture incorporates a dual-edge operation scheme, which further enhances energy efficiency by enabling computations on both rising and falling signal transitions.

Moreover, by harnessing the physical behavior of time-dependent gradual polarization switching in FeFETs, the same device can be used to implement activation functions in TD nvCIM systems. This feature introduces inherent memory trace dynamics, which are particularly beneficial for deep reinforcement learning. Remarkably, these functions can be realized using only a single FeFET device.[40] Overall, FeFET-based TD nvCIM represents a highly promising architecture for developing compact, low-power, and mixed-signal neural network accelerators optimized for energy-constrained edge AI systems (Figure 5.6d).

In nvCIM architectures, weight parameters stored in NVM can be directly accessed during inference, which poses significant security risks such as intellectual property (IP) theft and adversarial attacks. Therefore, integrating encryption mechanisms into nvCIM is essential for ensuring secure AI operations at the edge. Traditionally, secure nvCIM implementations rely on explicit decryption or key-controlled access, often requiring complex circuit designs and redundant storage, such as split WL SRAM cells or twin 2T+2RRAM (resistive RAM) architectures, resulting in increased area and power overheads. To address these challenges, a novel ferroelectric tunnel field-effect transistor (FeTFET) has been introduced and experimentally validated for implementing encryption-embedded nvCIM (Figure 5.6e).[41] This architecture enables *in-situ*

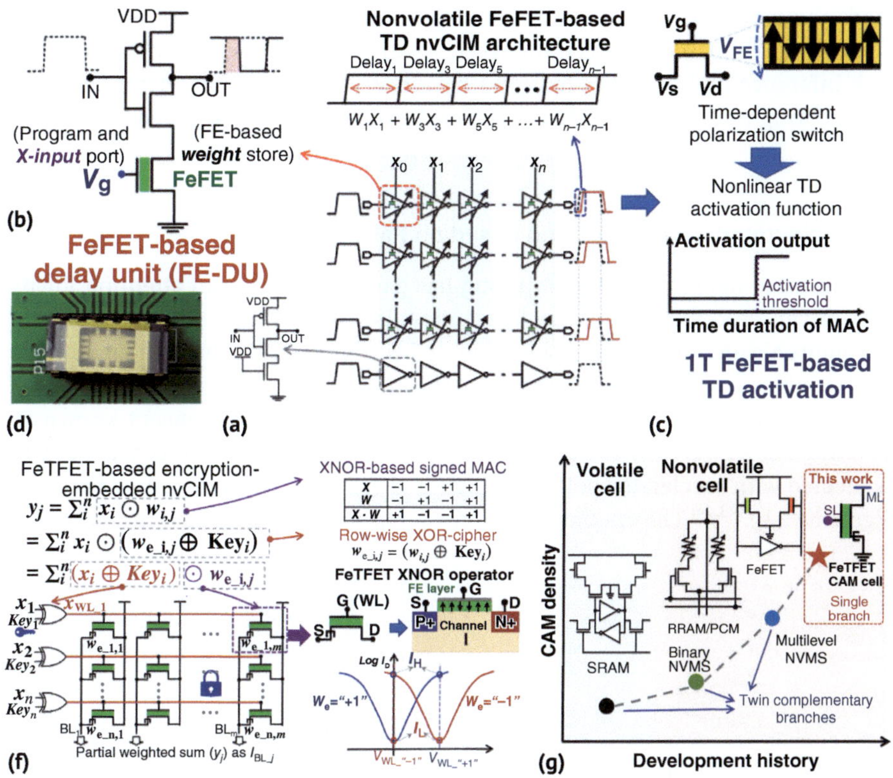

Figure 5.6 (a) FeFET-based TD nvCIM architecture incorporating FeFET-based delay unit (FE-DU) array and a reference delay chain. (b) FE-DU topology composed of a FeFET and an inverter for TD computation. (c) Operation principle of single-FeFET-based TD activation. *Source*: [39] / with permission of IEEE. (d) Hardware implementation of Fe-FinFET-based TD circuit. *Source*: [40] / with permission of IEEE. (e) FeTFET-based nvCIM embedding encryption with XOR-encrypted weight and XNOR-based signed MAC. *Source*: [41] / with permission of IEEE. (f) The FeTFET-based single-branch CAM cell for further improved CAM density. *Source*: [42] / with permission of IEEE.

key authentication and MAC computation within a unified framework, achieving both high area utilization and superior energy efficiency. By integrating XOR-based encryption directly into the XNOR MAC operation, the scheme eliminates the need for separate decryption stages. Local multiplications are performed directly on encrypted weights using an XNOR operator, simplifying circuit complexity.

The proposed FeTFET leverages nonvolatile polarization states to store encrypted weights and utilizes the ambipolar tunneling current of TFET to encode input signals. This unique combination allows the implementation of an XNOR function within a single FeTFET device, fabricated using a 14-nm FinFET-compatible process. It supports multilevel weight storage and encrypted MAC operations without requiring auxiliary decryption hardware or complementary storage cells.

Additionally, the FeTFET can serve as a compact CAM cell, performing pattern matching and similarity-based search using only one transistor, which avoids the twin-branch overhead typically found in conventional CAM designs (Figure 5.6f).[42] Based on this framework, secure and energy-efficient neural network inference and one-shot learning have been experimentally demonstrated, underscoring the promise of FeTFET-based nvCIM for trusted AI computing at the edge.

5.2.4 Neuromorphic Computing Based on FeFET

Brain-inspired neuromorphic computing has garnered significant attention for its event-driven processing model and exceptional energy efficiency. However, most neuromorphic architectures continue to rely heavily on CMOS-based components, which leads to considerable hardware overhead, especially as more advanced neuronal functions are integrated. To address this, a novel device concept known as the leaky-FeFET (L-FeFET) has been proposed, leveraging HfO_2-based FeFETs engineered with accelerated polarization degradation to emulate neuronal behavior (Figure 5.7a,b).[43,44] This method allows the realization of a compact, capacitor-less

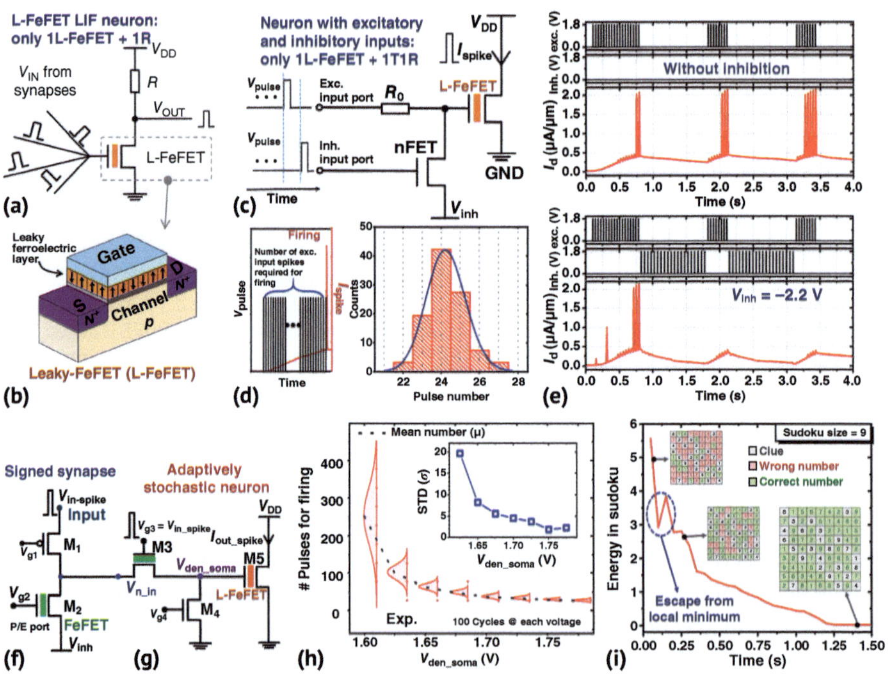

Figure 5.7 (a) Structure schematic of an excitatory spiking neuron. (b) Novel L-FeFET device enabling neuron behavior. (c) L-FeFET neuron with excitatory and inhibitory synaptic input. (d) Measured stochastic firing behavior showing Gaussian distributed pulse counts for neuron activation. (e) Experimental spike firing results of L-FeFET neuron. *Source*: [44] / with permission of IEEE. (f) FeFET-based signed synapse and (g) adaptively stochastic neuron design. *Source*: [45] / with permission of IEEE. (h) Measured adaptive reduction in neuron stochasticity. (i) FeFET-based SNN solving Sudoku CSP through simulated annealing. *Source*: [46] / with permission of IEEE.

leaky integrate-and-fire (LIF) neuron consisting of only a single transistor and one resistor. The L-FeFET mimics the "leaky" temporal integration characteristic of biological neurons while dramatically reducing circuit complexity.

Beyond basic LIF functionality, the L-FeFET's unique properties have also been exploited to implement more complex spiking behavior. A miniaturized spiking neuron capable of exhibiting both excitatory and inhibitory behaviors was realized using just three transistors. This was achieved by exploiting the bidirectional accumulation of ferroelectric polarization, which supports unsupervised learning within a highly compact hardware footprint (Figure 5.7c–e).[43]

Further advancement was achieved through device–circuit co-design, integrating an adaptively stochastic firing mechanism inspired by simulated annealing algorithms. This functionality leverages the inherent randomness of ferroelectric domain nucleation and switching, allowing neurons to probabilistically fire in a manner that supports energy-efficient global optimization. This strategy enabled the development of a fully FeFET-based spiking neural network (SNN) featuring improved area efficiency and scalability for neuromorphic computing tasks (Figure 5.7f–i).[45,46]

Collectively, these innovations underscore the transformative potential of FeFET technology in neuromorphic systems, offering compact, energy-efficient neuron implementations that support a wide range of cognitive computing functions with minimal circuit overhead.

5.3 FTJ-based Memory and Computing Applications

5.3.1 Merits and Limitations of HfO_2-based FTJs

FTJs constitute a new category of NVM devices that exploit quantum tunneling effect through a ferroelectric barrier. An FTJ typically comprises two metallic electrodes separated by an ultrathin ferroelectric film, generally 5–10 nm thick, enabling efficient electron tunneling.[47] When an external electric field is applied, the polarization direction of the ferroelectric barrier reverses, altering the potential energy profile across the junction. This reversal leads to a significant modulation in tunneling current, known as the tunneling electroresistance (TER) effect,[48] as the change in the polarization state modifies the effective tunneling barrier height and/or width, thus modulating the resistance of the device. This change in resistance serves as the basis for data encoding, enabling binary or even multilevel information storage. Importantly, the ferroelectric polarization remains stable in the absence of power, granting FTJs intrinsic NVM behavior suitable for long-term data retention.

As an emerging NVM concept, FTJs offer several notable benefits, including ultralow power consumption, fast switching characteristics, and the feasibility of multilevel storage.[49,50] In comparison to conventional flash memory, FTJs exhibit significantly faster write operations and lower power requirements. For instance, FTJs fabricated with an $Ag/BaTiO_3/Nb{:}SrTiO_3$ structure have demonstrated write current densities as low as $4 \times 10^3\,A/cm^2$ and switching speeds as fast as 0.6 ns.[51] This current density achieved in FTJs is approximately two orders of magnitude lower than that of other NVMs such as magnetic tunnel junctions (MTJs), resistive switching memory (RSM), and phase-change random-access memory (PCRAM), while also maintaining superior speed. These attributes make FTJs compelling candidates for high-speed memory and real-time CIM applications.

Despite their promising performance characteristics, FTJs still face several critical challenges that hinder their commercial deployment. A primary concern lies in the use of traditional perovskite ferroelectric materials, such as $BaTiO_3$, which offer favorable electrical properties but suffer from poor compatibility with silicon-based technologies. These incompatibilities stem from lattice mismatch, thermal expansion differences, interfacial chemical instability, and integration difficulties. Additionally, FTJs based on perovskite materials often require relatively high operating voltages in the range of 8–18 V, which restricts their suitability for modern low-power electronic systems.[52–54] To overcome these drawbacks, HfO_2-based ferroelectric materials have gained attention due to their excellent CMOS compatibility and relatively low processing temperatures. HfO_2-based FTJ devices can operate at significantly reduced voltages, and their ferroelectric behavior can be finely tuned through controlled doping and precise film thickness engineering. These optimizations enable improved device performance, with reported ON/OFF current ratios ranging from 10^2 to as high as 10^6, and write voltages reduced to approximately 3 V, greatly enhancing their practical appeal.[55]

Nonetheless, HfO_2-based FTJs are not without their own challenges. While they benefit from lower operating voltages, their switching speeds still lag behind volatile memory technologies such as SRAM, which are typically used in high-speed cache memory (e.g., level 1 [L1] and level 2 [L2] caches). Current HfO_2-based FTJ devices generally achieve switching speeds in the range of ~20 ns,[56] falling short of the sub-ns response times required for integration into high-performance processors or real-time computing environments. To overcome this limitation, ongoing studies aim to explore new material systems and device configurations that enable faster polarization switching and shorter response times to further improve FTJ performance.

Furthermore, while reducing the ferroelectric layer thickness can boost tunneling current and enhance ON-state conductance, it often compromises the polarization strength, thereby lowering the TER ratio and degrading MW.[57] Additionally, repeated switching cycles may introduce defect states and interface traps due to the ferroelectric–electrode interactions, which can interfere with charge transport and reduce device reliability over time.[58] These considerations emphasize the importance of continuous innovation in materials engineering, interface control, and structural design to fully unlock the potential of HfO_2-based FTJs for scalable, high-performance, NVM applications.

The application scope of FTJs extends beyond conventional NVM, with growing interest in their potential for neuromorphic computing. Neuromorphic architectures aim to emulate the brain's information processing mechanisms, in which signals are transmitted between neurons through synapses, functionally analogous to weight modulation in artificial neural networks (ANNs). In this context, FTJs serve as promising candidates for memristive synaptic devices, capable of emulating the dynamic plasticity of biological synapses.[59]

By modulating their resistance states through controlled polarization switching, FTJs can replicate key synaptic behaviors, enabling local weight storage and update operations directly within hardware. This capability opens a new avenue for implementing energy-efficient and highly parallel neural networks, essential for advancing edge AI and neuromorphic computing platforms.

While challenges remain, particularly in terms of material stability, switching speed, and endurance, the continued refinement of silicon-compatible ferroelectric materials, along with innovations in device architecture and circuit integration, positions FTJs as promising building blocks for future computing paradigms. Continued research into the physical mechanisms, material optimization, and system-level integration of these devices will be vital for realizing their full potential in next-generation storage-class memory and brain-inspired architectures.

5.3.2 Enhancing Performance and Reliability in HfO_2-based FTJs

Carrier modulation using semiconductor electrodes has shown promise in significantly enhancing the contrast in interface screening lengths, thereby boosting the resistance switching ratio beyond 100%. However, these semiconductor electrodes often introduce challenges such as parasitic interfacial layers and depolarization fields, which can compromise device stability and reliability. To mitigate these effects, researchers have explored symmetric energy-band engineering to stabilize the ferroelectric phase in HZO films. Specifically, introducing a ZrO_2 seed layer in combination with an Al_2O_3 barrier layer has been shown to enhance the switching ratio and suppress depolarization fields,[60] providing an effective means toward the realization of high-reliability FTJs.

Endurance remains a key limitation for ferroelectric devices, often degraded by internal bias fields generated through oxygen vacancy migration, which impedes domain switching. To further address these issues, La/Al co-doping in HfO_2 films has been identified as an effective approach to limit oxygen vacancy migration. This strategy increases P_r to about 30 $\mu C/cm^2$ and extends fatigue endurance up to 10^7 cycles, even at elevated temperatures up to 300 °C.[61] In addition, NH_3 plasma treatment at the TiN–HZO interface can reduce interfacial defect states and alleviate imprint phenomena.[62] The introduction of HfO_2–ZrO_2 superlattice structures further contributes to defect suppression by creating periodic potential barriers, thereby improving fatigue recovery by a factor of three and substantially extending device lifetime.[63]

Beyond binary switching, FTJs also support multilevel and polymorphic data storage. Multiferroic tunneling junctions (MFTJs), which leverage magnetoelectric coupling in heterostructures such as LSMO/$BaTiO_3$/LSMO, enable polymorphic storage capabilities.[64] In contrast, HfO_2-based FTJs can achieve up to 10 distinct resistance states within a single device by precisely controlling the amplitude and duration of applied voltage pulses. For ultrahigh-density applications, self-rectifying FTJ crossbar arrays have been proposed to reduce peripheral circuitry, minimize power consumption, and shrink cell size, hence facilitating dense integration.

To address the speed limitations in FTJ switching, La/Ta co-doping has been employed to finely control the oxygen vacancy concentration, thereby reducing the polarization switching time to 78 ns.[65] Moreover, combining ultra-scaled electrodes with a layered interface design has enabled record-fast switching speed of 780 ps under a 2 V bias, which approaches the fundamental speed limits of electronic switching.[66]

5.3.3 FTJ-enabled Neuromorphic Computing

The rapid growth of AI and the Internet of Things (IoT) has imposed increasingly stringent requirements on the performance and efficiency of image recognition hardware systems. In such applications, ranging from smartphones and drones to surveillance systems, low power consumption is a fundamental requirement.[67] Simultaneously, computational efficiency and rapid response are crucial, as tasks like image classification, video stream analysis, and object tracking require real-time processing capabilities. This is especially critical in latency-sensitive scenarios such as autonomous driving and medical diagnostics, where delays can lead to severe consequences.[68] To meet these requirements, HfO_2-based ferroelectric devices have emerged as highly promising candidates for next-generation, low-power, and high-speed image recognition architectures. These devices offer unique advantages, including fast switching, excellent CMOS compatibility, and scalability down to nanometer thicknesses, making them suitable for integration into neuromorphic and AI accelerator systems.

Recent research has focused on AFE HfO_2-based thin films, the materials that not only are compatible with standard CMOS fabrication but also exhibit distinct properties favorable for neuromorphic computing. As shown in Figure 5.8, researchers

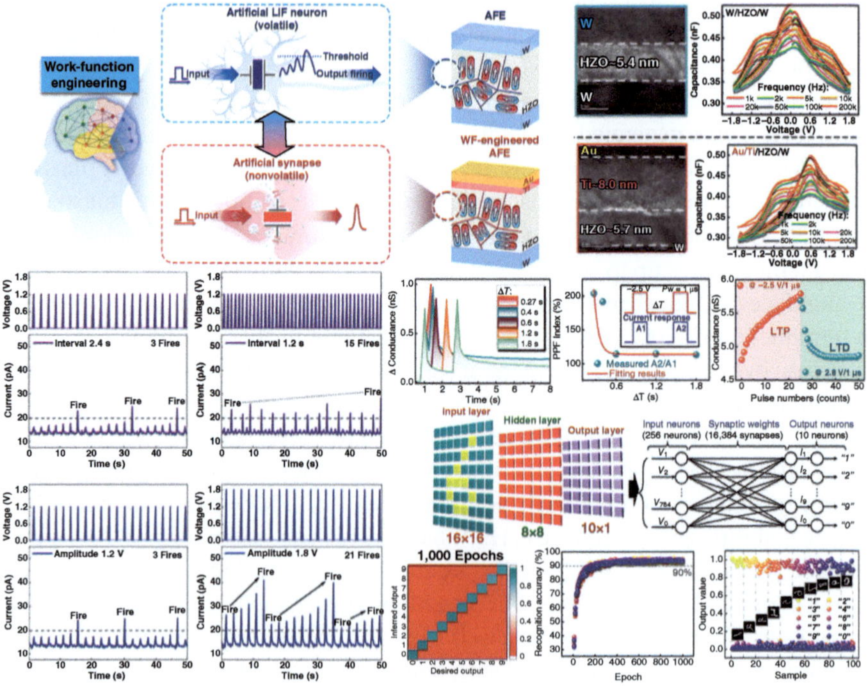

Figure 5.8 Reconfigurable HZO ferroelectric/AFE FTJ devices tailored for neuromorphic computing applications. *Source*: [69] / with permission of American Chemical Society.

have developed a novel artificial synapse/neuron device based on these AFE materials.[69] In configurations with symmetric electrodes, the device exhibits LIF behavior under pulsed stimuli, achieving remarkably low energy consumption of just 1.36 fJ/spike. By engineering asymmetric electrodes to introduce an internal bias field, researchers were able to shift the polarization hysteresis and realize two clearly distinguishable resistance states, enabling memory functionality. These devices successfully emulate key synaptic behaviors such as paired pulse facilitation (PPF) and long-term potentiation/depression (LTP/LTD), with energy consumption as low as 245 aJ/spike. Furthermore, ANNs constructed with these devices demonstrated effective handwritten digit recognition, achieving recognition rates exceeding 90%.

Importantly, AFE devices exhibit lower polarization currents and faster polarization dynamics compared to conventional ferroelectric counterparts, making them especially attractive for high-speed, low-power neuromorphic systems. These attributes highlight their potential to serve as fundamental building blocks for scalable neuromorphic computing architectures tailored to the growing demands of AI and edge computing applications.

5.4 DTCO for HfO_2-based Ferroelectric Devices

Ferroelectric memory technologies, including FeCAPs, FeFETs, and FTJs, offer significant advantages such as nonvolatility, multilevel storage capability, and extremely low energy consumption. These attributes position them as promising candidates for next-generation memory and computing systems. However, despite these benefits, ferroelectric devices still face significant challenges that hinder large-scale commercialization. In particular, process variation, D2D variability, and long-term reliability remain major concerns, especially in scenarios requiring massive integration and robustness under real-world operating conditions.

Compared to mature technologies like CMOS-based SRAM and DRAM, emerging ferroelectric devices currently lag in process maturity, yield stability, and ecosystem readiness. Nevertheless, cross-layer co-design methodologies that span device, circuit, and system levels have emerged as powerful strategies to bridge this gap. By holistically addressing the limitations across layers of abstraction, co-design approaches open up new design opportunities that can help compensate for device-level nonidealities.

From the architectural and circuit design perspectives, cross-layer co-optimization enables more flexible trade-offs between memory density, access latency, reliability, and energy efficiency. This expanded design space makes it possible to tailor ferroelectric memory solutions to diverse application requirements, accelerating their deployment in future computing systems where energy-efficient NVM is a key enabler.

5.4.1 Device–Technology Cross-layer Co-design for Ferroelectric Memory

At the circuit level, optimization efforts primarily target the memory array design, where cell architecture plays a critical role in determining the overall area efficiency, access latency, and energy consumption of ferroelectric memories. Multiple studies have adopted circuit–device co-design approaches to fully exploit the potential of ferroelectric memory technologies while addressing reliability-related challenges.[70–76]

For instance, a three-transistor (3T) symmetric FeFET memory cell has been proposed to support both row-wise and column-wise access operations (Figure 5.9a).[73] This design utilizes both the gate and the drain/source terminals of FeFET for writing, thereby enabling multidimensional access. With only one access NMOS per axis, the cell footprint is significantly reduced. The resulting compact and symmetric structure is particularly well suited for applications with multidimensional memory access patterns, such as matrix-based workloads in neural networks. It also opens new avenues for memory architecture research, as discussed in[74,75]. To enhance device reliability,

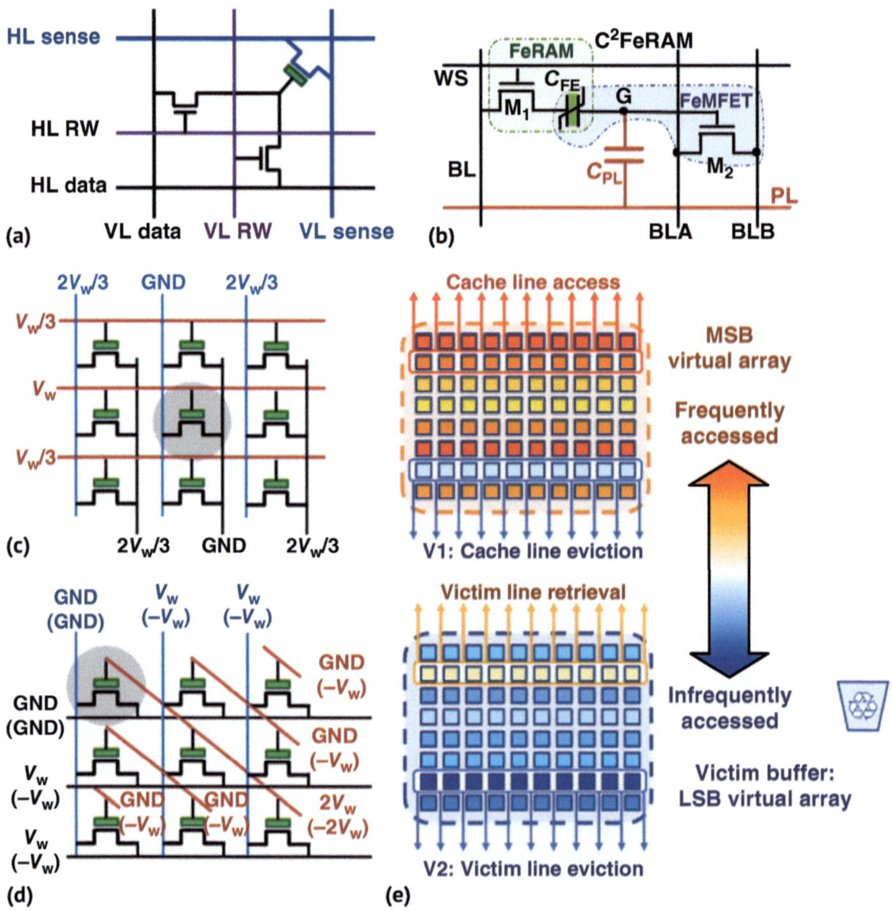

Figure 5.9 Co-design of ferroelectric memory across device, circuit, and architecture levels: (a) 3T compact symmetric cell and (b) reliable 2T2C cell. Panel (a): *Source*: [73] / with permission of ACM. Panel (b): *Source*: [76] / with permission of IEEE. (c) 1FeFET NAND cell with low disturbance. *Source*: [77] / with permission of IEEE. (d) A 1T NOR cell featuring low write disturbance. *Source*: [78] / with permission of IEEE. (e) MLC victim cache with low latency and high density. *Source*: [79] / with permission of IEEE.

particularly in endurance and read stability, a C^2FeRAM cell architecture has been introduced.[76] This design employs a 2T–2FeCAP configuration, where the read and write operations are decoupled: readout follows a FeFET-style approach, while writing uses a FeRAM-like mechanism (Figure 5.9b). This hybrid read–write strategy enables high write endurance, on par with traditional FeRAM, and minimizes read disturbance, addressing one of the key challenges in FeFET-based arrays. These innovations demonstrate how thoughtful circuit-level design, particularly when combined with a deep understanding of ferroelectric device behavior, can significantly improve memory performance, robustness, and integration efficiency.

Beyond individual cell innovations, array-level read/write schemes play a decisive role in defining the access characteristics and overall reliability of ferroelectric memory systems. To this end, several innovative architectures and adaptive techniques have been proposed to mitigate cell disturbances, optimize energy efficiency, and adapt to varying operating conditions.

For instance, a dense 1T FeFET NAND cell array was proposed (Figure 5.9c),[77] which eliminates the need for access transistors traditionally used for intercell isolation. Instead, the design suppresses disturbances to neighboring cells by carefully tuning the write voltage bias. This approach significantly improves area efficiency while maintaining acceptable write fidelity. Similarly, a 1T FeFET diagonal NOR array architecture has been introduced in[78] (Figure 5.9d), offering a compact layout with efficient interconnect utilization. This design is accompanied by a low-energy write scheme, enabled through circuit–device co-design that incorporates an accurate model of parasitic voltage drops within the array. This ensures reliable voltage delivery to each cell, further enhancing energy efficiency.

In addition to structural innovations, adaptive access schemes have been proposed to address the variability and sensitivity inherent in ferroelectric devices. For example,[80] proposed an adaptive read/write designed for MLC FeRAM, which dynamically adjusts sensing parameters to enhance the sensing margin and minimize access latency. Similarly,[1] presented a temperature-aware peripheral circuit that adaptively tunes the write voltage according to ambient temperature conditions. This technique significantly improves write endurance and reduces the bit error rate under varying thermal environments. Collectively, these circuit-level innovations underscore the importance of array-level co-optimization in improving performance, reliability, and energy efficiency in ferroelectric memory technologies. Such advancements pave the way for the practical deployment of ferroelectric memory technologies in future computing architectures.

From an architecture-level standpoint, cross-layer co-design must effectively leverage the unique physical characteristics of ferroelectric devices while exposing design trade-offs and performance constraints to lower abstraction layers, including device and circuit domains. Such an approach ensures that system-level goals are met through coordinated optimization across the entire design hierarchy.

In this context,[81] proposes a unified modeling framework for FeFET-based on-chip caches, specifically targeting 1FeFET AND-type array architectures. This model captures key behaviors such as read/write disturbance and P/E endurance limitations, starting from the device level and extending through full-array interactions. By enabling workload-driven evaluation of FeFET cache architectures, the model serves as a crucial bridge between device physics and architectural performance. It provides valuable insights for system designers, allowing them to identify reliability bottlenecks and guide optimization strategies through cross-layer feedback. In another example of architecture-guided co-design,[82] introduces an innovative MLC victim cache architecture that exploits the multilevel storage capability of ferroelectric and other emerging NVM technologies. This architecture capitalizes on the nonuniform access latency characteristics of MLC memory to improve cache efficiency. As illustrated in Figure 5.9e, only the most significant bits (MSBs), which can be accessed more quickly, are used for the primary cache data array, while the least significant bits (LSBs) are relegated to a secondary "victim" buffer.

This hierarchical access strategy ensures that latency-critical requests are served from fast MSBs, delivering response times comparable to single-level-cell (SLC) memory. At the same time, the MLC configuration reduces overall memory footprint, balancing performance with area efficiency. By tightly coupling architectural innovation with underlying device capabilities, this work highlights how cross-layer design can effectively unlock the full potential of ferroelectric memories in high-performance computing systems.

5.4.2 DTCO for FeFET-based CIM Accelerators

The intrinsic properties of FeFETs, such as their three-terminal configuration, high CMOS compatibility, and large ON/OFF current ratio, render them highly attractive for CIM architectures. In particular, FeFETs hold strong potential in analog CIM designs, where in-memory processing reduces data movement and significantly enhances computational parallelism.

As illustrated in Figure 5.10a, analog CIM architectures typically activate multiple memory rows simultaneously, allowing the output signals, which are either current or voltage, from the selected cells to be *in-situ* accumulated and subsequently digitized through ADCs. This in-memory accumulation eliminates the time and energy overheads caused by frequent data transfer between separate memory and processing units, while enabling massively parallel operations for neural network inference and matrix–vector multiplication (MVM). The transistor-like configuration of FeFETs is particularly well suited to implement AND/NAND-type memory arrays, which further improves memory density while preserving access flexibility. Additionally, the large ON/OFF current ratio characteristic of FeFETs contributes to enhanced computational fidelity by improving the distinguishability between logic states – an essential requirement for robust and accurate analog computations.

The first FeFET-based CIM architecture for MVM was reported in 2017, using the fundamental cell structure shown in Figure 5.10a.[30] Since then, extensive studies have explored how FeFET properties can be harnessed to enhance CIM efficiency across multiple performance dimensions. For example,[82] demonstrated an optimization of the resistive component in a 1T1R cell to achieve an exceptional energy

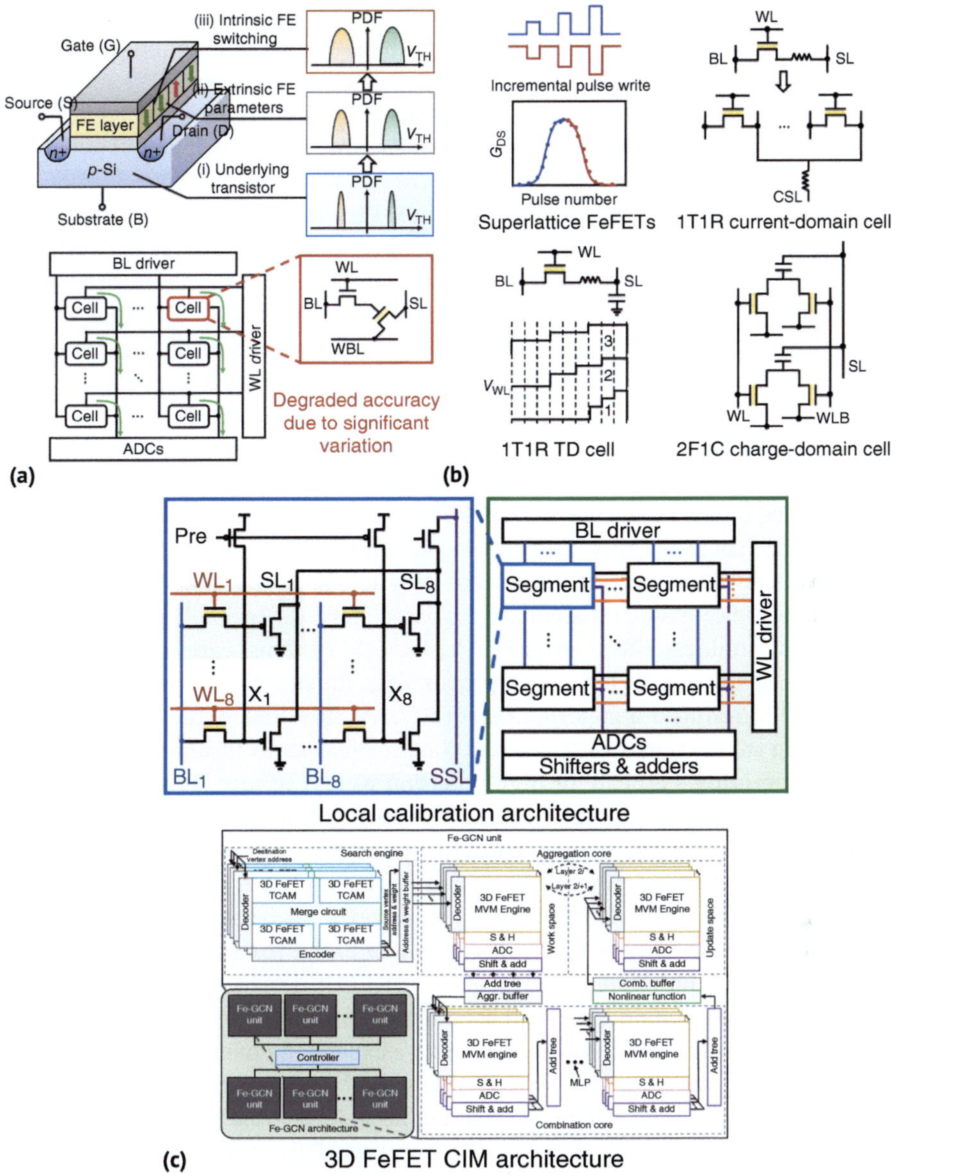

Figure 5.10 FeFET-based CIM architectures. (a) Key computation accuracy challenge caused by FeFET variability. *Source*: [83] / with permission of IEEE. (b) Existing device- and cell-level design solutions. *Source*: Refs. [30, 79, 84, 85]. (c) Current architecture-level optimization strategies. *Source*: Refs. [86, 87].

efficiency of 13,700 TOPS/W, setting a benchmark for low-power neuromorphic computing.

Building upon the excellent CMOS compatibility of FeFETs, researchers in[88] proposed a monolithic 3D integration approach, vertically stacking FeFETs to increase functional density without expanding the footprint. Similarly,[79] introduced the co-integration of FeFETs with RRAM to harness the strengths of both memory technologies, improving overall density and performance. More recently,[86] advanced this concept by implementing a three-dimensional FeFET NAND array, which significantly increases memory capacity and enables large-scale graph neural network (GNN) computation, as demonstrated in Figure 5.10c.

Despite these advancements, nonidealities in FeFET devices continue to pose challenges for reliable analog CIM operations. As depicted in Figure 5.10a, FeFETs exhibit significant D2D variability,[89] which can accumulate across rows and columns in the array, leading to substantial degradation in computation accuracy. This variability, if unaddressed, undermines the precision required for tasks such as neural inference or learning. Furthermore, MLC FeFETs are particularly susceptible to nonlinear and uneven state distributions, which complicate accurate weight mapping and retrieval. Therefore, a key challenge in FeFET-based CIM design lies in mitigating these nonideal effects, which include variation, nonlinearity, and limited endurance, while preserving the inherent advantages of FeFETs such as low power, high density, and three-terminal control. Cross-layer techniques encompassing device engineering, circuit-level compensation, and algorithm-level tolerance are essential to enable robust and high-accuracy CIM architectures based on FeFET technology.

To address nonideal effects in FeFET-based CIM systems, various device- and cell-level design strategies have been proposed, as illustrated in Figure 5.10b. For instance,[84] introduced a superlattice FeFET structure that improves switching linearity and supports up to 128 discrete conductance states through incremental pulse programming – crucial for high-precision weight storage. To suppress D2D variation,[90] proposed adding a current-limiting transistor in series with the FeFET, forming a 1T1R structure that narrows current distribution among cells. This idea was further optimized in[91] by sharing the limiting transistor across multiple FeFETs to enhance integration density. However, such a configuration inherently limits the feasibility of MLC operation due to the fixed current constraint. To overcome this limitation,[92] presents a TD 1T1R design that avoids analog current sensing altogether. Instead, it accumulates switching-induced charge over time and converts the result into a voltage on a capacitor, enabling accurate MLC readout without sacrificing density. Another work[85] proposed a two-FeFET one-capacitor (2F1C) charge-domain cell that employs differential storage and charge coupling to mitigate the impact of device variation. This concept has since been integrated into several FeFET-based CIM architectures to improve computational reliability[93,94]

Nevertheless, FeFET device innovations still face fabrication-related challenges and process variations, while cell-level circuit architectures must balance performance, density, and precision. To resolve these trade-offs, recent research[87,95] has proposed a calibrated shared local architecture, as illustrated in Figure 5.10c. In this design, MLC FeFETs are accessed within locally grouped blocks, where the analog outputs are calibrated on-chip before being converted into CMOS-compatible current signals for accumulation. This architecture not only enhances computational accuracy but also significantly reduces peripheral cost, as multiple memory cells share the same calibration and readout circuitry.

Through coordinated innovations at the device, cell, and architecture levels, FeFET-based CIM platforms are gradually meeting the requirements for high-density, energy-efficient, and reliable CIM, thereby opening new opportunities in edge AI and neuromorphic applications.

5.4.3 FeFET-enabled Hardware Security for IoT Edge Devices

In the IoT era, the widespread use of intelligent sensors and edge AI systems in domains such as smart homes, urban infrastructure, and healthcare has dramatically increased the volume of data exchanged between edge devices and cloud servers. These systems often handle sensitive personal or proprietary data, making them appealing targets for cyberattacks.[83] As illustrated in Figure 5.11a, vulnerabilities

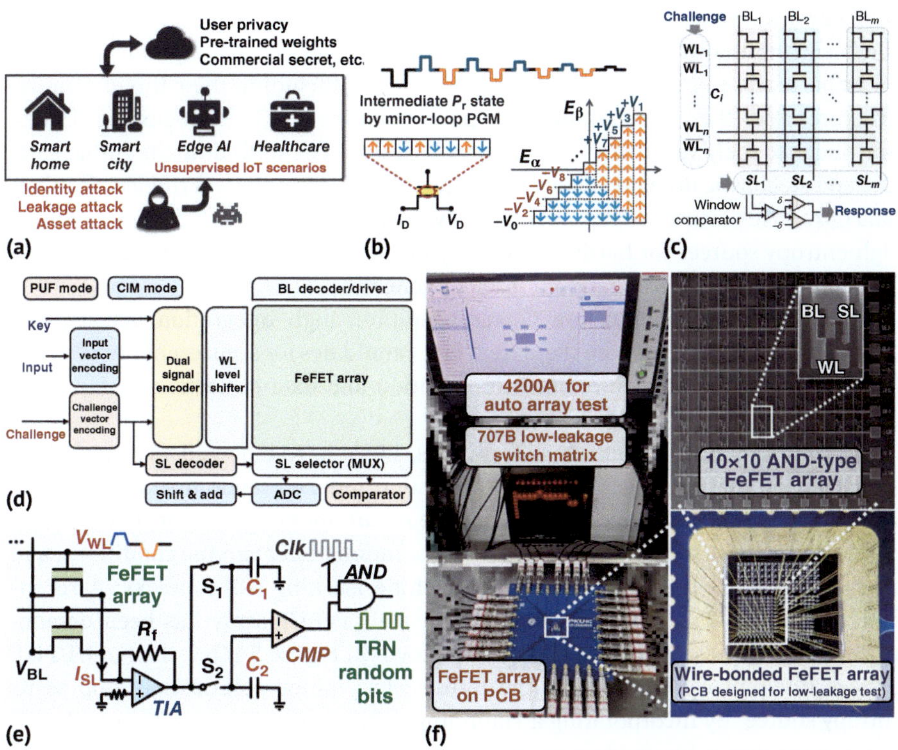

Figure 5.11 (a) Overview of multilevel security threats from edge to cloud. FeFET array-based strong PUF optimized for both reliability and security: (b) entropy source derived from random ferroelectric minor-loop fluctuations and (c) 1T AND-type FeFET array implementing a high-security, reliable, and strong PUF. (d) Macro-design of secure FeFET-based IMCE architecture. HE client integrating a multifunctional sampler and high-speed multiplier. *Source*: [96] / with permission of IEEE. (e) High-throughput FeFET TRNG utilizing fast gate perturbation and (f) VMV-NTT testing platform with wire-bonded 10×10 FeFET arrays. *Source*: [97] / with permission of IEEE.

may appear during data collection, transmission, and processing, creating serious risks of data leakage and system compromise.

Various forms of attacks can jeopardize cloud–edge ecosystems. Identity attacks involve impersonation, where adversaries mimic legitimate users to gain unauthorized access to cloud-stored data. Asset attacks focus on extracting information directly from edge memory via reverse engineering or brute-force techniques, targeting critical resources such as sensor outputs, communication parameters, or pretrained AI model weights. Leakage attacks exploit malware or intercept cloud–edge communication to siphon sensitive data, by either compromising cloud systems or eavesdropping on transmission channels.[98]

Given these challenges, ensuring robust data protection on resource-constrained IoT devices, where area and power budgets are limited, is of paramount importance. In this context, FeFET technology offers a promising solution. Its compact 1T AND-type memory arrays support CIM at low cost and with high efficiency, making it well suited for deployment in edge security modules. Compared to RRAM, FeFETs exhibit significantly lower write power and a higher ON/OFF current ratio, which contributes to substantial reductions in total energy consumption.[99]

Furthermore, the intrinsic properties of FeFETs, including their three-terminal structure and the incorporation of polycrystalline HfO_2 ferroelectric layers, contribute to their value in secure hardware design. The multidomain grain structure and stochastic domain switching behavior inherent in these ferroelectric materials introduce natural variations across devices and cycles, which can also serve as high-entropy sources for hardware security primitives such as physical unclonable functions (PUFs) and true random number generators (TRNGs).

Overall, FeFETs combine low-power operation, high integration density, and intrinsic stochasticity, making them excellent candidates for secure, energy-efficient IoT edge systems that require both computation and data protection within strict hardware constraints.

To counter identity attacks in IoT systems, PUFs have emerged as a hardware-level solution that leverages inherent fabrication-induced randomness to generate unique, device-specific fingerprints. The core requirements for an effective PUF are reliability, ensuring consistent fingerprint generation under varying conditions, and security, particularly resilience against machine learning (ML) attacks. A reconfigurable strong PUF architecture based on a 1T FeFET array has been demonstrated to meet these requirements.[100] As shown in Figure 5.11b,c, the FeFET PUF utilizes inherent stochasticity of the ferroelectric minor polarization loop as its entropy source. By incorporating a Hamming-distance-based comparator and nonlinear activation via a window comparator, this design enhances resistance to ML attacks while maintaining a low authentication error rate.

To protect against asset attacks, where local memory modules are targeted to extract sensitive data, XOR-based encryption strategies can be utilized. A complementary FeFET cell design, illustrated in Figure 5.11d, enables in-memory computing and encryption (IMCE) with strong data security.[96] This architecture combines memory and encryption directly in hardware, eliminating the traditional read–decrypt–compute cycle, which not only reduces latency but also significantly improves computational throughput and overall energy efficiency. In scenarios involving leakage attacks, where malicious agents attempt to extract data through malware or side-channel analysis, homomorphic encryption (HE) provides robust protection by allowing computations to occur directly on encrypted data without exposing the plaintext information. FeFET devices are particularly effective in reducing the computational overhead of HE at the edge. These architectures accelerate essential cryptographic operations such as polynomial random sampling and coefficient multiplication during both encryption and decryption processes.[97] As shown in Figure 5.11e,f, a fast gate-perturbation technique is used to modulate FeFETs in intermediate threshold voltage states, enabling the generation of high-throughput true random bitstreams, which serve as high-quality seeds for encryption protocols.

Moreover, hybrid-domain FeFET arrays have been developed to efficiently support ternary product computations required in vector–matrix–vector (VMV) multiplication, a critical operation in the Number Theoretic Transform (NTT) used for fast HE processing. These arrays enable low-latency, high-throughput multiplication, making FeFET a strong platform for secure and efficient data processing in future encrypted AI workloads and edge computing.

5.4.4 Ultralow-Power Computing Applications Enabled by FeFETs

CAM has become an important component for data-centric computing tasks such as ML, pattern recognition, and big data analytics. However, conventional SRAM-based CAM architectures suffer from limitations in scalability, static power consumption, and area.

To overcome these challenges, a novel multibit CAM (MCAM) design has been proposed and experimentally demonstrated using a FeFET-based NAND array architecture. By exploiting the MLC capability of FeFETs and applying complementary search line (SL) inputs, the MCAM cell achieves 2-bit-per-cell storage and search functionality. This significantly improves bit density, with the potential for further scalability through the stabilization of additional ferroelectric states in FeFET devices. Moreover, when MCAM cells are integrated into a three-dimensional vertical FeFET NAND structure, the system achieves extremely high density and superior energy efficiency at minimal cost, making it highly suitable for future large-scale data-intensive workloads.

Parallel search operations have been experimentally demonstrated on a 4×4 FeFET NAND array, supporting 4-bit query processing across an entire NAND block simultaneously. This highlights the architecture's high-throughput search capability and scalability, both of which are critical for data-centric systems.

Building upon this foundation, a novel data processing paradigm termed the Matching Univalue Segment Assimilating Nucleus (MUSAN) is proposed, which operates without convolution operations (Figure 5.12). The MUSAN method is designed to activate power consumption only during feature matching, substantially lowering energy usage. Match and mismatch results are distinguished by detecting high or low current levels, thereby eliminating the need for energy-intensive ADCs.

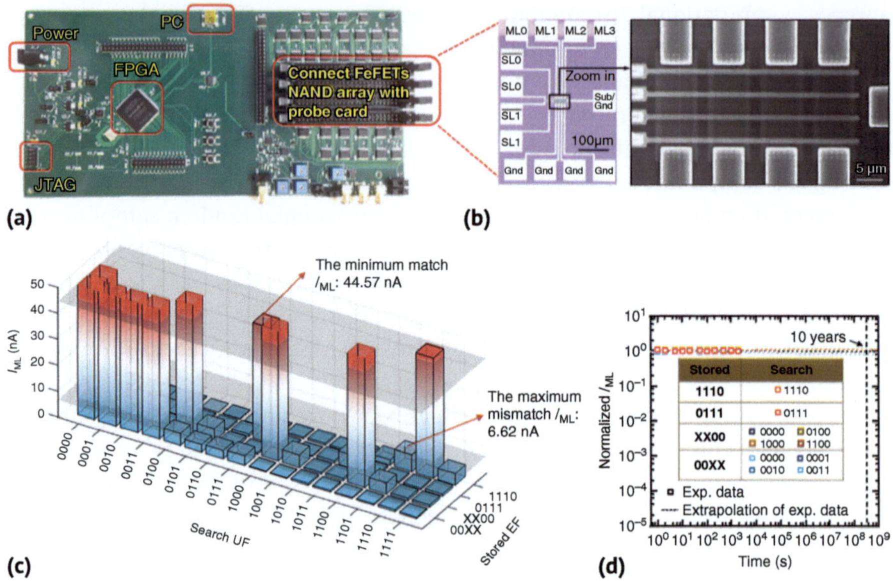

Figure 5.12 (a) Photograph of the MUSAN hardware PCB. (b) Microscopic image of a 4 × 4 FeFET NAND array. (c) Measured IML results for 16 search UF types across four stored EF types. (d) Retention behavior of the four stored EF states. *Source*: [101] / with permission of Springer Nature.

This not only simplifies the peripheral circuitry but also enhances speed and energy efficiency.

Furthermore, the MUSAN method exhibits strong resilience against performance degradation caused by device variability, making it particularly compatible with emerging NVM technologies. Hardware implementations of the FeFET-based MCAM system confirm its ultralow power consumption and high performance. Specifically, the proposed FeFET MCAM-based MUSAN edge detection architecture performs image edge extraction with exceptionally low energy consumption (~10 fJ per operation), operates without ADCs, and maintains zero accuracy loss. These advantages make it an outstanding solution for low-power edge computing scenarios, where energy, cost, and form factor are strictly constrained.[101]

Acknowledgments

The authors acknowledge support from the National Key Research and Development Project (Grant No. 2023YFB4402303), the Fundamental Research Funds for the Central Universities (Grant Nos. ZYTS25038 and YJSJ25013), the National Natural Science Foundation of China (Grant Nos. 62025402, 62374151, 62204226, 62274003, 61927901, 92164203, 92464205, 62474164, 92477135, 62174054, U21B2030, and

U24B6015), Major Program of Zhejiang Natural Science Foundation (Grant No. DT23F0402), and the 111 Project (Grant No. B18001).

References

1 Yang, J., Luo, Q., Xue, X. et al. (2023). A 9Mb HZO-based embedded FeRAM with 10^{12}-cycle endurance and 5/7ns read/write using ECC-assisted data refresh and offset-canceled sense amplifier. In: *Proceedings of the 2023 IEEE International Solid-State Circuits Conference (ISSCC).*

2 Ramaswamy, N., Calderoni, A., Zahurak, J. et al. (2023). NVDRAM: A 32Gb dual layer 3D stacked non-volatile ferroelectric memory with near-DRAM performance for demanding AI workloads. In: *Proceedings of the 2023 International Electron Devices Meeting (IEDM).*

3 Okuno, J., Kunihiro, T., Yonai, T. et al. (2023). A highly reliable 1.8 V 1 Mb $Hf_{0.5}Zr_{0.5}O_2$-based 1T1C FeRAM array with 3-D capacitors. In: *Proceedings of the 2023 International Electron Devices Meeting (IEDM).*

4 Francois, T., Coignus, J., Makosiej, A. et al. (2022). High-performance operation and solder reflow compatibility in BEOL-integrated 16-kb HfO_2: Si-based 1T-1C FeRAM arrays. *IEEE Transactions on Electron Devices* 69 (4): 2108–2114.

5 Chang, S., Neumann, C., Alpizar, B. et al. (2024). Reliable low-voltage FeRAM capacitors for high-speed dense embedded memory in advanced CMOS. In: *Proceedings of the 2024 IEEE Symposium on VLSI Technology and Circuits.*

6 Fu, Z., Wang, K., Xu, S. et al. (2023). Novel asymmetric operation scheme for HfO_2-based FeRAM based on reconstruction of ferroelectric dynamics impacts. *IEEE Electron Device Letters* 45: 20–23.

7 Okuno, J., Kunihiro, T., Konishi, K. et al. (2021). High-endurance and low-voltage operation of 1T1C FeRAM arrays for nonvolatile memory application. In: *Proceedings of the 2021 IEEE International Memory Workshop (IMW).*

8 Hur, J., Luo, Y., Tasneem, N. et al. (2021). Ferroelectric hafnium zirconium oxide compatible with back-end-of-line process. *IEEE Transactions on Electron Devices* 68: 3176–3180.

9 Huang, T., Li, Y., Chen, C. et al. (2023). Demonstration of robust breakdown reliability and enhanced endurance in gallium doped HfO_2 ferroelectric thin films. *IEEE Electron Device Letters* 44: 1476–1479.

10 Gong, T., Tao, L., Li, J. et al. (2021). $10^5\times$ Endurance improvement of FE-HZO by an innovative rejuvenation method for 1z node NV-DRAM applications. In: *Proceedings of the 2021 IEEE Symposium on VLSI Technology.*

11 Tahara, K., Toprasertpong, K., Hikosaka, Y. et al. (2021). Strategy toward HZO BEOL-FeRAM with low-voltage operation ($\leq$1.2 V), low process temperature, and high endurance by thickness scaling. In: *Proceedings of the 2021 IEEE Symposium on VLSI Technology.*

12 Toprasertpong, K., Tahara, K., Hikosaka, Y. et al. (2022). Low operating voltage, improved breakdown tolerance, and high endurance in $Hf_{0.5}Zr_{0.5}O_2$ ferroelectric

capacitors achieved by thickness scaling down to 4 nm for embedded ferroelectric memory. *ACS Applied Materials & Interfaces* 14: 51137–51148.

13 Guido, R., Wang, X., Xu, B. et al. (2024). Ferroelectric $Al_{0.85}Sc_{0.15}N$ and $Hf_{0.5}Zr_{0.5}O_2$ domain switching dynamics. *ACS Applied Materials & Interfaces* 16: 42415–42425.

14 Liao, A., Jerry, K., Hollander, M. et al. (2024). $4F^2$ stackable polysilicon channel access device for ultra-dense NVDRAM. In: *Proceedings of the 2024 IEEE Symposium on VLSI Technology and Circuits.*

15 Hwang, J., Lim, S., Kim, G. et al. (2022). Non-volatile majority function logic using ferroelectric memory for logic in memory technology. *IEEE Electron Device Letters* 43: 1049–1052.

16 Hsieh, E., Chang, J., Tang, T. et al. (2022). NVDimm-FE: A high-density 3D architecture of 3-bit/c 2TnCFE to break great memory wall with 10 ns of PGM-pulse, 10^{10} cycles of endurance, and decade lifetime at 103 °C. In: *Proceedings of the 2022 IEEE Symposium on VLSI Technology and Circuits.*

17 Chen, W. (2019). Selector-free cross-point memory architecture based on ferroelectric MFM capacitors. In: *Proceedings of the 2019 IEEE 11th International Memory Workshop (IMW).*

18 Fu, Z., Cao, S., Zheng, H. et al. (2023). First demonstration of hafnia-based selector-free FeRAM with high disturb immunity through design technology co-optimization. In: *Proceedings of the 2023 International Electron Devices Meeting (IEDM).*

19 Fu, Z., Cao, S., Zheng, H. et al. (2024). Hafnia-based high-disturbance-immune and selector-free cross-point FeRAM. *IEEE Transactions on Electron Devices* 71: 3358–3364.

20 Xu, W., Fu, Z., Wang, K. et al. (2023). A novel small-signal ferroelectric capacitance-based content addressable memory for area-and energy-efficient lifelong learning. *IEEE Electron Device Letters* 45: 24–27.

21 Luo, J., Song, B., Lin, Y. et al. (2024). Experimental demonstration of resonant adiabatic writing and computing in ferroelectric capacitive memory array for energy-efficient edge AI. In: *Proceedings of the 2024 International Electron Devices Meeting (IEDM).*

22 Xu, W., Fu, Z., Wang, K. et al. (2024). A novel small-signal ferroelectric memcapacitor based capacitive computing-in-memory for area-and energy-efficient quantized neural networks. In: *Proceedings of the 2024 8th IEEE Electron Devices Technology & Manufacturing Conference (EDTM).*

23 Fu, Z., Wang, K., Fu, B. et al. (2022). Novel energy-efficient hafnia-based ferroelectric processing-in-sensor with in-situ motion detection and four-quarter multiplication. In: *Proceedings of the 2022 International Electron Devices Meeting (IEDM).*

24 Gong, N. and Ma, T. P. (2017). A study of endurance issues in HfO_2-based ferroelectric field effect transistors: Charge trapping and trap generation. *IEEE Electron Device Letters* 39: 15–18.

25 Zhou, Y., Liang, Z., Luo, W. et al. (2022). Ferroelectric and interlayer co-optimization with in-depth analysis for high endurance FeFET. In: *Proceedings of the 2022 International Electron Devices Meeting (IEDM).*

26 Zhou, Y., Huang, W., Zhu, R. et al. (2024). A reliable 2 bit MLC FeFET with high uniformity and 10^9 endurance by gate stack and write pulse co-optimization. In: *Proceedings of the 2024 IEEE European Solid-State Electronics Research Conference (ESSERC).*

27 Zhou, Y., Shao, H., Zhu, R. et al. (2023). Hybrid-FE-layer FeFET with high linearity and endurance towards on-chip CIM by array demonstration. *IEEE Electron Device Letters* 45: 276–279.

28 Garg, C., Chauhan, N., Deng, S. et al. (2021). Impact of random spatial fluctuation in non-uniform crystalline phases on the device variation of ferroelectric FET. *IEEE Electron Device Letters* 42: 1160–1163.

29 Mulaosmanovic, H., Ocker, J., Müller, S. et al. (2017). Novel ferroelectric FET based synapse for neuromorphic systems. In: *Proceedings of the 2017 IEEE Symposium on VLSI Technology.*

30 Jerry, M., Chen, P. Y., Zhang, J. et al. (2017). Ferroelectric FET analog synapse for acceleration of deep neural network training. In: *Proceedings of the 2017 International Electron Devices Meeting (IEDM).*

31 Li, Z., Wu, J., Mei, X. et al. (2022). A 3D vertical-channel ferroelectric/anti-ferroelectric FET with indium oxide. *IEEE Electron Device Letters* 43: 1227–1230.

32 Feng, Y., Zhang, D., Sun, C. et al. (2024). First demonstration of BEOL-compatible 3D vertical FeNOR. In: *Proceedings of the 2024 IEEE Symposium on VLSI Technology and Circuits.*

33 Sun, C., Han, K., Samanta, S. et al. (2021). First demonstration of BEOL-compatible ferroelectric TCAM featuring a-IGZO Fe-TFTs with large memory window of 2.9 V, scaled channel length of 40 nm, and high endurance of 10^8 cycles. In: *Proceedings of the 2021 IEEE Symposium on VLSI Technology.*

34 Dutta, S., Ye, H., Chakraborty, W. et al. (2020). Monolithic 3D integration of high endurance multi-bit ferroelectric FET for accelerating compute-in-memory. In: *Proceedings of the 2020 IEEE International Electron Devices Meeting (IEDM).*

35 Yoo, J., Hong, J., Kim, H. et al. (2023). High-pressure H_2O post-annealing for improving reliability of LTPS and a-IGZO thin-film transistors within a coplanar structure. *Materials Science in Semiconductor Processing* 157: 107299.

36 Liang, Z., Tang, K., Dong, J. et al. (2021). A novel high-endurance FeFET memory device based on ZrO_2 anti-ferroelectric and IGZO channel. In: *Proceedings of the 2021 International Electron Devices Meeting (IEDM).*

37 Zhu, R., Zhou, Y., Sun, C. et al. (2024). Improved memory density and endurance by a novel 1T3C FeFET for BEOL multi-level cell memory. In: *Proceedings of the 2024 8th IEEE Electron Devices Technology and Manufacturing Conference (EDTM).*

38 Lu, T., Zhao, X., Liu, H. et al. (2023). Optimal weight models for ferroelectric synapses toward neuromorphic computing. *IEEE Transactions on Electron Devices* 70: 2297–2303.

39 Luo, J., Xu, W., Du, Y. et al. (2021). Energy- and area-efficient Fe-FinFET-based time-domain mixed-signal computing in memory for edge machine learning. In: *Proceedings of the 2021 International Electron Devices Meeting (IEDM).*

40 Luo, J., Fu, B., Du, Y. et al. (2023). Energy-efficient ferroelectric-FET-based agent with memory trace for enhanced reinforcement learning. *IEEE Electron Device Letters* 45: 264–267.

41 Luo, J., Shao, H., Fu, B. et al. (2022). Novel ferroelectric tunnel FinFET based encryption-embedded computing-in-memory for secure AI with high area-and energy-efficiency. In: *Proceedings of the 2022 International Electron Devices Meeting (IEDM).*

42 Luo, J., Xu, W., Fu, B. et al. (2022). A novel ambipolar ferroelectric tunnel FinFET based content addressable memory with ultra-low hardware cost and high energy efficiency for machine learning. In: *Proceedings of the 2022 IEEE Symposium on VLSI Technology and Circuits.*

43 Chen, C., Yang, M., Liu, S. et al. (2019). Bio-inspired neurons based on novel leaky-FeFET with ultra-low hardware cost and advanced functionality for all-ferroelectric neural network. In: *Proceedings of the 2019 IEEE Symposium on VLSI Technology.*

44 Luo, J., Liu, T., Fu, Z. et al. (2019). Capacitor-less stochastic leaky-FeFET neuron of both excitatory and inhibitory connections for SNN with reduced hardware cost. In: *Proceedings of the 2019 International Electron Devices Meeting (IEDM).*

45 Luo, J., Liu, T., Fu, Z. et al. (2023). Ferroelectric FET based signed synapses of excitatory and inhibitory connection for stochastic spiking neural network based optimizer. In: *Proceedings of 2023 7th IEEE Electron Devices Technology & Manufacturing Conference (EDTM).*

46 Luo, J., Liu, T., Fu, Z. et al. (2021). A novel ferroelectric FET-based adaptively-stochastic neuron for stimulated-annealing based optimizer with ultra-low hardware cost. *IEEE Electron Device Letters* 43: 308–311.

47 Ali, F., Ali, T., Lehninger, D. et al. (2022). Fluorite-structured ferroelectric and antiferroelectric materials: A gateway of miniaturized electronic devices. *Advanced Functional Materials* 32: 2201737.

48 Setter, N., Damjanovic, D., Eng, L. et al. (2006). Ferroelectric thin films: Review of materials, properties, and applications. *Journal of Applied Physics* 100: 051606.

49 Garcia, V., Fusil, S., Bouzehouane, K. et al. (2009). Giant tunnel electroresistance for non-destructive readout of ferroelectric states. *Nature* 460: 81–84.

50 Hu, W. J., Wang, Z., Yu, W. et al. (2016). Optically controlled electroresistance and electrically controlled photovoltage in ferroelectric tunnel junctions. *Nature Communications* 7: 10808.

51 Ma, C., Luo, Z., Huang, W. et al. (2020). Sub-nanosecond memristor based on ferroelectric tunnel junction. *Nature Communications* 11: 1439.

52 Chanthbouala, A., Garcia, V., Cherifi, R. et al. (2012). A ferroelectric memristor. *Nature Materials* 11: 860–864.

53 Luo, Z., Wang, Z., Guan, Z. et al. (2022). High-precision and linear weight updates by subnanosecond pulses in ferroelectric tunnel junction for neuro-inspired computing. *Nature Communications* 13: 699.

54 Boyn, S., Douglas, A. M., Blouzon, C. et al. (2016). Tunnel electroresistance in $BiFeO_3$ junctions: Size does matter. *Applied Physics Letters* 109: 232902.

55 Lee, K., Byun, J., Park, K. et al. (2022). Giant tunneling electroresistance in epitaxial ferroelectric ultrathin films directly integrated on Si. *Applied Materials Today* 26: 101308.

56 Berdan, R., Marukame, T., Ota, K. et al. (2020). Low-power linear computation using nonlinear ferroelectric tunnel junction memristors. *Nature Electronics* 3: 259–266.

57 Ryu, H., Wu, H., Rao, F. et al. (2019). Ferroelectric tunneling junctions based on aluminum oxide/zirconium-doped hafnium oxide for neuromorphic computing. *Scientific Reports* 9: 20383.

58 Cheema, S., Shanker, N., Hsu, C. et al. (2022). One nanometer HfO_2-based ferroelectric tunnel junctions on silicon. *Advanced Electronic Materials* 8: 2100499.

59 Gränsmark, E. (2023). *Ru and RuO_2 as bottom electrodes for HZO based FTJ's.* Master's thesis, Lunds University.

60 Liu, Y., Cao, Y., Zhu, H. et al. (2021). $HfZrO_x$-based ferroelectric tunnel junction with crested symmetric band structure engineering. *IEEE Electron Device Letters* 42: 1311–1314.

61 Li, Z. H., Tang, S. Q., Wang, T. Y. et al. (2025). Effect of lanthanum-aluminum co-doping on structure of hafnium oxide ferroelectric crystals. *Advanced Science* 12: 2410765.

62 Yuan, P., Wang, B., Yang, Y. et al. (2022). Enhanced remnant polarization (30 $\mu C/cm^2$) and retention of ferroelectric $Hf_{0.5}Zr_{0.5}O_2$ by NH_3 plasma treatment. *IEEE Electron Device Letters* 43: 1045–1048.

63 Kang, M., Peng, Y., Xiao, W. et al. (2024). HfO_2-ZrO_2 ferroelectric capacitors with superlattice structure: Improving fatigue stability, fatigue recovery, and switching speed. *ACS Applied Materials & Interfaces* 16: 2954–2963.

64 Yin, Y., Huang, W., Liu, Y. et al. (2015). Octonary resistance states in $La_{0.7}Sr_{0.3}MnO_3/BaTiO_3/La_{0.7}Sr_{0.3}MnO_3$ multiferroic tunnel junctions. *Advanced Electronic Materials* 1: 1500183.

65 Zhou, C., Ma, L., Feng, Y. et al. (2024). Enhanced polarization switching characteristics of HfO_2 ultrathin films via acceptor-donor co-doping. *Nature Communications* 15: 2893.

66 Song, Y., Yu, J., Wang, Z. et al. (2024). Ultrafast switching of ferroelectric HfO_2-ZrO_2 under low voltage with layered structure. *IEEE Electron Device Letters* 46: 12–15.

67 Wong, H. S. and Salahuddin, S. (2015). Memory leads the way to better computing. *Nature Nanotechnology* 10: 191–194.

68 Merolla, P. A., Arthur, J. V., Alvarez-Icaza, R. et al. (2014). A million spiking-neuron integrated circuit with a scalable communication network and interface. *Science* 345: 668–673.

69 Xu, K., Wang, T., Lu, C. et al. (2024). Novel two-terminal synapse/neuron based on an antiferroelectric hafnium zirconium oxide device for neuromorphic computing. *Nano Letters* 24: 11170–11178.

70 George, S., Ma, K., Aziz, A. et al. (2016). Nonvolatile memory design based on ferroelectric FETs. In: *Proceedings of the 2016 53nd ACM/EDAC/IEEE Design Automation Conference (DAC).*

71 George, S., Li, X., Liao, M. et al. (2018). Symmetric 2-D-memory access to multidimensional data. *IEEE Transactions on Very Large Scale Integration (VLSI) Systems* 26: 1040–1050.

72 Li, X., Wu, J., Ni, K. et al. (2019). Design of 2T/cell and 3T/cell nonvolatile memories with emerging ferroelectric FETs. *IEEE Design & Test* 36: 39–45.

73 Wu, J., Zhong, H., Ni, K. et al. (2019). A 3T/cell practical embedded nonvolatile memory supporting symmetric read and write access based on ferroelectric FETs. In: *Proceedings of the 2019 56th ACM/IEEE Design Automation Conference (DAC).*

74 George, S., Liao, M., Jiang, H. et al. (2018). MDACache: Caching for multi-dimensional-access memories. In: *Proceedings of the 2018 51st IEEE/ACM International Symposium on Microarchitecture (MICRO).*

75 Wang, P., Li, S., Sun, J. et al. (2018). RC-NVM: Enabling symmetric row and column memory accesses for in-memory databases. In: *Proceedings of the 2018 IEEE International Symposium on High Performance Computer Architecture (HPCA).*

76 Ma, X., Deng, S., Wu, J. et al. (2023). A 2-transistor-2-capacitor ferroelectric edge compute-in-memory scheme with disturb-free inference and high endurance. *IEEE Electron Device Letters* 44: 1088–1091.

77 Ni, K., Li, X., Smith, J. A. et al. (2018). Write disturb in ferroelectric FETs and its implication for 1T-FeFET AND memory arrays. *IEEE Electron Device Letters* 39: 1656–1659.

78 Xiao, Y., Xu, Y., Jiang, Z. et al. (2022). On the write schemes and efficiency of FeFET 1T NOR array for embedded nonvolatile memory and beyond. In: *Proceedings of the 2022 International Electron Devices Meeting (IEDM).*

79 Wu, J., Liao, T., Li, T. et al. (2023). Lowering latency of embedded memory by exploiting in-cell victim cache hierarchy based on emerging multi-level memory devices. In: *Proceedings of the 2023 IEEE/ACM International Conference on Computer Aided Design (ICCAD).*

80 Wu, J., Xu, Y., Xue, B. et al. (2020). Adaptive circuit approaches to low-power multi-level/cell FeFET memory. In: *Proceedings of the 2020 25th Asia and South Pacific Design Automation Conference (ASP-DAC).*

81 Genssler, P. R., Santen, V. M., Henkel, J. et al. (2022). On the reliability of FeFET on-chip memory. *IEEE Transactions on Computers* 71: 947–958.

82 Saito, D., Kobayashi, T., Koga, H. et al. (2021). Analog in-memory computing in FeFET-based 1T1R array for edge AI applications. In: *Proceedings of the 2021 IEEE Symposium on VLSI Circuits.*

83 Xiao, Y., Jia, Y., Liu, C. et al. (2019). Edge computing security: State of the art and challenges. *Proceedings of the IEEE* 107: 1608–1631.

84 Aabrar, K. A., Gomez, J., Kirtania, S. et al. (2021). BEOL compatible superlattice FerroFET-based high precision analog weight cell with superior linearity and symmetry. In: *Proceedings of the 2021 International Electron Devices Meeting (IEDM).*

85 Yin, G., Cai, Y., Wu, J. et al. (2021). Enabling lower-power charge-domain nonvolatile in-memory computing with ferroelectric FETs. *IEEE Transactions on Circuits and Systems II: Express Briefs* 68: 2262–2266.

86 Zhong, H., Zhu, Y., Luo, L. et al. (2023). Fe-GCN: A 3D FeFET memory based PIM accelerator for graph convolutional networks. In: *Proceedings of the 2023 IEEE Computer Society Annual Symposium on VLSI (ISVLSI).*

87 Li, T., Zhong, H., Xu, Y. et al. (2024). REMNA: Variation-resilient and energy-efficient MLC FeFET computing-in-memory using NAND flash-like read and adaptive control. In: *Proceedings of the 43rd IEEE/ACM International Conference on Computer-Aided Design (ICCAD).*

88 Chen, W., Huang, F., Qin, S. et al. (2022). 4 Bits/cell hybrid 1F1R for high density embedded non-volatile memory and its application for compute in memory. In: *Proceedings of the 2022 IEEE Symposium on VLSI Technology and Circuits.*

89 Ni, K., Gupta, A., Prakash, O. et al. (2020). Impact of extrinsic variation sources on the device-to-device variation in ferroelectric FET. In: *Proceedings of the 2020 IEEE International Reliability Physics Symposium (IRPS).*

90 Soliman, T., Müller, F., Kirchner, T. et al. (2020). Ultra-low power flexible precision FeFET based analog in-memory computing. In: *Proceedings of the 2020 International Electron Devices Meeting (IEDM).*

91 Soliman, T., Laleni, N., Kirchner, T. et al. (2022). FELIX: A ferroelectric FET based low power mixed-signal in-memory architecture for DNN acceleration. *ACM Transactions on Embedded Computing Systems* 21: 1–25.

92 Soliman, T., Chatterjee, S., Laleni, N. et al. (2022). First demonstration of in-memory computing crossbar using multi-level cell FeFET. *Nature Communications* 14: 6348.

93 Li, T., Zhong, H., Wu, J. et al. (2024). CafeHD: A charge-domain FeFET-based compute-in-memory hyperdimensional encoder with hypervector merging. In: *Proceedings of 2024 Design, Automation & Test in Europe Conference & Exhibition (DATE).*

94 Li, T., Guo, X., Li, T. et al. (2024). First demonstration of reconfigurable and low-power strong physical unclonable function empowered by FeFET cycle-to-cycle variation. Preprint (Version 1). Research Square.

95 Lee, M., Tang, W., Chen, Y. et al. (2023). Victor: A variation-resilient approach using cell-clustered charge-domain computing for high-density high-throughput MLC CiM. In: *Proceedings of the 2023 60th ACM/IEEE Design Automation Conference (DAC).*

96 Shao, H., Fu, B., Yang, J. et al. (2024). IMCE: An in-memory computing and encrypting hardware architecture for robust edge security. In: *Proceedings of 2024 Design, Automation & Test in Europe Conference & Exhibition (DATE).*

97 Shao, H., Zhou, Y., Ning, Z. et al. (2024). First demonstration of high throughput and reliable homomorphic encryption using FeFET arrays for resource-limited IoT clients. In: *Proceedings of the 2024 International Electron Devices Meeting (IEDM).*

98 Yang, K. (2019). Custom CMOS and post-CMOS crossbar circuits for resource-constrained hardware security primitives. In: *Proceedings of the 2019 IEEE International Electron Devices Meeting (IEDM).*

99 Khan, A. I., Keshavarzi, A., Datta, S. et al. (2020). The future of ferroelectric field-effect transistor technology. *Nature Electronics* 3: 588–597.

100 Shao, H., Zhou, Y., Huang, W. et al. (2023). A novel FeFET array-based PUF: Co-optimization of entropy source and CRP generation for enhanced robustness in IoT security. In: *Proceedings of the 2023 International Electron Devices Meeting (IEDM).*

101 Chen, J., Xu, J., Gu, J. et al. (2025). Low-power edge detection based on ferroelectric field-effect transistor. *Nature Communications* 16: 565.

6

1T1C HfO_2 FeRAM Materials

Hitoshi Saito and Takashi Eshita**

RAMXEED LIMITED, Yokohama, Japan

*Corresponding authors: saito.hitoshi@ramxeed.com; eshita@ramxeed.com

6.1 Introduction

In the currently used von Neumann architecture computing system, static random-access memory (SRAM), dynamic random-access memory (DRAM), and FLASH memory are currently the main types of memories. SRAM is the primary cache memory in the central processing unit (CPU) due to its highest speed among these three types of memory.[1] DRAM, with the second highest speed, offers advantages such as large memory capacity (e.g., 192Gb), random accessibility, low voltage (e.g., 1.05V), and free cycling endurance.[2] However, DRAM consumes significant power because it requires periodic refresh cycles due to its volatility.[3] FLASH memory, on the other hand, has the largest memory capacity (e.g., 1Tb), and is non-volatile.[4] However, it suffers from drawbacks such as low write and erase speeds (e.g., 100 μs), lack of random accessibility, limited cycling endurance (10^5cycles), and the need for high voltages (e.g., 6–12V) for writing and erasing, often requiring booster circuits.[5] Ferroelectric memory (Fe-M), specifically ferroelectric random-access memory (FeRAM or FRAM), offers superior characteristics compared to FLASH memory, such as high writing speed, random accessibility, and low writing voltage,[6,7] as well as high radiation immunity.[8,9]

HfO_2-based or fluorite-structured ferroelectric materials (FE-HfO_2), including impurity-doped HfO_2 and HfO_2–ZrO_2 (HZO) solid solutions, have attracted significant attention due to their potential to enable the development of high-density, high-speed FeRAM. This interest stems from Böscke et al.'s discovery of ferroelectricity in Si-doped HfO_2 in 2011.[10,11] HfO_2, a well-known high-k gate insulator for transistor, is compatible with CMOS fabrication processes, making it particularly attractive for memory applications. FE-HfO_2 has been reported to exhibit excellent electrical properties, such as large remnant polarization (P_r) above 10 μC/cm^2, low leakage current even at thicknesses under 10 nm, and low crystallization temperatures. These characteristics make HfO_2 a promising candidate for one-transistor, one-capacitor (1T1C) FeRAM (HfO_2 1T1C FeRAM) applications.

In comparison to conventional ferroelectrics such as lead zirconate titanate ($Pb(Zr,Ti)O_3$ or PZT) and strontium bismuth tantalate ($SrBi_2Ta_2O_9$ or SBT), which typically require thicker layers (> 50 nm), higher crystallization temperatures (> 700 °C), and noble metal electrodes, FE-HfO_2-based FeRAM (HfO_2 FeRAM) offers significant advantages in terms of processing ease and scalability.[7,12,13] A comparison of the electrical and structural properties of various memories, including emerging memories, such as magnetoresistive random-access memory (MRAM), resistive random-access memory (ReRAM), and phase-change random-access memory (PCRAM), is summarized in Table 6.1. The numerical values in this comparison are rearranged based on [5,14–17] and our internal data. HfO_2 FeRAM demonstrates clear advantages over other memories. However, several challenges remain concerning the electrical properties of FE-HfO_2 capacitors (HfO_2-FCs), such as wake-up behavior (an increase in P_r during polarization switching), imprint (shifts in coercive electric field, E_c, or voltage, V_c), breakdown (including both soft and hard breakdowns), and time-dependent dielectric breakdown (TDDB).[13] The E_c of FE-HfO_2 is approximately 1 MV/cm, which is about 10 times higher than that of PZT, and it is close to its breakdown electric field.[18,19]

In this chapter, we review the trends in HfO_2-FC materials used for 1T1C FeRAM, focusing on FE-HfO_2 materials and the electrodes, as well as the challenges associated with these materials. We also discuss the approaches taken to address these challenges, which include not only efforts in HfO_2 1T1C FeRAM but also fundamental experiments using metal–ferroelectric–metal (MFM) structures.

6.2 Fe-Ms and 1T1C FeRAM

6.2.1 Fe-Ms

Fe-Ms are typically categorized into three types[7,20]:

a. capacitor-type;
b. field-effect transistor (FET)-type (commonly referred to as FeFET); and
c. ferroelectric tunnel junction (FTJ)-type.

Figure 6.1 shows the schematic structures and circuits for these types. A typical capacitor-type FeRAM (or FRAM) is based on the 1T1C configuration, where the memory unit – i.e., the circuit used for 1-bit storage – consists of 1T1C. The 1T1C FeRAM circuit is typically designed similarly to DRAM,[21,22] enabling random access to specific data locations without the need for sequential processing. In addition to 1T1C FeRAM, other capacitor-type FeRAM configurations include the two-transistor, two-capacitor (2T2C) type, which offers more reliable operation and improved data storage information efficiency compared to the 1T1C type, even at

Table 6.1 A comparison of electrical and structural properties of various memories. The numerical values in this comparison have been rearranged based on [5,14–17] and our internal data. *F* in the "Cell size" column refers to the minimum feature size.

Memory	Memory principle	Boost circuit	Volatility	Cell size	Special material	Speed (ns)	Cycles
DRAM	Capacity	Not necessary	Volatile	$4–6F^2$	Necessary	<0.2	$>10^{15}$
SRAM	Logical latch	Not necessary	Volatile	$50–130F^2$	Not necessary	<0.1	$>10^{15}$
NAND FLASH	Trapped charge	Necessary	Nonvolatile	$< F^2$	Not necessary	100,000	10^5
NOR FLASH	Trapped charge	Necessary	Nonvolatile	$4F^2$	Not necessary	10,000	10^5
EEPROM	Trapped charge	Necessary	Nonvolatile	$10–17F^2$	Not necessary	100,000	10^6
MRAM	Magnetoresistance	Not necessary	Nonvolatile	$4–6F^2$	Necessary	<10	$>10^{14}$
ReRAM	Resistance	Not necessary	Nonvolatile	$4–6F^2$	Not necessary	<100	10^{11}
PCRAM	Resistance	Not necessary	Nonvolatile	$4F^2$	Necessary	100	$>10^9$
PZT FeRAM	Ferroelectricity	Not necessary	Nonvolatile	$12–25F^2$	Necessary	<50	10^{14}
FE-HfO_2 FeRAM	Ferroelectricity	Not necessary	Nonvolatile	$4–6F^2$	Not necessary	<50	10^{14}

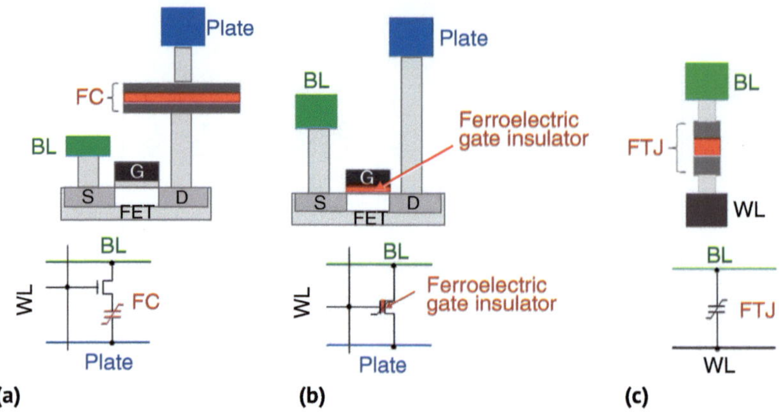

Figure 6.1 Schematic drawings of the structures and circuits for (a) capacitor-type, (b) FET-type (FeFET), and (c) FTJ memories.

the cost of a large area. This is because the sensing operation in 2T2C type compares complementary data (e.g., "1" versus "0" and vice versa), resulting in a more robust sensing margin. In contrast, the sensing operation in the 1T1C type compares a selected datum with a fixed reference signal. The 1T1C and 2T2C FeRAMs have been already commercially available for two decades and have most matured technology among the Fe-Ms.

Another variant is the SRAM-type FeRAM, which incorporates SRAM-like circuits with ferroelectric capacitors (FCs) to achieve nonvolatile memory operation,[23,24] as shown in Figure 6.2. SRAM-type FeRAM was first introduced in 1988 with two FCs. However, due to its poor endurance (10^6cycles), the so-called shadow SRAM was proposed.[25] It operates as an SRAM when external power is supplied, while storing data in FCs when external power is unavailable. Similar nonvolatile SRAM structures using magnetoresistive elements,[26] resistive switching elements,[27] or phase-change switching elements[28] instead of FCs have been proposed. A key challenge of nonvolatile SRAM is reducing its cell area.

6.2.2 FC Characteristics

6.2.2.1 *P–E* Hysteresis Curves

The relationship between polarization (P) and the external electric field (E) is described by a P–E hysteresis curve as shown in Figure 6.3. The polarization at $E = 0$ is called P_r (remnant polarization, as described earlier), which can take positive ($+P_r$) and negative ($-P_r$) values depending on the polarization history. For example, consider the following behaviors[7]:

- When the initial state is $+P_r$ (storing "0") at $E = 0$, polarization increases with increasing E (U-term) and then returns closely to $+P_r$ as E decreases back to zero (U_a-term).

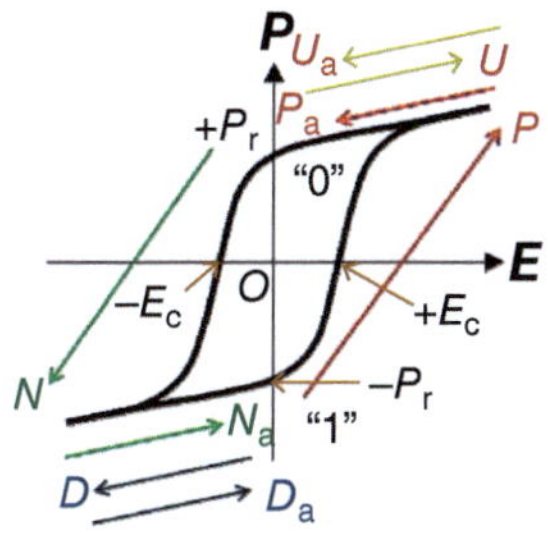

Figure 6.2 Schematic circuit drawing of SRM-type Fe-M.

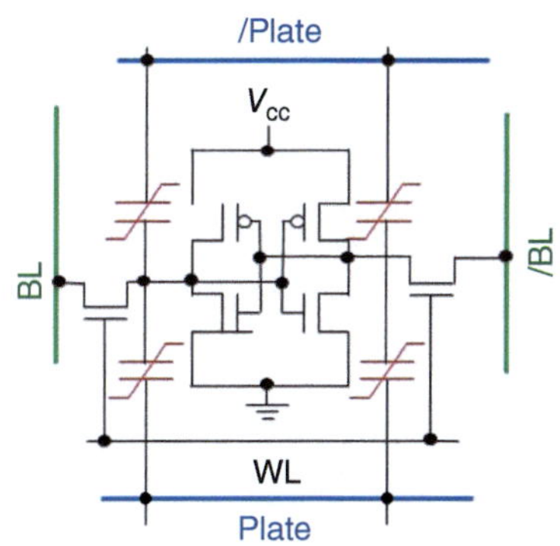

Figure 6.3 *P*–*E* hysteresis and polarization reversal and non-reversal terms.

- For the same initial $+P_r$ state, as E decreases further in the negative direction polarization also decreases (N-term) and eventually reaches $-P_r$ when E returns to zero (N_a-term).
- Conversely, when the initial state is $-P_r$ at $E = 0$, polarization decreases as the negative electric field increases (D-term) and remains at $-P_r$ as E returns to zero (D_a-term).
- Additionally, for a cell starting at $-P_r$ at $E = 0$, as E increases in the positive direction, polarization increases (P-term) and eventually returns to $+P_r$ as E decreases back to zero (P_a-term).

These changes in polarization behaviors are depicted in Figure 6.3. For Fe-M operation with a sufficient margin, $\pm P_r$ should be sufficiently large, and the *P*–*E* hysteresis should exhibit a square-like, symmetric shape. This means that polarization preferably changes abruptly in the *P*- and *N*-terms, and similarly in the *U*-, U_a-, P_a-, N_a-, *D*-, and D_a-terms.[29,30]

As an example of *P*–*E* hysteresis measurements, the FC's polarization is first set to $-P_r$ ("1") by applying a 1-ms-wide rectangular negative pulse. Then, triangular pulses with a frequency of, for example, 100 kHz are applied at temperatures ranging from −45 to 150 °C, to measure the polarization change in the *P*-term. After waiting for about 1 ms, the FC's polarization is set to $+P_r$ by applying a 1-ms-wide rectangular positive pulse. Following this, triangular pulses with a frequency of, for example, 100 kHz, are applied to the FC again to measure the polarization change

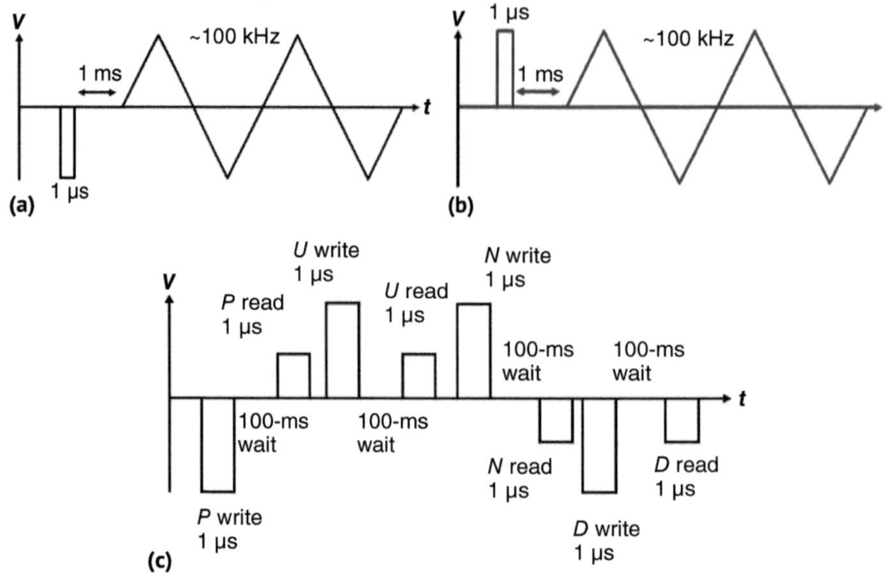

Figure 6.4 Voltage pulse profiles for (a) *P*–*E* hysteresis measurement starting from $-P_r$ ("1") write, (b) *P*–*E* hysteresis measurement starting from $+P_r$ ("0") write, and (c) PUND measurement.

in the *U*-term. These measurement pulses are shown in Figure 6.4. The positive-up negative-down (PUND) method[31] is also commonly used for evaluating P_r switching behavior, as depicted in Figure 6.3. It provides polarization behavior of FC that more closely resembles real FeRAM operation because it uses short pulses. A more precise evaluation scheme that mimics FeRAM using the high-*Z* readout scheme (see "Bit-Line Ground Sensing [BGS] Scheme" in Appendix A) reads the polarization at a voltage lower than the write (polarization) voltage.

6.2.2.2 Imprint, Depolarization, Wake-up, Fatigue, and Their Measurements

In this subsection, we describe polarization degradation observed on the *P–E* hysteresis, which is related to FeRAM failures.

6.2.2.2.1 Imprint

Imprint is observed when *P–E* hysteresis shifts from its original position to positive or negative *E*-direction. For example, as shown in Figure 6.5a, *P–E* hysteresis shifts in the positive direction after storing the "1" state, making it difficult to reverse the polarization direction. In other words, this shift in E_c toward the positive direction makes it difficult to rewrite a "0" from "1." In addition, a similar E_c shift that makes the *U*-term larger than usual reduces the difference in sensed voltages between "0" and "1."

6.2.2.2.2 Depolarization

Depolarization is observed when there is a decrease in P_r from original value, as shown in Figure 6.5b, and this effect is exacerbated at elevated temperature.

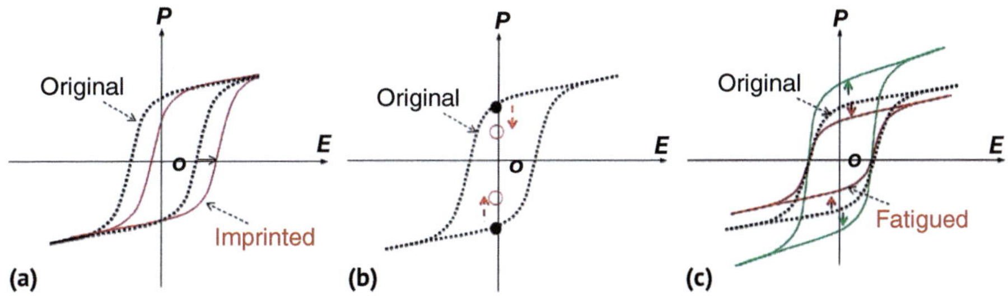

Figure 6.5 *P–E* hysteresis curves suffering from (a) imprint, (b) depolarization, and (c) wake-up and fatigue.

A depolarization field, typically appearing in structures such as ferroelectric–dielectric stacked structures (e.g., the gate structure of FeFETs), also contributes to depolarization when charge extinction occurs in the electrode.[32]

6.2.2.2.3 Wake-up and Fatigue

Wake-up is observed when there is an increase in polarization during switching cycles, whereas fatigue manifests as a gradual decrease in polarization with continued cycling, as shown in Figure 6.5c. Wake-up affects stable operation, while fatigue limits the rewriting endurance of FeRAM.[7,32]

6.2.2.2.4 Measurements

For 1T1C FeRAM, including the 2T2C type, a testing procedure has been established.[33] A retention test checks the ability of FC to retain the initially stored data ("same state" or SS). If the FC is degraded by depolarization, the data reading operation may fail during this test. An imprint test checks the ability of an FC to rewrite the stored data as the opposite data ("opposite state" or OS); for example, writing and storing "1" corresponding to the position "1," and then rewriting the stored "1" as "0." If the *P–E* hysteresis of FC shifts, the reading operation of the rewritten data may fail during this test, because rewriting the opposite data becomes difficult due to an increase in E_c. In this procedure, the baking temperature is set according to the activation energy of degradation, which is previously evaluated using the Arrhenius equation before baking.

A wake-up and fatigue endurance test checks the read/write function of FC after polarization cycling. The fatigue endurance of commercialized PZT-based 1T1C FeRAMs ranges from 10^{13} to 10^{15}. Since it takes a long time to cycle the polarization over 10^{14}, the fatigue endurance of FC is often tested using "voltage accelerating methods," which cycle the polarization of FC at a higher voltage than the actual operating voltage, and then predict the fatigue lifetime from the obtained acceleration factors.[34,35] To evaluate fatigue, wake-up, and endurance, changes in P_r during sequential polarization reversal (cycling) are measured using sequential bipolar pulses (e.g., 4 μs width and 2–4 MV/cm electric field), as shown in Figure 6.6a. It is worth mentioning dynamic imprint,[36] a type of imprint

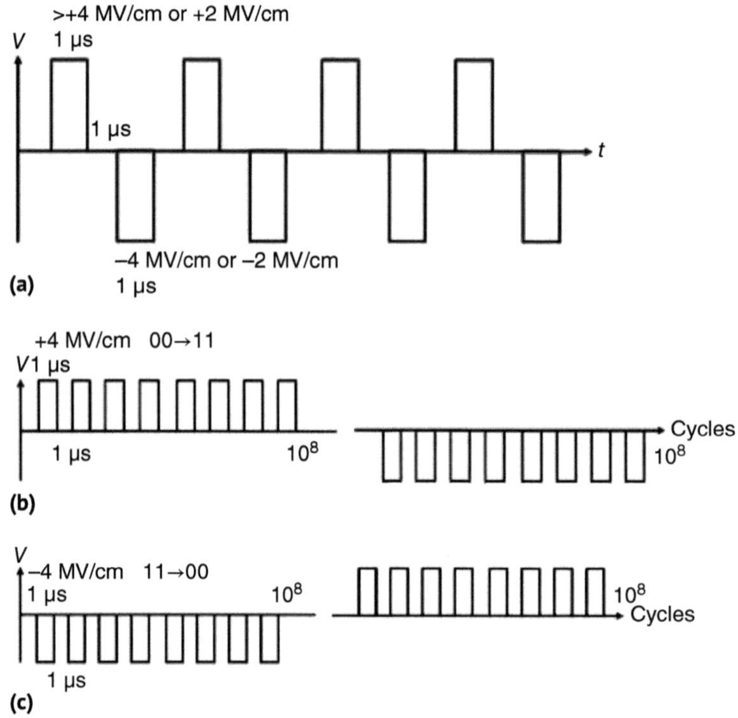

Figure 6.6 Voltage pulse profiles for measuring (a) wake-up and fatigue, (b) dynamic imprint after sequentially writing "0" and then writing "1," and (c) dynamic imprint after sequentially writing "1" and then writing "0."

phenomenon caused by polarization cycling. An evaluation method for assessing dynamic imprint involves applying sequential unipolar pulses, for example, with a 1 μs width and an electric field of about +4 MV/cm, at temperatures ranging from −45 to +150 °C. This measurement typically involves writing "0" for approximately 10^8 cycles, followed by writing "1" for approximately 10^8 cycles, and vice versa, as illustrated in Figure 6.6b,c, respectively.

6.2.2.2.5 Breakdown Electric Field and TDDB

The breakdown electric field is an important factor in determining the reliability of FeRAM. In particular, the polarization-cycling-induced soft breakdown has been carefully investigated in FE-HfO_2 due to the close values of E_c and breakdown field.[37] TDDB is another important factor for FeRAM reliability, as it measures the breakdown time under continuous DC bias applied to FC.

6.2.3 The 1T1C FeRAM Operation

This subsection describes the write and read operations of 1T1C FeRAM using Figure 6.7a–d.

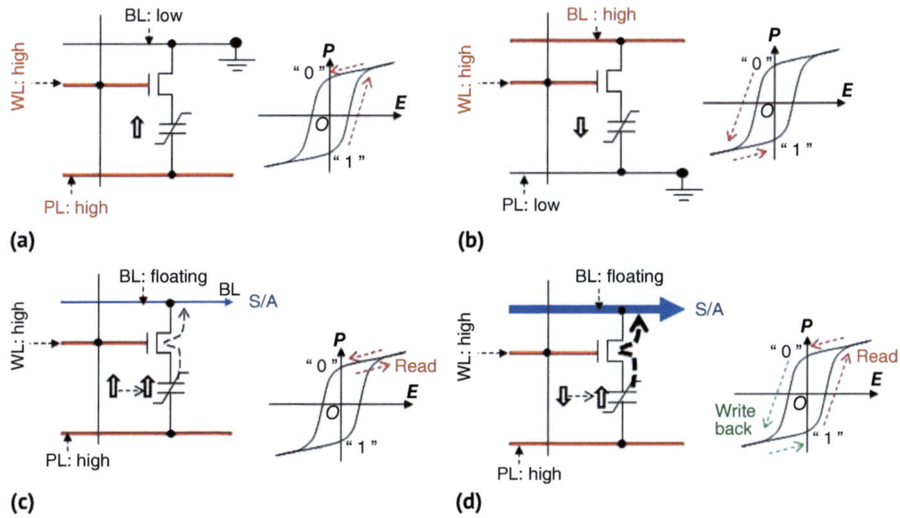

Figure 6.7 Unit circuit and *P*–*E* hysteresis of 1T1C FeRAM operating as (a) write "0," (b) write "1," (c) read "0," and (d) read "1."

6.2.3.1 Write Operation

The "0" write operation:

- set the word line (WL) to a high voltage;
- set the plate line (PL) to a high voltage;
- set the bit line (BL) to a low voltage (Figure 6.7a).

The "1" write operation:

- set WL to a high volage;
- set PL to a low voltage;
- set BL to a high voltage (Figure 6.7b).

6.2.3.2 Read Operation

For reading both for "0" and "1":

- set WL to a high voltage;
- set PL to a high voltage;
- connect BL to a sense amplifier (S/A) with floating (high-*Z* sensing);
- reading "1": polarization reversal charges flow to BL;
- reading "0": linear charges flow to BL.

The increase in BL voltage (V_{bl}) during a "1" read is typically larger than that during a "0" read because the polarization reversal charge is larger than the linear charge. The S/A connected to BL determines whether the stored value is "1" or "0" by comparing V_{bl} to a reference voltage generated by a designated FC or a preset voltage:

- After reading "0": the polarization direction does not change from its state before reading (Figure 6.7c).
- The polarization direction is reversed by a write-back operation to "1," a phenomenon known as "destructive reading" (Figure 6.7d).

6.2.3.3 Sensing Operation and C_s/C_b Optimization

During the sensing operation, the voltages applied to FC and BL, which are connected in series through PL, are divided according to the inverse ratio of the FC capacitance (C_s) and the BL capacitance (C_b). The S/A determines "0" or "1" by comparing V_{bl} with the reference voltage. A larger voltage difference between V_{bl} values obtained during "0" read and "1" read results in an increased sensing margin, enhancing FeRAM reliability. The V_{bl} is given by:

$$V_{bl} = Q_{SW}/C_b$$

where Q_{sw} is the emitted charge from FC to BL during sensing. A larger C_s/C_b ratio leads to a higher V_{bl} for a given Q_{sw}. However, a larger C_s/C_b also reduces Q_{sw}, because of a smaller application voltage to FC due to capacitive voltage division. Thus, we should optimize the C_s/C_b ratio to obtain the appropriate V_{bl} for sensing,[38] while also considering the nonlinear behaviors of C_s and Q_{sw} with applied voltage. In the actual sensing operation of FeRAM, the sensing margin strongly depends on the characteristics of FC, which can change due to temperature, polarization cycling history (such as fatigue and wake-up), and manufacturing process deviations. Several proposals for improving the sensing margin are described in Appendix A.

6.3 HfO_2 1T1C FeRAM

6.3.1 History of HfO_2 Fe-Ms

Müller et al. were the first to demonstrate that HZO could be applied to a trench capacitor with high-aspect-ratio structures, as used in DRAM. In 2013, they successfully integrated a 10-nm-thick HZO layer into an array of 1.6-μm-deep trenches with TiN electrodes and an aspect ratio of 13:1.[11] Initially. FE-HfO_2 was crystallized at temperatures around 800 °C shortly after its discovery. However, many efforts have been made to lower its crystallization temperature to below 500 °C, meeting the demand for HfO_2-FCs to be fabricated in the back end of line (BEOL), instead of before BEOL. These efforts aim to enhance memory density and reduce fabrication costs. Unfortunately, capacitor-type FeRAMs, including 1T1C FeRAM using FE-HfO_2, did not initially attract as much attention as FeFETs, since FeFETs were considered more advantageous for scalability. This trend changed when Okuno et al. demonstrated a 64-kbit 1T1C FeRAM using a HZO planar capacitor fabricated in BEOL in 2020.[39] This was followed by notable demonstrations, such as their improvements,[40–42] the development of a 16-kbit 1T1C FeRAM using Si-doped FE-HfO_2,[43,44] 8-Gb FeRAM with 3D trench FC using 5-nm-thick HZO,[45] FeRAM with 3D trench FE-HfO_2-FC,[46,47] and FE-HfO_2 combined with a low-leakage oxide transistor using Sn-doped InGaZnO (IGZO).[48] More recently, Ramaswamy et al. demonstrated a 3D stacked nonvolatile Fe-M with memory density of 32Gb, utilizing a $4F^2$ architecture with 48-nm pitch, where F and $4F^2$ present the feature size and the minimum feature area, respectively. Their fabricated FeRAM was successfully tested from −40 to 95 °C along with 10-year reliability.[49]

6.3.2 Advantages of FE-HfO_2 FeRAM

The advantages of FE-HfO_2 FeRAMs, including HZO FeRAM, lie not only in the conventional electrode materials they use, such as TiN, but also in the ferroelectric layer thickness, which is less than 10nm, making it easier to pattern both the capacitor and 3D pillar structure. In contrast, PZT capacitors use noble metal electrodes, such as Ir and Pt, and have a ferroelectric layer thicker than 50nm. These advantages enable HfO_2 1T1C FeRAM to have increased memory density. The memory density integrated into a 5×5mm^2 die is 16Mb, as shown in Figure 6.8a,[50] and can be estimated to reach up to 41Mb with the adoption of a pillar structure, as depicted in Figure 6.8b. However, the memory density of HfO_2 1T1C FeRAM can exceed 11Gb in the same square die, as shown in Figure 6.8c.

FeRAM with FE-HfO_2 also is known to have excellent radiation immunity. Huang et al. demonstrated that the electrical properties of FeRAM using Y-doped HfO_2, such as leakage current, permittivity, P_r, endurance, and fatigue, were not affected by ^{60}Co γ-ray irradiation up to 12.96 Mrad (Si).[9] Zhang et al. investigated the radiation effects on Al-doped HfO_2 with various configurations (TiN/HfAlO/TiN, Pt/HfAlO/TiN, and Ta/HfAlO/TiN) and found that the leakage current (I) increased with voltage in the range from zero to the forward coercive voltage (V_{c+}), and then exhibited a reduction above V_{c+}. After a dose of 1Mrad (Si), this indicated a weakened built-in electric field or a decreased depolarization field induced by ionizing radiation.[51] Kang et al. found that the ferroelectric properties improved by bombarding hafnium oxide films with a beam of helium ions, as the ion bombardment created oxygen vacancies (V_Os) and strain changes from helium implantation, which pushed more of the polycrystalline samples into the ferroelectric orthorhombic phase (o-phase).[52]

6.3.3 A Comparison of Polarization Properties Between PZT and HZO

In this subsection, we compare the polarization properties of PZT[7,50] and HZO,[18,53–55] including P_r, the voltage dependence of polarization, and cycling endurance. The P–E hysteresis measured on the PZT FC at 90 °C with a 2-MHz-frequency pulse at voltages ranging from 1.2 to 2.2V, and that measured at 1.2V on the 4–6.5-nm-thick HZO FCs are shown in Figure 6.9a,b, respectively. The P_r of HZO FC obtained at 1.2 V measurement appears smaller than that of PZT. The voltage dependence of

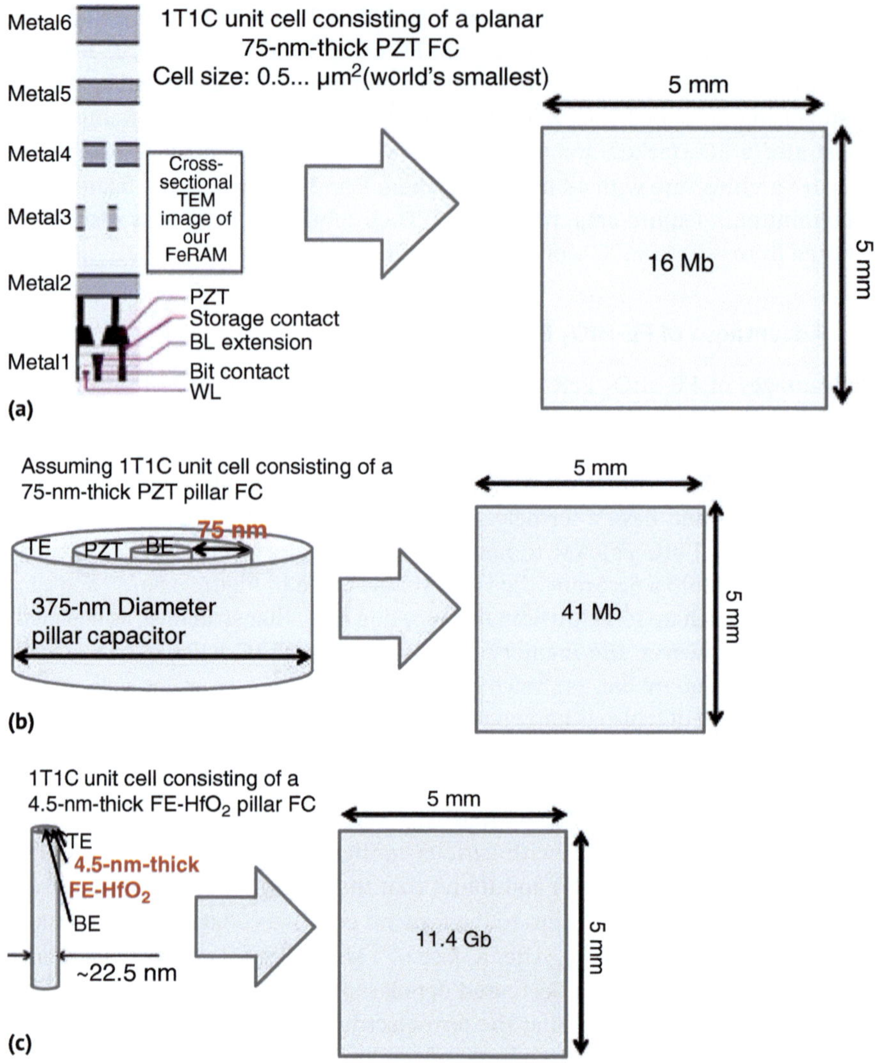

Figure 6.8 Memory density evaluations of 1T1C FeRAM for integration within a 5×5 mm^2 die using (a) a planar PZT capacitor, (b) a cylindrical PZT capacitor, and (c) a cylindrical HfO_2-FC.

polarization of PZT FC, measured with 20 and 100 ns read/write pulses, and that measured on the 4–9.5-nm-thick HZO FC at the crystallization temperature range of 400–500 °C are shown in Figure 6.9c,d, respectively. The polarization rise of HZO below 0.6 V seems slightly more gradual than that of PZT. The cycling endurances of PZT and HZO FCs are shown in Figure 6.9e,f, respectively. The projected endurance of HZO FCs,[18] shown in Figure 6.9f, is 10^{14}cycles, which is much smaller than that of PZT FC of 10^{17}cycles.[50] Therefore, based on these data, challenges for HZO, such as relatively low P_r, slow polarization rise at low voltage, and endurance, still remain.

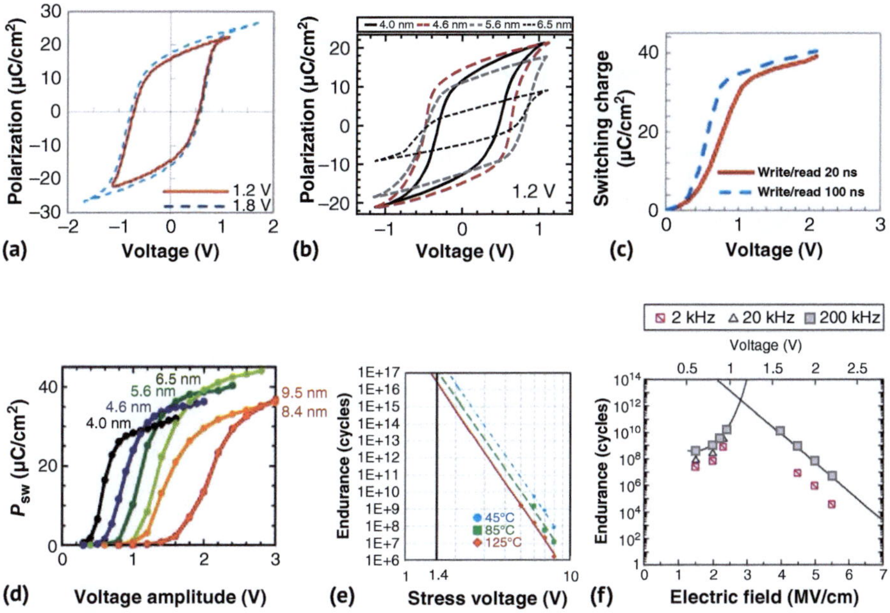

Figure 6.9 *P–E* hysteresis of HZO obtained under (a) varying applied voltage and (b) varying HZO thickness. Switching charge dependences of HZO FC measured by (c) varying write/read pulse widths and (d) varying temperature. Endurance estimations for (e) PZT capacitors and (f) HZO capacitors.

6.4 HfO_2 1T1C FeRAM and Its FC Materials

6.4.1 HfO_2-FC Element Materials for 1T1C FeRAM

6.4.1.1 FE-HfO_2

Unlike the perovskite structure of PZT crystal, FE-HfO_2 has a fluorite-type structure under atmospheric conditions, with its most stable phase at room temperature being the monoclinic phase (m-phase). PZT and FE-HfO_2 crystalline structures and their atomic shifts causing polarization are depicted in Figure 6.10a,b, respectively. The m-phase of HfO_2 can be transformed into a tetragonal phase (t-phase) or cubic phase at high temperatures, or into a ferroelectric o-phase under very high pressure. The ferroelectricity of HfO_2 (FE-HfO_2) is achieved by stabilizing its o-phase, typically through impurity doping or solid solution with ZrO_2.[56,57] Many impurities have been found to stabilize the o-phase of FE-HfO_2 since the first report of Si-doped HfO_2 in 2011.[10] Xu et al. discovered an interesting fact that the maximum switching polarization value was almost the same, independent of the dopant elements, while doping sensitivity to polarization varied quantitatively among the investigated dopants (Sc, Y, Al, Nb, Si, Ge, and N).[58] They explained that both oxygen vacancy and the size effects of the dopants are responsible for these different doping sensitivity coefficients. According to their experimental data on the relationship between polarization and dopant concentration, Zr concentration in HfO_2 was the least sensitive to polarization

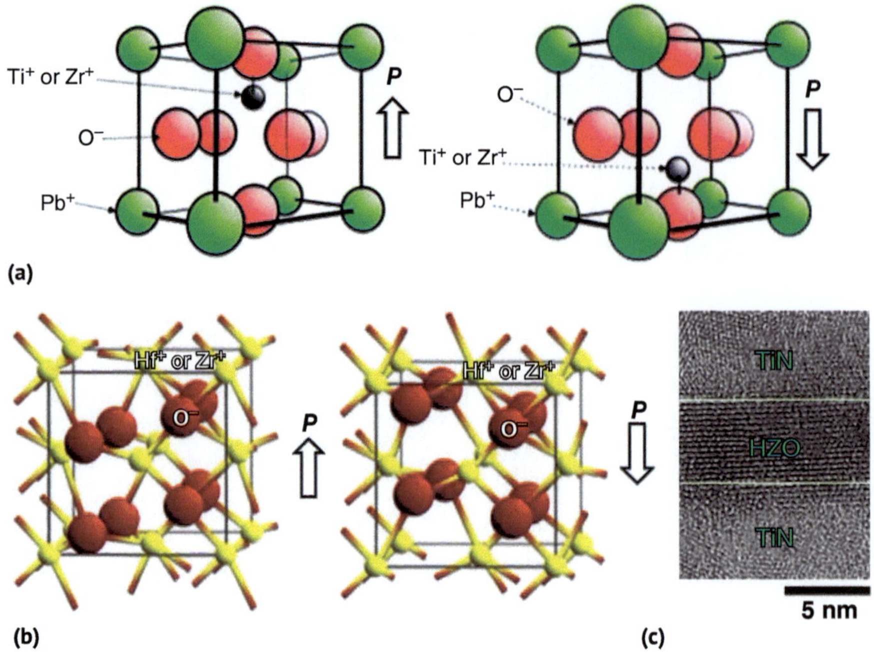

Figure 6.10 Schematic illustrations of unit cells: (a) PZT with upward (left) and downward (right) polarizations, (b) FE-HfO_2 with upward (left) and downward (right) polarizations, and (c) TEM image of a TiN/HZO/TiN capacitor.

among the dopants investigated. Considering these findings, along with Müller et al.'s first successful demonstration of 1T1C FeRAM using HZO, it is reasonable to conclude that HZO is typically used for 1T1C FeRAM[11]. Thus, HZO has been widely used not only for 1T1C FeRAM but also for other Fe-Ms.[11,59,60] A transmission electron microscope (TEM) image of TiN/4.6-nm-thick HZO (crystallized at 400 °C)/TIN is shown in Figure 6.10c.[53]

6.4.1.2 HZO

Müller et al. investigated the composition- and temperature-dependent ferroelectric phase transition of HZO. They found that Hf-rich HZO exhibited ferroelectricity and a predominantly monoclinic structure, whereas Zr-rich HZO, which exhibited antiferroelectricity, crystallized into a t-phase. Ferroelectricity appeared in the mixed oxide and was attributed to a tetragonal to orthorhombic transformation during cooling.[62] Park et al. studied the hydrogen-induced degradation of $Hf_{0.5}Zr_{0.5}O_2$ using "forming gas," a N_2 gas containing a few percent of H_2, and found that the degradation degree of HZO was significantly lower than those of other ferroelectrics.[63] They also found that $Hf_{0.5}Zr_{0.5}O_2$ films became more antiferroelectric (AFE)-like with decreasing film thickness in the pristine state, whereas all the $Hf_{0.5}Zr_{0.5}O_2$ films exhibited ferroelectric characteristics after the wake-up process.[64] Li et al. explored the effect of ZrO_2 content on the switching speed of 10-nm-thick HZO capacitors and demonstrated that the switching times for 80% partial polarization decreased with decreasing ZrO_2 content, even though E_c increased (an increase in E_c is expected to make switching polarization more difficult). This was because lower ZrO_2 contents led to higher active fields and a less uniform distribution of local fields.[65]

6.4.1.3 Electrode Materials

For the electrodes of HfO_2-FC, TiN has been widely used for both the top electrode (TE) and bottom electrode (BE) because it is a well-established material in Si-based semiconductor fabrication processes and is highly compatible with FE-HfO_2.[10,11] However, partial oxidation of TiN during crystallization annealing can degrade the ferroelectric properties of FE-HfO_2. As a result, oxide electrodes or oxidation-resistant metals are sometimes employed for the electrodes of HfO_2-FCs, as described in a later subsection.

6.4.2 Formation of FE-HfO_2 Layers

There are typically three types of deposition techniques for FE-HfO_2: chemical vapor deposition (CVD), physical vapor deposition (PVD), and chemical solution deposition (CSD).

6.4.2.1 ALD

Atomic-layer deposition (ALD), a specialized form of CVD, is one of the most common deposition techniques. It sequentially deposits single atomic layers using vaporized precursor materials in a reaction chamber, taking advantage of the self-limiting surface reaction.[66] An ALD system and its deposition procedure are shown in Figure 6.11. ALD has been widely adopted for FE-HfO_2 deposition because it can meet the miniaturization demands by precisely controlling deposition conditions, such as stoichiometry, thickness, and doping concentration. Furthermore, it can deposit very thin layers conformally, even on 3D trench capacitors with high aspect ratios.[67,68] Müller et al.[11] applied ALD FE-HfO_2, a 10-nm-thick $Hf_{0.5}Zr_{0.5}O_2$, to 1T1C FeRAM for the first time on TiN BE. Okuno et al. also demonstrated 1T1C

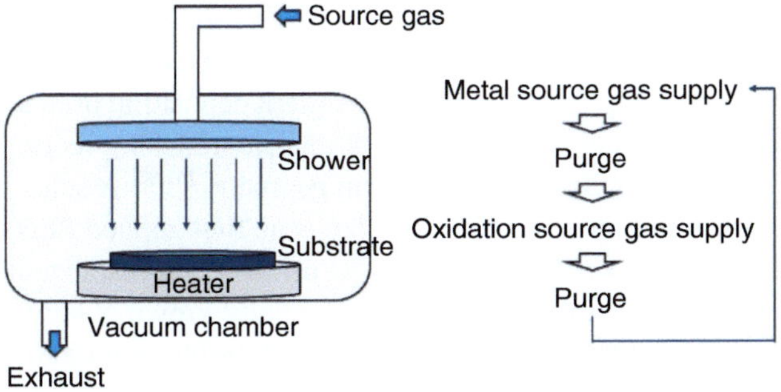

Figure 6.11 Schematic diagram of ALD system and an example of the ALD procedure.

FeRAM using a TiN/11-nm-thick HZO/TiN stack. They crystallized HZO via ALD under 500 °C fabricated 1T1C.[39] François et al. deposited 1% Si-doped HfO_2 on sputter-deposited TiN at 300 °C and crystallized it at 350 °C.[43]

Apart from the direct application of ALD to 1T1C FeRAM, many interesting fundamental investigations have been reported on FE-HfO_2 grown by ALD and its electrical properties in MFM structures. In ALD for FE-HfO_2, the oxidizing conditions during growth play an important role in determining phase transitions, electrical properties, and reliability,[66,69–74] although this is not only the case for ALD but also for other deposition techniques. Low oxidation conditions generally lead to the formation of the o-phase, resulting in high P_r, similar to the crystallization conditions of the ferroelectric PZT.[75] Materano et al. reported that low O_3 dosing during HZO growth induced a higher amount of o-phase and larger P_r in HfO_2.[69] Pal et al. demonstrated that 6.9-nm-thick undoped HfO_2 grown via ALD revealed ferroelectricity with $P_r \approx 13.5\ \mu C/cm^2$ by lowering the oxidation during growth, which suppressed the m-phase.[70]

Plasma-enhanced ALD (PEALD) has also been employed for FE-HfO_2 deposition. Chernikova et al. demonstrated that La-doped HfO_2 via low-temperature (330 °C) PEALD revealed 3×10^8 endurance after 650 °C annealing.[76] Hur et al. found a trade-off between conventional ALD (thermal ALD [THALD]) and PEALD; while the THALD-grown HZO maintained higher cycling endurance, PEALD-grown HZO demonstrated more stable characteristics over cycling with reduced wake-up and fatigue effects, in addition to better tolerance against breakdown under high electric fields.[77]

6.4.2.2 MOCVD

Metal–organic CVD (MOCVD), a common CVD variant which typically does not control deposition at the atomic layer level unlike ALD, has also been adopted for FE-HfO_2 film formation.[78,79] According to the first report on FE-HfO_2, a 7–10-nm-thick Si-doped HfO_2 was deposited via MOCVD using tetrakis-(ethylmethylamino)-hafnium (TEMA-Hf) and tetrakis-dimethylamino-silane (4DMAS) metal–organic precursors as the oxidants. The film was subsequently crystallized at 1,000 °C.[10]

6.4.2.3 PLD

Pulsed laser deposition (PLD) is one of the PVD techniques that are widely used to grow thin films, primarily for fundamental research. PLD uses pulsed laser radiation, usually in the ultraviolet region, to irradiate a solid-state or liquid target, resulting in ablation of material from the target surface and subsequent deposition onto a substrate. This technique offers advantages such as stoichiometric transfer, precise control over the growth rate, and flexibility in the ablation geometry.[61,80] A schematic of the PLD system is shown in Figure 6.12.[61] Shimizu et al. first applied PLD to deposit a 20-nm-thick Y-doped HfO_2 (Y:HfO_2) at 700 °C on an yttria-stabilized zirconia substrate in 2015.[81] They reported that their epitaxially grown Y:HfO_2 exhibited much higher temperatures (above 450 °C) for the orthorhombic-to-tetragonal structural phase transition than the reported maximum temperature of 200 °C to obtain ferroelectricity.[82] Song et al. investigated HZO growth via plasma-enhanced

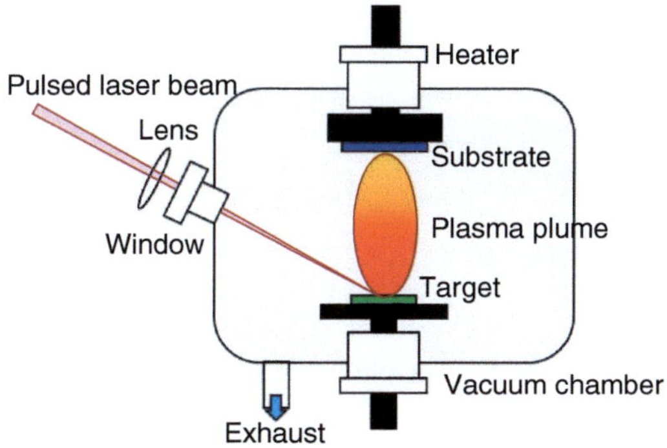

Figure 6.12 Schematic diagram of PLD system.

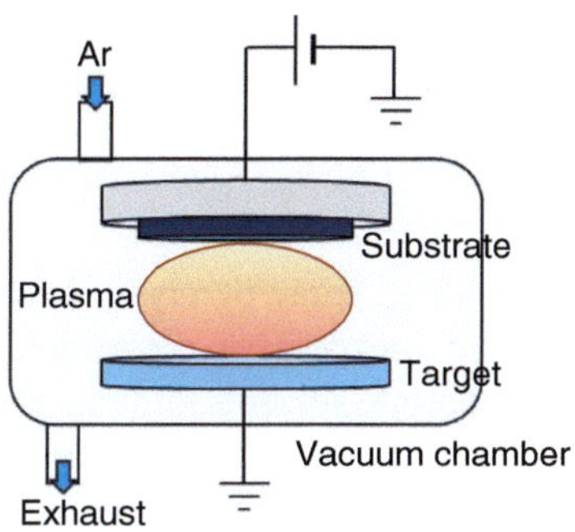

Figure 6.13 Schematic diagram of sputter deposition system.

PLD (PEPLD) using Ar and O_2 gases and found that low plasma energy and low O_2 pressure allowed a significant increase in ferroelectric polarization, up to about 30 $\mu C/cm^2$.[83]

6.4.2.4 Sputter Deposition

Another important PVD technique is sputtering, which has matured as a technique for semiconductor device fabrication. Sputter deposition, as depicted in the schematic in Figure 6.13, bombards a target material (the source material) with high-energy ions, typically from an inert gas like argon. This bombardment causes atoms from the target to be ejected and deposited onto a substrate within a vacuum chamber. Olsen et al. fabricated TiN/Y-doped HfO_2/TiN stacks on Si, incorporating yttrium into the HfO_2 layers by simultaneously sputtering from Y_2O_3 and HfO_2 sources in 2012.[84] Lee et al. reported that oxygen-deficient conditions were necessary during sputter deposition to suppress grain growth, while annealing with O_2 gas was critical to avoid defects and leakage problems when fabricating an 11-nm-thick HZO film on a 50-nm-thick TiN BE by radio-frequency sputtering with HfO_2 and ZrO_2 targets, followed by annealing at 550–650 °C. They demonstrated that the optimized fabrication condition, with Pt (30 nm)/TiN (5 nm)/HZO (11 nm)/TiN (50 nm), resulted in a P_r of about 20 $\mu C/cm^2$.[85] Ryu et al. investigated the effects

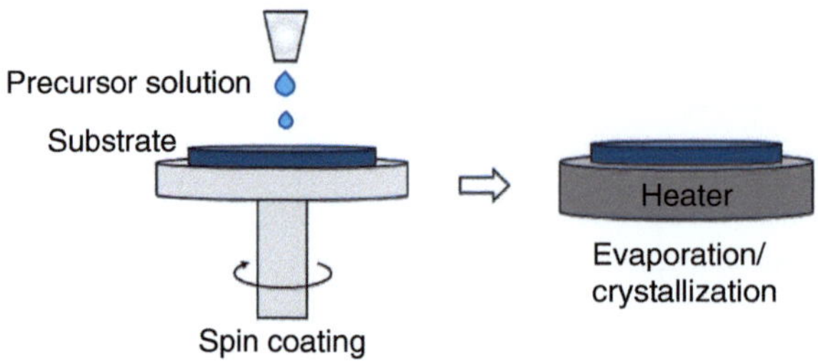

Figure 6.14 Schematic diagram of CSD system and an example of the CSD procedure.

of oxygen partial pressure (PO_2) on sputter-deposited HZO thin films and HZO film thickness in Pt/HZO/TiN capacitors. They found that a P_r of 24.8 μC/cm² for the optimized HZO growth condition of a 20-nm-thick HZO film deposited at a PO_2 of 0% was achieved, and P_r decreased with increasing PO_2 due to oxygen reduction in the amount of ferroelectric o-phase.[86]

6.4.2.5 CSD

CSD, including the so-called sol–gel deposition, spin coating, spray pyrolysis, and metal–organic decomposition (MOD), is a flexible and inexpensive deposition method. In this process, precursors are dissolved in appropriate solvents and mixed in stoichiometric ratios to yield the desired composition of the final film, which is deposited on substrates by spin coating, spraying, and other techniques. The desired film is then obtained by heating and decomposing the deposited precursor.[87] The procedure is depicted in Figure 6.14. Starschich et al. fabricated 17–70-nm-thick 5.3% Y:HfO_2 films using CSD and crystallized them at 700 °C in 2014. They found that the P_r of 13 μC/cm² obtained on their Pt/Y:HfO_2/Pt capacitor was mostly independent of the yttrium composition, attributed to their crystalline procedure, which involved step-by-step annealing after each spin coating of the solution.[88] Nakayama et al. investigated the crystallization behaviors of $Hf_{0.5}Zr_{0.5}O_2$ via CSD, crystallizing at 600, 700, and 800 °C. They found that $Hf_{0.5}Zr_{0.5}O_2$ crystallized into the t-phase at 600°C or a mixture of o- and m-phases at 700°C and above. At 800°C, the microstructure degraded, resulting in a maximum P_r of 8 μC/cm² obtained from the Pt/700 °C crystallized $Hf_{0.5}Zr_{0.5}O_2$/Pt capacitor.[89]

6.4.2.6 Formation of Electrode Materials

Sputter-, ALD-, or CVD-deposited TiN has been commonly used as TE and BE for 1T1C HfO_2-based FeRAM, likely because it is a mature material in semiconductor fabrication.[90] However, TiAlN, NbN,[90] TaN,[91] W,[92] Mo,[93,94] Ru,[95] and oxide materials[96,97] have also been used in fundamental investigations. In HfO_2-FCs with nitride electrodes such as TiN and TaN, decomposed nitrogen atoms form Hf–N bonds, while the remaining oxidized electrode material persists throughout the annealing process. As a result, the concentration of V_O, which will be described later, in FE-HfO_2 near the electrode increases.[98]

6.5 Challenges and Improvements of FE-HfO_2 for 1T1C FeRAM

6.5.1 Oxygen Vacancy-related Degradation

V_O generally exists in oxide materials due to crystalline configurational entropy and the total enthalpy balance. Reliability problems such as wake-up (increase in polarization), fatigue,[99–101] imprint (V_c shift)[102,103] as shown in Figure 6.15,[90] and soft breakdown or TDDB[104,105] appear in HfO_2-FCs during polarization switching. These issues have been related to the redistribution and accumulation of V_O near the electrode interface in FE-HfO_2.[106–108]

In the case of FE-HfO_2 with nitride electrodes, redox reactions at the electrode–FE-HfO_2 interface increase the concentration of V_O near the interface.[66,89] Alcala et al. reported that asymmetric V_O concentration generated during the annealing process, even with a symmetric TiN/HZO/TiN FC structure, depends on the fabrication process sequence. This leads to an internal electric field due to differences in charge at the TE and BE interfaces.[90] In comparison to conventional ferroelectrics like PZT and SBT, which typically require thicker layers (> 50 nm), higher crystallization temperatures (> 700 °C), and noble metal electrode, HfO_2 FeRAM offers significant advantages in terms of processing ease and scalability.[7,12,13] Grimley et al. investigated the wake-up phenomenon in their prepared TiN/Gd-doped HZO/TiN FC and concluded that the wake-up regime arises due to changes in bulk and interfacial structures: the bulk undergoes a phase transformation from monoclinic to

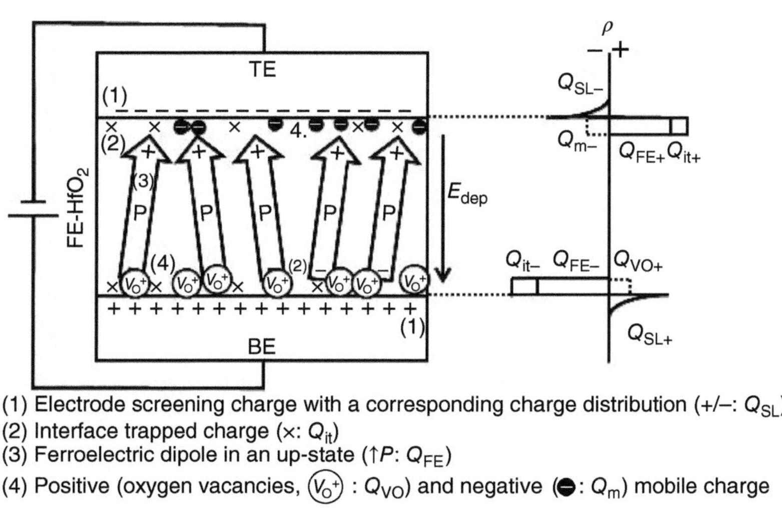

Figure 6.15 A cross-sectional view of an HfO_2-FC showing the various charges present under zero external bias but with a fully poled ferroelectric layer, along with the corresponding charge density plot: electrode screening charges with their respective distributions (+/−: Q_{SL}), interface trapped charges (× : Q_{it}), ferroelectric dipoles in the up-state (↑P : Q_{FE}), and positive (oxygen vacancies $[V_O{}^+]$: Q_{VO}) and negative (•: Q_m) mobile charges. *Source*: [90] / with permission of John Wiley & Sons.

orthorhombic grains, while the interfaces show changes and a diminishment of a nonuniform, defect-rich tetragonal HfO_2 layer near the electrodes.[109] Yuan et al. demonstrated that injection-induced charged V_Os were at the heart of imprint in HZO by experimentally investigating their fabricated TiN/$Hf_{0.5}Zr_{0.5}O_2$/TiN structure and performing first-principles calculations.[102]

Wei et al. investigated the TDDB of TiN/10-nm ALD HZO/TiN by combining TDDB measurements and first-principles calculations, finding that annealing temperature had a large impact on the initial defect or V_O concentrations. The breakdown paths mainly occur in the amorphous regions, and the temporarily recovered ferroelectricity during cycling can be explained by considering the unstable breakdown paths generated at the grain boundary.[104] Chouprik et al. investigated retention loss in their fabricated TiN (TE)/10-nm ALD HZO/W (BE) structure and found that retention loss is caused by the accumulation of positively charged V_Os at the interfaces with capacitor electrodes.[105]

6.5.2 Addressing the Challenges

The following addressing methods are schematically depicted in Figure 6.16a–e and the main topics are summarized in Table 6.2. Important studies are introduced as follows. V_Os at the FE-HfO_2–electrode interface cause reliability issues such as fatigue, wake-up, imprint, and leakage current. Therefore, interface engineering is crucial for improving 1T1C HfO_2 FeRAM.

6.5.2.1 Thermally Stable Electrode and Interlayer

As described earlier, since V_O is generated in FE-HfO_2 near the TiN electrode by oxygen reduction, thermally stable metals or metal oxides, such as TaN,[91] W,[92] Mo,[93,94] Ru,[95] MoO_x,[96] RuO_2,[97] and Ru[95,110] as depicted in Figure 6.16a, have reportedly been adopted for HfO_2-FC, resulting in an enhancement of P_r and

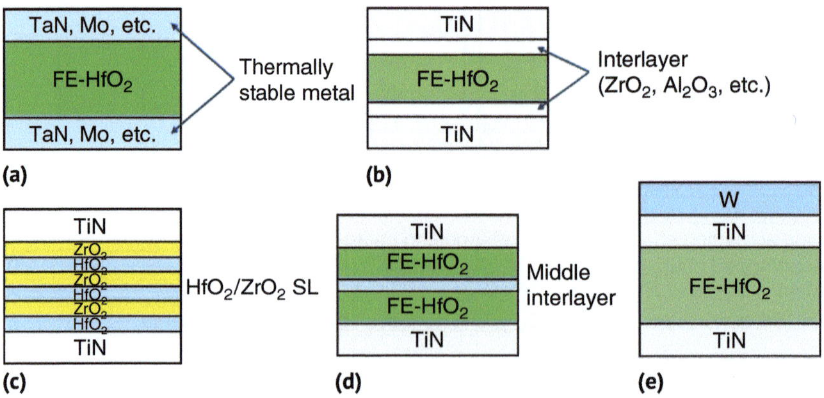

Figure 6.16 Cross-sectional diagrams of HfO_2-FCs illustrating methods for improving FE-HfO_2 or HZO capacitors through (a) thermally stable electrodes, (b) interlayers inserted between FE-HfO_2 and electrodes, (c) SL structures, (d) interlayers placed within FE-HfO_2, and (e) stress engineering.

Table 6.2 Summary of key topics regarding challenges and improvements for FE-HfO_2 in 1T1C FeRAM.

Improvement category	Key technology	Improvement effects										Mechanism	Reference	Title
		P_r	P_r (thin FE-HfO_2)	E	Fatigue lifetime	Wake-up	Imprint	Retention lifetime	Leakage current	Breakdown lifetime	Breakdown voltage			
Thermally stable electrode	MoO_x (O_x/Ar 15% ratio) BE	✓							✓			MoO_x electrode minimizes the thickness of the non-ferroelectric dead layer	[96]	Impact of molybdenum oxide electrode on the ferroelectricity of doped-hafnia oxide capacitors
Thermally stable electrode	Ru/TiN_x TE	✓			✓							Ru prevents TiN_x oxidization, and TiN_x controls oxygen and oxygen vacancy which enhances o-phase formation	[110]	Low polarization loss of long endurance on scavenged Ru-based electrode ferroelectric $Hf_{0.5}Zr_{0.5}O_2$ by optimizing TiN_x interfacial capping layer and its fatigue mechanism

(*Continued*)

Table 6.2 (Continued)

Improvement category	Key technology	Improvement effects										Mechanism	Reference	Title
		P_r	P_r (thin FE-HfO_2)	E	Fatigue lifetime	Wake-up	Imprint	Retention lifetime	Leakage current	Breakdown lifetime	Breakdown voltage			
Interlayer	1-nm LaO_2 interlayer between BE and HZO				✓							LaO interlayer suppresses the interaction between HZO/Mo and formation of oxygen vacancies inside HZO	[115]	Interfacial layer selection methodology for customized ferroelectric memories
Interlayer	AlON interlayer (TiN TE/1-nm AlON/10-nm HZO/1-nm HfO_2/TiN BE) post-deposition RTA (PDA) at 400 °C					✓						AlON interlayer enhanced P_r, PDA reduces oxygen vacancy	[116]	Improved reliability for back-end-of-line compatible ferroelectric capacitor with 3bits/cell storage capability by interface engineering and post deposition annealing

Optimization of FE-HfO_2 deposition	ALD HZO with optimum (1–5 s) O_3 pulse				✓			Increase of polar o-phase	[119]	Impact of oxygen vacancy content in ferroelectric HZO films on the device performance
Optimization of FE-HfO_2 deposition	Low-temperature HZO deposition at 200 °C	✓					✓	Low-temperature deposition causes large HZO grain leading to smaller number of charge defects	[129]	Performance improvement by modifying deposition temperature in $HfZrO_x$ ferroelectric memory
Optimization of FE-HfO_2 deposition	Low annealing temperature (400 °C) RTA for long time	✓	✓					Large (> 30 nm) and uniform grain size of HZO	[130]	Hafnia-based high-disturbance-immune and selector-free cross-point FeRAM
Optimization of FE-HfO_2 deposition	Solid-phase epitaxy of ALD HZO on W (W BE→ALD HZO→W→RTA→W etching→ALD HZO→W TE→RTA)	✓	✓	✓				Decrease in m-phase and enhancement of oxygen vacancy effect within the small-grain HZO layers	[131]	Grain size reduction of ferroelectric HZO enabled by solid phase epitaxy (SPE): Working principle, experimental demonstration, and theoretical understanding
Pretreatment	NH_3 plasma treatment both for BE and TE surfaces		✓	✓		✓		Reduction of oxygen vacancy in HZO, especially the treatment at the BE interface	[123]	Impact of plasma treatment on reliability performance for $HfZrO_x$-based metal-ferroelectric-metal capacitors

(*Continued*)

Table 6.2 (Continued)

Improvement category	Key technology	Improvement effects										Mechanism	Reference	Title
		P_r	P_r (thin FE-HfO_2)	E	Fatigue lifetime	Wake-up	Imprint	Retention lifetime	Leakage current	Breakdown lifetime	Breakdown voltage			
Pretreatment	NH_3 plasma nitridation for both BE and TE surfaces									✓		NH_3 plasma nitridation suppressing oxygen vacancy generation results in a significantly improved voltage stress immunity in the HZO-based capacitors	[124]	Role of nitrogen in ferroelectricity of $Hf_xZr_{1-x}O_2$-based capacitors with metal-ferroelectric-insulator-metal structure
Pretreatment	NH_3 plasma treatment over TiN BE before ALD HZO	✓						✓				A higher ferroelectric o-phase formation	[125]	Enhanced remnant polarization (30 μC/cm²) and retention of ferroelectric $Hf_{0.5}Zr_{0.5}O_2$ by NH_3 plasma treatment

Annealing process	HPA (10 °C/s) and rapid cooling (RC, 200 atm)		✓			✓	HPA–RC induce high tensile stress resulting in high fraction of stable o-phase	[128]	Sub-5-Å EOT $Hf_xZr_{1-x}O_2$ for next-generation DRAM capacitors using morphotropic phase boundary and high-pressure (200 atm) annealing with rapid cooling process
Annealing process	HPA (200 atm, 600 °C), with 1:3 Hf:Zr ratio	✓		✓			Reduction in the effective thickness of non-ferroelectric film. Reduction of oxygen vacancy implying a phase transition from t-phase of non-ferroelectric layer to ferroelectric o-phase	[126]	Influence of high-pressure annealing conditions on ferroelectric and interfacial properties of Zr-rich $Hf_xZr_{1-x}O_2$ capacitors
Annealing process	240 °C, 70 atm, C_2H_6O, 30 nm as secondary annealing			✓	✓		C_2H_6O provides the hydrogen ions as deoxidizers to help cure the defects and reduce dangling bonds in the non-ferroelectric layers	[127]	Improved symmetry of ferroelectric switching in HZO based MFM capacitors enabled by high pressure annealing

(Continued)

Table 6.2 (Continued)

Improvement category	Key technology	Improvement effects										Mechanism	Reference	Title
		P_r	P_r (thin FE-HfO_2)	E	Fatigue lifetime	Wake-up	Imprint	Retention lifetime	Leakage current	Breakdown lifetime	Breakdown voltage			
La doping	Optimizing Hf:La ratio (Hf:La = 15:1)	✓									✓	La elements can reduce the energy difference between the o- and m-phases, favoring the formation of o-phase	[134]	Effect of doping concentration on intrinsic ferroelectric properties of HfLaO-based ferroelectric memory
La doping	2.3 mol.% La doping				✓		✓		✓			Trap energy level reduction	[136]	Reliability improvement of ferroelectric $Hf_{0.5}Zr_{0.5}O_2$ thin films by lanthanum doping for FeRAM applications

Intercalation	Intercalating separated dual-atomic-thin Ti (DL-Ti) doping-layer HZOs	✓		✓		✓	DL-Ti scavenges oxygen and decreases the concentration of pinning site.	[135]	Significant reliability improvement by inducing dual atomic-thin titanium intercalation layers in $Hf_{0.5}Zr_{0.5}O_2$ films
SL	1.5-nm ZrO_2 (seed) + SL (1-nm HfO_2 + 1-nm ZrO_2) × five times		✓				SL HZO improves HZO quality (reduction of oxygen vacancy) and suppresses TiO_xN_{1-x} layers formation at HZO–Ti BE interface	[143]	Investigation of HfO_2 and ZrO_2 nanolamination thickness on superlattice HZO FeRAMs exhibiting enhanced remnant polarization and improved endurance
SL	Thick sublayers (≥1 nm)				✓		Small grain size and comb-like band structure	[139]	Ferroelectric HfO_2/ZrO_2 superlattices with improved leakage at bias and temperature stress
Stress engineering	Inserting a dielectric (TiO_2) interlayer and elucidating the influence of interlayer on the relatively thick HZO films (TiN BE, TiN+W TE)	✓					The tensile stress enhances o-phase formation from t-phase	[145]	Stress engineering as a strategy to achieve high ferroelectricity in thick hafnia using interlayer

(*Continued*)

Table 6.2 (Continued)

Improvement category	Key technology	Improvement effects										Mechanism	Reference	Title
		P_r	P_r (thin FE-HfO_2)	E	Fatigue lifetime	Wake-up	Imprint	Retention lifetime	Leakage current	Breakdown lifetime	Breakdown voltage			
Stress engineering	Ti BE and TiN+W TE, with different thermal expansion coefficients for each, inducing mechanical tensile stress	✓										Promotes the formation of o-phase	[147]	The influence of top and bottom metal electrodes on ferroelectricity of hafnia
Stress engineering	Interconnecting metal (W) with TiN/HZO/TiN to induce mechanical tensile stress	✓		✓				✓				Intense compressive stress enhances o-phase formation	[148]	Stress effects of interconnecting metals on back-end-of-line compatible $Hf_{0.5}Zr_{0.5}O_2$ ferroelectric capacitors

mitigating the wake-up effect by preventing V_O generation or FE-HfO_2 reduction. Zhang et al. reported that a rough interface between TaN and HZO enhanced P_r. However, it also led to an increase in interface traps and leakage current, as well as the degradation of endurance.[111]

It has also been reported that thin oxide layers, such as ZrO_2,[112,113] Al_2O_3,[114] SiO_2, LaO,[115] and AlON,[116] enhance P_r by preventing the formation of a dead layer with inferior ferroelectricity, as depicted in Figure 6.16b. Additionally, the HfO_2 seed layer on HZO promotes o-phase generation by facilitating the diffusion/migration of oxygen atoms between the seed layer and the substrate or vice versa.[117] Jeong et al. found that their fabricated TiN/HZO/TaO/TiN structure, using high-pressure annealing (HPA), exhibited a high imprint field (1.65 MV/cm), which influenced the switching, interfacial, and ferroelectric properties.[118]

6.5.2.2 Optimization of FE-HfO_2 Deposition, Pretreatment, and Annealing Processes

Various effective improvements have been reported in FE-HfO_2 deposition and post-deposition annealing. In the case of 1T1C FeRAM, FC is desired to be fabricated at temperatures lower than 400 °C in the BEOL process, so controlling V_O concentration is important. Dose optimization of O_3 as an oxidation gas during ALD deposition of HZO[119,120] and crystallization annealing[121] has reportedly improved the FC characteristics, especially polarization, as well as reliability. Lee et al. developed a new annealing method for HZO crystallization at 250 °C and 10–5 Torr. They successively obtained good electric properties, $2P_r$ and $2E_c$ values of 36.7 μC/cm^2 and 2.38 MV/cm, respectively, on an Mo/HZO/Mo capacitor.[122] NH_3 plasma pretreatment before FE-HfO_2 deposition has been reported to be effective in improving the ferroelectricity of FE-HfO_2. This treatment reduces V_O,[123,124] restrains the interfacial layer (IL), and enhances the formation of o-phase by reducing the intrinsic TiO_xN_y dead layer.[125] HPA for crystallization of FE-HfO_2 has also been found effective in improving endurance, wake-up behavior,[126,127] and leakage current[128] by reducing the non-ferroelectric IL.

It has been reported that the grain size of FE-HfO_2, controlled by annealing or deposition temperatures, significantly influences the electrical properties of HfO_2-FC. Chen et al.[129] and Fu et al.[130] reported that larger grain sizes, such as 80 nm and over 30 nm, improved P_r. In contrast, Zhang et al.[131] demonstrated that smaller grain sizes, less than 10 nm, also enhanced P_r.

6.5.2.3 La Doping and Intercalation

La doping has been reported to be highly effective in decreasing E_c and leakage current as well as increasing P_r for HZO[132] and HfO_2,[133] due to the inhibition of m-phase generation. Silva et al.[133] investigated the effects of La content on the structural and ferroelectric properties of epitaxial La-doped HfO_2 thin films using experimental and theoretical methods. They found that the increase in P_r and decrease in E_c were due to the inhibition of m-phase formation, as well as a reduction in domain wall motion and polarization switching energy. Liu et al. found that the P_r of La-doped HfO_2-FC increased with increasing La concentration until the Hf:La atomic ratio reached 15:1. This improvement was attributed to a reduction in the energy difference between the o- and m-phases.[134] Tai et al. demonstrated that dual-Ti intercalation in the HZO capacitor enhanced P_r and improved endurance due to oxygen scavenging and a decrease in the concentration of pinning sites caused by the dual-Ti layers or TiO_2.[135] Mehmood et al. reported that trap energy level in HZO decreased from 0.5 to 0.2V with La doping, which led to improvements in cycling endurance, a reduction of leakage current, and enhanced imprint resistance.[136]

6.5.2.4 Superlattices and Middle Interlayers

HfO_2/ZrO_2 superlattices (SLs), as depicted in Figure 6.16c, which consist of alternating stacks of a few atomic layers of HfO_2 and ZrO_2, whereas HZO typically consists of a compound of HfO_2 and ZrO_2 via CVD and CSD, or alternating single atomic layers of HfO_2 and ZrO_2 via ALD, have been reported to improve polarization properties, such as fatigue resistance[137,138] and leakage current.[139] Peng and co-workers investigated a TaN/HfO_2 (1.2nm)–ZrO_2 (1.2nm) SL (with a total thickness of 10 nm)/TaN capacitor and found that their capacitor exhibited an endurance of 5×10^{12} cycles. Fatigue could be fully or partially recovered with a ~30-s break or higher field cycling due to a decrease in charge trapping/detrapping, which reduced defect density.[138] Grain boundary engineering using middle interlayers of Al_2O_3,[140] AlN,[141] Mo, and W,[142] as depicted in Figure 6.16d, has reportedly improved the P_r properties of HZO, likely due to the suppression of m-phase formation in $Hf_{0.5}Zr_{0.5}O_2$. Hsieh et al. demonstrated that an FC consisting of a TiN electrode, an HZO SL with ZrO_2 nanolaminations (NLs), and an ZrO_2 seed layer achieved an excellent $2P_r$ of 43.3 μC/cm^2. They explained that the periodic SL structure of HZO and HfO_2 NLs, acting as diffusion barrier layers, effectively suppressed O and Ti diffusion at the TiN–HZO interfaces. This, in turn, inhibited the formation of thin TiO_xN_{1-x} layers and a V_O-rich layer at the TiN–HZO interfaces.[143]

6.5.2.5 Stress Engineering

Stress engineering applied to Si-doped HfO_2[144] using a four-point bending tool, to HZO with a TiO_2 insertion layer,[145] and to HZO via a thermal expansion mismatch using W/TiN, as depicted in Figure 6.16e, as well as surface roughness control of the electrode and HZO,[146] has been effective in enhancing the P_r of FE-HfO_2. Kruv et al.[144] reported that this enhancement is due to in-plane compressive strain, which promotes polarization alignment parallel to the electric field, leading to a $2P_r$ increase. It has been reported that applying stress through the thermal expansion difference between TE and a capping layer,[146–148] or by using an insertion layer such as TiO_2 or SiO_2,[145] promotes the formation of o-phase in HZO FCs. This, in turn, results in an increase in P_r.

6.5.2.6 AFEs

FE-HfO_2 suffers from polarization-switching-induced problems such as wake-up and fatigue. ZrO_2, an AFE material, is a candidate to mitigate these problems, as reported by Pešić et al.[149,150] They demonstrated nonvolatile capacitors using AFE ZrO_2 by employing electrodes with different work function values, thereby introducing a built-in bias within the AFE stack, which creates two stable nonvolatile states. Their fabricated AFE FC exhibited superior endurance compared to FE-HfO_2-based memories.[150] Chang et al. demonstrated a deep-trench capacitor using AFE HZO (Hf/Zr ratio not described) that exhibited low-voltage operation at 1.8 V and a high endurance of 10^{12} cycles.[151] Chen et al. also demonstrated a capacitor with a 6-nm $Hf_{0.2}Zr_{0.8}O_2$, exhibiting a high endurance of 10^{12} cycles under full polarization switching conditions at an electric field of 4.5 MV/cm.[152] Hsiang et al. proposed unipolar recovery of FC for the first time, which completely restores a fatigued AFE capacitor to its initial state. This was proven with an endurance of 5.6×10^{11} cycles on their fabricated FC using AFE $Hf_{0.1}Zr_{0.9}O_2$. Although AFE capacitors have high endurance and low operating voltage, their P_r and fatigue improvement are insufficient for FeRAM.[153] Lo et al. proposed a capacitor comprising a stack of 5-nm-thick AFE $Hf_{0.25}Zr_{0.75}O_2$ on 5-nm-thick FE $Hf_{0.6}Zr_{0.4}O_2$ to combine the advantages of AFE and FE materials. The bilayer HZO exhibited the benefits of both antiferroelectricity and ferroelectricity, including a high dielectric constant (58), a high P_r (approximately 15 $\mu C/cm^2$), a low coercive field, and excellent endurance ($> 10^{10}$ cycles).[154]

6.5.2.7 Scaling Issues

Electric properties and reliability parameters, such as P_r, wake-up, imprint, and retention, associated with FC scaling, both in terms of lateral area and FE-HfO_2 thickness, are important. Alcala et al. investigated these factors in a 1T1C FeRAM with Si-doped HfO_2-FCs fabricated using 130-nm technology. They found that lateral (areal) scaling of the devices showed no detrimental effects on wake-up, fatigue, imprint, or retention behavior, compared to larger devices, likely due to the absence of edge effects. However, an improvement in field cycling endurance was observed, possibly due to a net reduction in defects within the ferroelectric layer, simply as a result of having a smaller area, which reduces the probability of breakdown events.[155] Tahara et al. found that the thickness–temperature relationship indicated a clear trade-off between thickness scaling and the crystallization temperature of their fabricated ferroelectric $Hf_{0.5}Zr_{0.5}O_2$, with thickness varying from 9.5 down to 2.8 nm.[53] They demonstrated that a 4-nm-thick $Hf_{0.5}Zr_{0.5}O_2$ crystallized at 500 °C exhibited excellent ferroelectricity $\left(2P_r > 25\ \mu C/cm^2\right)$, low operating voltage (0.7–1.2 V), and high read/write endurance (projected to 10^{14} cycles).

6.6 Future Trend of HfO_2 1T1C FeRAM

The recent surge of data-centric applications, such as artificial intelligence (AI) and the Internet of Things (IoT), has significantly burdened conventional computing systems to increase processing speed while minimizing energy consumption. This has created a demand for memories with nonvolatile characteristics, high operation speed, and low energy consumption. The 1T1C HfO_2 FeRAM will play an important role in small systems at the edge of main IoT systems, helping to mitigate data processing issues. Given the recent progress in memory density for 1T1C HfO_2 FeRAM, which is similar to the memory density of DRAM (as referenced in [49]), it will accommodate the demand in the memory hierarchy of von Neumann architecture-based computing systems. From another perspective, conventional von Neumann computing is approaching its limitations due to enormous energy consumption and the constraints of Moore's Law.[156] As a result, new types of computing systems have been proposed. Takagi and co-workers proposed a reservoir computing system using HZO/Si FeFETs and TiN/ HZO/TiN MFM capacitors for AI applications. They demonstrated that MFM capacitors with an HZO thickness of less than 5 nm could achieve low crystallization temperatures, excellent ferroelectricity, low operating voltage, and high read/write endurance by performing sufficient wake-up operations on the thin HZO films.[54,55] In-memory computing is another promising non-von Neumann computing approach, expected to enable highly energy-efficient AI engines – potentially improving compute efficiency by a factor of 1,000 – particularly for edge intelligence applications. Chen et al. demonstrated NOR and NOT logic gates consisting of TiN/Al-doped HfO_2/Pt/TiN capacitors.[157]

If the reliability challenges of HfO_2-based Fe-Ms including 1T1C FeRAM can be fully addressed, the currently existing DRAM and flash memory will be replaced with Fe-Ms. Furthermore, Fe-Ms could become the primary memory solution in non-von Neumann architecture-based computing systems, paving the way for the next generation of computing technology.

Appendix A

Sensing circuits should be robust enough to account for semiconductor processing variations across the die and the inherent device imperfections of FCs. There have been reports of these elaborate circuits.[158] In this appendix, we introduce the following three sensing schemes.

A.1 Column Sensing Scheme

Figure A.1 compares two reference schemes: the conventional row sensing scheme[159] and the column sensing scheme.[160] The conventional row sensing scheme uses a reference cell for each column. Assuming that the memory array consists of n rows, the FC used for each reference cell is influenced by polarization cycling (such as wake-up and fatigue) n times more than the memory cell. In contrast, the column sensing scheme uses a reference cell for each row. In this scheme, the FCs used for both the reference cell and the memory cell experience polarization cycling the same number of times. Therefore, the column sensing scheme is more robust than the row sensing scheme, although it sacrifices cost in slightly more complex circuitry.

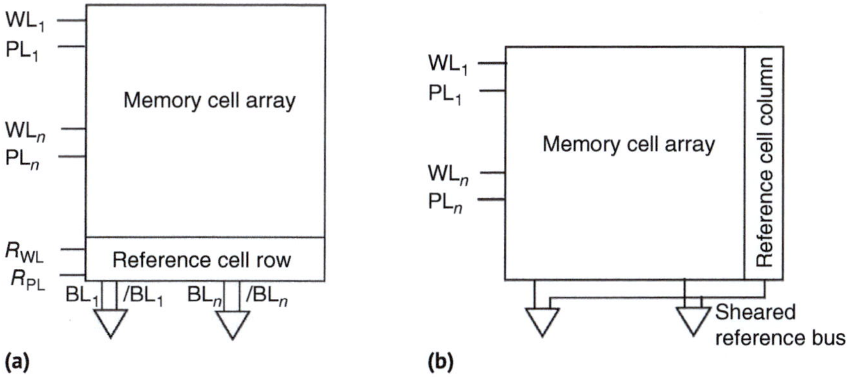

Figure A.1 Schematic illustration of the (a) conventional row sensing scheme and (b) column sensing scheme.

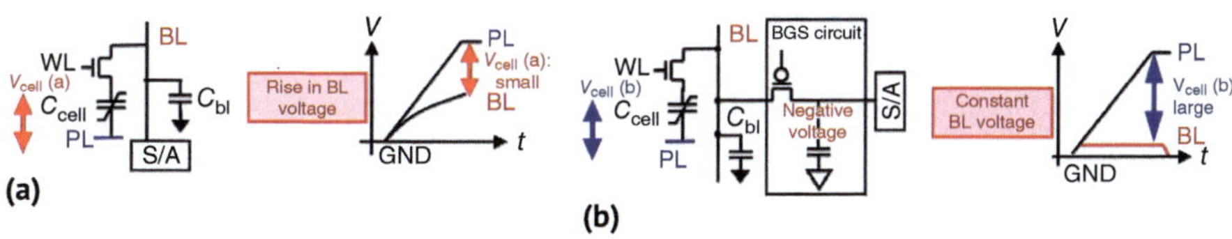

Figure A.2 The 1T1C FeRAM sensing methods and corresponding voltage charts for PL and BL: (a) the conventional high-*Z* sensing scheme and (b) the BGS scheme.

A.2 BGS Scheme

In the conventional high-*Z* sensing scheme, as shown in Figure A.2a, a cell FC connected to BL cannot be fully applied with V_{dd}, power supply voltage, due to the floating BL. This results in a marginal sensing performance, especially when the P_r of cell FC decreases due to depolarization, imprint, or fatigue. In the BGS scheme,[161] the charges output from cell FC to BL are absorbed by a capacitor in the BGS circuit, which is incorporated between BL and S/A as shown in Figure A.2b. This configuration ensures that the BL potential is maintained at ground level during sensing, allowing the cell FC to be fully applied with V_{dd}. Therefore, the BGS scheme provides a more robust sensing margin, although the power consumption is slightly increased by the BGS circuit.

A.3 Overwrite S/A and Programmable Ferroelectric Reference Generation Schemes

This architecture aims to achieve fast, reliable circuit operation of 1T1C FeRAM. McAdams et al. demonstrated 64-Mb 1T1C FeRAM with 30-ns access time, consuming less than 20 mA of operating current, while yielding a memory density of over 1.1 Mb/mm^2.[162]

References

1 Schenk, T., Pešić, M., Slesazeck, S. et al. (2020). Memory technology – a primer for material scientists. *Reports on Progress in Physics* 83: 086501. https://doi.org/10.1088/1361-6633/ab8f86.

2 Park, M. J., Lee, J., Cho, K. et al. (2022). A 192-Gb 12-high 896-GB/s HBM3 DRAM with a TSV auto-calibration scheme and machine-learning-based layout optimization. *IEEE Journal of Solid-State Circuits* 58: 256–269. https://doi.org/10.1109/ISSCC42614.2022.9731562.

3 Bhati, I., Chang, M. T., Chishti, Z. et al. (2016). DRAM refresh mechanisms, penalties, and trade-offs. *IEEE Transactions on Computers* 65: 108–121. https://doi.org/10.1109/TC.2015.2417540.

4 Lee, J. W., Na, D., Kavala, A. et al. (2020). A 1.8 Gb/s/pin 16Tb NAND flash memory multi-chip package with F-chip of Toggle 4.0 specification for high performance and high capacity storage systems. In: *Proceedings of the 2020 IEEE Symposium on VLSI Circuits*. IEEE. https://doi.org/10.1109/VLSICircuits18222.2020.9163052.

5 IEEE (2023). *International Roadmap for Devices and Systems: Mass Data Storage*. Piscataway, NJ: IEEE Computer Society. https://doi.org/10.60627/4hs9-2098.

6 Scott, J. F., Araujo, C. A., and McMillan, L. D. (1989). Ferroelectric memory applications. In: *Proceedings of IEEE Ultrasonics Symposium*. IEEE. pp. 299–308. https://doi.org/10.1109/ULTSYM.1989.66999.

7 Eshita, T., Wang, W., and Hikosaka, Y. (2023). Ferroelectric memory. In: *Encyclopedia of Materials: Electronics: Volume 1*. Elsevier Science. pp. 218–240. https://doi.org/10.1016/B978-0-12-819728-8.

8 Zanata, M., Wrachien, N., and Cester, A. (2008). Ionizing radiation effect on ferroelectric nonvolatile memories and its dependence on the irradiation temperature. *IEEE Transactions on Nuclear Sciences* 55: 3237–3245. https://doi.org/10.1109/TNS.2008.2006052.

9 Huang, F., Wang, Y., Liang, X. et al. (2017). HfO_2-based highly stable radiation-immune ferroelectric memory. *IEEE Electron Device Letters* 38: 330–333. https://doi.org/10.1109/LED.2017.2653848.

10 Böscke, T. S., Müller, J., Bräuhaus, D. et al. (2011). Ferroelectricity in hafnium oxide thin films. *Applied Physics Letters* 99: 102903. https://doi.org/2011 doi:10.1063/1.3634052.

11 Müller, S., Böscke, T. S., Müller, S. et al. (2013). Ferroelectric hafnium oxide: A CMOS-compatible and highly scalable approach to future ferroelectric memories. In: *Proceedings of the 2013 IEEE International Electron Devices Meeting (IEDM)*. https://doi.org/10.1109/IEDM.2013.6724605.

12 Mikolajick, T., Schroeder, U., and Slesazeck, S. (2020). The past, the present, and the future of ferroelectric memories. *IEEE Transactions on Electron Devices* 67: 1434–1443. https://doi.org/10.1109/TED.2020.2976148.

13 Park, J. Y., Yang, K., Lee, D. H. et al. (2020). A perspective on semiconductor devices based on fluorite structured ferroelectrics from the materials–device integration perspective. *Journal of Applied Physics* 128: 240904. https://doi.org/10.1063/5.0035542.

14 Aswathy, N. and Sivamangai, N. M. (2021). Future nonvolatile memory technologies: Challenges and applications. In: *Proceedings of the 2021 2nd International Conference on Advances in Computing, Communication, Embedded and Secure Systems (ACCESS)*. https://doi.org/10.1109/ACCESS51619.2021.9563288.

15 Schenk, T. and Mueller, S. (2021). A new generation of memory devices enabled by ferroelectric hafnia and zirconia. In: *Proceedings of the 2021 IEEE International Symposium on Applications of Ferroelectrics (ISAF)*. https://doi.org/10.1109/ISAF51943.2021.9477377.

16 Marinella, M. J. (2021). Radiation effects in advanced and emerging nonvolatile memories. *IEEE Transactions on Nuclear Science* 68: 546–572. https://doi.org/10.1109/TNS.2021.3074139.

17 Banerjee, W. (2020). Challenges and applications of emerging nonvolatile memory devices. *Electronics* 9: 1029. https://doi.org/10.3390/electronics9061029.

18 Toprasertpong, K., Tahara, K., Hikosaka, Y. et al. (2022). Interfaces, low operating voltage, improved breakdown tolerance, and high endurance in $Hf_{0.5}Zr_{0.5}O_2$ ferroelectric capacitors achieved by thickness scaling down to 4 nm for embedded ferroelectric memory. *ACS Applied Materials & Interfaces* 14: 51137–51148. https://doi.org/10.1021/acsami.2c15369.

19 Zagni, N., Puglisi, F. M., Pavan, P. et al. (2023). Reliability of HfO_2-based ferroelectric FETs: A critical review of current and future challenges. *Proceedings of the IEEE* 111: 158–184. https://doi.org/10.1109/JPROC.2023.3234607.

20 Esaki, L., Laibowitz, R. B., and Stiles, P. J. (1971). Polar switch. *IBM Technical Disclosure Bulletin* 13: 2161–2162.

21 Evans, J. T. and Womack, R. (1988). An experimental 512-bit non-volatile memory with ferroelectric storage cell. *IEEE Journal of Solid-State Circuits* 23: 1171–1175. https://doi.org/10.1109/4.5940.

22 Eaton, S. S., Butler, D. B., Parris, M. et al. (1988). A ferroelectric non-volatile memory. In: *Proceedings of the 1988 IEEE International Solid-State Circuits Conference: Digest of Technical Papers*. IEEE. pp. 130–131. https://doi.org/10.1109/ISSCC.1988.663665.

23 Dimmler, K. and Eaton, S. S., Jr. (1987). Memory cell with volatile and non-volatile portions having ferroelectric capacitors. Patent Number US4809225. Ramtron Corporation.

24 Masui, S., Ninomiya, T., Oura, M. et al. (2003). A ferroelectric memory-based secure dynamically programmable gate array. *IEEE Journal of Solid-State Circuits* 38: 715–725. https://doi.org/10.1109/JSSC.2003.810034.

25 Takeuchi, K., Kobayashi M., and Hiramoto, T. (2019) "A Feasibility Study on Ferroelectric Shadow SRAMs Based on Variability-Aware Design Optimization," In: IEEE Journal of the Electron Devices Society, vol. 7, pp. 1284–1292. doi. 10.1109/JEDS.2019.294956.

26 Zhao, W., Belhaire, E., Chappert, C. et al. (2008). Spintronic device based non-volatile low standby power SRAM. In: *Proceedings of the 2008 IEEE Computer Society Annual Symposium on VLSI*. IEEE. pp. 40–45. https://doi.org/10.1109/ISVLSI.2008.11.

27 Dilillo, L., Girard, P., Pravossoudovitch, S. et al. (2004). Resistive-open defects in embedded-SRAM core cells: Analysis and march test solution. In: *Proceedings of the 13th Asian Test Symposium*. https://doi.org/10.1109/ATS.2004.75.

28 Yamamoto, S., Shuto, Y., and Sugahara, S. (2009). Nonvolatile SRAM (NV-SRAM) using functional MOSFET merged with resistive switching devices. In: *Proceedings of the 2009 IEEE Custom Integrated Circuits Conference*. https://doi.org/10.1109/CICC.2009.5280761.

29 Traynor, S. D., Hadnagy, T. D., and Kammerdiner, L. (2017). Device integration issues and testing capacitor test simulation of retention and imprint characteristics for ferroelectric memory operation. *Integrated Ferroelectrics* 16: 63–76. https://doi.org/10.1080/10584589708013030.

30 Kawashima, S., Fukushi, I., Morita, K. et al. (2007). A reliable 1T1C FeRAM using a thermal history tracking 2T2C dual reference level technique for a smart card application chip. *IEICE Transactions on Electronics* E90C: 1941–1948. https://doi.org/10.1093/ietele/e90-c.10.1941 E90-C.

31 Chowdhury, U., Goswami, S., Bhattacharya, D. et al. (2016). Determination of intrinsic ferroelectric polarization in lossy improper ferroelectric systems. *Applied Physical Letters* 109: 092902. https://doi.org/10.1063/1.4961988.

32 Ma, T. P. and Han, J. (2002). Why is nonvolatile ferroelectric memory field-effect transistor still elusive? *IEEE Electron Device Letters* 23: 386–388. https://doi.org/10.1109/LED.2002.1015207.

33 Rodriguez, J., Remack, K., Gertas, J. et al. (2010). Reliability of ferroelectric random access memory embedded within 130nm CMOS. In: *Proceedings of the 2010 IEEE International Reliability Physics Symposium*. https://doi.org/10.1109/IRPS.2010.5488738.

34 Mihara, T., Watanabe, H., and de Araujo, P. (1994). Polarization fatigue characteristics of sol-gel ferroelectric $Pb(Zr_{0.4}Ti_{0.6})O_3$ thin-film capacitors. *Japanese Journal of Applied Physics* 33: 3996. https://doi.org/doi10.1143/JJAP.33.3996.

35 Polcawich, R. G., Feng, C., Vanatta, P. et al. (2000). Highly accelerated lifetime testing (HALT) of lead zirconate titanate (PZT) thin films. In: *Proceedings of the 2000 12th IEEE International Symposium on Applications of Ferroelectrics*. https://doi.org/10.1109/ISAF.2000.941571.

36 Chernikova, A. G. and Markeev, A. M. (2021). Dynamic imprint recovery as an origin of the pulse width dependence of retention in $Hf_{0.5}Zr_{0.5}O_2$-based capacitors. *Applied Physics Letters* 119: 032904. https://doi.org/10.1063/5.0057188.

37 Schroeder, U., Weinreich, W., Erben, E. et al. (2009). Detailed correlation of electrical and breakdown characteristics to the structural properties of ALD grown HfO_2- and ZrO_2-based capacitor dielectrics. *ECS Transactions* 25: 357. https://doi.org/10.1149/1.3205070.

38 Kawashima, S. and Cross, J. S. (2009). FeRAM. In: Zhang, K. (ed.). *Embedded Memories for Nano-Scale VLSIs*. Springer. pp. 279–328. https://doi.org/10.1007/978-0-387-88497-4_8.

39 Okuno, J., Kunihiro, T., Konishi, K. et al. (2020). SoC compatible 1T1C FeRAM memory array based on ferroelectric $Hf_{0.5}Zr_{0.5}O_2$. In: *Proceedings of the 2020 IEEE Symposium on VLSI Technology*. https://doi.org/10.1109/VLSITechnology18217.2020.9265063.

40 Okuno, J., Kunihiro, T., Konishi, K. et al. (2021). 1T1C FeRAM memory array based on ferroelectric HZO with capacitor under bitline. *IEEE Journal of the Electron Devices Society* 10: 29–34. https://doi.org/10.1109/JEDS.2021.3129279.

41 Okuno, J., Kunihiro, T., Konishi, K. et al. (2021). High-endurance and low-voltage operation of 1T1C FeRAM arrays for nonvolatile memory application. In: *Proceedings of the 2021 IEEE International Memory Workshop (IMW)*. https://doi.org/10.1109/IMW51353.2021.9439595.

42 Okuno, J., Kunihiro, T., Konishi, K. et al. (2022). Reliability study of 1T1C FeRAM arrays with $Hf_{0.5}Zr_{0.5}O_2$ thickness scaling. *IEEE Journal of the Electron Devices Society* 10: 778–783. https://doi.org/10.1109/JEDS.2022.3187101.

43 Francois, T., Coignus, J., Makosiej, A. et al. (2021). 16kbit HfO_2:Si-based 1T-1C FeRAM arrays demonstrating high performance operation and solder reflow compatibility. In: *Proceedings of the 2021 IEEE International Electron Devices Meeting (IEDM)*. https://doi.org/10.1109/IEDM19574.2021.9720640.

44 François, T., Coignus, J., Makosiej, A. et al. (2022). High-performance operation and solder reflow compatibility in BEOL-integrated 16-kb HfO_2: Si-based 1T-1C FeRAM arrays. *IEEE Transactions on Electron Devices* 69: 2108–2114. https://doi.org/10.1109/TED.2021.3138360.

45 Sung, M., Rho, K., Kim, J. et al. (2021). Low voltage and high speed 1×nm 1T1C FE-RAM with ultra-thin 5nm HZO. In: *Proceedings of the 2021 IEEE International Electron Devices Meeting (IEDM)*. https://doi.org/10.1109/IEDM19574.2021.9720545.

46 Okuno, J., Kunihiro, T., Yonai, T. et al. (2023). A highly reliable 1.8 V 1 Mb $Hf_{0.5}Zr_{0.5}O_2$-based 1T1C FeRAM array with 3-D capacitors. In: *Proceedings of the 2023 International Electron Devices Meeting (IEDM)*. https://doi.org/10.1109/IEDM45741.2023.10413661.

47 Wu, J., Mo, F., Saraya, T. et al. (2021). Monolithic integration of oxide semiconductor FET and ferroelectric capacitor enabled by Sn-doped InGaZnO for 3-D embedded RAM application. *IEEE Transactions on Electron Devices* 68: 6617–6622. https://doi.org/10.1109/TED.2021.3111145.

48 Endo, M., Numata, S., Ohshima, K. et al. (2022). A c-axis aligned crystalline IGZO FET and a 0.06-μm^2 HfO_2-based capacitor 1T1C FeRAM with high voltage tolerance and 10-ns write time. In: *Proceedings of the 2022 International Electron Devices Meeting (IEDM)*. https://doi.org/10.1109/IEDM45625.2022.10019506.

49 Ramaswamy, N., Calderoni, A., Zahurak, J. et al. (2023). NVDRAM: A 32Gb dual layer 3D stacked non-volatile ferroelectric memory with near-DRAM performance for demanding. In: *Proceedings of the 2023 International Electron Devices Meeting (IEDM)*. https://doi.org/10.1109/IEDM45741.2023.10413848.

50 Saito, H., Sugimachi, T., Nakamura, K. et al. (2015). A triple-protection structured COB FRAM with 1.2-V operation and 1017-cycle endurance. In: *Proceedings of the 2015 International Memory Workshop*. https://doi.org/10.1109/IMW.2015.7150275.

51 Zhang, W., Wan, G., Lin, Y. et al. (2024). Investigation of ionizing radiation mechanisms on HfO_2-based ferroelectric thin-film memories with various configurations. *Journal of Materials Science: Materials in Electronics* 35: 296. https://doi.org/10.1007/s10854-024-12033-5.

52 Kang, S., Jang, W. S., Morozovska, A. N. et al. (2022). Highly enhanced ferroelectricity in HfO_2-based ferroelectric thin film by light ion bombardment. *Sciences* 376: 731–738. https://doi.org/10.1126/science.abk319.

53 Tahara, K., Toprasertpong, K., Hikosaka, Y. et al. (2021). Strategy toward HZO BEOL-FeRAM with low-voltage operation ($\leq$ 1.2 V), low process temperature, and high endurance by thickness scaling. In: *Proceedings of the 2021 Symposium on VLSI Technology*.

54 Nako, E., Toprasertpong, K., Nakane, R. et al. (2020). Proposal and experimental demonstration of reservoir computing using $Hf_{0.5}Zr_{0.5}O_2$/Si FeFETs for neuromorphic applications. In: *Proceedings of the 2020 IEEE Symposium on VLSI Technology*. https://doi.org/10.1109/VLSITechnology18217.2020.9265110.

55 Takagi, S., Toprasertpong, K., Tahara, K. et al. (2021). HfZrO-based ferroelectric devices for lower power AI and memory applications. *ECS Transactions* 104: 17. https://doi.org/10.1149/10404.0017ecst.

56 Huan, T. D., Sharma, V., Rossetti, G. A., Jr. et al. (2014). Pathways towards ferroelectricity in hafnia. *Physical Review B* 90: 064111. https://doi.org/10.1103/PhysRevB.90.064111.

57 Materlik, R., Künneth, C., and Kersch, A. (2015). The origin of ferroelectricity in $Hf_{1-x}Zr_xO_2$: A computational investigation and a surface energy model. *Journal of Applied Physics* 117: 134109. https://doi.org/10.1063/1.4916707.

58 Xu, L., Nishimura, T., Shibayama, S. et al. (2017). Kinetic pathway of the ferroelectric phase formation in doped HfO_2 films. *Journal of Applied Physics* 122: 124104. https://doi.org/10.1063/1.5003918.

59 Ryu, H., Wu, H., Rao, F. et al. (2019). Ferroelectric tunneling junctions based on aluminum oxide/zirconium-doped hafnium oxide for neuromorphic computing. *Scientific Reports* 9: 20383. https://doi.org/10.1038/s41598-019-56816-x.

60 Park, M. H., Kim, H. J., Kim, Y. J. et al. (2013). Evolution of phases and ferroelectric properties of thin $Hf_{0.5}Zr_{0.5}O_2$ films according to the thickness and annealing temperature. *Applied Physics Letters* 102: 242905. https://doi.org/10.1063/1.4811483.

61 Willmott, P. R. and Huber, J. R. (2000). Huber "pulsed laser vaporization and deposition." *Reviews of Modern Physics* 72: 315–328. https://doi.org/10.1103/RevModPhys.72.315.

62 Müller, J., Böscke, T. S., Schröder, U. et al. (2012). Ferroelectricity in simple binary ZrO_2 and HfO_2. *Nano Letters* 12: 4318–4323. https://doi.org/10.1021/nl302049k.

63 Park, M. H., Kim, H. J., Kim, Y. J. et al. (2013). Effect of forming gas annealing on the ferroelectric properties of $Hf_{0.5}Zr_{0.5}O_2$ thin films with and without Pt electrodes. *Applied Physics Letters* 102: 112914. https://doi.org/10.1063/1.4798265.

64 Park, M. H., Kim, H. J., Kim, Y. J. et al. (2015). Study on the size effect in $Hf_{0.5}Zr_{0.5}O_2$ films thinner than 8 nm before and after wake-up field cycling. *Applied Physics Letters* 107: 192907. https://doi.org/10.1063/1.4935588.

65 Li, Y., Li, J., Liang, R. et al. (2019). Switching dynamics of ferroelectric HfO_2-ZrO_2 with various ZrO_2 contents. *Applied Physics Letters* 114: 142902. https://doi.org/10.1063/1.5093793.

66 Hsain, H. A., Lee, Y., Lancaster, S. et al. (2022). Role of oxygen source on buried interfaces in atomic-layer deposited ferroelectric hafnia–zirconia thin films. *ACS Applied Materials & Interfaces* 14: 42232–42244. https://doi.org/10.1021/acsami.2c11073.

67 George, S. M. (2010). Atomic layer deposition: An overview. *Chemical Reviews* 110: 111–131. https://doi.org/10.1021/cr900056b.

68 Johnson, R. W., Hultqvist, A., and Bent, S. F. (2014). A brief review of atomic layer deposition: From fundamentals to applications. *Materials Today* 17: 236–246. https://doi.org/10.1016/j.mattod.2014.04.026.

69 Materano, M., Mittmann, T., Lomenzo, P. D. et al. (2020). Influence of oxygen content on the structure and reliability of ferroelectric $Hf_xZr_{1-x}O_2$ layers. *ACS Applied Materials & Interfaces* 2: 3618–3626. https://doi.org/10.1021/acsaelm.0c00680.

70 Pal, A., Narasimhan, V. K., Weeks, S. et al. (2017). Enhancing ferroelectricity in dopant-free hafnium oxide. *Applied Physics Letters* 110: 022903. https://doi.org/10.1063/1.4973928.

71 Richter, C., Schenk, T., Schroeder, U. et al. (2014). Film properties of low temperature HfO_2 grown with H_2O, O_3, or remote O_2-plasma. *Journal of Vacuum Science & Technology A* 32: 01A117. https://doi.org/10.1116/1.4842675.

72 Nigro, R. L., Schilirò, E., Mannino, G. et al. (2020). Comparison between thermal and plasma enhanced atomic layer deposition processes for the growth of HfO_2 dielectric layers. *Journal of Crystal Growth* 539: 125624. https://doi.org/10.1016/j.jcrysgro.2020.125624.

73 Alcala, R., Richter, C., Materano, M. et al. (2021). Influence of oxygen source on the ferroelectric properties of ALD grown $Hf_{1-x}Zr_xO_2$ films. *Journal of Physics D: Applied Physics* 54: 035102. https://doi.org/10.1088/1361-6463/abbc98.

74 Kim, S. J., Mohan, J., Kim, H. S. et al. (2019). Effect of hydrogen derived from oxygen source on low-temperature ferroelectric TiN/$Hf_{0.5}Zr_{0.5}O_2$/TiN capacitors. *Applied Physics Letters* 115: 182901. https://doi.org/10.1063/1.5126144.

75 Wang, W., Nomura, K., Yamaguchi, H. et al. (2017). Control of La-doped Pb(Zr,Ti)O_3 crystalline orientation and its influence on the properties of ferroelectric random access memory. *Japanese Journal of Applied Physics* 56: 10PF14. https://doi.org/10.7567/jjap.56.10pf14.

76 Chernikova, A. G., Kuzmichev, D. S., Negrov, D. V. et al. (2016). Ferroelectric properties of full plasma-enhanced ALD TiN/La:HfO_2/TiN stacks. *Applied Physics Letters* 108: 242905. https://doi.org/10.1063/1.4953787.

77 Hur, J., Tasneem, N., Choe, G. et al. (2020). Direct comparison of ferroelectric properties in $Hf_{0.5}Zr_{0.5}O_2$ between thermal and plasma-enhanced atomic layer deposition. *Nanotechnology* 31: 505707. https://doi.org/10.1088/1361-6528/aba5b7.

78 Shimizu, T. (2018). Ferroelectricity in HfO_2 and related ferroelectrics. *Journal of the Ceramic Society of Japan* 126: 667–674. https://doi.org/10.2109/jcersj2.18104.

79 Shimizu, T., Yokouchi, T., Oikawa, T. et al. (2015). Contribution of oxygen vacancies to the ferroelectric behavior of $Hf_{0.5}Zr_{0.5}O_2$ thin films. *Applied Physics Letters* 106: 112904. https://doi.org/10.1063/1.4915336.

80 Ogugua, S. N., Ntwaeaborwa, O. M., and Swart, H. C. (2020). Latest development on pulsed laser deposited thin films for advanced luminescence applications. *Coatings* 10: 1078. https://doi.org/10.3390/coatings10111078.

81 Shimizu, T., Katayama, K., Kiguchi, T. et al. (2015). Growth of epitaxial orthorhombic $YO_{1.5}$-substituted HfO_2 thin film. *Applied Physics Letters* 107: 032910. https://doi.org/10.1063/1.4927450.

82 Mueller, S., Müller, J., Schroeder, U. et al. (2012). Reliability characteristics of ferroelectric $Si:HfO_2$ thin films for memory applications. *IEEE Transactions on Device Material Reliability* 13: 93–97. https://doi.org/10.1109/TDMR.2012.2216269.

83 Song, T., Solanas, R., Qian, M. et al. (2022). Large enhancement of ferroelectric polarization in $Hf_{0.5}Zr_{0.5}O_2$ films by low plasma energy pulsed laser deposition. *Journal of Materials Chemistry C* 10: 1084–1089. https://doi.org/10.1039/D1TC05387F.

84 Olsen, T., Schröder, U., Müller, S. et al. (2012). Co-sputtering yttrium into hafnium oxide thin films to produce ferroelectric properties. *Applied Physics Letters* 101: 082905. https://doi.org/10.1063/1.4747209.

85 Lee, Y. H., Kim, H. J., Moon, T. et al. (2017). Preparation and characterization of ferroelectric $Hf_{0.5}Zr_{0.5}O_2$ thin films grown by reactive sputtering. *Nanotechnology* 28: 305703. https://doi.org/10.1088/1361-6528/aa7624.

86 Ryu, T. H., Min, D. H., and Yoon, S. M. (2020). Comparative studies on ferroelectric switching kinetics of sputtered $Hf_{0.5}Zr_{0.5}O_2$ thin films with variations in film thickness and crystallinity. *Journal of Applied Physics* 128: 074102. https://doi.org/10.1063/5.0013487.

87 Schwartz, R. W., Schneller, T., and Waser, R. (2004). Chemical solution deposition of electronic oxide films. *Comptes Rendus Chimie* 7: 433–461. https://doi.org/10.1016/j.crci.2004.01.007.

88 Starschich, S., Griesche, D., Schneller, T. et al. (2014). Chemical solution deposition of ferroelectric yttrium-doped hafnium oxide films on platinum electrodes. *Applied Physics Letters* 104: 202903. https://doi.org/10.1063/1.4879283.

89 Nakayama, S., Funakubo, H., and Uchida, H. (2018). Crystallization behavior and ferroelectric property of HfO_2–ZrO_2 films fabricated by chemical solution deposition. *Japanese Journal of Applied Physics* 57: 11UF06. https://doi.org/10.7567/JJAP.57.11UF06.

90 Alcala, R., Materano, M., Lomenzo, P. D. et al. (2023). The electrode-ferroelectric interface as the primary constraint on endurance and retention in HZO-based ferroelectric capacitors. *Advanced Functional Materials* 33: 2303261. https://doi.org/10.1002/adfm.202303261.

91 Lomenzo, P. D., Takmeel, Q., Zhou, C. et al. (2015). TaN interface properties and electric field cycling effects on ferroelectric Si-doped HfO_2 thin films. *Journal of Applied Physics* 117: 134105. https://doi.org/10.1063/1.4916715.

92 Xu, P., Yan, S., Zhu, Y. et al. (2023). Effects of different metal electrodes on the ferroelectric properties of HZO thin films. *Journal of Materials Science: Materials in Electronics* 34: 1915. https://doi.org/10.1007/s10854-023-11303-y.

93 Lee, Y., Kim, S. H., Jeong, H. W. et al. (2023). Mitigation of field-driven dynamic phase evolution in ferroelectric $Hf_{0.5}Zr_{0.5}O_2$ films by adopting oxygen-supplying electrode. *Applied Surface Science* 648: 158948. https://doi.org/10.1016/j.apsusc.2023.158948.

94 Chen, K. T., Liao, C. Y., Lo, C. et al. (2019). Improvement on ferroelectricity and endurance of ultra-thin $HfZrO_2$ capacitor with molybdenum capping electrode. In: *Proceedings of the 2019 Electron Devices Technology and Manufacturing Conference (EDTM).* https://doi.org/10.1109/EDTM.2019.8731272.

95 Cao, R., Song, B., Shang, D. et al. (2019). Improvement of endurance in HZO-based ferroelectric capacitor using Ru electrode. *IEEE Electron Device Letters* 40: 1744–1747. https://doi.org/10.1109/LED.2019.2944960.

96 Zhao, R., Liu, T., Zhao, X. et al. (2022). Impact of molybdenum oxide electrode on the ferroelectricity of doped-hafnia oxide capacitors. *IEEE Transactions on Electron Devices* 69: 1492–1496. https://doi.org/10.1109/TED.2021.3138843.

97 Goh, Y., Cho, S. H., Park, S. H. K. et al. (2020). Crystalline phase-controlled high-quality hafnia ferroelectric with RuO_2 electrode. *IEEE Transactions on Electron Devices* 67: 3431–3434. https://doi.org/10.1109/TED.2020.2998444.

98 Li, Y., Liang, R., Xiong, B. et al. (2019). $TiN_x/Hf_{0.5}Zr_{0.5}O_2/TiN_x$ ferroelectric memory with tunable transparency and suppressed wake-up effect. *Applied Physics Letters* 114: 052902. https://doi.org/10.1063/1.5083231.

99 Starschich, S., Menzel, S., and Böttger, U. (2016). Evidence for oxygen vacancies movement during wake-up in ferroelectric hafnium oxide. *Applied Physics Letters* 108: 032903. https://doi.org/10.1063/1.4940370.

100 Park, M. H., Kim, H. J., Kim, Y. J. et al. (2015). Study on the internal field and conduction mechanism of atomic layer deposited ferroelectric $Hf_{0.5}Zr_{0.5}O_2$ thin films. *Journal of Materials Chemistry C* 3: 6291–6300. https://doi.org/10.1039/c5tc01074h.

101 Starschich, S., Menzel, S., and Böttger, U. (2017). Pulse wake-up and breakdown investigation of ferroelectric yttrium doped HfO_2. *Journal of Applied Physics* 121: 154102. https://doi.org/10.1063/1.4981893.

102 Yuan, P., Mao, G. Q., Cheng, Y. et al. (2022). Microscopic mechanism of imprint in hafnium oxide-based ferroelectrics. *Nano Research* 15: 3667–3674. https://doi.org/10.1007/s12274-021-4047-y.

103 Banerjee, W., Kashir, A., and Kamba, S. (2022). Hafnium oxide (HfO_2) – A multifunctional oxide: A review on the prospect and challenges of hafnium oxide in resistive switching and ferroelectric memories. *Small* 18: 2107575. https://doi.org/10.1002/smll.202107575.

104 Wei, W., Zhang, W., Wang, F. et al. (2020). Deep insights into the failure mechanisms in field-cycled ferroelectric $Hf_{0.5}Zr_{0.5}O_2$ thin film: TDDB characterizations and first-principles calculations. In: *Proceedings of the 2020 International Electron Devices Meeting (IEDM)*. https://doi.org/10.1109/IEDM13553.2020.9371932.

105 Chouprik, A., Kondratyuk, E., Mikheev, V. et al. (2021). Origin of the retention loss in ferroelectric $Hf_{0.5}Zr_{0.5}O_2$-based memory devices. *Acta Materialia* 204: 116515. https://doi.org/10.1016/j.actamat.2020.116515.

106 Nagata, T., Haemori, M., Yamashita, Y. et al. (2010). Oxygen migration at Pt/HfO_2/Pt interface under bias operation. *Applied Physics Letters* 97: 082902. https://doi.org/10.1063/1.3483756.

107 Nukala, P., Ahmadi, M., Wei, Y. et al. (2021). Reversible oxygen migration and phase transition in hafnia-based ferroelectric devices. *Science* 372: 630–635. https://doi.org/10.1126/science.abf3789.

108 Rushchanskii, K. Z., Blügel, S., and Ležaić, M. (2021). Ordering of oxygen vacancies and related ferroelectric properties in $HfO_{2-\delta}$. *Physical Review Letters* 127: 087602. https://doi.org/10.1103/PhysRevLett.127.087602.

109 Grimley, E. D., Schenk, T., Sang, X. et al. (2016). Structural changes underlying field-cycling phenomena in ferroelectric HfO_2 thin films. *Advanced Electronic Materials* 2: 1600173. https://doi.org/10.1002/aelm.201600173.

110 Senapati, A., Lou, Z. F., Lee, J. Y. et al. (2024). Low polarization loss of long endurance on scavenged Ru-based electrode ferroelectric $Hf_{0.5}Zr_{0.5}O_2$ by optimizing TiN_x interfacial capping layer and its fatigue mechanism. *IEEE Electron Device Letters* 45: 673–676. https://doi.org/10.1109/LED.2024.3368422.

111 Zhang, Y., Peng, Y., Liu, F. et al. (2023). Interface-roughness effect on polarization switching and reliability performance of doped-HfO_2 capacitors. *IEEE Transactions on Electron Devices* 70: 3788–3793. https://doi.org/10.1109/TED.2023.3279064.

112 Lee, S. J., Kim, M. J., Lee, T. Y. et al. (2019). Effect of ZrO_2 interfacial layer on forming ferroelectric $Hf_xZr_yO_z$ on Si substrate. *AIP Advances* 9: 125020. https://doi.org/10.1063/1.5124402.

113 Chen, Y. F., Hu, C. W., Kao, Y. C. et al. (2023). Wake-up free ferroelectric capacitor with quadruple-level storage by inserting ZrO_2 interlayer and bottom layer in. *IEEE Electron Device Letters* 44: 400–403. https://doi.org/10.1109/LED.2023.3235715.

114 Chen, H., Tang, L., Liu, L. et al. (2021). Significant improvement of ferroelectricity and reliability in $Hf_{0.5}Zr_{0.5}O_2$ films by inserting an ultrathin Al_2O_3 buffer layer. *Applied Surface Science* 542: 148737. https://doi.org/10.1016/j.apsusc.2020.148737.

115 Lee, H. J., Moon, T., Nam, S. et al. (2024). Interfacial layer selection methodology for customized ferroelectric memories. *IEEE Transactions on Electron Devices* 71: 1907–1912. https://doi.org/10.1109/TED.2024.3360009.

116 Peng, H. K., Lai, T. C., Kao, Y. C. et al. (2022). Improved reliability for back-end-of-line compatible ferroelectric capacitor with 3 bits/cell storage capability by interface engineering and post deposition annealing. *IEEE Electron Device Letters* 43: 2180–2183. https://doi.org/10.1109/LED.2022.3218253.

117 Gaddam, V., Das, D., and Jeon, S. (2020). Insertion of HfO_2 seed/dielectric layer to the ferroelectric HZO films for heightened remanent polarization in MFM capacitors. *IEEE Transactions on Electron Devices* 67: 745–750. https://doi.org/10.1109/TED.2019.2961208.

118 Jeong, Y., Gaddam, V., Goh, Y. et al. (2023). Oxygen vacancy control as a strategy to enhance imprinting effect in hafnia ferroelectric devices. *IEEE Transactions on Electron Devices* 70: 354–359. https://doi.org/109/TED.2022.3223886.

119 Mittmann, T., Materano, M., Chang, S. C. et al. (2020). Impact of oxygen vacancy content in ferroelectric HZO films on the device performance. In: *Proceedings of the 2020 IEEE International Electron Devices Meeting (IEDM)*. https://doi.org/10.1109/IEDM13553.2020.9372097.

120 Jang, H., Kashir, A., Schenk, T. et al. (2024). Demonstration of 1 V reliable FeRAM operation: V_c engineering using quasi-chirality of $Hf_{1-x}Zr_xO_2$ in a nanolaminate structure. *ACS Applied Materials & Interfaces* 16: 55627–55636. https://doi.org/10.1021/acsami.4c08641.

121 Park, H. S., Choi, S., Kim, K. D. et al. (2024). Optimization of the 4 nm-thick $Hf_{1-x}Zr_xO_2$ film with low operating voltage and high endurance for ferroelectric random access memory. *ACS Applied Materials & Interfaces* 6: 6826–6836. https://doi.org/10.1021/acsaelm.4c01216.

122 Lee, Y., Kim, U. J., Kim, K. et al. (2024). Boosting non-volatile memory performance with exhalative annealing: A novel approach to low-temperature crystallization of hafnia based ferroelectric. *Materials Today Nano* 28: 100546. https://doi.org/10.1016/j.mtnano.2024.100546.

123 Chen, K. Y., Chen, P. H., Kao, R. W. et al. (2017). Impact of plasma treatment on reliability performance for $HfZrO_x$-based metal-ferroelectric-metal capacitors. *IEEE Electron Device Letters* 39: 87–90. https://doi.org/10.1109/LED.2017.2771390.

124 Hsieh, D. R., Lee, C. C., and Chao, T. S. (2022). Role of nitrogen in ferroelectricity of $Hf_xZr_{1-x}O_2$-based capacitors with metal-ferroelectric-insulator-metal structure. *IEEE Transactions on Electron Devices* 69: 2074–2079. https://doi.org/10.1109/TED.2022.3153263.

125 Yuan, P., Wang, B., Yang, Y. et al. (2022). Enhanced remnant polarization (30 $\mu C/cm^2$) and retention of ferroelectric $Hf_{0.5}Zr_{0.5}O_2$ by NH_3 plasma treatment. *IEEE Electron Device Letters* 43: 1045–1048. https://doi.org/10.1109/LED.2022.3178867.

126 Das, D., Buyantogtokh, B., Gaddam, V. et al. (2021). Influence of high-pressure annealing conditions on ferroelectric and interfacial properties of Zr-Rich $Hf_xZr_{1-x}O_2$ capacitors. *IEEE Transactions on Electron Devices* 68: 1996–2002. https://doi.org/10.1109/TED.2021.3061963.

127 Wang, Q., Zhang, Y., Yang, P. et al. (2022). Improved symmetry of ferroelectric switching in HZO based MFM capacitors enabled by high pressure annealing. *IEEE Transactions on Electron Devices* 69: 103–108. https://doi.org/10.1109/TED.2021.3131403.

128 Das, D., Buyantogtokh, B., Gaddam, V. et al. (2022). Sub 5 Å-EOT $Hf_xZr_{1-x}O_2$ for next-generation DRAM capacitors using morphotropic phase boundary and high-pressure (200 atm) annealing with rapid cooling process. *IEEE Transactions on Electron Devices* 69: 103–108. https://doi.org/10.1109/TED.2021.3131403.

129 Chen, W. C., Tan, Y. F., Lin, S. K. et al. (2021). Performance improvement by modifying deposition temperature in $HfZrO_x$ ferroelectric memory. *IEEE Transactions on Electron Devices* 68: 3838–3842. https://doi.org/10.1109/TED.2021.3093256.

130 Fu, Z., Cao, S., Zheng, H. et al. (2024). Hafnia-based high-disturbance-immune and selector-free cross-point FeRAM. *IEEE Transactions on Electron Devices* 71: 3358–3364. https://doi.org/10.1109/TED.2024.3369569.

131 Zhang, D., Wu, J., Kong, Q. et al. (2023). Grain size reduction of ferroelectric HZO enabled by solid phase epitaxy (SPE): Working principle, experimental demonstration, and theoretical understanding. *IEEE Transactions on Electron Devices* 70: 6665–6672. https://doi.org/10.1109/TED.2023.3317007.

132 Kozodaev, M. G., Chernikova, A. G., Korostyev, E. V. et al. (2019). Mitigating wakeup effect and improving endurance of ferroelectric HfO_2-ZrO_2 thin films by careful La-doping. *Journal of Applied Physics* 125: 034101. https://doi.org/10.1063/1.5050700.

133 Silva, A., Fina, I., Sánchez, F. et al. (2023). Unraveling the ferroelectric switching mechanisms in ferroelectric pure and La doped HfO_2 epitaxial thin films. *Materials Today Physics* 34: 101064. https://doi.org/10.1016/j.mtphys.2023.101064.

134 Liu, Y., Wang, T., Xu, K. et al. (2024). Effect of doping concentration on intrinsic ferroelectric properties of HfLaO-based ferroelectric memory. *IEEE Electron Device Letters* 45: 376–379. https://doi.org/10.1109/LED.2023.3347920.

135 Tai, L., Wei, W., Sang, P. et al. (2023). Significant reliability improvement by inducing dual atomic-thin titanium intercalation layers in $Hf_{0.5}Zr_{0.5}O_2$ films. *IEEE Electron Device Letters* 44: 753–756. https://doi.org/10.1109/LED.2023.3263294.

136 Mehmood, F., Mikolajick, T., and Schroeder, U. (2020). Reliability improvement of ferroelectric $Hf_{0.5}Zr_{0.5}O_2$ thin films by lanthanum doping for FeRAM applications. In: *Proceedings of the 2020 Device Research Conference*. https://doi.org/10.1109/DRC50226.2020.9150531.

137 Lehninger, D., Prabhu, A., Sünbül, A. et al. (2023). Ferroelectric HfO_2/ZrO_2 superlattices with enhanced polarization, tailored coercive field, and improved high temperature reliability. *Advanced Physics Research* 2: 2200108. https://doi.org/10.1002/apxr.202200108.

138 Peng, Y., Xiao, W., Liu, Y. et al. (2022). HfO_2-ZrO_2 superlattice ferroelectric capacitor with improved endurance performance and higher fatigue recovery capability. *IEEE Electron Device Letters* 43: 216–219. https://doi.org/10.1109/LED.2021.3135961.

139 Lehninger, D., Sünbül, A., Olivo, R. et al. (2023). Ferroelectric HfO_2/ZrO_2 superlattices with improved leakage at bias and temperature stress. In: *Proceedings of the 2023 International Memory Workshop*. https://doi.org/10.1109/IMW56887.2023.10145927.

140 Kim, H. J., Park, M. H., Kim, Y. J. et al. (2014). Grain size engineering for ferroelectric $Hf_{0.5}Zr_{0.5}O_2$ films by an insertion of Al_2O_3 interlayer. *Applied Physics Letters* 105: 192903. https://doi.org/10.1063/1.4902072.

141 Lee, S., Kim, G., Lee, S. et al. (2024). First demonstration of thermally stable Zr:HfO_2 ferroelectrics via inserting AlN interlayer. *IEEE Electron Device Letters* 45: 1578–1581. https://doi.org/10.1109/LED.2024.3424973.

142 Nguyen, A. D., Song, S. U., Nguyen, A. H. T. et al. (2024). Metal interlayer insertion for grain engineering in ferroelectric $Hf_{0.5}Zr_{0.5}O_2$. *IEEE Electron Device Letters* 71: 6647–6651. https://doi.org/10.1109/TED.2024.3447614.

143 Hsieh, D. R., Hong, Z. Y., Yeh, W. J. et al. (2024). Investigation of HfO_2 and ZrO_2 nanolamination thickness on superlattice HZO FeRAMs exhibiting enhanced remnant polarization and improved endurance. In: *Proceedings of the 2024 IEEE Silicon Nanoelectronics Workshop (SNW)*. https://doi.org/10.1109/SNW63608.2024.10639221.

144 Kruv, A., McMitchell, S. R. C., Clima, S. et al. (2021). Impact of mechanical strain on wakeup of HfO_2 ferroelectric memory. In: *Proceedings of the 2021 IEEE International Reliability Physics Symposium*. https://doi.org/10.1109/IRPS46558.2021.9405159.

145 Joh, H., Jung, T., and Jeon, S. (2021). Stress engineering as a strategy to achieve high ferroelectricity in thick hafnia using interlayer. *IEEE Transactions on Electron Devices* 68: 2538–2542. https://doi.org/10.1109/TED.2021.3068246.

146 Tsai, S. H., Fang, Z., Wang, X. et al. (2022). Stress-memorized HZO for high-performance ferroelectric field effect memtransistor. *ACS Applied Electronic Materials* 4: 1642–1650. https://doi.org/10.1021/acsaelm.1c01321.

147 Lee, Y., Goh, Y., Hwang, J. et al. (2021). The influence of top and bottom metal electrodes on ferroelectricity of hafnia. *IEEE Transactions on Electron Devices* 68: 523–528. https://doi.org/10.1109/TED.2020.3046173.

148 Jiang, P., Yang, Y., Wei, W. et al. (2023). Stress effects of interconnecting metals on back-end-of-line compatible $Hf_{0.5}Zr_{0.5}O_2$ ferroelectric capacitors. *IEEE Electron Device Letters* 44: 602–605. https://doi.org/10.1109/LED.2023.3248103.

149 Pešić, M., Knebel, S., Hoffmann, M. et al. (2016). How to make DRAM non-volatile? Anti-ferroelectrics: A new paradigm for universal memories. In: *Proceedings of the 2016 IEEE International Electron Devices Meeting (IEDM)*. https://doi.org/10.1109/IEDM.2016.7838398.

150 Pešić, M., Li, T., Lecce, V. D. et al. (2018). Built-in bias generation in anti-ferroelectric stacks: Methods and device applications. *IEEE Journal of the Electron Devices Society* 6: 1019–1025. https://doi.org/10.1109/JEDS.2018.2825360.

151 Chang, S. C., Haratipour, N., Shivaraman, S. et al. (2020). Anti-ferroelectric HfxZr$_1$-xO_2 capacitors for high-density 3-D embedded-DRAM. In: *Proceedings of the 2020 IEEE International Electron Devices Meeting (IEDM)*. https://doi.org/10.1109/IEDM13553.2020.9372011.

152 Chen, D., Zhong, S., Dong, Y. et al. (2022). Antiferroelectric phase evolution in $Hf_xZr_{1-x}O_2$ thin film toward high endurance of non-volatile memory devices. *IEEE Electron Device Letters* 43: 2065–2068. https://doi.org/10.1109/LED.2022.3217813.

153 Hsiang, K. Y., Chen, Y. C., Chang, F. S. et al. (2022). Novel opposite polarity cycling recovery (OPCR) of $HfZrO_2$ antiferroelectric-RAM with an access scheme toward unlimited endurance. In: *Proceedings of the 2022 International Electron Devices Meeting (IEDM)*. https://doi.org/10.1109/IEDM45625.2022.10019508.

154 Lo, C., Chen, C. K., Chang, C. F. et al. (2022). High endurance and low fatigue effect of bilayer stacked antiferroelectric/ferroelectric $Hf_xZr_{1-x}O_2$. *IEEE Electron Device Letters* 43: 224–227. https://doi.org/dop:10.1109/LED.2021.3134178.

155 Alcala, R., Materano, M., Lomenzo, P. D. et al. (2022). BEOL integrated ferroelectric HfO_2-based capacitors for FeRAM: Extrapolation of reliability performance to use conditions. *IEEE Journal of Electron Devices Society* 10: 907–912. https://doi.org/10.1109/JEDS.2022.3198138.

156 Leiserson, C. E., Thompson, N. C., Emer, J. S. et al. (2020). There's plenty of room at the top: What will drive computer performance after Moore's law? *Science* 368: eaam9744. https://doi.org/10.1126/science.aam9744.

157 Chen, P. Y., He, Z. Y., Cha, M. Y. et al. (2021). Hafnium oxide-based ferroelectric devices for computing-in-memory applications. *Physica Status Solidi A* 218: 2000635. https://doi.org/10.1002/pssa.202000635.

158 Sheikholeslami, A. and Gulak, P. G. (2000). A survey of circuit innovations in ferroelectric random-access memories. *Proceedings of the IEEE* 88: 667–689. https://doi.org/10.1109/5.849164.

159 Sumi, T., Moriwaki, N., Nakane, G. et al. (2002). A 256 kb nonvolatile ferroelectric memory at 3V and 100 ns. In: *Proceedings of IEEE International Solid-State Circuits Conference – ISSCC '94*. https://doi.org/10.1109/CICC.2002.1012777.

160 Siu, J. W. K., Eslami, Y., Sheikholeslami, A. et al. (2002). A 16 kb 1T1C FeRAM test chip using current-based reference scheme. In: *Proceedings of the IEEE 2002 Custom Integrated Circuits Conference*. https://doi.org/10.1109/CICC.2002.1012777.

161 Kawashima, S., Endo, T., Yamamoto, A. et al. (2002). Bitline GND sensing technique for low-voltage operation FeRAM. *IEEE Journal of Solid-State Circuits* 37: 592–598. https://doi.org/10.1109/4.997852.

162 McAdams, H. P., Acklin, R., Blake, T. et al. (2004). A 64-Mb embedded FRAM utilizing a 130-nm 5LM Cu/FSG logic process. *IEEE Journal of Solid-State Circuits* 39: 667–677. https://doi.org/10.1109/JSSC.2004.825241.

7

3D Ferroelectric Capacitor Memories for Data-centric Computing

*Jiahui Duan and Kai Ni**

College of Engineering, University of Notre Dame, Notre Dame, IN, USA

*Corresponding author: kni@nd.edu

7.1 Introduction

In the age of rapid data generation, driven by advancements in artificial intelligence (AI), edge computing, and large-scale data analytics, traditional computing systems are increasingly struggling to meet the performance and energy efficiency requirements of modern workloads. These applications are marked by substantial data volumes, frequent memory access, and often low arithmetic intensity, placing unprecedented pressure on established computing architectures. At the core of this challenge lies the von Neumann architecture, which physically segregates the processing and memory components. This division creates the "memory wall" – a bottleneck resulting from the latency and energy overhead of continuous data transfers between computation and storage units.[1] As data-centric applications grow in scale and complexity, the expense of this data movement becomes a pivotal factor in system performance and power consumption, making traditional architectures progressively less effective.

To overcome this fundamental bottleneck, new computing paradigms and memory technologies must be considered. Among these, compute-in-memory (CIM) has emerged as a promising architectural solution.[2] By performing computation directly where the data resides – within the memory arrays themselves – CIM architectures effectively eliminate the need for frequent data transfers between the processor and memory. This scheme not only mitigates the memory wall but also takes advantage of the intrinsic parallelism and locality of memory structures, leading to significantly improved throughput and energy efficiency.[2–6] However, realizing the full potential of CIM requires memory technologies that can combine superior energy efficiency, reliable data retention, fast switching, high endurance, and compatibility with existing semiconductor manufacturing processes.

Emerging nonvolatile memory (NVM) technologies offer a viable path forward. Among the various candidates, ferroelectric capacitor (FeCap)-based memories have garnered significant attention due to their unique set of properties that make them particularly well suited for data-centric applications.[7–11] These memories harness the nonvolatile spontaneous polarization of ferroelectric materials that can be reversed by an external electric field. The bistable polarization states can be used to store binary information in a nonvolatile manner, while partial polarization states can support multibit storage. Compared to other NVMs such as resistive RAM (ReRAM) or phase-change memory (PCM), FeCap memories offer faster switching speeds, lower write voltages, high endurance, and enhanced energy efficiency due to the electric field-driven polarization switching mechanisms of FeCap.

Historically, ferroelectric memories development was centered around perovskite-based materials such as lead zirconate titanate (PZT) and strontium bismuth tantalate (SBT).[2,12–15] These materials demonstrated strong remanent polarization and fast switching capabilities, enabling early implementations of ferroelectric random-access memory (FeRAM). Despite their favorable electrical characteristics, these perovskite ferroelectrics faced several critical limitations. Their integration into CMOS platforms proved challenging due to poor material compatibility, high annealing temperatures, and the need for complex deposition processes for mobile elements such as Pb and Bi.[15] Additionally, their scalability was limited as their ferroelectric properties degraded at nanometer-scale thicknesses,[16,17] impeding their use in highly scaled semiconductor nodes. As a result, while perovskite-based FeRAM found niche applications, it failed to achieve widespread adoption in mainstream computing systems.

A paradigm shift occurred with the discovery of ferroelectricity in doped hafnium oxide (HfO_2) thin films.[18] This exciting finding revitalized interest in ferroelectric memory technologies by addressing many of the drawbacks of prior materials. HfO_2-based ferroelectrics exhibit robust ferroelectric properties at thicknesses below 10 nm,[19–26] and retain ferroelectricity at even 1 nm,[22] making them inherently scalable for advanced technology nodes. Furthermore, these materials are fully compatible with CMOS processes, requiring relatively low thermal budgets, and amenable to atomic-layer deposition (ALD), a mature and industry-standard thin-film fabrication technique. Importantly, HfO_2 and its doped variants, such as hafnium–zirconium oxide ($Hf_{1-x}Zr_xO_2$ or HZO), demonstrate stable remanent polarization and high endurance over billions of cycles, making them exceptionally well suited for NVM and CIM applications.

The discovery of HfO_2-based ferroelectricity has opened up new avenues for realizing compact, energy-efficient, and high-performance FeCap-based memory structures, which is also the only commercially produced ferroelectric memory so far, used in applications such as IC cards and medical instruments.[27,28] These include pure capacitor arrays as well as hybrid memory cells that integrate access transistors for improved performance. Architectures such as pure FeCap crossbar arrays, 1T-1C (one-transistor one-capacitor), 1T-*n*C (one-transistor *n*-capacitor), 2T-1C (two-transistor one-capacitor), and 2T-*n*C (two-transistor *n*-capacitor) FeRAM cells have been proposed and demonstrated for various applications.[29–33] These structures exploit the switching behavior of FeCaps to enable in-memory implementation of multiply-and-accumulate

(MAC) operations, threshold logic, vector–matrix multiplication (VMM), and even synaptic learning behaviors. Such capabilities are especially valuable for AI and ML workloads, which rely heavily on large-scale matrix computations and parallel processing.

Moreover, HfO_2-based FeCap memories present a promising route toward three-dimensional (3D) integration, allowing vertical stacking of memory and logic layers to enhance both density and performance – key to meeting the demands of high-density computing and data storage. The compatibility of ferroelectric HfO_2 with back-end-of-line (BEOL) fabrication processes enables the direct integration of memory layers atop logic circuits, paving the way for tightly integrated CIM architectures. This strategy substantially minimizes interconnect latency, enables fine-grained parallel processing, and improves energy efficiency across data-intensive tasks, including neural network inference, signal processing, and edge AI applications.

In this chapter, we focus on the transformative potential of FeCaps based on HfO_2, which have emerged as highly promising candidates for next-generation memory and computing technologies. These devices, owing to the unique polarization switching behavior of doped HfO_2 thin films, offer a scalable, CMOS-compatible, and nonvolatile platform that not only supports efficient data storage but also enables in-memory computation. FeCap-based memories, such as pure FeCap crossbar arrays, 1T-1C, 1T-*n*C, 2T-1C, and 2T-*n*C FeRAMs, and 1T-1C FeMFET configurations as shown in Figure 7.1, exploit the switching behavior of ferroelectric materials to perform fundamental arithmetic and logic operations directly within memory arrays, thereby significantly reducing data movement and computational overhead. This capability is especially valuable in the context of data-intensive applications where conventional von Neumann architectures face severe bottlenecks due to the separation of processing and memory units. By leveraging the dense integration potential of HfO_2-based ferroelectrics and their suitability for 3D stacking, these memory architectures pave the way for compact, high-throughput, and energy-efficient CIM systems. In this chapter, we present a detailed exploration of the physical mechanisms underlying these capacitor-based memories, review their integration strategies and architectural variants, and examine the associated challenges in material reliability

	FeCap	1T-1C FeRAM	1T-*n*C FeRAM	1T-1C FeMFET	2T-1C FeRAM	2T-*n*C FeRAM
Structure						
Write/read path	Same	Same	Same	Same	Different	Different
Bit-cost 3D integration	Good	Moderate	Good	Moderate	Bad	Good
Power consumption	Low	Moderate	Moderate	Moderate	Moderate	Moderate
Disturbance management	Bad	Good	Bad	Good	Good	Bad

Figure 7.1 Ferroelectric memory cells. Key performance and reliability metrics comparing various ferroelectric memory candidates, including write voltage, retention, write endurance, read endurance, sense margin, and scalability for charge-based 1T-1C FeRAM, capacitive-based 1T-1C FeRAM, 1T-1C FeMFET, 2T-1C FeRAM, and 2T-*n*C FeRAM.

and circuit implementation. Furthermore, we highlight their compelling application in CIM, showcasing how FeCap-based memories can accelerate data processing and overcome the fundamental limitations of traditional computing paradigms.

7.2 Classification of FeCap-based Memories

To harness the full potential of HfO_2-based FeCaps in memory and CIM applications, various memory cell architectures have been developed, each offering distinct trade-offs in performance, scalability, and circuit complexity. This section introduces and classifies the primary types of FeCap-based memory structures – namely, the crossbar one-capacitor (1C) FeCap array, 1T-1C, 1T-nC, 2T-1C, and 2T-nC FeRAMs, and 1T-1C FeMFET architectures – highlighting their operational principles.

7.2.1 Crossbar FeCap

One basic implementation of ferroelectric memory is the crossbar array architecture based solely on FeCaps, commonly referred to as the 0T-1C structure. This architecture consists of a two-dimensional grid formed by intersecting word lines (WLs) and bit lines (BLs), where each intersection defines a memory cell containing a single FeCap that stores binary data via its polarization state. In the absence of selection devices (e.g., diodes or transistors), the passive crossbar array relies on the voltage drop between WLs and BLs to address specific memory cells.

During a write operation, a write voltage V_{write} pulse is applied at the intersection of activated WL and BL, resulting in a voltage drop across the targeted FeCap. This voltage must exceed the coercive voltage (V_c) of the ferroelectric material to induce polarization switching, thereby storing a logical "1" or "0." A key challenge in this architecture arises from half-selected cells, which are connected to either the active WL or the active BL but not both. These cells experience partial voltage excitation that, over repeated operations, may cause unintentional polarization switching. To mitigate this issue, voltage biasing schemes such as $V_{\text{write}}/3$ or $V_{\text{write}}/2$ are employed.[34] These techniques ensure that only the selected cell experiences the full write voltage, while half-selected and unselected cells are subjected to sub-V_c levels, thus preserving their polarization states. Under these biasing strategies, the full write voltage is dropped only across the selected FeCap, resulting in a reduced effective write voltage requirement. However, even with such inhibition bias schemes, under repetitive excitation, disturbance to polarization can still occur and additional disturbance mitigation techniques are needed, which will be discussed in more detail in Section 7.2.2.

During a read operation, a read voltage V_{read} is applied to the selected WL, and the resulting displacement current from FeCap is detected through the BL. Also, the unselected cells are also applied $V_{\text{read}}/3$ or $V_{\text{read}}/2$ to prevent polarization switching during read operation. Since ferroelectric readout is inherently destructive, the original data must be rewritten following each read cycle to maintain data integrity.

Then the difference between data "0" and "1" can be sensed as shown in Figure 7.2. The sense margin for this read operation can be quantified as:

$$\Delta V_{\mathrm{BL}} = \frac{\Delta P_{\mathrm{r}}}{C_{\mathrm{BL}}} \propto A_{\mathrm{Cap}} \tag{7.1}$$

where ΔP_{r} is the change in remanent polarization due to switching, C_{BL} is the BL capacitance, and A_{Cap} is the area of the FeCap. Given that each memory cell consists

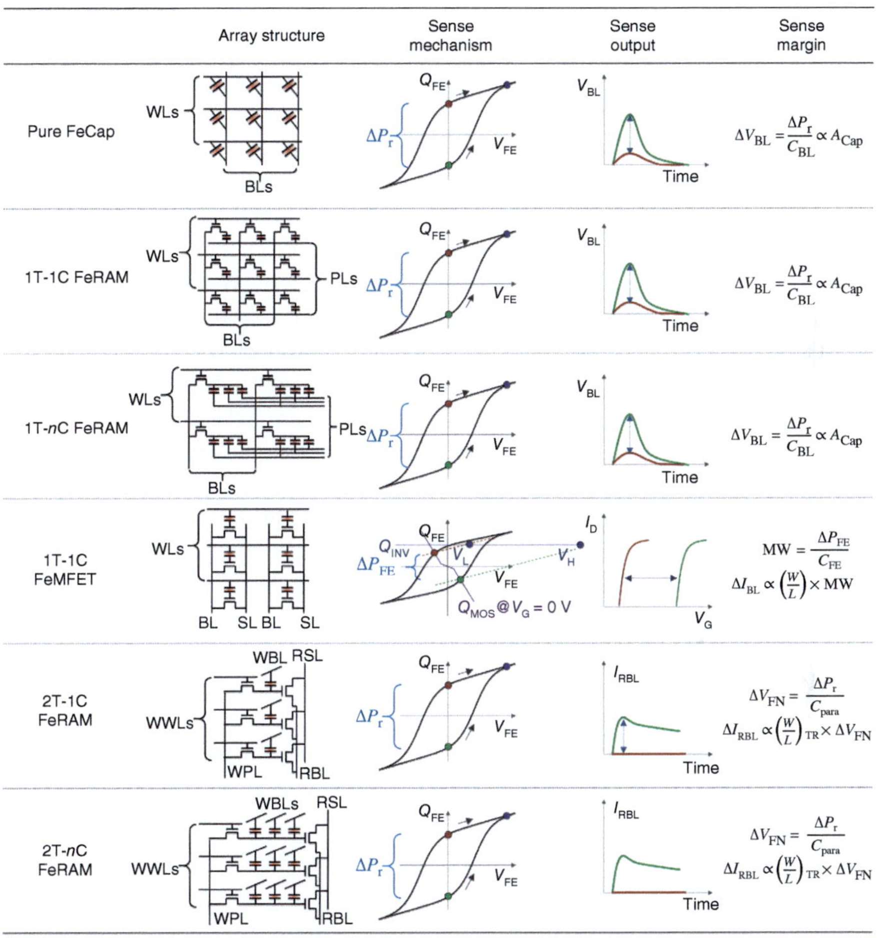

Figure 7.2 Summary of sensing mechanisms of different ferroelectric memory cells. The charge-based and capacitance-based pure FeCap, 1T-1C, and 1T-*n*C FeRAMs share the same write and read paths, while charge-based sensing directly reads the polarization current when a high voltage is applied at BL. For these three FeRAM designs, the sense margins are both proportional to the area of the capacitor. The write path for 1T-1C FeMFET is from the gate to the source/drain, while the read path is from source to drain. The sense margin is determined by the *W/L* ratio and also can be understood from the load line of MOSCAP and polarization–voltage curves of MFM capacitor. The write paths of 2T-1C and 2T-*n*C FeRAMs are both from WPLs to WBLs, which are controlled by the write transistor. And the read paths are from WBLs to RBLs.

of only a single FeCap, this architecture is highly compatible with 3D integration which offers high memory density, making it attractive for applications where area efficiency is critical.

7.2.2 1T-1C FeRAM

The 1T-1C FeRAM cell adopts a DRAM-like structure, comprising one transistor (1T) and one FeCap. The transistor's gate is connected to WL, which controls access to the cell upon activation. BL links to the transistor's drain/source terminal, serving as the primary data line for write and sensing, while the plate line (PL) connects to the opposing terminal of FeCap, supplying the reference voltage required for write and read operations.

During a write operation, a positive voltage pulse is applied to BL or PL, depending on whether a logical "0" or "1" is being stored. Concurrently, WL is driven to the supply voltage (V_{DD}) to activate the access transistor, enabling the voltage to polarize FeCap. The applied pulse must exceed the coercive voltage (V_c) of the ferroelectric material to ensure reliable polarization switching between the remaining states (P_{r+} or P_{r-}). The presence of the access transistor provides electrical isolation between adjacent cells, preventing unintended disturbances in unselected memory cells.

For the read operation, a sensing voltage V_{read} is applied to PL while the access transistor is also activated, which results in a charge displacement in FeCap that manifests as a detectable current on BL. Similar to crossbar FeRAM architectures, this readout mechanism is destructive, necessitating a write-back operation to restore the original data. While this ensures data integrity, it introduces additional power overhead, which may be a limiting factor in low-power applications. Figure 7.2 shows the sense mechanism and sense margin for the 1T-1C read operation. Sense margin can also be expressed as:

$$\Delta V_{BL} = \frac{\Delta P_r}{C_{BL}} \propto A_{Cap} \tag{7.2}$$

Due to the additional access transistor per cell, the overall memory density of the 1T-1C architecture is lower than that of the crossbar FeCap array.

7.2.3 1T-*n*C FeRAM

An alternative to the conventional 1T-1C FeRAM architecture is the 1T-*n*C design, in which a single access transistor is shared among multiple FeCaps, grouped into a subarray within the overall memory matrix.[35–37] This configuration enhances memory density by reducing the number of transistors per memory cell, offering an effective trade-off between area efficiency and control over memory operations. During a write operation, the shared access transistor is enabled by applying a voltage to WL, allowing a write voltage pulse to be applied across the selected FeCap. The amplitude of this pulse must exceed the coercive voltage (V_c) of the ferroelectric material to switch its polarization state and store binary information

("1" or "0"). To prevent inadvertent polarization switching in unselected FeCaps within the group, voltage biasing schemes such as $V_{\text{write}}/2$ or $V_{\text{write}}/3$ are employed, ensuring that only the target cell experiences the full write voltage.

The read operation follows a procedure similar to that of the 1T-1C architecture. A sensing voltage V_{read} is applied to the target PL, and the resulting displacement current, due to polarization switching, is sensed through BL. As with 1T-1C FeRAM, the read process is inherently destructive, necessitating a restore cycle following each read operation to preserve data integrity. Given the comparable read and write procedures, the write voltage requirements and sense margin characteristics remain similar to those of the 1T-1C architecture. The sense margin is given by:

$$\Delta V_{\text{BL}} = \frac{\Delta P_{\text{r}}}{C_{\text{BL}}} \propto A_{\text{Cap}} \tag{7.3}$$

Owing to the shared access transistor, the memory density of the 1T-nC architecture falls between those of the high-density crossbar FeCap array and the lower-density 1T-1C FeRAM.

7.2.4 1T-1C FeMFET

In a 1T-1C FeMFET, the FeCap[38–42] can be either implemented in the BEOL or integrated with the Si MOSFET in the front-end-of-line (FEOL) process, and it is electrically connected to the gate of the MOSFET, rather than the drain in the 1T-1C FeRAM cell. This separation provides significant design flexibility, particularly in optimizing the area ratio between FeCap and the MOSFET gate – a key factor in controlling the memory window and ensuring reliable switching behavior. Reducing the FeCap size relative to the MOSFET area maximizes the voltage drop across the ferroelectric layer, facilitating efficient polarization switching at lower write voltages. Additionally, the 1T-1C FeMFET improves endurance, as the isolated MFM capacitor absorbs most of the write voltage, minimizing stress on the transistor. This reduces charge injection, suppresses defect generation, and mitigates cycle-dependent degradation, ultimately extending device lifetime. However, the introduction of a floating metal in the FeMFET can degrade retention due to leakage from the floating node between the MOSFET and the MFM capacitor, leading to charge loss and instability over time.

In this 1T-1C FeMFET cell, WL, BL, and source line (SL) are connected to the gate, drain, and source terminals, respectively. To program the memory, voltages on WL and necessary inhibition bias schemes applied to BL and SL are exerted to set the device to the low threshold voltage (LVT) or high threshold voltage (HVT) while keeping the unselected cells intact. After polarization is programmed, sensing of the FeFET is typically conducted in the current domain where a proper read voltage between the LVT and HVT is applied to WL, which could induce different drain currents for the LVT and HVT states. It should be mentioned here that the write voltage is dropped at the FeCap and a MOSCAP in series. So, the write voltage is higher than the previously introduced FeCap-based memories.

The read operation of 1T-1C FeMFET can be understood by examining the load line of the semiconductor channel and the ferroelectric polarization hysteresis loops. Since the semiconductor channel cannot provide screening charge for all the polarizations, the ferroelectric will likely operate in minor hysteresis loops. After programming, and when the WL voltage returns to ground, there are two intersection points between the channel charge and the hysteresis loop, corresponding to the LVT and HVT states, as shown in Figure 7.2. It is assumed that during sensing, the polarization remains fixed, thus enabling nondestructive read such that the linear ferroelectric capacitance supports the read operation. In this way, the memory window and the sense margin can be expressed as:

$$\mathrm{MW} = \frac{\Delta P_{\mathrm{FE}}}{C_{\mathrm{FE}}} \tag{7.4}$$

$$\Delta I_{\mathrm{BL}} \propto \frac{W}{L} \times \mathrm{MW} \tag{7.5}$$

where MW is the voltage separation between the LVT and HVT states and the current sense margin is estimated based on the assumption that HVT yields negligible current while the LVT state works at the linear region during sensing. Because this margin is not proportional to the area of the transistor, but the W/L ratio, the scalability is better than those of above-mentioned ferroelectric memory devices.

7.2.5 2T-1C FeRAM

The 2T-1C FeRAM architecture can also overcome the limitations of earlier area-dependent sense margin memory structures, while keeping the low write voltage of 1T-1C FeRAM.[29,30,32,43] This design incorporates two transistors per memory cell alongside a single FeCap, thereby improving overall performance and robustness. The first transistor serves as the write access device, controlled by WL, while the second transistor functions as the readout or gain transistor, amplifying the stored charge during the read operation. During a write cycle, the write word line (WWL) is biased to the supply voltage (V_{DD}), turning on the write access transistor. This enables the application of a voltage pulse across FeCap via the write bit line (WBL) and write plate line (WPL). The applied voltage must exceed the coercive voltage (V_{c}) to switch the polarization of the ferroelectric layer, thereby storing a binary "1" or "0." The write mechanism mirrors that of the 1T-1C FeRAM architecture.

For the read operation, the write access transistor is deactivated by grounding the WWL, isolating the write path. A sensing voltage is then applied to WBL, which induces a difference in the stored node potential for different stored states due to polarization switching. This variation influences the current flow through the read transistor, allowing the stored data to be detected through the read bit line (RBL) and read source line (RSL). The sense margin can be quantitatively described by:

$$\Delta V_{\mathrm{FN}} = \frac{\Delta P_{\mathrm{r}}}{C_{\mathrm{para}}} \tag{7.6}$$

$$\Delta I_{\mathrm{RBL}} \propto \left(\frac{W}{L}\right)_{\mathrm{TR}} \times \Delta V_{\mathrm{FN}} \tag{7.7}$$

where C_{para} is the parasitic capacitance at the floating node, and $(W/L)_{TR}$ is the width-to-length ratio of the read transistor. Although this dual-transistor configuration enhances read accuracy and sense margin, it also significantly reduces memory density due to the increased silicon area per cell.

7.2.6 2T-*n*C FeRAM

The 2T-nC FeRAM architecture integrates the high-density advantage of shared capacitor structures, such as 1T-nC, with the sensing precision enabled by dual-transistor designs like 2T-1C. In this architecture, multiple FeCaps share a pair of write and read transistors, achieving a balance between density and performance.[44,45] The write operation is similar to that in 2T-1C FeRAM. The write access transistor is activated by applying V_{DD} to WWL, allowing a voltage pulse to be applied across the selected FeCap via BL and PL. This pulse must exceed the coercive voltage (V_c) to alter the polarization state and store data. To prevent unintentional switching in unselected capacitors, biasing schemes such as $V_{write}/2$ or $V_{write}/3$ are used to ensure subthreshold voltages on nontargeted cells.

During the read operation, the write path is disabled by grounding the WWL, and a read voltage V_{read} is applied to the target PL, while the unselected read cells are biased at $V_{read}/2$ or $V_{read}/3$. This voltage induces a potential change at the floating node, which modulates the current through the read transistor. The readout is performed via RBL and RSL. The sense margin is governed by the same expressions used in 2T-1C FeRAM:

$$\Delta V_{FN} = \frac{\Delta P_r}{C_{para}} \tag{7.8}$$

$$\Delta I_{RBL} \propto \left(\frac{W}{L}\right)_{TR} \times \Delta V_{FN} \tag{7.9}$$

By leveraging shared transistors for both write and read operations, the 2T-nC architecture achieves improved memory density compared to 2T-1C, while retaining the benefits of amplified and reliable sensing.

7.3 Optimization Guidance

7.3.1 Nondestructive Readout

In all the FeCap-based memory designs described so far, except for 1T-1C FeMFET, the read operation involves applying a sufficiently high voltage to switch the polarization, leading to a destructive read process. This significantly increases the requirements on endurance of the ferroelectric and also increases the read latency due to additional write-back step. To avoid that, recent studies have proposed nondestructive readout (NDRO) methods for 1C FeCap, 1T-1C, and 1T-nC FeRAM designs.[46–50] Typically, these methods utilize capacitive sensing to achieve NDRO. Instead of relying on transient currents from polarization switching, NDRO involves measuring the small-signal capacitance of FeCap under controlled bias conditions. Due to asymmetric polarization dynamics and domain wall contributions, the capacitance varies at low or zero DC bias depending on its polarization state. This difference enables nondestructive sensing of stored binary information without inducing significant polarization switching.

By superimposing a small alternating current (AC) signal on a low or zero DC bias and detecting the resulting AC current, the effective capacitance is extracted. Two distinct capacitance values correspond to two polarization states – high-capacitance state (HCS) and low-capacitance state (LCS) – which serve as the memory states for binary data storage, which is shown in Figure 7.3. Then, the sense margin of this NDRO method can be expressed as:

$$\mathrm{MW}(V_{\mathrm{read}}) = |\varepsilon_+ - \varepsilon_-| \times \frac{\varepsilon_0 A_{\mathrm{FE}}}{t_{\mathrm{FE}}} \int \frac{\partial V}{\partial t} dt \tag{7.10}$$

Here, ε_+ and ε_- denote the relative dielectric constants for HCS and LCS, respectively, while t_{FE} represents the thickness of the ferroelectric layer. This method offers true NDRO, benefiting from enhanced read endurance, lower power consumption, and seamless integration with CMOS BEOL processes. However, a key challenge remains in the limited difference between HCS and LCS (often with capacitance ratios < 1.3), constraining the sensing margin and complicating multilevel data storage and reliable readout under process variations and aging effects. For MFM FeCap, the asymmetric CV can be employed to increase the ON/OFF ratio between HCS and LCS.[48] However, the ratio is still quite low. Then, the MFS FeCapis used; Si electrode[47,51–53] or a BEOL-process-compatible oxide-semiconductor inserting layer[46,54] is used in the structure.

Due to their larger sense margins, 2T-1C and 2T-nC FeRAM architectures can utilize a sufficiently distinguishable read voltage at the PL node – without fully switching the polarization – to enable NDRO, albeit with some sacrifice in sense margin. This behavior can be analyzed using load-line analysis, as shown in Figure 7.3. When WBL voltages are applied, the polarization charge changes for

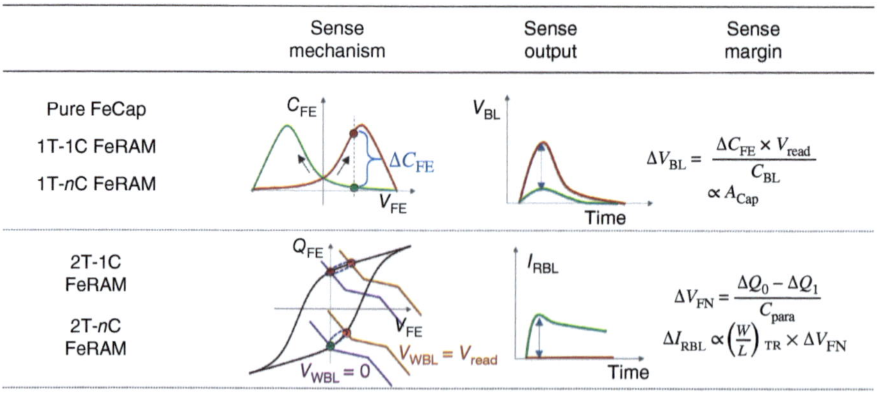

Figure 7.3 NDRO mechanism and capacitive sensing mechanism for 1C, 1T-1C, and 1T-nC FeRAMs and small read voltage sensing for 2T-1C and 2T-nC FeRAMs.

state "0" (ΔQ_0) are greater than those for state "1" (ΔQ_1), resulting in a floating-node voltage difference ΔV_{FN}:

$$\Delta V_{FN} = \frac{\Delta Q_0 - \Delta Q_1}{C_{para}} \tag{7.11}$$

The resulting RBL current sense margin is analogous to that of the 2T-1C FeRAM case, as described in Equation (7.9). In this configuration, multiple FeCaps are connected to the floating node, potentially increasing the parasitic capacitance (C_{para}) compared to the 2T-1C FeRAM case. However, assuming a fixed number of capacitors, the sense margin remains relatively constant with geometric scaling, offering improved scalability over 1T-1C FeRAM designs.

Another noteworthy point is that following the application of the read pulse, some of the switched polarization may partially revert, thereby increasing the number of sustainable read cycles. This phenomenon is reflected in the small minor hysteresis loops shown in the load-line analysis. If the read operation is confined within these minor loops, it can be considered effectively disturbance-free. Nevertheless, in practical scenarios, repeated reads may lead to cumulative polarization switching, eventually driving the device beyond the minor loop and resulting in read-induced degradation.

7.3.2 Disturbance Management

In FeRAM architectures such as 1T-1C and 2T-1C FeRAMs, the unselected memory cells are fully isolated during read and write operations, providing excellent immunity to disturbance. This isolation ensures that only the targeted cell experiences the necessary switching voltage, leaving neighboring cells unaffected. However, in configurations like 1C, 1T-*n*C, and 2T-*n*C, where multiple FeCaps share common access paths, unselected cells can still experience partial voltage stress – even when employing half-selection biasing schemes such as $V_{DD}/2$ or $V_{DD}/3$. These partial voltages can gradually induce unintended polarization switching over time, leading to reliability concerns and data retention degradation. To mitigate these disturbance effects, integrating a selector device with the FeCap has proven to be an effective solution. Highly nonlinear selector devices – such as Ovonic threshold switches (OTSs) or back-to-back diodes[55–58] – can maintain a high-resistance state under low voltages, effectively suppressing ferroelectric voltage drop and preventing unintentional switching. For example, back-to-back diodes offer bidirectional conduction and rectifying behavior, suppressing ferroelectric voltage drop under inhibition biases and further improving disturbance resilience. By incorporating such selector devices, the overall reliability and endurance of high-density FeRAM arrays can be significantly enhanced.

7.4 CIM Applications and 3D Integration

Beyond their traditional role as storage elements, FeCap-based memories demonstrate strong potential for integration into CIM frameworks – a transformative computing paradigm that rethinks conventional system architectures by embedding computational capabilities directly within memory arrays. By enabling *in-situ* data processing, CIM drastically reduces or eliminates the overhead associated with memory access, leveraging the intrinsic locality and massive parallelism of memory structures. This architectural shift offers significant benefits in terms of throughput, energy efficiency, and scalability, especially for workloads characterized by extensive data movement and low arithmetic intensity. As the limitations of the von Neumann architecture become increasingly evident in the era of big data and AI, CIM emerges as a compelling solution for energy-efficient, high-performance computing.

CIM architectures can be categorized into six types: XYZ-CIM, XZ-CIM, XY-CIM, Z-CIM, X-CIM, and O-CIM.[6] These categories depend on whether the inputs (X, Y, etc.) originate from memory cells and whether the output (Z) is re-stored in a memory cell after computation, as illustrated in Figure 7.4. XYZ-CIM takes both X and Y from memory, computes Z, and stores the result back in memory. XZ-CIM accesses X from a memory cell and Y from elsewhere (typically WL or BL), computes Z, and then stores Z in memory. XY-CIM retrieves both X and Y from memory, computes Z, but does not store Z in memory. Z-CIM takes X and Y out of memory, computes Z, and stores the result in memory. X-CIM takes X from memory and Y from elsewhere, computes Z, and outputs Z without storing it in memory. O-CIM does not use memory for inputs nor does it store computation results in memory; instead, memory cells serve purely as computational platforms or logic modules.

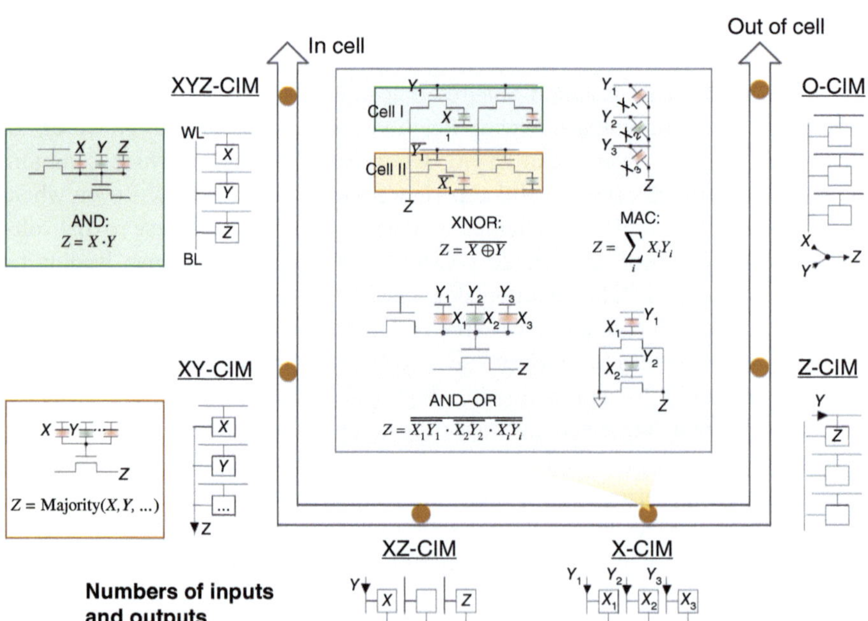

Figure 7.4 CIM classes sorted by input and output locations. The reported FeCap-based memories can involve XYZ-, XY-, and X-type CIM classes. Most of the FeCap-based memories can be used as X-type to perform MAC, AND–OR, and XNOR operations. FeMFET can be used as XY-CIM class by employing 1T-nC majority gate FeMFET. XYZ-CIM requires all inputs and outputs to be stored in memory cells, so 2T-nC FeRAM can be adopted.

Most FeCap-based memory designs are naturally compatible with the X-CIM configuration, where one operand (X) is stored in FeCap and the other operand (typically Y) is the applied voltage at WL during computation. For example, a 12×12 FeCap crossbar array has been demonstrated to perform MAC operations,[59] which is a nondestructive operation using capacitive sensing. In this implementation, the weights are first programmed into the FeCaps, and different input voltages – representing the variable x – are applied to WLs. The resulting total accumulated charge across the FeCaps and, thus, the output voltage (V_{out}) can be expressed as:

$$Q = \sum C_{FE_i} V_{FE_i} \tag{7.12}$$

$$V_{out} = \frac{Q}{C_{ref}} \tag{7.13}$$

where C_{FE} is the capacitance of the ferroelectric layer, and V_{FE} is the voltage drop across FeCap. However, due to the nonlinear and voltage-dependent nature of ferroelectric capacitance, this simplified expression is an approximation. A more accurate representation requires integrating the differential capacitance over the voltage range, leading to the continuous form:

$$Q = \int_0^{V_{FE}} C_{FE}(V)dV \tag{7.14}$$

This integral accounts for the nonlinear polarization response of the ferroelectric material and provides a more faithful estimation of the total charge stored, especially in high-precision analog or MAC-based applications.

For crossbar FeCaps, a promising 3D integration method involves sequential stacking of multiple crossbar layers, as shown in Figure 7.5. This configuration consists of vertically stacked two-dimensional arrays of FeCaps, with each layer formed by intersecting WLs and BLs at cross-points. Sequential stacking maximizes memory density while maintaining efficient access to individual cells. However, low-temperature BEOL processing (typically below 400 °C) is required to prevent damage to FeCap layers and underlying CMOS circuits. This thermal constraint limits material choices to those with high stability and minimal diffusion. As stacking layers increase, signal routing complexity grows. Vertical vias must be precisely aligned and isolated to prevent crosstalk, ensuring selective addressing of each layer. Advanced layout strategies and potential use of 3D selective devices such as diodes or transistors can mitigate disturbance issues and improve system robustness. Despite these challenges, sequentially stacked crossbar FeCap arrays hold great promise for ultra-dense, low-power NVM though overcoming material, thermal, and routing obstacles remains critical for practical deployment.

Another viable candidate for X-CIM implementation is the 1T-1C FeMFET,[42] which is nondestructive. In this structure, the input X, which may also represent a weight, is stored in FeCap through polarization programming. During the computation phase, another input is applied to WL, and the output drain current reflects the logical AND between the programmed state and the applied input voltage. Also, Figure 7.5 shows the sequential stacking 3D integration, which can be adopted to increase the density of this array, which faces the same thermal issue and routing obstacles and requires careful material selection and layout design.

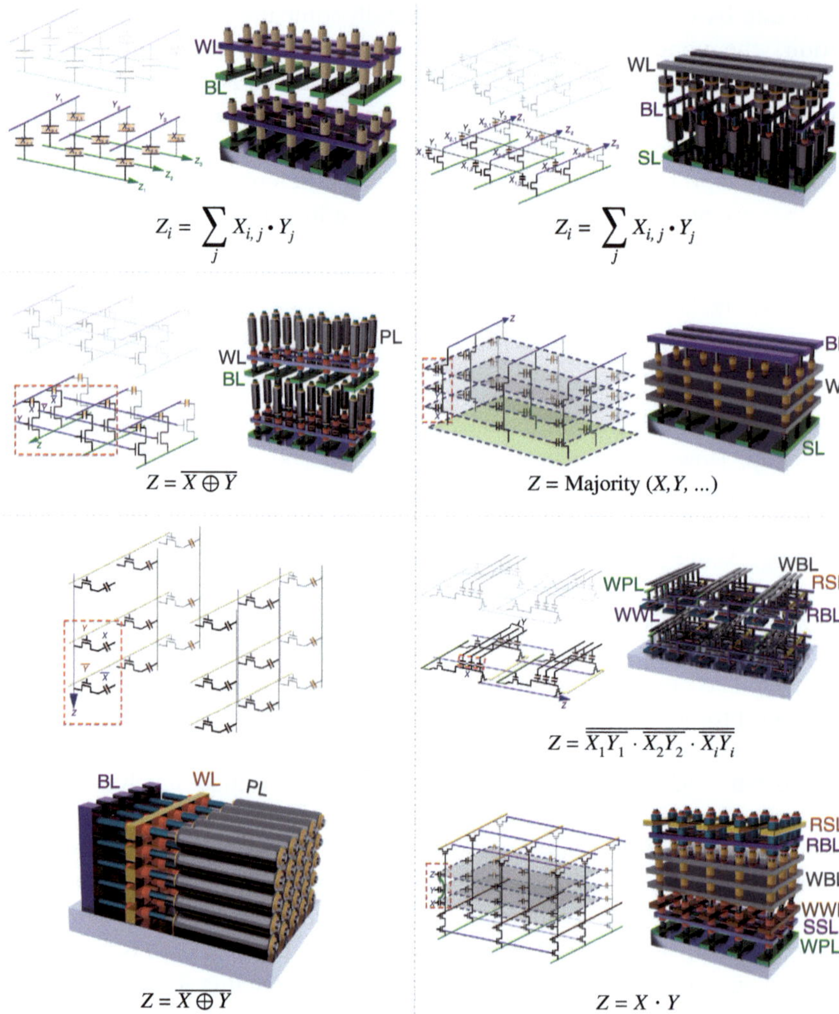

Figure 7.5 The 3D integration method of FeCap-based memories. All these FeCap memories have the capability for vertical stacking in 3D integration, which significantly increases the density of the FeCap-based memory.

Dual-row 1T-1C FeRAM cells can also perform XNOR operations in an X-CIM configuration, which is destructive. In this case, two oppositely polarized FeCaps are used to store the input data X.[60] During the computation step, both WL_1 and WL_2 are applied opposite voltage V_{DD} and ground, which represents input Y, while the two PLs biased at V_{DD}. The resulting voltage on BL reflects charge sharing between the two FeCaps, effectively representing the logical XNOR result of the inputs, which can be used as the ternary content-addressable memory (TCAM) cell. The 3D integration of 1T-1C FeRAM in Figure 7.5 offers a potential path to outperform conventional 3D DRAM, delivering higher memory density and performance. One approach uses sequential stacking, similar to $4F^2$ RAM, where planar 1T-1C

FeRAM arrays are repeated across multiple layers. This design employs vertical cylindrical FeCaps and gate-all-around MOSFET access transistors, increasing surface area for higher density. A recent milestone demonstrated a 32-Gb 1T-1C FeRAM with dual-layer stacking,[11] using dual-gate poly-Si-channel access transistors above CMOS circuitry. This architecture exhibits near-unlimited endurance and excellent retention across temperatures. Alternatively, vertical stacking akin to 3D DRAM[61–67] through top-down parallel processing that achieves high density and performance can also be adopted. Each layer integrates a horizontal GAA access transistor with metal–ferroelectric–semiconductor or metal–ferroelectric–metal capacitors. The nonvolatility of ferroelectric materials allows cost-effective channel materials such as poly-Si or amorphous oxide semiconductors instead of single-crystal silicon. Two routing variants exist: one with vertical BLs and horizontal WLs minimizes BL coupling and ensures consistent sensing delay while enhancing sense margin with layer scaling; the other with horizontal BLs and vertical WLs requires tighter layer pitches and more precise integration, posing greater challenges. Varying contact heights across layers can degrade sense margin, particularly in upper tiers, necessitating further optimization. Improvements may come from channel-all-around (CAA) access transistors, such as those using indium gallium zinc oxide (IGZO) channels demonstrated in 3D DRAM which offer bit-cost scalability and long retention, making them promising for 3D 1T-1C FeRAM.[68]

The 2T-*n*C FeRAM architecture is also compatible with the X-CIM framework. A 2T-4C configuration is reported to implement an in-memory AND–OR logic gate,[69] which can achieve a quasi-nondestructive operation. In this design, input Y is controlled by the voltage level on the WBL, representing a logic "0" when grounded and a logic "1" when driven by a positive voltage. Input X is preprogrammed into FeCaps. To suppress the voltage-sharing effect among FeCaps, an additional compensation capacitor is incorporated, which is consistently biased with a negative voltage. The operation of this logic structure consists of two distinct stages. During the programming stage, input X is written into the memory cell through the application of appropriate WBL and WPL voltages, mediated by the write transistor. In the subsequent computation stage, input A is applied through WPL, and the resulting AND–OR logic outcome is inferred from the drain current of the read transistor.

Although the additional pair of write and read transistors in 2T-*n*C FeRAM could offset density gain, it could be compensated by 3D integration, like layer-by-layer sequential stacking. The 2T-*n*C FeRAM is also compatible with parallel integration of vertical string configurations.[30] These two 3D stacking methods are both shown in Figure 7.5. In the vertical 2T-*n*C design, the number of stacked layers directly corresponds to the number of MFM capacitors per cell. Yet, as layer count increases, electrical performance may degrade due to parasitic effects on the floating node. The array resembles a vertical NAND structure, with transistors positioned at the top and bottom and memory cells in between. However, unlike NAND, the string's core consists of a metal conductor rather than a semiconductor channel, placing cells in parallel rather than in series. Each vertical string incorporates one write transistor and one or more string-select transistors at its base, enabling selective WPL activation for FeCap programming. At the top, a read (or gain) transistor amplifies charge current to detect polarization states. To ensure precise write operations, inhibition biasing schemes are critical: target cells must receive the full write voltage, while unselected cells experience a fraction (e.g., half or a third) of this voltage. This is achieved by carefully biasing WBLs and internal nodes via WWL, SSL, and WPL combinations. During reads, all WWLs and SSLs are grounded to deactivate write transistors, while the target WBL, RBL, and RSL are activated for sensing.

For XY-CIM-type memory cell, where all the inputs are stored in memory cells, the FeMFET can be adopted by employing a 1T-*n*C FeMFET structure, which serves as a majority gate to implement nondestructive majority logic.[43] Here, multiple FeCaps are connected to one MOSFET. In this configuration, an LVT associated with polarization-down is encoded as Z = "1," while a high V_{TH} associated with polarization-up is encoded as Z = "0." And the inputs X, Y, and so on are stored in FeCaps. The AND operation is conducted locally at each MFM gate, which is enabled by applying read voltage at WLs. The results of these local AND operations contribute to the floating gate potential, and the collective outcome determines the channel current. A three-input programmable majority gate is illustrated in Figure 7.4, where the channel conducts a high current only if the majority of the AND outputs are logic "1," thereby realizing the desired majority functionality. This 1T-*n*C majority gate FeMFET can achieve high density using 3D NAND-like integration strategy as shown in Figure 7.5.

The 2T-*n*C FeRAM architecture can be extended to implement XYZ-CIM.[44] Here, two FeCaps store the input values, and during computation, both are biased at the read voltage. The third FeCap, intended to store the logic output, is biased with a negative voltage. This configuration enables the AND/OR result of the two inputs to be directly written into the third FeCap. Although negative biasing is typically avoided in practical circuit implementations, this method highlights the potential of 2T-*n*C FeRAM for realizing advanced XYZ-CIM applications.

7.5 Conclusion

In conclusion, FeCap-based memories, particularly those utilizing doped HfO_2, present a transformative solution for next-generation computing. Their CMOS compatibility, nanoscale scalability, and analog programmability enable efficient CIM architectures that overcome the von Neumann bottleneck. By integrating computation directly within memory arrays, these devices significantly reduce energy consumption and latency while accelerating data-intensive operations like AI inference. While challenges remain in reliability and circuit optimization, HfO_2-based ferroelectric memories offer a scalable pathway toward high-performance, energy-efficient computing. Their potential for 3D integration further enhances their promise for future memory-centric systems, positioning them as a key enabler of advanced computing paradigms.

Acknowledgments

The authors would like to acknowledge the financial support from the US Department of Energy, Office of Science, Office of Basic Energy Sciences Energy Frontier Research Centers program (DESC0021118), SUPREME, one of the SRC/DARPA JUMP 2.0 center, and National Science Foundation (No. 2346953).

References

1 Mutlu, O., Ghose, S., Gómez-Luna, J. et al. (2019). Processing data where it makes sense: Enabling in-memory computation. *Microprocessors and Microsystems* 67: 28–41. https://doi.org/10.1016/j.micpro.2019.01.009.

2 Sebastian, A., Le Gallo, M., Khaddam-Aljameh, R. et al. (2020). Memory devices and applications for in-memory computing. *Nature of Nanotechnology* 15 (7): 529–544. https://doi.org/10.1038/s41565-020-0655-z.

3 Indiveri, G. and Liu, S. C. (2015). Memory and information processing in neuromorphic systems. *Proceedings of the IEEE* 103 (8): 1379–1397. https://doi.org/10.1109/JPROC.2015.2444094.

4 Tye, N. J., Hofmann, S., and Stanley-Marbell, P. (2023). Materials and devices as solutions to computational problems in machine learning. *Nature Electronics* 6 (7): 479–490. https://doi.org/10.1038/s41928-023-00977-1.

5 Le Gallo, M., Sebastian, A., Mathis, R. et al. (2018). Mixed-precision in-memory computing. *Nature Electronics* 1 (4): 246–253. https://doi.org/10.1038/s41928-018-0054-8.

6 Sun, Z., Kvatinsky, S., Si, X. et al. (2023). A full spectrum of computing-in-memory technologies. *Nature Electronics* 6 (11): 823–835. https://doi.org/10.1038/s41928-023-01053-4.

7 Gong, T., Tao, L., Li, J. et al. (2021). $10^5\times$ Endurance improvement of FE-HZO by an innovative rejuvenation method for 1z node NV-DRAM applications. In: *Proceedings of the 2021 Symposium on VLSI Technology*. IEEE. pp. 1–2.

8 Liao, P. J., Chang, Y. K., Lee, Y. H. et al. (2021). Characterization of fatigue and its recovery behavior in ferroelectric HfZrO. In: *Proceedings of the 2021 Symposium on VLSI Technology*. IEEE. pp. 1–2.

9 Okuno, J., Kunihiro, T., Konishi, K. et al. (2020). SoC compatible 1T1C FeRAM memory array based on ferroelectric $Hf_{0.5}Zr_{0.5}O_2$. In: *Proceedings of the 2020 IEEE Symposium on VLSI Technology*. IEEE. pp. 1–2. https://doi.org/10.1109/VLSITechnology18217.2020.9265063.

10 Francois, T., Coignus, J., Makosiej, A. et al. (2021). 16kbit HfO_2:Si-based 1T-1C FeRAM arrays demonstrating high performance operation and solder reflow compatibility. In: *Proceedings of the 2021 IEEE International Electron Devices Meeting (IEDM)*. IEEE. pp. 33.1.1–33.1.4. https://doi.org/10.1109/IEDM19574.2021.9720640.

11 Ramaswamy, N., Calderoni, A., Zahurak, J. et al. (2023). NVDRAM: A 32Gb dual layer 3D stacked non-volatile ferroelectric memory with near-DRAM performance for demanding AI workloads. In: *Proceedings of the 2023 International Electron Devices Meeting (IEDM)*. IEEE. pp. 1–4. https://doi.org/10.1109/IEDM45741.2023.10413848.

12 Martin, L. W. and Rappe, A. M. (2016). Thin-film ferroelectric materials and their applications. *Nature Review Materials* 2 (2): 16087. https://doi.org/10.1038/natrevmats.2016.87.

13 Setter, N. (2006). Ferroelectric thin films: Review of materials, properties, and applications. *Journal of Applied Physics* 100 (5): 051606. https://doi.org/10.1063/1.2336999.

14 Mikolajick, T., Park, M. H., BegonLours, L. et al. (2023). From ferroelectric material optimization to neuromorphic devices. *Advanced Materials* 35 (37): 2206042. https://doi.org/10.1002/adma.202206042.

15 Kim, K. H., Karpov, I., Olsson, R. H. et al. (2023). Wurtzite and fluorite ferroelectric materials for electronic memory. *Nature Nanotechnology* 18 (5): 422–441. https://doi.org/10.1038/s41565-023-01361-y.

16 Bokov, A. A. and Ye, Z. G. (2006). Recent progress in relaxor ferroelectrics with perovskite structure. *Journal of Materials Science* 41 (1): 31–52. https://doi.org/10.1007/s10853-005-5915-7.

17 Waser, R., Dittmann, R., Staikov, G. et al. (2009). Redox-based resistive switching memories – nanoionic mechanisms, prospects, and challenges. *Advanced Materials* 21 (25–26): 2632–2663. https://doi.org/10.1002/adma.200900375.

18 Böscke, T. S., Müller, J., Bräuhaus, D. et al. (2011). Ferroelectricity in hafnium oxide thin films. *Applied Physics Letters* 99 (10): 102903. https://doi.org/10.1063/1.3634052.

19 Pal, A., Narasimhan, V. K., Weeks, S. et al. (2017). Enhancing ferroelectricity in dopant-free hafnium oxide. *Applied Physics Letters* 110 (2): 022903. https://doi.org/10.1063/1.4973928.

20 Lee, H. J., Lee, M., Lee, K. et al. (2020). Scale-free ferroelectricity induced by flat phonon bands in HfO_2. *Science* 369 (6509): 1343–1347. https://doi.org/10.1126/science.aba0067.

21 Lancaster, S., Slesazeck, S., and Mikolajick, T. (2024). On the thickness scaling of ferroelectric hafnia. *IEEE Transactions on Materials for Electron Devices* 1: 36–48. https://doi.org/10.1109/TMAT.2024.3423665.

22 Cheema, S. S., Kwon, D., Shanker, N. et al. (2020). Enhanced ferroelectricity in ultrathin films grown directly on silicon. *Nature* 580 (7804): 478–482. https://doi.org/10.1038/s41586-020-2208-x.

23 Cai, Z., Toprasertpong, K., Liu, Z. et al. (2024). Understanding HZO thickness scaling in Si FeFETs: Low operating voltage, fast wake-up, and suppressed charge trapping. *IEEE Transactions on Electron Devices* 71 (6): 3633–3639. https://doi.org/10.1109/TED.2024.3386508.

24 Cai, Z., Toprasertpong, K., Takenaka, M. et al. (2023). HZO scaling and fatigue recovery in FeFET with low voltage operation: Evidence of transition from interface degradation to ferroelectric fatigue. In: *Proceedings of the 2023 IEEE Symposium on VLSI Technology and Circuits (VLSI Technology and Circuits)*. IEEE. pp. 1–2. https://doi.org/10.23919/VLSITechnologyandCir57934.2023.10185295.

25 Xu, L., Shibayama, K. Izukashi, T. et al. (2016). General relationship for cation and anion doping effects on ferroelectric HfO_2 formation. In: *Proceedings of the 2016 IEEE International Electron Devices Meeting (IEDM)*. IEEE. pp. 25.2.1–25.2.4. https://doi.org/10.1109/IEDM.2016.7838477.

26 Khan, A. I., Keshavarzi, A., and Datta, S. (2020). The future of ferroelectric field-effect transistor technology. *Nature Electronics* 3 (10): 588–597. https://doi.org/10.1038/s41928-020-00492-7.

27 Raj, B., Tripathi, S. L., Chaudhary, T. et al. (eds.) (2024). Front matter. In: *Integrated Devices for Artificial Intelligence and VLSI* (1st ed.). Wiley. pp. 125–126. https://doi.org/10.1002/9781394205158.fmatter.

28 Chen, H. (2024). Non-volatile photonic-electronic memory via 3D monolithic ferroelectric-silicon ring resonator. *Light: Science & Applications* 13 (1): 271. https://doi.org/10.1038/s41377-024-01625-9.

29 Hur, J., Luo, Y. C., Wang, Z. et al. (2021). A technology path for scaling embedded FeRAM to 28nm with 2T1C structure. In: *Proceedings of the 2021 IEEE International Memory Workshop (IMW)*. IEEE. pp. 1–4. https://doi.org/10.1109/IMW51353.2021.9439624.

30 Deng, S., Xiao, Y., Zhao, Z. et al. (2023). Comparative advantages of 2T-nC FeRAM in empowering high density 3D ferroelectric capacitor memory. In: *Proceedings of the 2023 International Electron Devices Meeting (IEDM)*. IEEE. pp. 1–4. https://doi.org/10.1109/IEDM45741.2023.10413855.

31 Okuno, J., Kunihiro, T., Konishi, K. et al. (2021). High-endurance and low-voltage operation of 1T1C FeRAM arrays for nonvolatile memory application. In: *Proceedings of the 2021 IEEE International Memory Workshop (IMW)*. IEEE. pp. 1–3. https://doi.org/10.1109/IMW51353.2021.9439595.

32 Luo, Y. C., Hur, J., Wang, Z. et al. (2022). A technology path for scaling embedded FeRAM to 28 nm and beyond with 2T1C structure. *IEEE Transactions on Electron Devices* 69 (1): 109–114. https://doi.org/10.1109/TED.2021.3131108.

33 Yoo, J., Song, H., Lee, H. et al. (2023). Recent research for HZO-based ferroelectric memory towards in-memory computing applications. *Electronics* 12 (10): 2297. https://doi.org/10.3390/electronics12102297.

34 Fu, Z., Cao, S., Zheng, H. et al. (2023). First demonstration of hafnia-based selector-free FeRAM with high disturb immunity through design technology co-optimization. In: *Proceedings of the 2023 International Electron Devices Meeting (IEDM)*. IEEE. pp. 1–4. https://doi.org/10.1109/IEDM45741.2023.10413887.

35 Myeong, I., Kim, H., Kim, S. et al. (2024). Strategies for a wide memory window of ferroelectric FET for multilevel ferroelectric VNAND operation. *IEEE Electronic Device Letters* 45 (7): 1185–1188. https://doi.org/10.1109/LED.2024.3400983.

36 Lim, E., Park, Y., Youn, C. et al. (2025). 3D vertical ferroelectric capacitors with excellent scalability. *Nano Letters* 25 (6): 2166–2172. https://doi.org/10.1021/acs.nanolett.4c05121.

37 Florent, K., Lavizzari, S., Di Piazza, L. et al. (2017). First demonstration of vertically stacked ferroelectric Al doped HfO_2 devices for NAND applications. In: *Proceedings of the 2017 Symposium on VLSI Technology*. IEEE. pp. T158–T159. https://doi.org/10.23919/VLSIT.2017.7998162.

38 Zheng, Z., Zhang, D., Jiao, L. et al. (2024). BEOL-compatible MFMIS ferroelectric/anti-ferroelectric FETs – Part II: Mechanism with load line analysis and scaling strategy. *IEEE Transactions on Electron Devices* 71 (9): 5325–5331. https://doi.org/10.1109/TED.2024.3421184.

39 Wang, X., Sun, C., Zheng, Z. et al. (2024). Comprehensive experiments and modeling applicable for ferroelectric transistors with an MFMIS structure and a wide range of area ratios: Unveiling the operation mechanisms. *IEEE Transactions on Electron Devices* 71 (9): 5332–5338. https://doi.org/10.1109/TED.2024.3421180.

40 Wang, X., Zheng, Z., Jiao, L. et al. (2024). Unveiling the intricate dynamic characteristics of FeFETs with a MFMIS structure: Experiment and modeling. In: *Proceedings of the 2024 IEEE International Electron Devices Meeting (IEDM)*. IEEE. pp. 1–4. https://doi.org/10.1109/IEDM50854.2024.10873561.

41 Zheng, Z., Jiao, L., Zhang, D. et al. (2024). BEOL-compatible MFMIS ferroelectric/anti-ferroelectric FETs – Part I: Experimental results with boosted memory window. *IEEE Transactions on Electron Devices* 71 (3): 1827–1833. https://doi.org/10.1109/TED.2023.3326116.

42 Ni, K., Smith, J. A., Grisafe, B. et al. (2018). SoC logic compatible multi-bit FeMFET weight cell for neuromorphic applications. In: *Proceedings of the 2018 IEEE International Electron Devices Meeting (IEDM)*. IEEE. pp. 13.2.1–13.2.4. https://doi.org/10.1109/IEDM.2018.8614496.

43 Deng, S., Benkhelifa, M., Thomann, S. et al. (2022). Compact ferroelectric programmable majority gate for compute-in-memory applications. In: *Proceedings of the 2022 International Electron Devices Meeting (IEDM)*. IEEE. pp. 36.7.1–36.7.4. https://doi.org/10.1109/IEDM45625.2022.10019400.

44 Slesazeck, S., Ravsher, T., Havel, V. et al. (2019). A 2TnC ferroelectric memory gain cell suitable for compute-in-memory and neuromorphic application. In: *Proceedings of the 2019 IEEE International Electron Devices Meeting (IEDM)*. IEEE. pp. 38.6.1–38.6.4. https://doi.org/10.1109/IEDM19573.2019.8993663.

45 Xiao, Y., Deng, S., Zhao, Z. et al. (2023). Quasi-nondestructive read out of ferroelectric capacitor polarization by exploiting a 2TnC cell to relax the endurance requirement. *IEEE Electron Device Letters* 44 (9): 1436–1439. https://doi.org/10.36227/techrxiv.22570033.v1.

46 Kim, T. H., Luo, Y. C., Phadke, O. et al. (2024). Engineering nvCap from FEOL to BEOL with ferroelectric small-signal non-destructive read. In: *Proceedings of the 2024 IEEE International Memory Workshop (IMW)*. IEEE. pp. 1–4. https://doi.org/10.1109/IMW59701.2024.10536973.

47 Zhou, Z., Jiao, L., Kong, Q. et al. (2023). Non-destructive-read 1T1C ferroelectric capacitive memory cell with BEOL 3D monolithically integrated IGZO access transistor for 4F2 high-density integration. In: *Proceedings of the 2023 IEEE Symposium on VLSI Technology and Circuits (VLSI Technology and Circuits)*. IEEE. pp. 1–2. https://doi.org/10.23919/VLSITechnologyandCir57934.2023.10185243.

48 Mukherjee, S., Bizindavyi, J., Clima, S. et al. (2023). Capacitive memory window with non-destructive read in ferroelectric capacitors. *IEEE Electron Device Letters* 44 (7): 1092–1095. https://doi.org/10.1109/LED.2023.3278599.

49 Mukherjee, S., Bizindavyi, J., Luo, Y. C. et al. (2023). Pulse-based capacitive memory window with high non-destructive read endurance in fully BEOL compatible ferroelectric capacitors. In: *Proceedings of the 2023 International Electron Devices Meeting (IEDM)*. IEEE. pp. 1–4. https://doi.org/10.1109/IEDM45741.2023.10413879.

50 Mukherjee, S., Bizindavyi, J., Clima, S. et al. (2024). Improved capacitive memory window for non-destructive read in HZO-based ferroelectric capacitors with incorporation of semiconducting IGZO. In: *Proceedings of the 2024 IEEE International Electron Devices Meeting (IEDM)*. IEEE. pp. 1–4. https://doi.org/10.1109/IEDM50854.2024.10873459.

51 Zhou, Z., Jiao, L., Zhou, J. et al. (2023). Inversion-type ferroelectric capacitive memory and its 1-Kbit crossbar array. *IEEE Transactions on Electron Devices* 70 (4): 1641–1647. https://doi.org/10.1109/TED.2023.3243556.

52 Zhou, Z., Leming, J., Zhou, J. et al. (2022). Experimental demonstration of an inversion-type ferroelectric capacitive memory and its 1 kbit crossbar array featuring high CHCS/CLCS, fast speed, and long retention. In: *Proceedings of the 2022 IEEE Symposium on VLSI Technology and Circuits (VLSI Technology and Circuits)*. IEEE. pp. 357–358. https://doi.org/10.1109/VLSITechnologyandCir46769.2022.9830291.

53 Zhou, Z., Zhou, J., Wang, X. et al. (2020). A metal-insulator-semiconductor non-volatile programmable capacitor based on a $HfAlO_x$ ferroelectric film. *IEEE Electron Device Letters* 41 (12): 1837–1840. https://doi.org/10.1109/LED.2020.3035276.

54 Lee, J., Zhang, C., Shon, M. et al. (2024). BEOL-compatible non-volatile capacitive synapse with ALD W-doped In_2O_3 semiconductor layer. In: *Proceedings of the 2024 IEEE International Electron Devices Meeting (IEDM)*. IEEE. pp. 1–4. https://doi.org/10.1109/IEDM50854.2024.10873481.

55 Musisi-Nkambwe, M., Afshari, S., Barnaby, H. et al. (2021). The viability of analog-based accelerators for neuromorphic computing: A survey. *Neuromorphic Computing and Engineering* 1 (1): 012001. https://doi.org/10.1088/2634-4386/ac0242.

56 Syed, G. S., Le Gallo, M., and Sebastian, A. (2025). Phase-change memory for in-memory computing. *Chemical Reviews* 125: 5163–5194. https://doi.org/10.1021/acs.chemrev.4c00670.

57 Hsieh, C. C., Chang, Y. F., Chen, Y. C. et al. (2017). Highly non-linear and reliable amorphous silicon based back-to-back Schottky diode as selector device for large scale RRAM arrays. *ECS Journal of Solid State Science and Technology* 6 (9): N143–N147. https://doi.org/10.1149/2.0041709jss.

58 Li, Y., Gong, Q., Li, R. et al. (2014). A new bipolar RRAM selector based on anti-parallel connected diodes for crossbar applications. *Nanotechnology* 25 (18): 185201. https://doi.org/10.1088/0957-4484/25/18/185201.

59 Luo, Y. C., Hur, J., Wang, T. H. et al. (2021). Experimental demonstration of non-volatile capacitive crossbar array for in-memory computing. In: *Proceedings of the 2021 IEEE International Electron Devices Meeting (IEDM)*. IEEE. pp. 1–4. https://doi.org/10.1109/IEDM19574.2021.9720508.

60 Li, J., Xie, M., and Xiong, S. (2024). A compute-in-memory macro based on complementary 2T2C FeRAM cell for BNNs. In: *Proceedings of the 2024 IEEE 17th International Conference on Solid-State & Integrated Circuit Technology (ICSICT)*. IEEE. pp. 1–3. https://doi.org/10.1109/ICSICT62049.2024.10831829.

61 Choi, K. S., Kim, S. H., Seo, J. W. et al. (2024). A three dimensional DRAM (3D DRAM) technology for the next decades. In: *Proceedings of the 2024 IEEE Symposium on VLSI Technology and Circuits (VLSI Technology and Circuits)*. IEEE. pp. 1–2. https://doi.org/10.1109/VLSITechnologyandCir46783.2024.10631471.

62 Chen, W. C., Lue, H. T., Wu, M. H. et al. (2023). A highly pitch-scalable capacitor-less 3D DRAM using cross-bar selection with gate-controlled thyristor (GCT) featuring high endurance and free read-disturb. In: *Proceedings of the 2023 International Electron Devices Meeting (IEDM)*. IEEE. pp. 1–4. https://doi.org/10.1109/IEDM45741.2023.10413828.

63 Wu, X., Upton, L. R., Hsu, P. K. et al. (2024). A comparative study of HBL and VBL 3D DRAM: signal margin, bit-cell density, and scalability. In: *Proceedings of the 2024 IEEE Silicon Nanoelectronics Workshop (SNW)*. IEEE. pp. 19–20. https://doi.org/10.1109/SNW63608.2024.10639232.

64 Chen, W. C., Lue, H. T., Wu, M. Y. et al. (2022). A 3D stackable DRAM: Capacitor-less three-wordline gate-controlled thyristor (GCT) RAM with >40 μA current sensing window, >1010 endurance, and 3-second retention at room temperature. In: *Proceedings of the 2022 International Electron Devices Meeting (IEDM)*. IEEE. pp. 26.3.1–26.3.4. https://doi.org/10.1109/IEDM45625.2022.10019464.

65 Han, J. W., Park, S. H., Jeong, M. Y. et al. (2023). Ongoing evolution of DRAM scaling via third dimension – vertically stacked DRAM. In: *Proceedings of the 2023 IEEE Symposium on VLSI Technology and Circuits (VLSI Technology and Circuits)*. IEEE. pp. 1–2. https://doi.org/10.23919/VLSITechnologyandCir57934.2023.10185290.

66 Hsu, P. K., Sharda, J., Wu, X. et al. (2025). Monolithic 3D stackable DRAM. *IEEE Nanotechnology Magazine* 19 (2): 7–16. https://doi.org/10.1109/MNANO.2025.3533815.

67 Huang, M., Si, S., He, Z. et al. (2023). A 3D stackable 1T1C DRAM: Architecture, process integration and circuit simulation. In: *Proceedings of the 2023 IEEE International Memory Workshop (IMW)*. IEEE. pp. 1–4. https://doi.org/10.1109/IMW56887.2023.10145931.

68 Lee, F. M., Tseng, P. H., Lin, Y. Y. et al. (2024). Bit-cost-scalable 3D DRAM architecture and unit cell first demonstrated with integrated gate-around and channel-around IGZO FETs. In: *Proceedings of the 2024 IEEE Symposium on VLSI Technology and Circuits (VLSI Technology and Circuits)*. IEEE. pp. 1–2. https://doi.org/10.1109/VLSITechnologyandCir46783.2024.10631386.

69 Xiao, Y., Xu, Y., Deng, S. et al. (2023). A compact ferroelectric 2T-(n+1)C cell to implement AND-OR logic in memory. In: *Proceedings of the 2023 IEEE Computer Society Annual Symposium on VLSI (ISVLSI)*. IEEE. pp. 1–6. https://doi.org/10.1109/ISVLSI59464.2023.10238503.

8

Basic Mechanism of Si-channel HfO_2-FeFET and Its Reliability

Reika Ichihara and Masumi Saitoh*

Frontier Technology Research and Development Institute, Kioxia Corporation, Yokohama-shi, Kanagawa, Japan

*Corresponding author: reika.ichihara@kioxia.com

8.1 Introduction

The ferroelectric field-effect transistor (FeFET) is a nonvolatile memory based on threshold voltage (V_{th}) modulation due to spontaneous polarization (P_s) reversal in ferroelectric layer used as a gate dielectric. Since the discovery of ferroelectricity in thin HfO_2,[1–3] HfO_2-based FeFET (HfO_2-FeFET) has attracted much attention because of its high-speed/low-voltage operation and CMOS compatibility.

FeFET utilizes the difference in gate voltages (V_g) realizing the same polarization charge, which is the sum of dielectric polarization charges determined by electric field and dielectric permittivity and polarization charges originating in P_s, in the different memorized states with up-state P_s and down-state P_s. Here, P_s changes from down (up) state to up (down) state by applying positive (negative) voltage between the gate electrode and the channel, resulting in V_{th} modulation in negative (positive) direction. That is, the polarities of operating voltages and the directions of resulting V_{th} modulation are opposite. Therefore, to detect V_{th} of both memorized states, we must apply read voltages with the polarity reversing P_s from its current state to the opposite state. This relationship makes a somewhat special situation during the read operation as follows. Figure 8.1 explains this situation by using an example when trying to detect the flat-band voltages (V_f) for both memorized states. If the flat-band situation, where the polarization charge is zero, is achieved solely by the change in polarization charges determined by permittivity and electric field, V_f expressed by blue in Figure 8.1 are needed. However, considering the general characteristics of ferroelectric HfO_2 that $2P_s$ is $20\mu C/cm^2$ or more and the coercive electric field is about $2 MV/cm$, these V_f values exceed coercive voltages (V_c). Therefore, the reversal

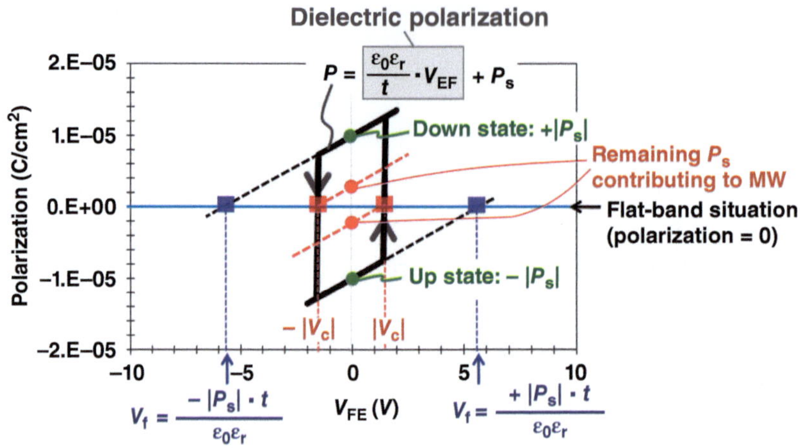

Figure 8.1 Explanation of determinant factor of MW in HfO_2-FeFET using the general polarization–voltage characteristic of FE-HfO_2. P_s, t, and E_c are assumed to be 10 μC/cm^2, 10 nm, and 1.5 MV/cm, respectively. ε_r is assumed to be 20.

of a part of P_s is mandatory to detect V_f. The same explanation also holds true when detecting V_{th}. Consequently, the memory window (MW), which is the difference in measurable V_{th} for both states with up-state P_s and down-state P_s, is determined by $2V_c$, not by the maximum value of P_s. This is the operation principle of the ideal FeFET.[4]

However, in the real FeFET devices, in addition to P_s reversal, charge trapping/de-trapping also affects V_{th}[5–14] resulting in the situation different from the ideal model described above. Therefore, it is necessary to understand the behavior of both P_s reversal and charge trapping/de-trapping for FeFET development.

This chapter provides the recent understandings of basic operation mechanism and reliability for the real HfO_2-FeFET with Si channel, in which the interfacial layer (IL) SiO_2 is inevitably formed, based on the experimental analysis of both P_s reversal and charge trapping/de-trapping.

8.2 Phenomena Occurring in FeFET

8.2.1 Experimental Extraction of Spontaneous Polarization and Trapped Charges

So far, many studies have reported charge trapping/de-trapping in FeFET based on the observation of undesirable or unexpected V_{th} behaviors in FeFET.[5–14] For example,[6] reported the rapid V_{th} reduction (increase) in short time domain after program (erase) operation. Here, program (erase) operation refers to the operation applying positive (negative) voltages between the gate electrode and the channel. Because the direction of this rapid V_{th} drift is opposite to that expected for P_s loss, it is considered that electrons (holes) are trapped temporarily during program (erase) operation and are released immediately. However, it is difficult to know the quantitative amounts of both P_s reversal and charge trapping/de-trapping in FeFET just from V_{th} behavior, because there are countless combinations of P_s reversal and charge trapping/de-trapping that can explain V_{th} behavior.

In 2019, Toprasertpong et al. have reported the amounts of both P_s reversal and charge trapping/de-trapping generated by FeFET switching extracted based on quasi-static split C–V measurement in addition to the investigation of V_{th} modulation.[15] They have clarified that a surprisingly huge amount of charge trapping of about $10^{14}/cm^2$ occurs at the same time as P_s reversal of about $\pm 10 \mu C/cm^2$, which cannot be imagined just from V_{th} behavior.

For FeFET memory, a convenient analysis method that can extract the amounts of both P_s reversal and charge trapping/de-trapping in various memory operating processes including not only program (erase) but also the retention period, has been developed,[16] which will be explained briefly below.

Based on the assumption that all trapped charges (Q_t) are located at the interface between the ferroelectric layer and IL as shown in Figure 8.2, the changes of both P_s and Q_t (ΔP_s and ΔQ_t) are expressed using experimentally measurable parameters Q_m, ΔV_{th}, and $C_{before}(V)$ as described in Equations (1)–(4),[16]

$$\Delta P_s = Q_m - \frac{\varepsilon_r \varepsilon_0 \Delta V_{th}}{EOT_f} + \Delta Q_{Si} \tag{8.1}$$

$$\Delta Q_t = Q_m + \Delta Q_{Si} \tag{8.2}$$

$$\Delta V_{th} \geq 0 : \Delta Q_{Si} = \int_{-\Delta V_{th}}^{0} C_{before}(V) dV \tag{8.3}$$

$$\Delta V_{th} < 0 : \Delta Q_{Si} = -\int_{0}^{-\Delta V_{th}} C_{before}(V) dV \tag{8.4}$$

Here, $Q_m, \Delta V_{th}$, and $C_{before}(V)$ are the total amount of charges moving in and out during the operation, the difference in V_{th} before and after the operation, and the differential capacitance $(C) - V_g$ characteristic just before the operation. EOT_f, $\varepsilon_r, \varepsilon_0$, and ΔQ_{Si} are the oxide equivalent thickness of FE-HfO_2, relative permittivity of the oxide, vacuum permittivity, and the change of carrier density in Si channel induced by the operation. Positive (negative) ΔP_s means the change from up (down) state to down (up) state, and positive (negative) ΔQ_t means electron (hole) trapping or hole (electron) de-trapping.

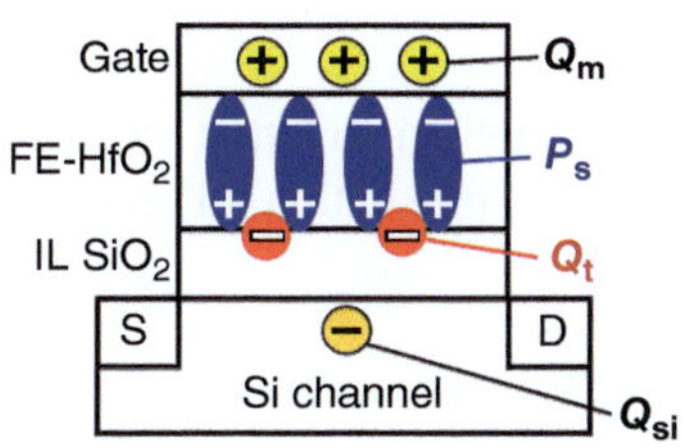

Figure 8.2 Definitions of parameters.

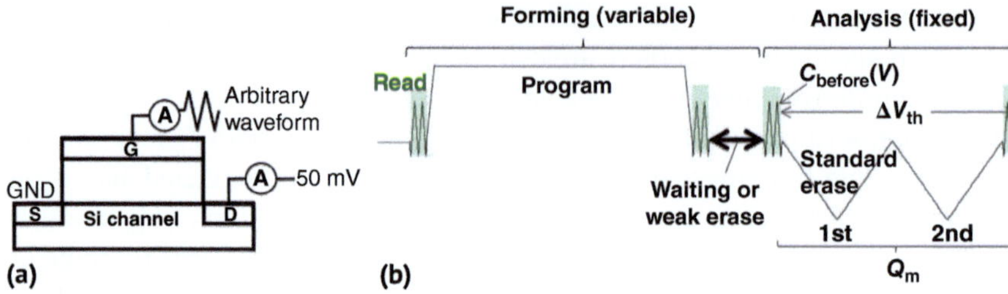

Figure 8.3 Experimental scheme to measure parameters required for the extraction of both ΔP_s and ΔQ_t during memory operation: (a) measurement setup and (b) typical voltage waveform applied to the gate.

The measurements of Q_m, ΔV_{th}, and $C_{before}(V)$, which are parameters required for the extraction of ΔP_s and ΔQ_t, are achieved by the experimental scheme shown in Figure 8.3. The drain current I_d and the gate current I_g are monitored simultaneously while the arbitrary voltage waveform consisting of "Forming" and "Analysis" and DC bias are applied to the gate and the drain, respectively, as shown in Figure 8.3a. In the waveform "Analysis," the parameters required to extract ΔP_s and ΔQ_t generated by a standard erase operation are obtained as follows. First, ΔV_{th} is extracted by comparing the $I_d - V_g$ characteristics during read operations before and after a standard erase operation. Second, Q_m is extracted by integrating transient I_g other than the leakage current, flowing during a standard erase operation, over time. If the nonnegligible Q_m is observed during read operation, that is also taken into consideration in the analysis. And C_{before} (V) is extracted by analyzing the $I_g - V_g$ characteristic during read operation performed just before a standard erase operation as described in Figure 8.3b. By using these three parameters, ΔP_s and ΔQ_t generated by a standard erase operation are extracted based on Equations (8.1)–(8.4). Here, ΔP_s and ΔQ_t caused by a standard erase operation reflect the status generated by "Forming" performed just before applying "Analysis" waveform. For example, if the program in "Forming" becomes weaker, ΔP_s and ΔQ_t induced by the following standard erase operation in "Analysis" also become smaller. Also, for example, if the status partially returns to the erased state during waiting after program operation in "Forming," ΔP_s and ΔQ_t induced by the following standard erase operation become smaller. Therefore, by changing the conditions of the waveform "Forming," ΔP_s and ΔQ_t corresponding to various operating processes including time evolution are extracted.

8.2.2 Phenomena Classification

From the method explained in Section 8.2.1, it has become evident that various phenomena occur in FeFET, which will be introduced one by one in this subsection.

8.2.2.1 Read-induced Electron Trapping

Figure 8.4a shows the typical $I_g - V_g$ characteristic during read operation in FET with paraelectric HfO_2 and the IL SiO_2. In addition to charging and discharging originating in metal–insulator–semiconductor capacitance, current peak indicating electron trapping appears in forward sweep. On the other hand, there is no clear hysteresis in $I_d - V_g$ characteristic shown in Figure 8.4b measured simultaneously with $I_g - V_g$ characteristic. This is because electron trapping occurs at the same time as inversion.

Similar electron trapping during read operation is also observed in FeFET as shown in Figure 8.5a. Because this electron trapping occurs in every read operation regardless of the device status, measurable V_{th} of both programmed and erased states are always somewhat increased by this read-induced electron trapping. Although this phenomenon does not affect MW, which is V_{th} difference between programmed state and erased state as schematically shown in Figure 8.5b, read-induced electron trapping should be taken into consideration when discussing the V_{th} behavior itself as explained later.

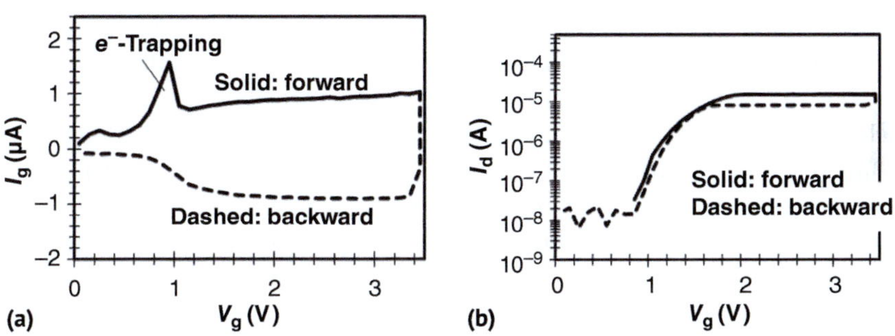

Figure 8.4 (a) $I_g - V_g$ characteristic and (b) $I_d - V_g$ characteristic during read operation in paraelectric HfO_2 device. The data of the region with a constant I_d is not measured correctly because I_d is outside the current measurement range. *Source*: [17] / with permission of IEEE.

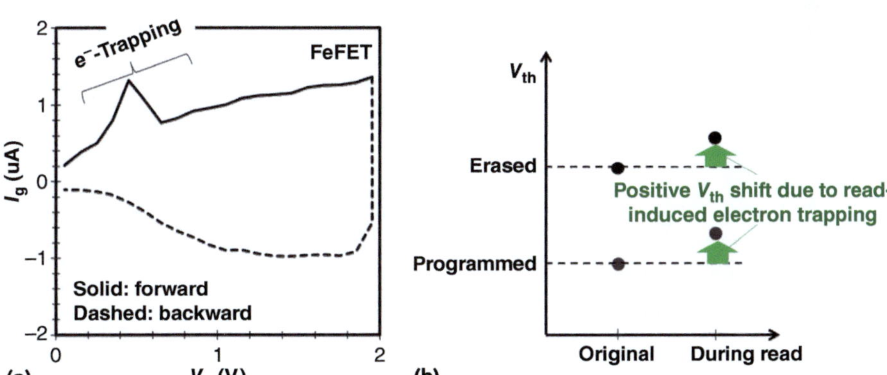

Figure 8.5 (a) $I_g - V_g$ characteristic during read in FeFET. *Source*: [17] / with permission of IEEE. (b) Schematic image of the impact of read-induced electron trapping on V_{th} for both programmed state and erased state.

Regarding the spatial position, the analysis of cycling dependence of read-induced electron trapping suggests that this phenomenon occurs within the IL SiO_2.[17]

8.2.2.2 Program-induced Phenomena

Figure 8.6 shows the typical time evolutions of both P_s and Q_t after program operation extracted by the method explained in Section 8.2.1. Q_t larger than P_s is generated temporarily, but within 10s, a part of Q_t disappears more rapidly than P_s. After that, the state is stable until 1,000s followed by the synchronized decrease of P_s and Q_t. Based on this result, the charges generated by program operation can be classified into two components. One is the unstable component mainly composed of Q_t, which disappears immediately after program operation. The other is the stable component with both P_s and Q_t. In the following, the details of both components are provided.

8.2.2.2.1 Unstable Trapped Charges

Similar to read-induced electron trapping, the electron trapping caused by program operation and the subsequent rapid electron de-trapping are not specific to FeFET but also observed in FET with paraelectric HfO_2/SiO_2 gate dielectric. In the case of paraelectric device, not only charge density but also its mean spatial position is experimentally extracted[18] thanks to the absence of P_s as described by the equations in Figure 8.7a. Figure 8.7b,c show the time evolutions of density and mean spatial position of remaining trapped charges at each timing after program operation. Trapped electrons are released rapidly within 1s after program operation as well as a part of trapped electrons in FeFET. In this de-trapping period, the mean position of the residual Q_t. is always located near the interface between HfO_2 and SiO_2. From these results, it is suggested that unstable trapped electrons observed in FeFET are related to the defects at HfO_2–SiO_2 interface which are not specific to ferroelectric HfO_2.

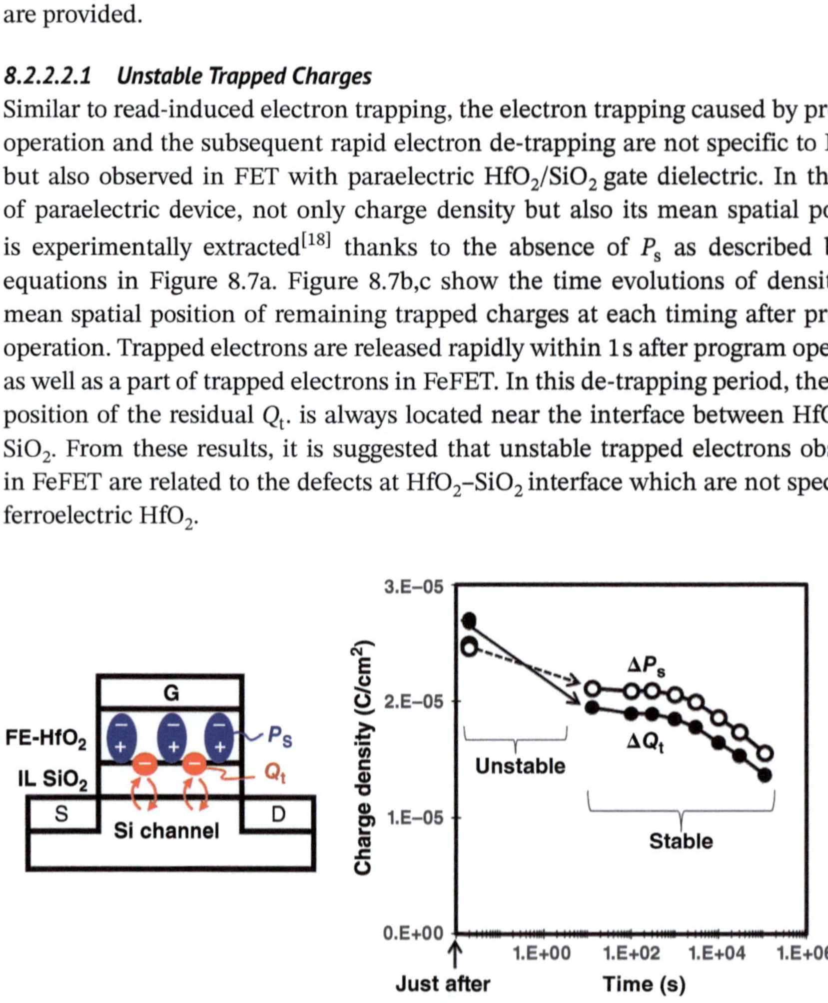

Figure 8.6 Time evolutions of both P_s and Q_t after program operation, which are extracted by the method described in Section 8.2.1. *Source:* [16] / with permission of IEEE.

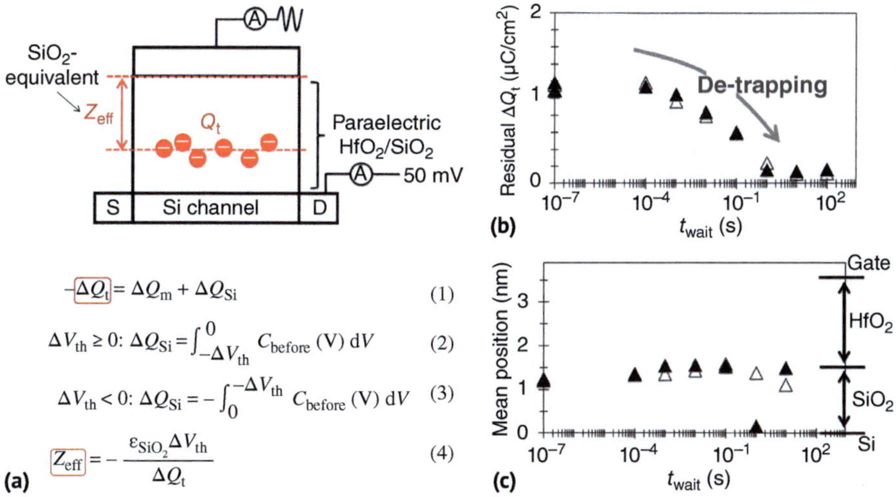

Figure 8.7 (a) Extraction of both ΔQ_t and their mean special position Z_{eff} using paraelectric FET. (b) Time evolution of residual ΔQ_t after program operation. (c) Time evolution of mean spatial position of residual trapped charges. *Source*: [17] / with permission of IEEE.

8.2.2.2.2 Stable Spontaneous Polarization and Trapped Charges

Figure 8.8 shows the relationship between ΔQ_t and ΔP_s of the stable component, which remains even after de-trapping of the unstable Q_t unrelated to ferroelectricity, for various conditions such as different program pulses and different types of dopants in FE-HfO_2. ΔQ_t, which is almost 90% of ΔP_s, always exists regardless of the conditions of the operation and the device. Although the reason for "90%" is unknown at this moment, the strong correlation between P_s and Q_t is clearly demonstrated.

The high electric field applied to IL SiO_2 when P_s reverses in FE-HfO_2 is considered the driving force for Q_t being correlated to P_s. Because IL SiO_2 has to realize the same polarization as that of FE-HfO_2 including P_s by only dielectric polarization, the extremely high electric field is generated in IL SiO_2 self-consistently. Considering the typical value of P_s such as $\pm 25\ \mu C/cm^2$ for FE-HfO_2, the electric field of IL SiO_2 could be over 50 MV/cm. It should be noted that the actual electric field in IL SiO_2 directly below FE-HfO_2 domain is expected to be even larger, because the measurable P_s is just the average value of P_s in HfO_2 film including both FE phase and paraelectric phase. Under such a high electric field, it is inevitable that Q_t, which cancels the electric field generated by P_s, is easily injected from the Si channel resulting in the subsequent reduction of electric field of IL SiO_2.

8.2.2.3 Overall Picture

As schematically shown in Figure 8.9, phenomena that occur in FeFET can be classified into three components. First one is read-induced electron trapping. Second one is program-induced electron trapping. These two phenomena occur regardless

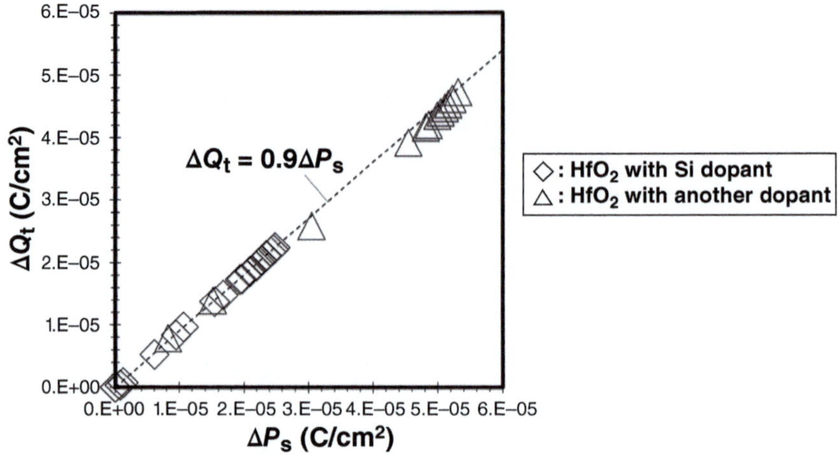

Figure 8.8 Relationships between ΔQ_t and ΔP_s of the stable component for various conditions. The unstable component is excluded by waiting for appropriate time after program operation. *Source*: [16] / with permission of IEEE.

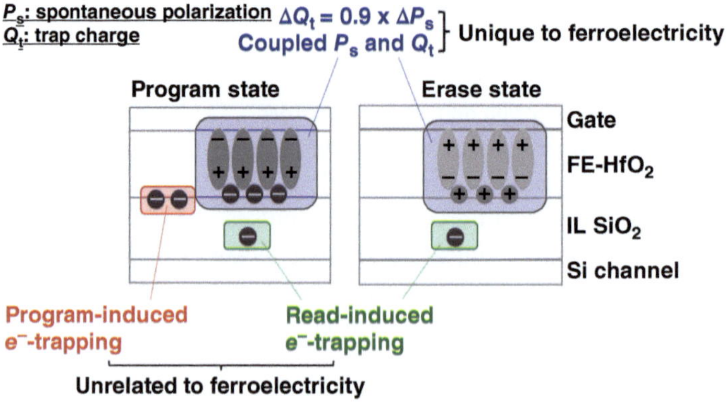

Figure 8.9 Classification of phenomena occurring in FeFET.

of whether HfO_2 is ferroelectric or paraelectric. The third one is stable P_s and huge amount of Q_t coupled with each other, which is unique to FeFET.

The relationships between these components and FeFET operations will be discussed in the following sections.

8.3 Basic Operation of Real FeFET

In this section, the role of huge amount of Q_t coupled with P_s, which is highlighted in blue in Figure 8.10, in the basic operation of FeFET is described.

As explained in Section 8.2.2.2.2, a huge amount of Q_t is inevitable because the extremely high electric field in IL SiO_2 generated by P_s reversal unavoidably induces charge injection from the Si channel. Such a Q_t alters the behavior of P_s during read

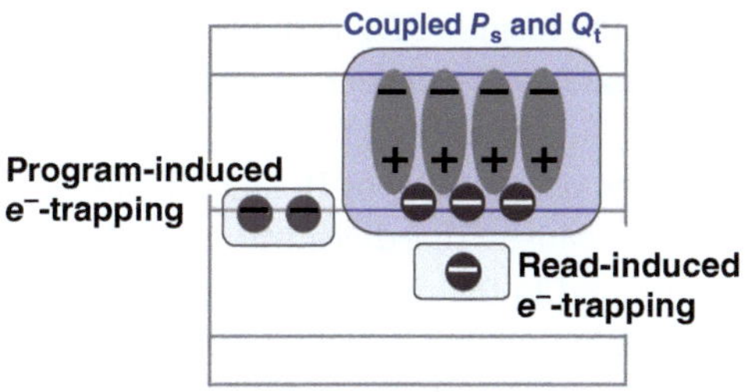

Figure 8.10 Coupled P_s and Q_t, under consideration in Section 8.3.

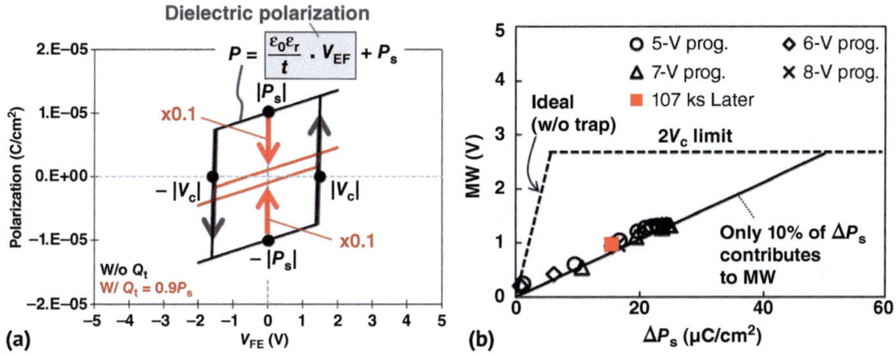

Figure 8.11 (a) Comparison between effective *P–V* characteristic with Q_t of 0.9 P_s and original *P–V* characteristic without Q_t. (b) Relationship between MW and ΔP_s. *Source*: [16] / with permission of IEEE.

operation. Unlike the ideal model mentioned in the introduction, in the real devices with a huge amount of Q_t coupled with P_s, V_{th} can be detected by only the change of dielectric polarization without P_s reversal despite sufficiently large P_s. This is because most of the electric field generated by P_s is compensated by Q_t resulting in the situation equivalent to much smaller P_s as schematically shown by red line in Figure 8.11a. As a result, MW, which is the difference in V_{th} between programmed state and erased state, is not yet limited by V_c despite sufficiently large P_s of more than 10 $\mu C/cm^2$, as shown in Figure 8.11b.

Here, is an ideal situation, where charge trapping does not occur and the partial reversal of P_s is mandatory for read operation, really the ideal we should aim for? Regarding this matter, it should be noted that the ideal FeFET without Q_t has the following potential issues. First, IL SiO_2 would break without Q_t, because the extremely high electric field induced by P_s is continuously applied to IL SiO_2. Second, based on the common understanding that P_s in a single domain can only be stable in either up state or down state, partial reversal of P_s, which is necessary for read operation in the ideal situation, is hard to occur in reality. Although the gradual change of P_s is observed experimentally, this is just the result of area normalization of various domains of up-state P_s and down-state P_s. That is, the measured gradual change of P_s does not mean the gradual change of electric field in the local region unless a metal layer is inserted between ferroelectric layer and IL SiO_2.

The root cause of these potential problems in the ideal FeFET without Q_t is excessively large P_s of FE-HfO_2 that is disproportionate to the dielectric permittivity of the materials in the gate stack including both IL SiO_2 and FE-HfO_2 itself, as follows. Because P_s is too large compared to the dielectric permittivity of IL SiO_2, the excessive electric field applied to IL SiO_2 probably causes the dielectric breakdown. And besides, because P_s is also too large compared to the dielectric permittivity of FE-HfO_2, a partial disappearance of P_s in a single domain, which is hard to occur in reality, is required for read operation. Huge amount of Q_t coupled with P_s has solved these issues consequently, since Q_t effectively realizes ferroelectric layer with smaller P_s that matches the dielectric permittivity of materials in the gate stack. Correlation of P_s and a huge amount of Q_t is considered not only inevitable but essential for the real HfO_2-FeFET operation with IL SiO_2.

8.4 Reliability Issues

The phenomena indicated in Figure 8.9 have impacts on FeFET reliability in various ways, which will be described in this section.

8.4.1 Data Retention

8.4.1.1 Short-term Characteristics

Figure 8.12 shows V_{th} behaviors after program and erase operations. While V_{th} after erase operation is relatively stable, that after program operation shows a complicated behavior. First, V_{th} shifts to positive after program operation, which is opposite to the expected direction for FeFET. After that, V_{th} decreases in short time domain, which is similar to what has been reported in[6]. These temporal positive V_{th} shifts and the subsequent rapid V_{th} reductions originate in program-induced unstable

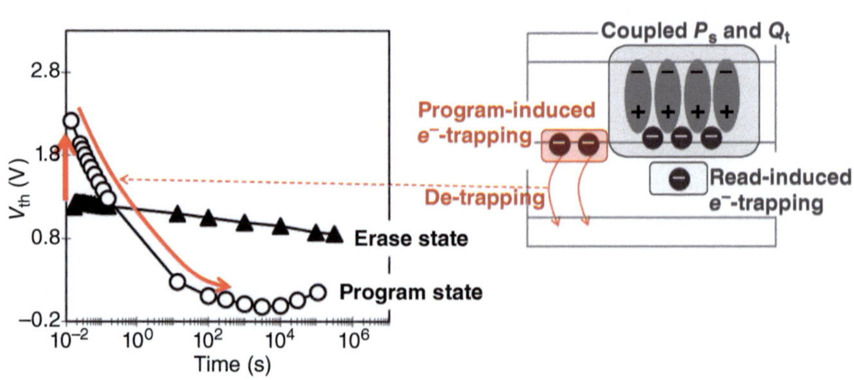

Figure 8.12 V_{th} behaviors after program and erase operations. Program-induced electron trapping and subsequent electron de-trapping that cause temporal positive V_{th} shift and the subsequent rapid reduction of V_{th} after program are highlighted by red in the illustration. *Source*: [16] / with permission of IEEE.

trapped electrons and their de-trapping, as highlighted by red in the illustration in Figure 8.12. This kind of unstable V_{th} behavior is often referred to as "Read-After-Write-Delay"[19–22] as this phenomenon leads to longer read latency. While qualitatively similar phenomenon is widely observed, significantly different electron emission rates have been reported,[6,16,19–23] indicating that engineering of the IL could reduce this problem. As a solution through the operation, it has been reported that weak pulse with opposite polarity applied after the operation effectively pushes away the excessive trapped charges.[14,23,24]

8.4.1.2 Long-term Characteristics

In long time domain, V_{th} of programmed state increases with time resulting in MW closure as shown by the red arrows in Figure 8.13a. In Figure 8.13b, the relationships between P_s and Q_t, which remain at each time in this time domain, are shown by red symbols. It is clearly shown that P_s and Q_t decrease together maintaining their relationship. Strong correlation between P_s and Q_t, which has already become evident in Figure 8.8, is confirmed again.

Figure 8.14 shows the schematic image of energy band diagram of programmed state in long-term retention period after de-trapping of unstable trapped electrons. As widely known, depolarization field, which causes P_s reversal, is applied to FE-HfO_2. Therefore, partial reversal of P_s during long-term retention process is a natural phenomenon.[25] On the other hand, IL SiO_2 is subjected to an electric field in the direction that accelerates electron injection from the Si channel. However, in fact, electrons are de-trapped with P_s reversal. From these results, it is strongly suggested that P_s reversal, which is caused by depolarization field, occurs first triggering the change of Q_t. Reduction of depolarization field is the key for improving long-term retention characteristics.

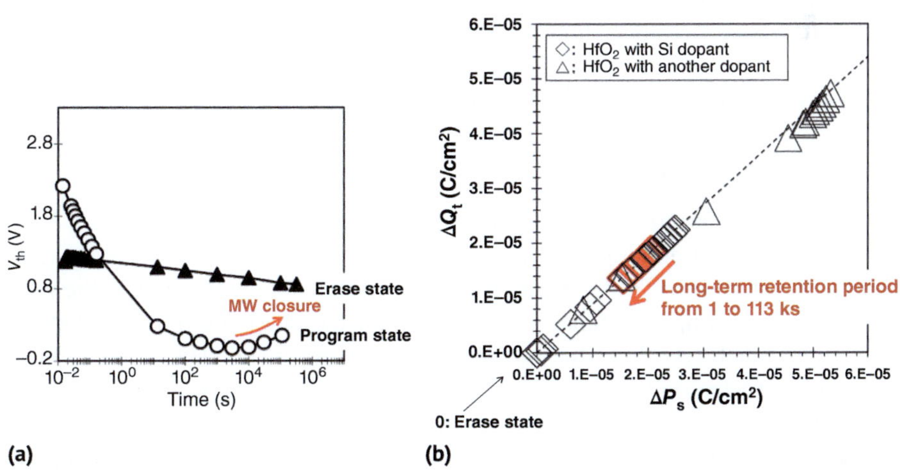

Figure 8.13 (a) Time evolutions of V_{th} of programmed state and erased state. (b) Relationship between residual Q_t and P_s during long-term retention process. *Source*: [16] / with permission of IEEE.

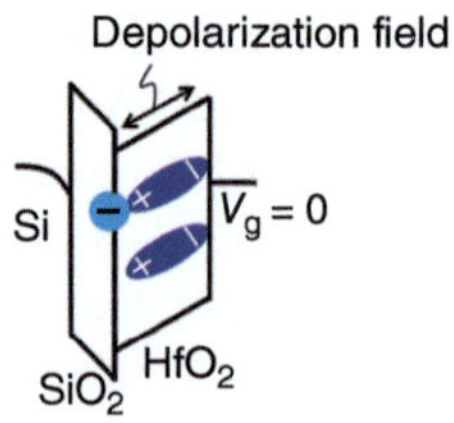

Figure 8.14 Schematic image of energy band diagram of programmed state in the long-term retention process after de-trapping of excessive electrons. *Source*: [16] / with permission of IEEE.

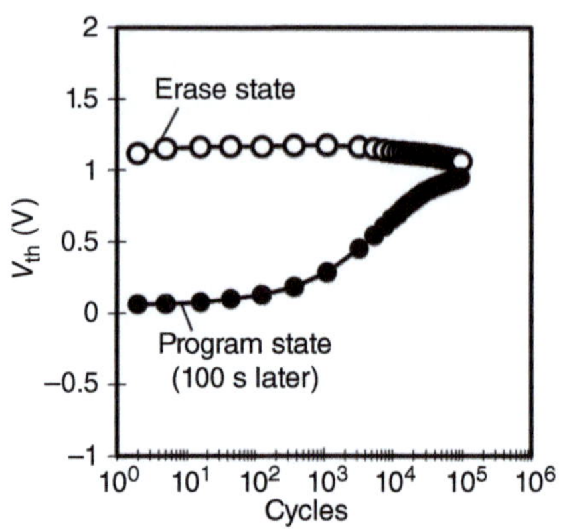

Figure 8.15 Cycling dependence of V_{th} for both programmed state and erased state. For programmed state, V_{th} after eliminating excessive trapped electrons by waiting for 100 s is shown. *Source*: [17] / with permission of IEEE.

Here, it should be noted that, contrary to the drastic changes of both P_s and Q_t during long-term retention process as shown in Figure 8.13b, the change in V_{th} in this time domain is trivial as shown in Figure 8.13a. This is because P_s and Q_t, which have the opposite effects on V_{th} modulation, decrease together. What happens in FeFET is very difficult to predict just from V_{th} behavior.

8.4.2 Cycling Endurance

8.4.2.1 Various Phenomena Affecting V_{th} Behavior During Cycling

The limited cycling endurance is recognized as a major problem in FeFET that needs to be addressed.[26] In general, it is believed that charge trapping is the root cause of poor cycling endurance in FeFET, because cycling endurance of FeFET is generally much less than that of metal–ferroelectric–metal capacitor in which fatigue of ferroelectric layer dominates the cycling characteristic.[27, 28] However, there is not just one type of charge trapping, but many different types as explained in Section 8.2.2. In this subsection, an example, where each type of charge trapping affects cycling endurance in different ways, is provided.

Figure 8.15 is an example of cycling dependence of V_{th} for both programmed state and erased state. Due to the characteristic that V_{th} of erased state remains almost unchanged while that of programmed state increases, it may seem that the amount of program-induced electron trapping has increased with cycling. However, the actual situation is not so straightforward, as explained below.

Figure 8.16a shows the cycling dependence of charges moving in and out during read operation (Q_{m_read}), which is an indicator of the density of trap sites capturing electrons during read operation. Q_{m_read} increases with cycling indicating that cycling stress generates new trap sites as schematically shown in Figure 8.16b. Because these trap sites capture electrons during read operation before V_{th} is detected as explained in Section 8.2.2.1, V_{th} for both programmed state and erased state are elevated together with cycling.

On the other hand, the amount of unstable electrons trapped during program decreases with cycling as shown in Figure 8.17a. Since the extracted value is the information of reversible charges, this result strongly suggests that a part of electrons trapped during program become fixed through cycling. These fixed electrons also increase the value of V_{th} for both programmed state and erased state similar to the read-induced electron trapping as illustrated in Figure 8.17b, because they always exist regardless of the device status.

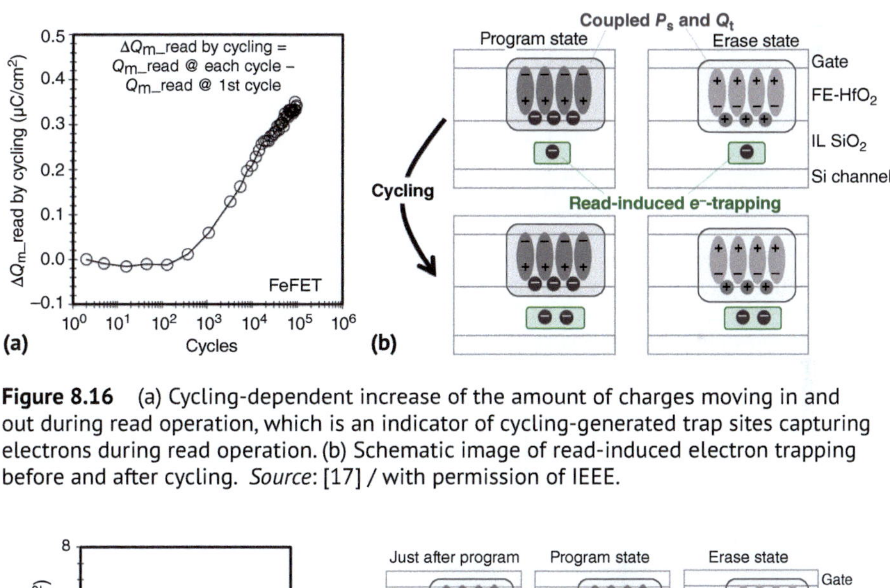

Figure 8.16 (a) Cycling-dependent increase of the amount of charges moving in and out during read operation, which is an indicator of cycling-generated trap sites capturing electrons during read operation. (b) Schematic image of read-induced electron trapping before and after cycling. *Source*: [17] / with permission of IEEE.

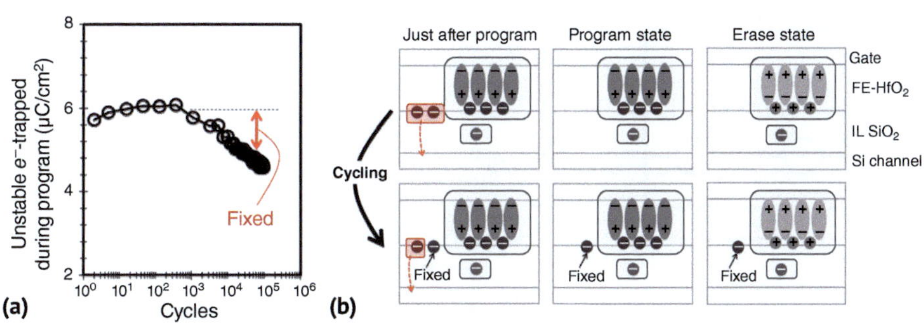

Figure 8.17 (a) Cycling dependence of the amount of unstable electrons trapped during program. (b) Schematic image of fixed electrons before and after cycling. *Source*: [17] / with permission of IEEE.

Figure 8.18a shows V_{th} modulation due to the increase of read-induced electron trapping, program-induced fixed electrons, and their total. Nonnegligible impact on V_{th} is confirmed. By eliminating this V_{th} modulation, corrected V_{th} behavior during cycling, which is greatly different from the measured trend shown again in Figure 8.18b, is obtained as shown in Figure 8.18c. In the corrected characteristic, V_{th} of erased state drastically decreases with cycling, whereas that of programmed state is relatively stable. This result indicates that additional holes are trapped during erase operation and are released during program operation after cycling. This is the direct cause of MW closure with cycling in this case.

The root cause of additional hole trapping during erase operation after cycling can be explained as shown in Figure 8.19. After cycling, there are many fixed electrons at the interface between ferroelectric HfO_2 and IL SiO_2 as explained before. These fixed electrons modulate IL SiO_2 electric field to accelerate hole injection from Si channel resulting in additional hole trapping during erase operation.

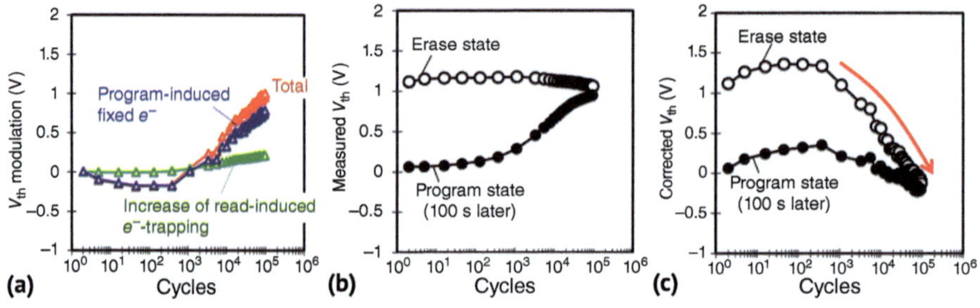

Figure 8.18 (a) V_{th} modulation due to the increase of read-induced electron trapping, program-induced fixed electrons, and their total. (b) Measured cycling dependence of V_{th} for programmed state and erased state. (c) Corrected V_{th} behavior during cycling without the components shown in panel (a). *Source*: [17] / with permission of IEEE.

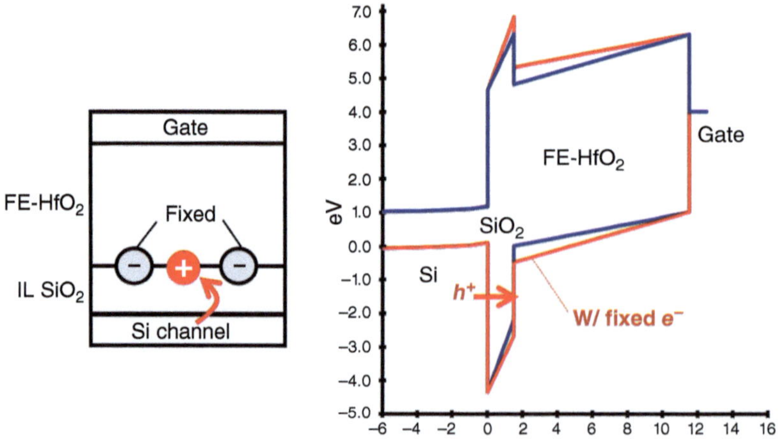

Figure 8.19 Explanation for additional hole trapping caused by fixed electrons at the interface between FE-HfO_2 and IL SiO_2. *Source*: [17] / with permission of IEEE.

In this manner, V_{th} behavior during cycling is influenced by various phenomena. Although the balance of phenomena could change depending on the conditions of devices and operations, it would be worth mentioning that the actual phenomena occurring during cycling are difficult to imagine just from V_{th} behavior, similar to the case of long-term data retention explained in Section 8.4.1.2.

8.4.2.2 Endurance Improvement by Intentional Charge Trapping

Although various phenomena affect cycling characteristics as explained above, it is believed that they are commonly associated with the degradation of IL SiO_2. This is because IL SiO_2 is subjected to extremely strong electric field at the same time as P_s reversal as mentioned in Section 8.2.2. To reduce the electric field across IL SiO_2 effectively, TCIL (trap-controlled interfacial layer) with metal doping realizing intentional and controllable charge trapping is utilized.[24] Figure 8.20a shows cross-sectional TEM image of the gate stack of FeFET with TCIL and EDX intensity of the metal around TCIL. Metal regions, which act as trap sites, are introduced at the interface between IL and FE-HfO_2. Figure 8.20b shows the cycling dependence of MW for the devices with and without TCIL for different metal doses. Compared to the device without TCIL, those with TCIL show much less shrinkage of MW during cycling in exchange for a decrease of initial MW.

The improvement in cycling endurance by introducing TCIL is attributed to the reduction of the electric field across IL due to charges captured at metal trap sites in TCIL, which compensates for the electric field generated by P_s reversal as schematically shown by dashed circle in Figure 8.21.

On the other hand, this technique has a negative aspect of reducing initial MW before cycling degradation as mentioned before, because charge trapping compensates V_{th} modulation by P_s. To overcome this trade-off, charge-control pulse (CCP) method shown in Figure 8.22a is adopted. In this method, the pulse with negative voltage is applied before read operation to promote de-trapping of program-induced trapped electrons. Figure 8.22b shows CCP voltage dependence of V_{th} for both programmed state and erased state. CCPs with voltages from −2.5 to −4 V effectively decrease V_{th} of programmed state indicating that electron de-trapping is dominant,

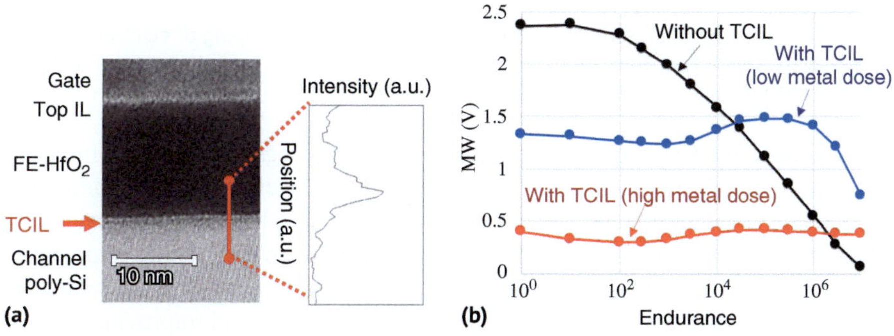

Figure 8.20 (a) Cross-sectional TEM image of FeFET with TCIL and EDX intensity of the metal around TCIL. (b) Endurance cycling dependence of MW for FeFETs with and without TCIL. *Source*: [24] / with permission of IEEE.

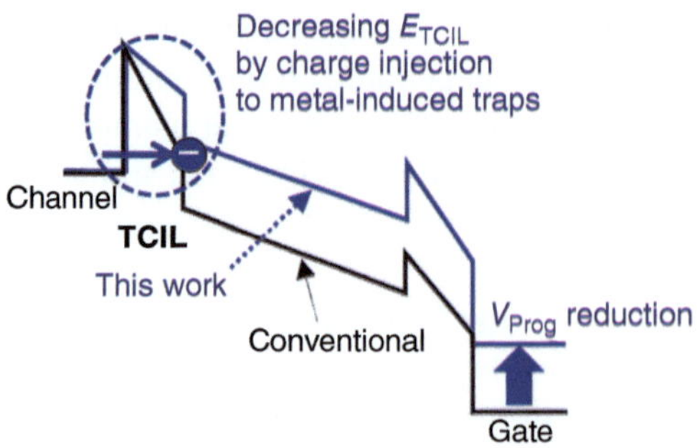

Figure 8.21 Schematic image explaining that larger amount of trapped charges at the interface between FE-HfO_2 and IL decreases the electric field across IL. *Source*: [24] / with permission of IEEE.

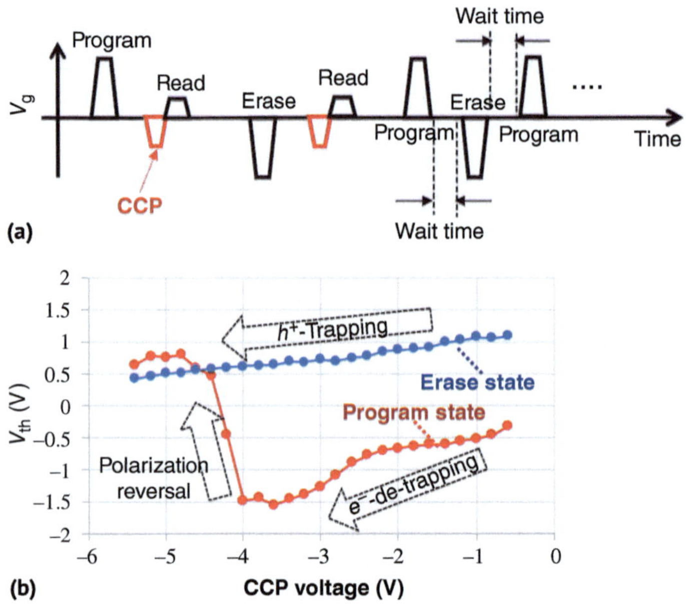

Figure 8.22 (a) Operation sequence with CCP intended for de-trapping of electrons trapped at the metal trap sites. (b) CCP voltage dependence of V_{th} for both programmed state and erased state. *Source*: [24] / with permission of IEEE.

whereas CCPs with voltages greater than −4V in the negative direction increase V_{th} of programmed state indicating that undesirable P_s reversal becomes dominant. Although V_{th} of erased state also decreases with increasing CCP voltage suggesting additional hole trapping, this is much less compared to electron de-trapping in programmed state. Consequently, MW is expanded by adopting the optimal CCP condition, achieving both suppressing initial MW reduction and improving cycling endurance.

In this way, control of charge trapping which is not coupled with P_s is one of the key factors for higher cycling endurance.

8.4.2.3 Recovery of Cycling-induced Degradation

As described in Section 8.4.2.1, cycling stress generates new trap sites capturing electrons during read operation resulting in elevated values of V_{th} in both programmed state and erased state. Figure 8.23a shows charges moving in and out during read operation, which is an indicator of the density of trap sites capturing electrons during read operation, before and after cycling, and its time evolution after cycling. The density of trap sites increases with cycling as explained before but decreases over time after cycling. That is, cycling-generated trap sites do not exist indefinitely, but disappear with time as illustrated in Figure 8.23b. Contrary to the generation of electron trap sites resulting in positive shift of V_{th}, disappearance of electron trap sites causes negative shift of V_{th}. It means that, in the retention period after cycling, V_{th} behavior is determined not only by P_s reversal and charge trapping/de-trapping, but also by the disappearance of trap sites capturing electrons during read operation. To discuss the reliability after cycling, it is needed to take note of the recovery of cycling-induced degradation.

8.4.3 Disturb Immunity

The behavior of charge trapping also affects disturb immunity.[30] Figure 8.24a shows a schematic image of memory array with FeFET. In the typical array operation scheme, half of the operating voltage is applied to the cells which share the wires with selected cell. Such half-voltage called "disturb" is applied to the cells repeatedly for various intervals resulting in undesirable V_{th} shift. Therefore, it is important to know the impacts of interval on V_{th} shift caused by disturb accumulation.

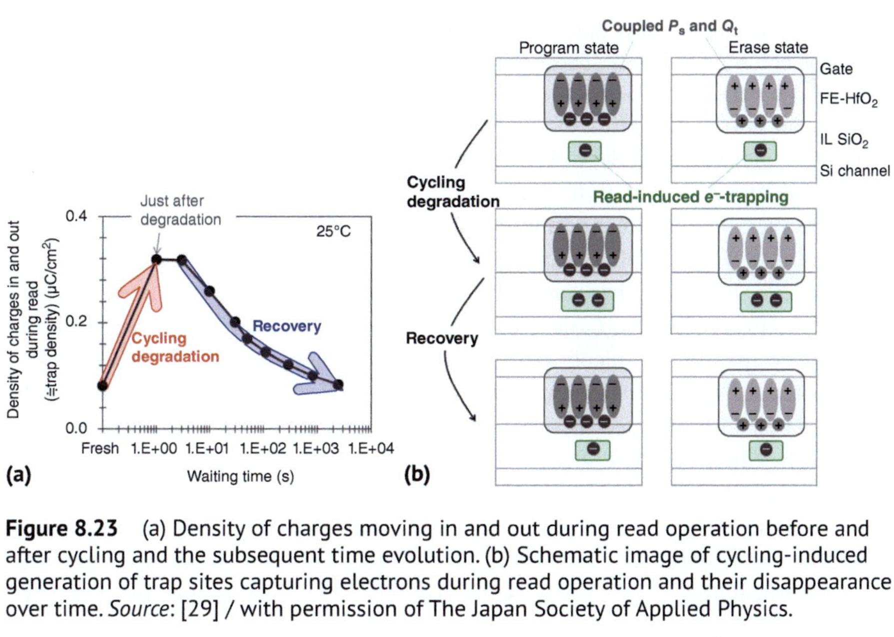

Figure 8.23 (a) Density of charges moving in and out during read operation before and after cycling and the subsequent time evolution. (b) Schematic image of cycling-induced generation of trap sites capturing electrons during read operation and their disappearance over time. *Source*: [29] / with permission of The Japan Society of Applied Physics.

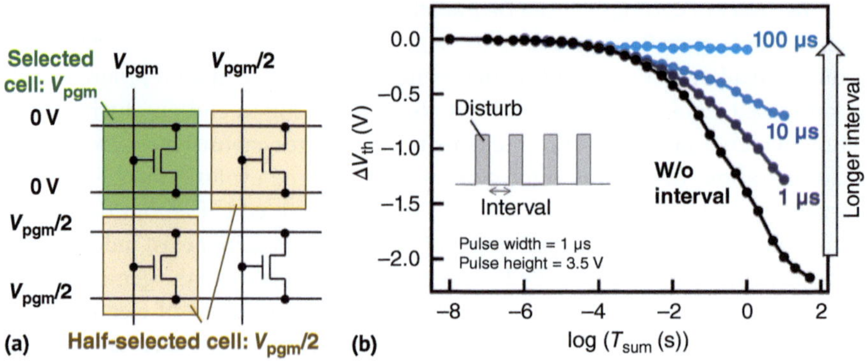

Figure 8.24 (a) Typical array operation scheme. (b) Dependence of V_{th} on total disturb time for various intervals between subsequent disturb pulses. *Source*: [30] / with permission of IEEE.

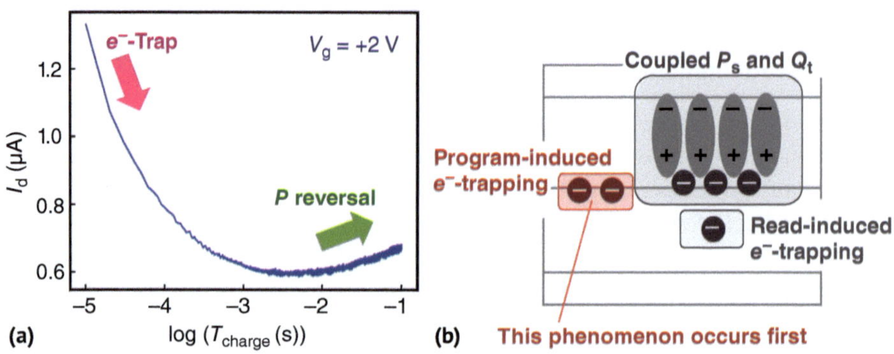

Figure 8.25 (a) Disturb stress time (T_{charge}) dependence of I_d in erased state. (b) Phenomena occurring in FeFET. *Source*: [30] / with permission of IEEE.

Figure 8.24b shows the shift of V_{th} in erased state caused by program disturb accumulation for various intervals. Longer interval significantly suppresses V_{th} shift caused by disturb accumulation. Such a strong impact of interval on disturb immunity can be explained by the modulation of electric field across FE-HfO_2 due to electron trapping and de-trapping as explained next.

Figure 8.25a shows the time evolution of I_d of erased state under a program disturb of 2V. Under disturb, I_d decreases first, then switches to an increase. This characteristic means that electron trapping not coupled with P_s, which is highlighted by red in Figure 8.25b, occurs first followed by P_s reversal. If program disturb is accumulated without interval, more and more trapped electrons are accumulated at the interface between IL SiO_2 and FE-HfO_2 before P_s reversal. These trapped electrons modulate the electric field across FE-HfO_2 to the direction that enhances P_s reversal, as shown by red line in energy band diagram in Figure 8.26a. However, as explained in Section 8.2.2, these trapped electrons are not stable, and they are released immediately if the stress is interrupted. Therefore, longer interval between subsequent disturb stresses decreases the total amount of trapped electrons at the interface between IL SiO_2 and FE-HfO_2 and electric field across FE-HfO_2,

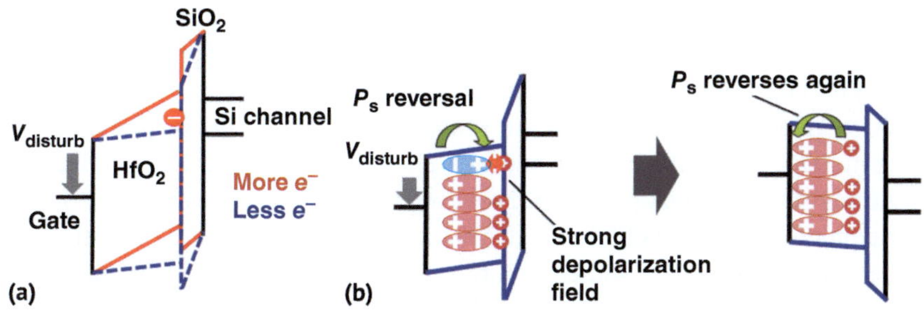

Figure 8.26 (a) Modulation of electric field across each layer depending on the density of electrons trapped at the interface between IL SiO_2 and FE-HfO_2. (b) Explanation of the instability of reversed P_s without sufficient electrons at the interface between IL SiO_2 and FE-HfO_2. *Source*: [30] / with permission of IEEE.

suppressing undesirable disturb-induced P_s reversal. Even if P_s reverses under disturb stress without sufficient trapped charges at the interface between IL SiO_2 and FE-HfO_2, P_s is unstable due to strong depolarization field resulting in the re-reversal of P_s before read operation and apparent absence of disturb effect, as illustrated in Figure 8.26b. From the perspective of disturb immunity, the operation that continues to apply disturb on the specific group of devices should be avoided. Like this, the dynamics of charge trapping and de-trapping even affects the reliability of array operation.

8.5 Summary

This chapter has presented the recent understandings of basic operation mechanism and reliability for HfO_2-FeFET with Si channel, in which the IL SiO_2 is inevitably formed, based on the experimental analysis of both spontaneous polarization reversal and charge trapping/de-trapping. One important message is that huge amount of charge trapping always compensates for a large part of spontaneous polarization. Such charge trapping makes the operation of FeFET, which utilizes ferroelectric layer with large spontaneous polarization that is disproportionate to dielectric permittivity of gate stack layers, realistic. The coupled spontaneous polarization reversal and charge trapping/de-trapping are the basis of FeFET and they determine the base value of MW. In addition to this basic phenomenon, there are two types of charge trapping unrelated to ferroelectricity. First one is electron trapping which occurs during every read operation. This read-induced electron trapping raises the base value of V_{th} of both programmed and erased states. Second one is program-induced electron trapping. These trapped electrons are not stable and start to be released immediately after program operation, causing V_{th} instability during short-term retention process. With cycling, a part of electrons trapped during program become fixed. These fixed electrons modulate SiO_2 electric field followed by additional hole trapping during erase leading to MW closure. And besides, cycling generates new trap sites capturing electrons during read operation resulting in an increase of V_{th} values for both programmed and erased states. These newly generated trap sites capturing electrons during read operation disappear with time making FeFET characteristics after cycling more complicated.

The above-mentioned components are probably not yet the entire phenomena occurring in FeFET. And the balance between phenomena may vary depending on the conditions of device and operations. However, it is worth emphasizing that multiple components coexist in FeFET and all have impacts on V_{th} of FeFET in different ways. Due to such complexity, sometimes, the real situation is totally different from what we can simply imagine just from V_{th} behavior. Continual effort to accurately understand the phenomena really occurring in the device is necessary to realize highly reliable FeFET.

References

1 Böscke, T. S., Müller, J., Bräuhaus, D. et al. (2011). Ferroelectricity in hafnium oxide thin films. *Applied Physics Letters* 90: 102903. https://doi.org/10.1063/1.3634052.

2 Müller, J., Böscke, T. S., Schröder, U. et al. (2012). Ferroelectricity in simple binary ZrO_2 and HfO_2. *Nano Letters* 12: 4318–4323. https://doi.org/10.1021/nl302049k.

3 Mueller, S., Mueller, J., Singh, A. et al. (2012). Incipient ferroelectricity in Al-doped HfO_2 thin films. *Advanced Functional Materials* 22: 2412–2417. https://doi.org/10.1002/adfm.201103119.

4 Mueller, J., Slesazeck, S., and Mikolajick, T. (2019). Ferroelectric field effect transistor. In: Schroeder, U., Hwang, C. S., and Funakubo, H. (eds.). *Woodhead Publishing Series in Electronic and Optical Materials: Ferroelectricity in Doped Hafnium Oxide: Materials, Properties and Devices*. Woodhead Publishing. pp. 451–471. https://doi.org/10.1016/B978-0-08-102430-0.00022-X.

5 Yurchuk, E., Müller, J., Müller, S. et al. (2016). Charge-trapping phenomena in HfO_2-based FeFET-type nonvolatile memories. *IEEE Transactions on Electron Devices* 63 (9): 3501–3507. https://doi.org/10.1109/TED.2016.2588439.

6 Ni, K., Sharma, P., Zhang, J. et al. (2018). Critical role of interlayer in $Hf_{0.5}Zr_{0.5}O_2$ ferroelectric FET nonvolatile memory performance. *IEEE Transactions on Electron Devices* 65 (6): 2461–2469. https://doi.org/10.1109/TED.2018.2829122.

7 Gong, N. and Ma, T.-P. (2018). A study of endurance issues in HfO_2-based ferroelectric field effect transistors: Charge trapping and trap generation. *IEEE Electron Device Letters* 39 (1): 15–18. https://doi.org/10.1109/LED.2017.2776263.

8 Pesic, M., Padovani, A., Slesazeck, S. et al. (2018). Deconvoluting charge trapping and nucleation interplay in FeFETs: Kinetics and Reliability. In: *Proceedings of the 2018 IEEE International Electron Devices Meeting (IEDM)*. IEEE. pp. 25.1.1–25.1.4. https://doi.org/10.1109/IEDM.2018.8614492.

9 Kleimaier, D., Mulaosmanovic, H., Dünkel, S. et al. (2021). Demonstration of a p-type ferroelectric FET with immediate read-after-write capability. *IEEE Electron Device Letters* 42 (12): 1774–1777. https://doi.org/10.1109/LED.2021.3118645.

10 Tian, F., Zhao, S., Xu, H. et al. (2021). Impact of interlayer and ferroelectric materials on charge trapping during endurance fatigue of FeFET with TiN/

$Hf_xZr_{1-x}O_2$/interlayer/Si (MFIS) gate structure. *IEEE Transactions on Electron Devices* 68 (11): 5872–5878. https://doi.org/10.1109/TED.2021.3114663.

11 Higashi, Y., Ronchi, N., Alam, M. N. K. et al. (2021). Impact of charge trapping and depolarization on data retention using simultaneous *P–V* and *I–V* in HfO_2-based ferroelectric FET. *IEEE Transactions on Electron Devices* 68 (9): 4391–4396. https://doi.org/10.1109/TED.2021.3096510.

12 Deng, S., Zhao, Z., Kim, Y. S. et al. (2022). Unraveling the dynamics of charge trapping and de-trapping in ferroelectric FETs. *IEEE Transactions on Electron Devices* 69 (3): 1503–1511. https://doi.org/10.1109/TED.2022.3143485.

13 Zhao, S., Tian, F., Xu, H. et al. (2022). Experimental extraction and simulation of charge trapping during endurance of FeFET with TiN/HfZrO/SiO_2/Si (MFIS) gate structure. *IEEE Transactions on Electron Devices* 69 (3): 1561–1567. https://doi.org/10.1109/TED.2021.3139285.

14 Liao, C. Y., Lin, C. Y., Lee, Z. X. et al. (2022). Mechanisms of instability retention for ferroelectric field effect transistors with $HfZrO_2$ gate stack scaling down. *Applied Physics Letters* 121 (25): 252902. https://doi.org/10.1063/5.0111592.

15 Toprasertpong, K., Takenaka, M., and Takagi, S. (2019). Direct observation of interface charge behaviors in FeFET by quasi-static split C-V and Hall techniques: Revealing FeFET operation. In: *Proceedings of the 2019 IEEE International Electron Devices Meeting (IEDM)*. IEEE. pp. 23.7.1–23.7.4. https://doi.org/10.1109/IEDM19573.2019.8993664.

16 Ichihara, R., Suzuki, K., Kusai, H. et al. (2020). Re-examination of Vth window and reliability in HfO_2 FeFET based on the direct extraction of spontaneous polarization and trap charge during memory operation. In: *Proceedings of the 2020 IEEE Symposium on VLSI Technology*. IEEE. pp. 1–2. https://doi.org/10.1109/VLSITechnology18217.2020.9265055.

17 Ichihara, R., Higashi, Y., Suzuki, K. et al. (2021). Accurate picture of cycling degradation in HfO_2-FeFET based on charge trapping dynamics revealed by fast charge centroid analysis. In: *Proceedings of the 2021 IEEE International Electron Devices Meeting (IEDM)*. IEEE. pp. 6.3.1–6.3.4. https://doi.org/10.1109/IEDM19574.2021.9720516.

18 Fujii, S., Yasuda, N., Fujiki, J. et al. (2018). A new method to extract the charge centroid in the program operation of metal-oxide-nitride-oxide semiconductor memories. *Japanese Journal of Applied Physics* 49 (4S): 04DD06. https://doi.org/10.1143/JJAP.49.04DD06.

19 Cai, P., Zhu, T., Duan, J. et al. (2022). Deep understanding of reliability in Hf-based FeFET during bipolar pulse cycling: Trap profiling for read-after-write delay and memory window degradation. In: *Proceedings of the 2022 International Electron Devices Meeting (IEDM)*. IEEE. pp. 32.2.1–32.2.4. https://doi.org/10.1109/IEDM45625.2022.10019441.

20 Wang, Z., Tasneem, N., Hur, J. et al. (2021). Standby bias improvement of read after write delay in ferroelectric field effect transistors. In: *Proceedings of the 2021 IEEE International Electron Devices Meeting (IEDM)*. IEEE. pp. 19.3.1–19.3.4. https://doi.org/10.1109/IEDM19574.2021.9720502.

21 Hoffmann, M., Tan, A. J., Shanker, N. et al. (2022). Fast read-after-write and depolarization fields in high endurance n-type ferroelectric FETs. *IEEE Electron Device Letters* 43 (5): 717–720. https://doi.org/10.1109/LED.2022.3163354.

22 Myeong, I., Lim, S., Kim, T. et al. (2024). A comprehensive study of read-after-write-delay for ferroelectric VNAND. In: *Proceedings of the 2024 IEEE International Reliability Physics Symposium (IRPS).* IEEE. pp. 9B.3-1–9B.3-6. https://doi.org/10.1109/IRPS48228.2024.10529486.

23 Zhao, G., Wei, W., Tai, L. et al. (2022). Suppressing interfacial layer degradation in $Hf_{0.5}Zr_{0.5}O_2$-based FeFETs using a pre-erase strategy during program/erase cycling. In: *Proceedings of the 2022 IEEE Silicon Nanoelectronics Workshop (SNW).* IEEE. pp. 1–2. https://doi.org/10.1109/SNW56633.2022.9889043.

24 Suzuki, K., Sakuma, K., Yoshimura, Y. et al. (2023). High-endurance FeFET with metal-doped interfacial layer for controlled charge trapping and stabilized polarization. In: *Proceedings of the 2023 International Electron Devices Meeting (IEDM).* IEEE. pp. 1–4. https://doi.org/10.1109/IEDM45741.2023.10413699.

25 Yoon, S.-J., Min, D.-H., Moon, S.-E. et al. (2020). Improvement in long-term and high-temperature retention stability of ferroelectric field-effect memory transistors with metal–ferroelectric–metal–insulator–semiconductor gate-stacks using Al-doped HfO_2 thin films. *IEEE Transactions on Electron Devices* 67 (2): 499–504. https://doi.org/10.1109/TED.2019.2961117.

26 Zagni, N., Puglisi, F. M., Pavan, P. et al. (2023). Reliability of HfO_2-based ferroelectric FETs: A critical review of current and future challenges. *Proceedings of the IEEE* 111 (2): 158–184. https://doi.org/10.1109/JPROC.2023.3234607.

27 Fengler, F., Park, M. H., Schenk, T. et al. (2019). Field cycling behavior of ferroelectric HfO_2-based capacitors. In: Schroeder, U., Hwang, C. S., and Funakubo, H. (eds.). *Woodhead Publishing Series in Electronic and Optical Materials: Ferroelectricity in Doped Hafnium Oxide: Materials, Properties and Devices.* Woodhead Publishing. pp. 381–398. https://doi.org/10.1016/B978-0-08-102430-0.00017-6.

28 Pešić, M., Larcher, L., and Schroede, U. (2019). Modeling of field cycling behavior of ferroelectric hafnia-based capacitors. In: Schroeder, U., Hwang, C. S., and Funakubo, H. (eds.). *Woodhead Publishing Series in Electronic and Optical Materials: Ferroelectricity in Doped Hafnium Oxide: Materials, Properties and Devices.* Woodhead Publishing. pp. 399–411. https://doi.org/10.1016/B978-0-08-102430-0.00018-8.

29 Schlykow, V., Suzuki, K., Yoshimura, Y. et al. (2024). Recovery of cycling-induced degradation of interfacial SiO_2 in HfO_2-FeFET and its impact on retention characteristics. *Japanese Journal of Applied Physics* 63: 02SP43. https://doi.org/10.35848/1347-4065/ad1255.

30 Hamai, T., Suzuki, K., Ichihara, R. et al. (2023). Novel operation scheme for suppressing disturb in HfO_2-based FeFET considering charge-trapping-coupled polarization dynamics. In: *Proceedings of the 2023 IEEE International Reliability Physics Symposium (IRPS).* IEEE. pp. 1–5. https://doi.org/10.1109/IRPS48203.2023.10118125.

9

Reliability of the Hafnia-based Ferroelectric Memory

*Wei Wei, Xuedong Zhao, and Qing Luo**

State Key Laboratory of Fabrication Technologies for Integrated Circuits, Key Laboratory of Microelectronics Devices and Integrated Technology, Institute of Microelectronics of the Chinese Academy of Sciences, Beijing, China

*Corresponding author: luoqing@ime.ac.cn

9.1 Introduction

Since the discovery of ferroelectric properties in doped-HfO_2 thin films, these materials have transitioned from serving as high-κ dielectrics to being recognized as promising ferroelectric materials, attracting widespread attention not only for elucidating the fundamental mechanisms underlying ferroelectricity but also for evaluating their potential as next-generation, high-density memory devices.[1,2] HfO_2-based ferroelectric materials offer several advantages, including excellent scalability, low operating voltage, fast read/write speed, and compatibility with CMOS processes. These materials can be utilized in the development of various ferroelectric devices, such as ferroelectric capacitors (FeCAPs), ferroelectric field-effect transistors (FeFETs), and ferroelectric tunnel junctions (FTJ). These devices can be applied to embedded memory and hold promise for applications in nonvolatile dynamic random access memory (DRAM) and storage-class memory (SCM).

Despite these benefits, endurance remains a significant challenge for HfO_2-based ferroelectric devices. Depending on the specific fabrication processes and operating conditions, the endurance of HfO_2-based devices typically ranges from 10^6 to 10^8 cycles. Although this figure surpasses that of NAND flash, it is nonetheless constrained when compared to the 10^{15} endurance cycles exhibited by traditional perovskite ferroelectrics. Furthermore, the destructive readout nature of ferroelectric devices also intensifies the issue of limited endurance. Therefore, attaining higher endurance is essential for applications in nonvolatile DRAM and SCM.

In broader terms, endurance constitutes a key aspect of device reliability and closely relates to other reliability metrics, such as retention, with potential trade-offs between them. Additionally, ferroelectric materials exhibit unique phenomena – including imprint and field cycling effects – that can profoundly influence device performance and, consequently, storage reliability. In summary, the reliability of HfO_2-based ferroelectric devices has been a widely discussed issue and remains an urgent challenge requiring further investigation and innovation.

9.2 Reliability of HfO_2-based FeCAPs

9.2.1 Dielectric Breakdown

Before remanent polarization (P_r) drops to zero, HfO_2-based metal–ferroelectric–metal (MFM) capacitors generally exhibit premature dielectric breakdown, which can be classified into soft and hard breakdown as evidenced by the current-time characteristics. From a macroscopic perspective, HfO_2-based films subjected to rapid annealing are polycrystalline, consisting of various crystalline phases (i.e., orthorhombic, tetragonal, and monoclinic (m) phases). As a result, several leakage paths may form across the films when a dielectric breakdown occurs.

Traditional approaches to studying dielectric breakdown behavior commonly rely on electrical measurement methods. A systematic study by Wei et al. identifies three dominant failure modes in $Hf_{0.5}Zr_{0.5}O_2$ (HZO) ferroelectric films.[3] They point out that the soft breakdown occurs through the formation of reconfigurable filaments at grain boundaries, while the hard breakdown occurs through the formation of unreconfigurable filaments in amorphous regions, as shown in Figure 9.1. The time-dependent dielectric breakdown (TDDB) measurements indicate that the annealing

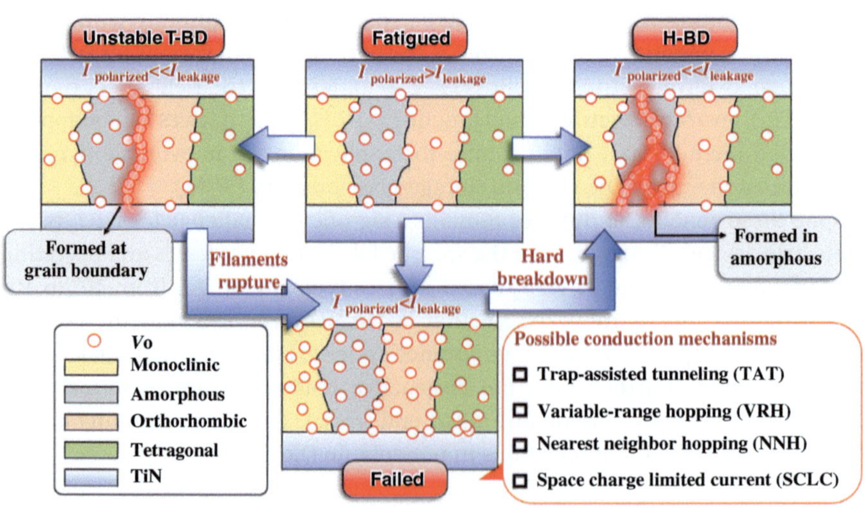

Figure 9.1 Possible microstructures corresponding to different stages and potential conduction mechanisms of the "failed" state. The red lines indicate the conducting filaments after dielectric breakdown. *Source*: [3] / with permission of IEEE.

temperature has a large impact on the initial defect concentration, while having a weak effect on the trap generation rate.

Besides the polycrystalline nature of HfO_2-based ferroelectric films, the primary factor that leads to premature dielectric breakdown and consequently limited endurance during electric field cycling is the much higher ratio of the coercive field (E_C) to the breakdown electric field (E_{BD}). To address this issue, one can either increase the E_{BD} or decrease the E_C.

There are several approaches to increasing the E_{BD}, such as reducing the thickness of films.[4] One effective method proposed by Xu et al. is to insert an Al_2O_3 layer within the HZO film, as shown in Figure 9.2a,b.[5] Figure 9.2c presents the operation voltage extrapolation for a 10-year lifetime at 63% failure of MFM, MFIM, and MFIFM devices. This intermediate layer not only significantly increases the breakdown electric field but also extends the operational lifetime of the devices. With this structure, the device can achieve an impressive endurance of 10^{10} cycles, far surpassing traditional designs, as shown in Figure 9.2d. The mechanism behind this improvement is the ability of the Al_2O_3 layer to disrupt the continuity of grain boundaries within the HZO layer. By interrupting these grain boundaries, the Al_2O_3 layer effectively prevents the formation of conductive pathways, which would

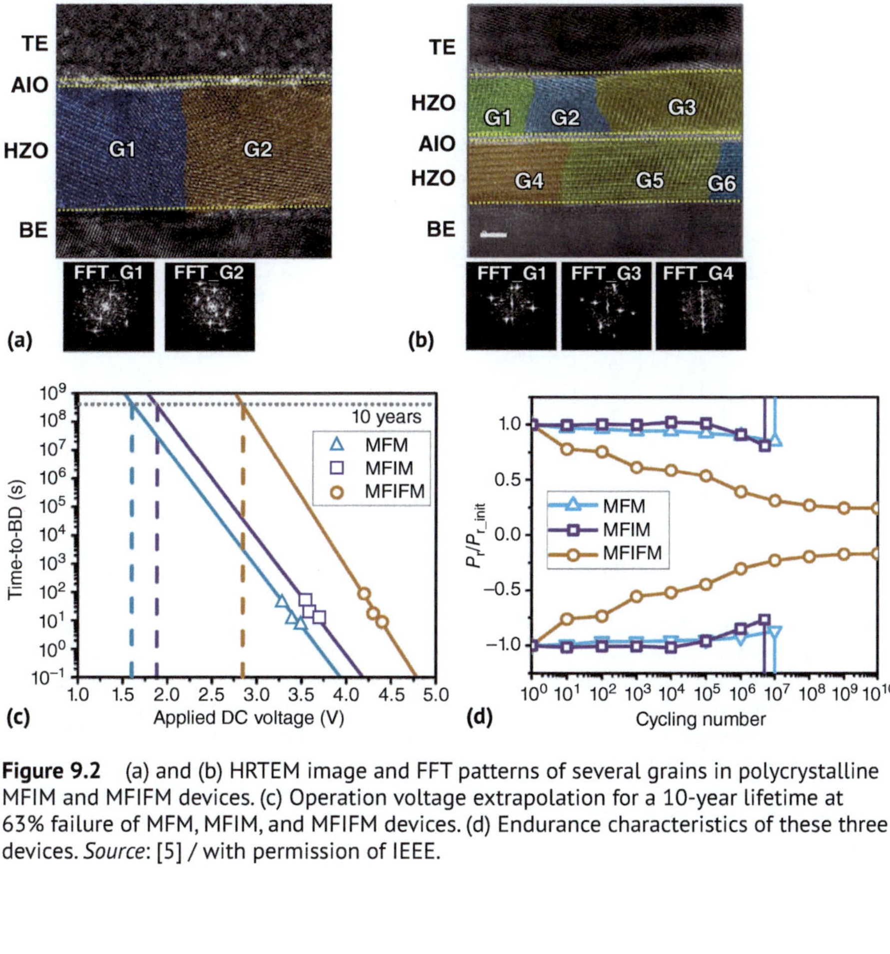

Figure 9.2 (a) and (b) HRTEM image and FFT patterns of several grains in polycrystalline MFIM and MFIFM devices. (c) Operation voltage extrapolation for a 10-year lifetime at 63% failure of MFM, MFIM, and MFIFM devices. (d) Endurance characteristics of these three devices. *Source*: [5] / with permission of IEEE.

otherwise connect the top and bottom electrodes. As a result, leakage current is suppressed, greatly enhancing the overall stability and endurance of the device.

In addition to enhancing the E_{BD}, another effective approach is to reduce the E_C, though it remains a highly challenging task. From a microscopic perspective, such a large E_C stems from the much higher switching barrier (1.38 eV) of dipoles in orthorhombic HfO_2 with the *Pca2*$_1$ space group compared to that of $PbTiO_3$ (0.024 eV).[6] There are many theoretical studies, which are based on first-principles calculations, that concentrate on reducing the E_C of HfO_2-based materials, such as dopants, strains, and other factors.[7]

Due to the separation of dipoles by spacers in the *Pca2*$_1$ ferroelectric phase of HfO_2, the polarization switching process is characterized by a hopping mechanism, resulting in a high polarization switching barrier and a high E_C. Therefore, it is necessary to identify materials where dipoles can switch polarization continuously to address this issue. In recent years, the rhombohedral (r) phase has been identified as one such material. However, as it is also unstable, exploring methods to enhance the stability remains necessary.[8–14] Wang et al. discovered a CMOS-compatible rhombohedral ferroelectric $Hf(Zr)_{1+x}O_2$ material rich in Hf(Zr).[15] Excess Hf(Zr) atoms embedded within the lattice of the thin film can stabilize the ferroelectric rhombohedral crystal structure, as shown in Figure 9.3a,b. Density functional theory (DFT)

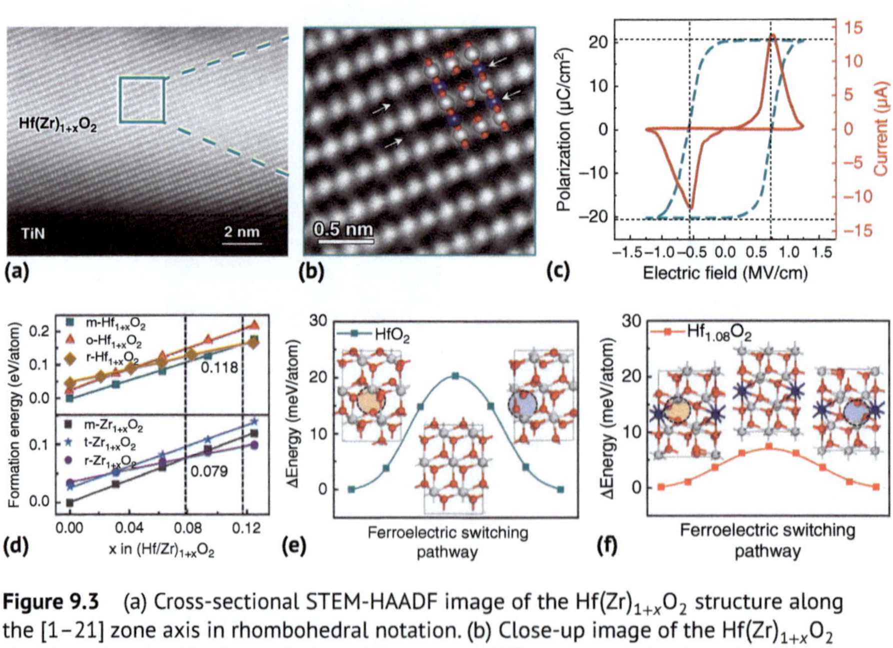

Figure 9.3 (a) Cross-sectional STEM-HAADF image of the $Hf(Zr)_{1+x}O_2$ structure along the [1−21] zone axis in rhombohedral notation. (b) Close-up image of the $Hf(Zr)_{1+x}O_2$ structure showing the intercalation of the excess Hf(Zr) atoms (highlighted by the white arrows). An atomic model is overlaid to guide the eyes. (c) PUND measurements of $Hf(Zr)_{1+x}O_2$ capacitors under an electric field of frequency of 1 KHz. (d) Formation energies of $Hf_{1+x}O_2$ in r-, o-, and m-phases as a function of x (upper) and $Zr_{1+x}O_2$ in r-, t-, and m-phases as a function of x (lower). (e, f) Energy landscape showing the ferroelectric switching process of HfO_2 and $Hf_{1.08}O_2$ with d_{111} of 3.11 Å, respectively. *Source*: [15] / with permission of American Association for the Advancement of Science - AAAS.

calculations reveal that the formation energy of the r-phase is lower than those of the orthorhombic (o) phase and m-phase in Hf(Zr)-rich Hf(Zr)O_2 films when the ratio of Hf(Zr) to O is larger than 1.079:2, which is more conducive to the formation of the r-phase, as shown in Figure 9.3d. More importantly, Hf(Zr) interstitial atoms enhance the ferroelectricity of the thin film and reduce its ferroelectric polarization switching barrier (Figure 9.3e,f), resulting in an ultralow coercive electric field (0.65MV/cm) and a small saturation polarization field (1.25MV/cm) (Figure 9.3c). The device lifetime of more than 10^{12} cycles is achieved at saturation polarization. The discovery of this material may contribute to realizing low-cost and long-life memory chips.

9.2.2 Electric Field Cycling Behaviors

Conventional ferroelectric films exhibit fatigue behavior, characterized by a decrease in the value of $2P_r$ with increasing electric field cycles. In contrast, HfO_2-based ferroelectric films show an initial increase in $2P_r$ value, which is called "wake-up," reaching a maximum before decreasing, indicating "fatigue." Such electric field cycling behaviors are observed in various HfO_2-based ferroelectric films with different dopants and are strongly dependent on annealing temperature, electrode materials, and deposition techniques. The change in $2P_r$ with electric field cycles may lead to misreading of stored information. Therefore, understanding the underlying mechanisms and seeking practical solutions are essential.

The wake-up and fatigue behaviors observed in HfO_2-based ferroelectric films are mainly attributed to two mechanisms. One focuses on the redistribution and generation of oxygen vacancy (Vo) defects, while the other emphasizes the phase transition between the paraelectric and ferroelectric phases.

Pešić et al. conducted experiments to study the electric field cycling behaviors of Sr:HfO_2 FeCAPs.[16] They found that the number of switching cycles needed to wake up the sample decreases significantly as temperature increases. However, the evolution of P_r at elevated temperatures also exhibits enhanced fatigue, as shown in Figure 9.4a,b. Their STEM images at the interface of Sr:HfO_2/electrode in pristine, wake-up, and fatigue states, as shown in Figure 9.4c, indicate that the phase transition and the change in the polarization direction of the o-phase are responsible for the wake-up and fatigue, respectively. In addition, their electrical experiments indicated that the leakage current and defects concentrations remain constant during the wake-up state and then continuously increase during the fatigue state. Guided by these observations, they proposed an underlying mechanism for wake-up and fatigue behaviors, as illustrated in Figure 9.4d. During the wake-up process, interfacial defects redistribute and diffuse from the interfaces to the bulk. Defects diffusing into the m-phase grains in the bulk promote phase transitions from the m/t-phase to ferroelectric o-phase, resulting in the rise of $2P_r$. During the fatigue process, defect generation becomes dominant. The generated defects pin the domains, leading to unswitched ferroelectric grains. Meanwhile, some of the o-phase grains rotate, causing the polarization direction to become perpendicular to the electric field. These combined effects contribute to the reduction of P_r.

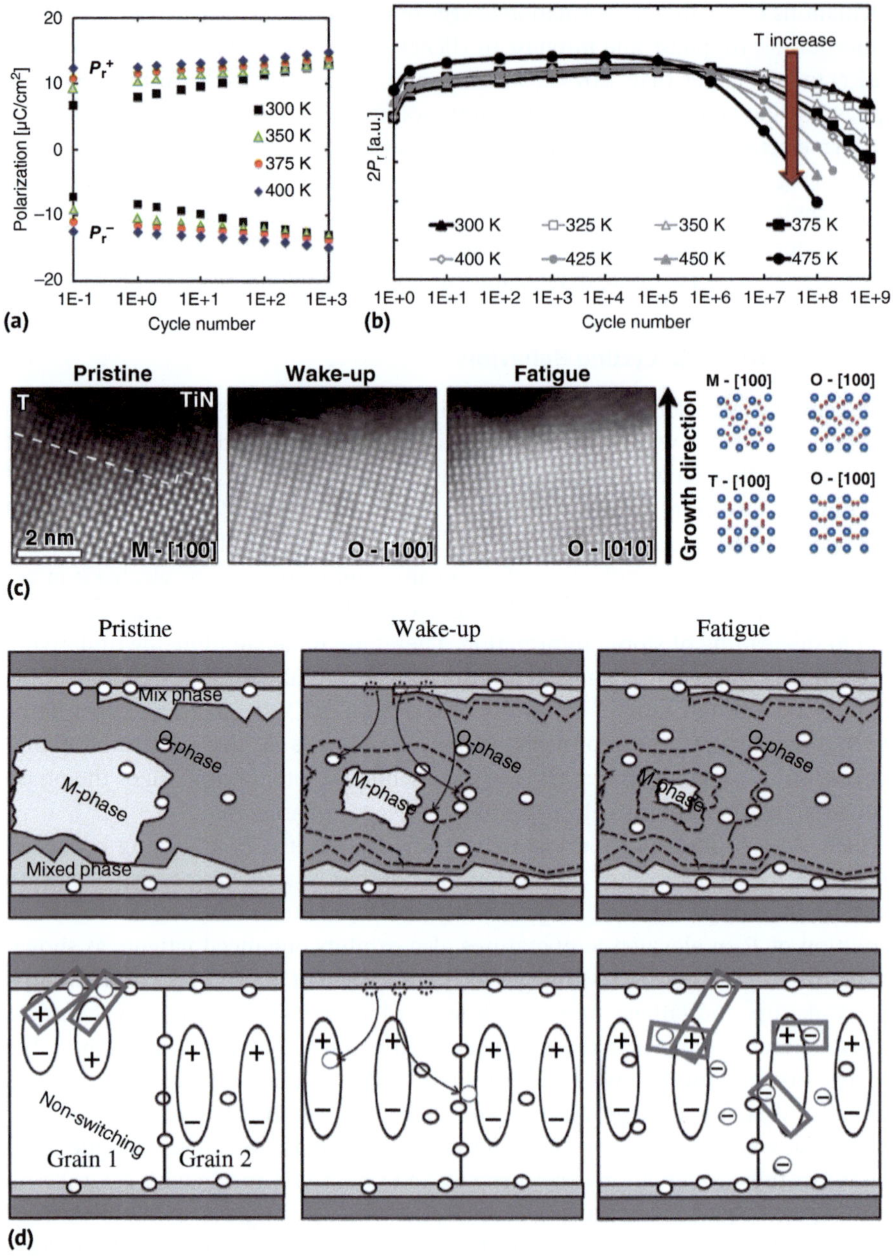

Figure 9.4 (a) Thermal wake-up dependence on elevated temperatures; (b) MW evolution as a function of device cycling and temperature. Increased temperature results in an earlier fatigue onset; (c) HAADF-STEM images of doped HfO_2 in pristine, wake-up, and fatigue conditions, showing relaxation of the bulk monoclinic and tetragonal symmetry at the electrode interfaces toward orthorhombic phase; (d) proposed model of the three different stages of the FeCap lifetime at pristine, wake-up, and fatigue conditions. *Source*: [16] / with permission of John Wiley & Sons.

While wake-up can be addressed by controlling the defects within the HfO_2-based films, the fatigue remains insufficiently resolved. However, an alternative method has been discovered to temporarily recover the degraded $2P_r$, thereby improving the overall endurance.[17,18] Gong et al. found that applying a high voltage to devices with degraded P_r effectively restores the P_r to its pre-degradation level.[18] They discovered that when HfO_2-based capacitors undergo cycling under low electric fields, the concentration of interfacial defects remains unchanged, as shown in Figure 9.5a. In addition, the aberration-corrected electron microscopy revealed that the random distribution of interface charges relates to the randomly oriented polarization directions pointing up or down within a single grain during the fatigue stage, as illustrated in Figure 9.5b. This demonstrates that changes in the charged states of interface defects can cause P_r degradation under low electric fields. Based on this principle, a method involving the periodic application of continuous high-voltage and low-voltage operations was proposed, as shown in Figure 9.5c,d. This approach restores the charged states of interface defects, enabling the periodic recovery of residual polarization and achieving an ultra-high fatigue endurance of 10^{14} cycles, thereby meeting the application requirements of nonvolatile memory (Figure 9.5e). This approach has also been further explored. Li et al. observed a re-wakeup phenomenon during the fatigue stage under high-temperature field cycling.[19]

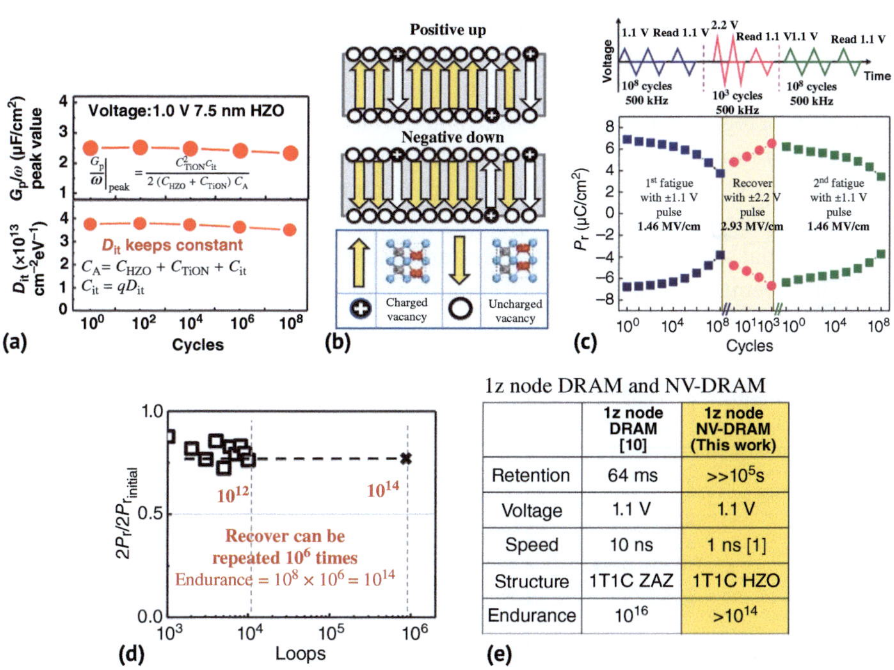

Figure 9.5 (a) Interface defects under low electric fields as a function of read/write cycles; (b) charged states of interface defects leading to low-field P_r failure; (c) operational algorithm for P_r recovery; (d) and (e) validation and comparison of nonvolatile DRAM with endurance exceeding 10^{14} cycles. *Source*: [18] / with permission of The Japan Society of Applied Physics.

They found that re-wakeup temperatures significantly affect P_r fatigue, with devices exhibiting excellent anti-fatigue characteristics after high-temperature re-wakeup. In summary, the high-voltage recovery strategy can substantially enhance the endurance of HfO_2-based ferroelectric devices.

9.2.3 Retention

Retention refers to the ability of a ferroelectric film to maintain its predefined polarization state over time. This involves writing a polarization state into the film, periodically reading it, and assessing any degradation. Specifically, it can be characterized as same-state (SS), new-same-state (NSS), and opposite-state (OS) retention, as shown in Figure 9.6.[20]

In FeCAPs with destructive readout, every read operation inevitably affects the device's polarization state. Retention loss can result from depolarization caused by factors like thermal effects and depolarization fields or from imprint, where the accumulation of internal bias in the ferroelectric device progressively restricts its ferroelectric response under a constant readout voltage.

Additionally, measuring the retention characteristics in ferroelectric devices is rather time-consuming. Therefore, one commonly used approach to accelerate degradation is to measure at elevated temperatures.

Due to the low surface-to-thickness ratio of the film in bulk ferroelectric materials, the alignment of defect dipoles within the film becomes a primary factor contributing to the imprint effect. In contrast, for thin-film ferroelectric materials, the interfacial region between the surface and the electrode plays a dominant role.[21] The depolarization field generated by polarization charges in either material must be screened to stabilize a specific ferroelectric orientation. In thin-film MFM capacitors, this screening is typically achieved by free electrons within the electrodes. However, a minimal spatial separation between the external screening charges and the polarization charges remains even under ideal conditions, resulting in incomplete screening.

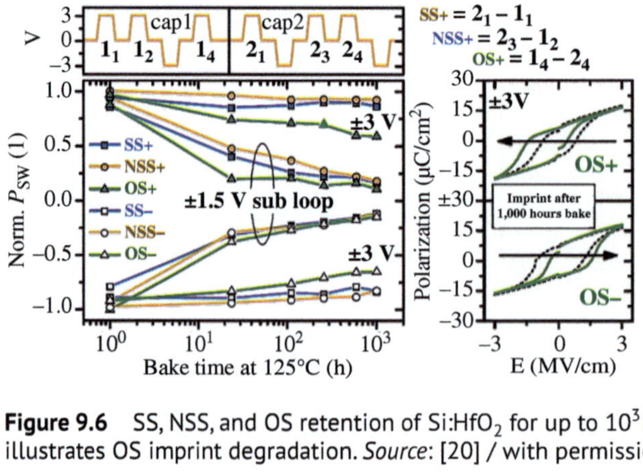

Figure 9.6 SS, NSS, and OS retention of Si:HfO_2 for up to 10^3 h bake at 125 °C. P-E illustrates OS imprint degradation. *Source*: [20] / with permission of IEEE.

This causes a small residual depolarization field to persist, generating an electric field in the region and leading to charge injection. Consequently, the trapped charges accumulate over time, resulting in the imprint effect.

Additionally, during the fabrication process of devices, chemical interactions between the electrode and ferroelectric material may produce parasitic layers with distinct electrical properties. These layers exhibit a lower dielectric constant and may possess a higher defect density, further exacerbating the imprint effect. Under such conditions, elevated electric fields can accelerate charge injection, and more charges may be trapped.

Yuan et al. investigated retention loss in HZO capacitors caused by imprint and highlighted the role of charged *V*os.[22] They found that initial downward polarization leads to a more pronounced imprint effect than upward polarization, which they attributed to the asymmetric distribution of vacancies, with a higher concentration observed at the upper interface, as shown in Figure 9.7a–c. They concluded that the imprint effect in HfO_2-based ferroelectric materials stems primarily from charge injection, electron release from oxygen vacancies, and changes in the built-in electric field. Based on their findings, they proposed a pre-shipment baking process to eliminate irreversible imprint effects, as shown in Figure 9.7d,e. They recommended applying unipolar voltage pulses during operation to recover both coercive field and P_r strength, restoring up to 90% of the P_r. This approach enables HfO_2-based FeCAPs to achieve retention characteristics of up to 10 years, thus effectively improving device reliability and extending the operational lifespan of HfO_2-based ferroelectric memories.

9.3 Reliability of HfO_2-based FeFET

9.3.1 Endurance

In the initial stages of FeFET development, the ferroelectric layer is integrated into the gate stack of a MOSFET structure, replacing the gate dielectrics (metal oxide). The native SiO_2 layer on Si forms a metal–ferroelectric–insulator–semiconductor (MFIS) gate stack. However, the low dielectric constant of SiO_2 causes a significant voltage drop across the layer. This can lead to dielectric breakdown of the SiO_2 layer, causing reliability issues. Furthermore, the SiO_2 layer reduces the memory window (MW). Early studies on FeFETs demonstrated that the MW increases as the thickness of the ferroelectric layer increases and the interlayer (IL) decreases.[23]

Yurchuk et al. performed an extensive study of the endurance characteristics in FeFETs using capacitor and transistor structures, establishing that the endurance degradation in HfO_2-FeFET relates to the degradation of the transistor gate stack, specifically the interfacial layer (IL) rather than to the ferroelectric layer itself.[24] Tests on Si:HfO_2-based MFIS-FETs revealed an initial increase in the MW up to 10^3 cycles, attributed to the "wake-up" effect frequently observed in ferroelectric materials. Beyond 10^3 cycles, however, the MW exhibited an almost linear decay with the number of cycles (Figure 9.8a). In contrast, MFM capacitor structures displayed no fatigue in P_r and coercive voltage (V_C) even after up to 10^6 cycles (Figure 9.8b), further corroborating that

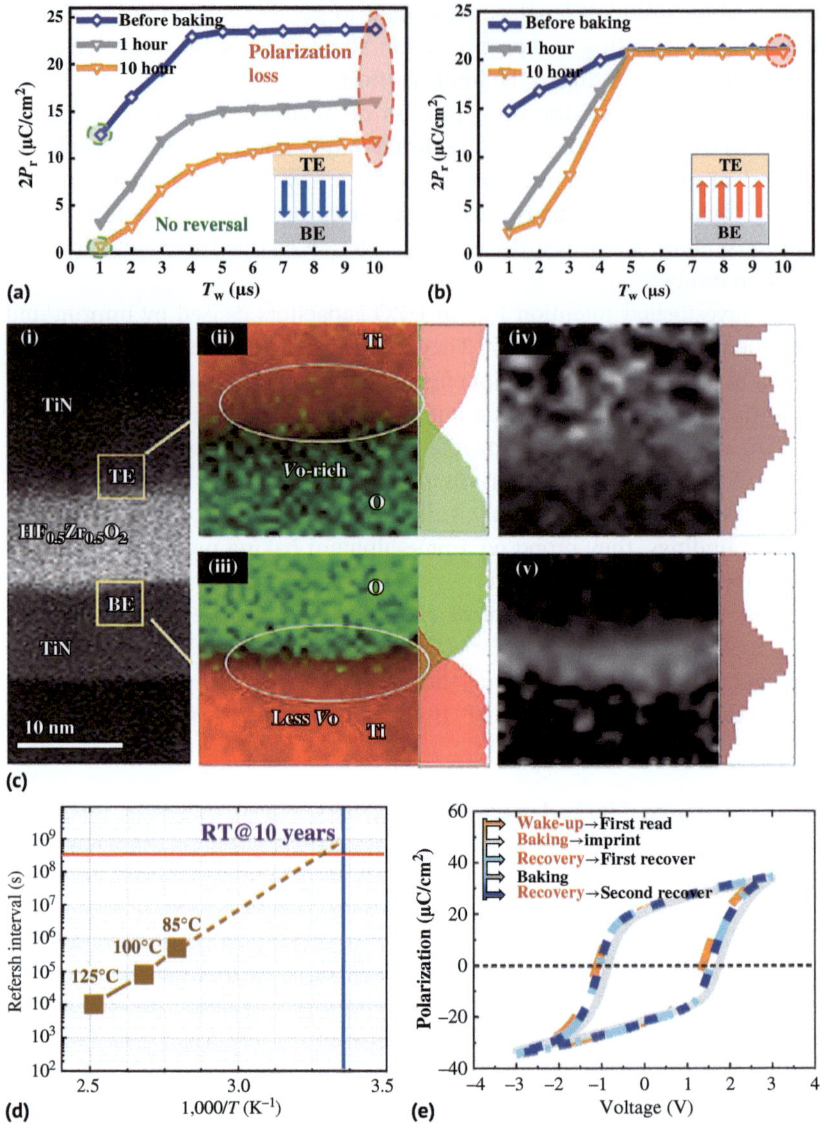

Figure 9.7 Imprint test results with a small voltage (2V). Before baking, the cell was programmed either to (a) the down-polarization state or (b) the up-polarization state. When a relatively small writing pulse was used (amplitude: 2V), the E_C shift for the down-state capacitor was much more severe than the up-state capacitor, together with a great decrease of P_r value by ~50%. (c) High-angle annular dark-field (HAADF) image of the TiN/HZO/TiN sample: (i) the yellow rectangles inside the image mark the area of electron energy loss spectroscopy (EELS) mapping in the top and bottom interfaces, respectively. (ii) and (iii) are normalized EELS intensity mappings of Ti (red) and O (green) from the top and bottom interfaces, respectively; vertical profiles of the mappings are shown on the right side. (iv) and (v) show the multiplications of the intensities of Ti and O (ITi·IO) from each mapping. (d) extrapolation of room temperature imprint–recovery measurement by the failure time, from 85, 100, and 125 °C data; (e) P–V loops of imprint–recovery test, where the V_c shift between 1st recovery to 2nd recovery is negligible. *Source*: [22] / with permission of Springer Nature.

the endurance of FeFETs is constrained by the gate stack degradation. After 10^3 cycles, an increase in the gate current and interfacial trap density was detected, corresponding to the onset of MW degradation (Figure 9.8c). The high-density trap generation impairs the insulating properties of the interfacial layer, resulting in increased gate leakage current and, ultimately, FeFET endurance degradation (Figure 9.8d).

To understand the endurance behaviors of FeFETs, multiple defect extraction techniques have been employed to characterize and analyze gate defects. Yurchuk et al. used a single-pulse $I_D - V_G$ technique to accurately monitor the rapid charge trapping dynamics in the gate dielectric layer, thereby revealing its impact on MW degradation. They also utilized the charge-pumping technique to observe the degradation of the SiON interfacial layer under cycling stress.[24] Gong and Ma identified two mechanisms of MW degradation: reversible threshold voltage (V_{th}) drift caused by charge trapping during early cycles and irreversible subthreshold swing (SS) degradation caused by interface trap generation after 10^4 cycles (Figure 9.9a,b).[25] As shown in Figure 9.9c, Ni et al. evaluated the effects of charge release dynamics

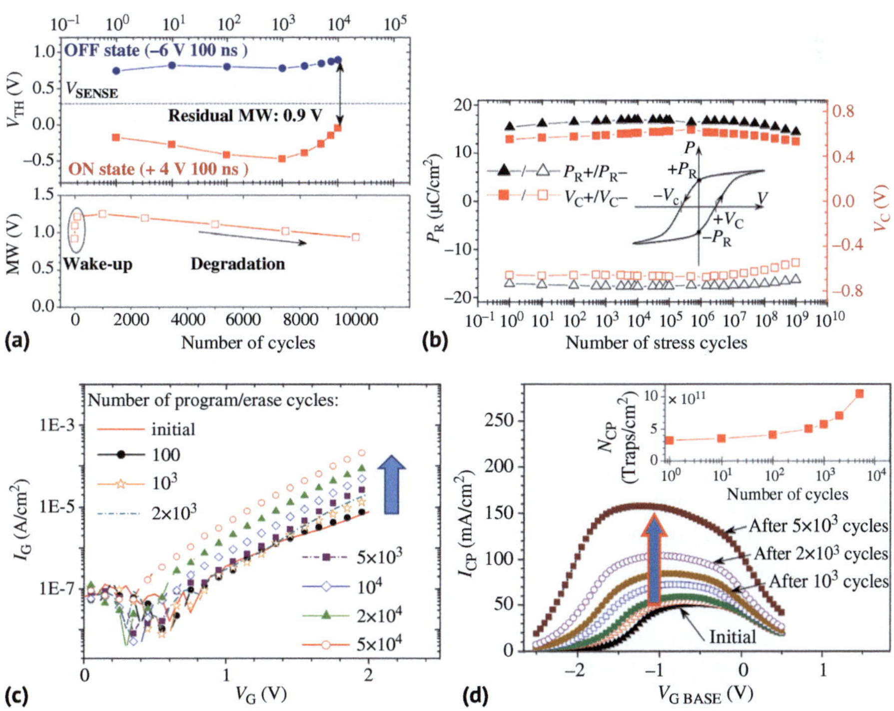

Figure 9.8 (a) Typical endurance characteristics of Si:HfO_2-based MFIS-FETs exposed to alternating program/erase pulses. V_{th} in the "ON" and "OFF" states (upper panel) and a corresponding MW (lower panel) are plotted versus the number of program/erase cycles. (b) P_r and V_C as functions of stress cycles of an MFM capacitor. (c) Evolution of gate leakage current characteristics (I_G - V_G) of MFIS-FETs with continuous program/erase cycling. (d) Variable base level CP characteristics with increasing number of program/erase cycles. Insert shows the interface trap density, extracted from the CP characteristics, as a function of cycle number. *Source*: [24] / with permission of IEEE.

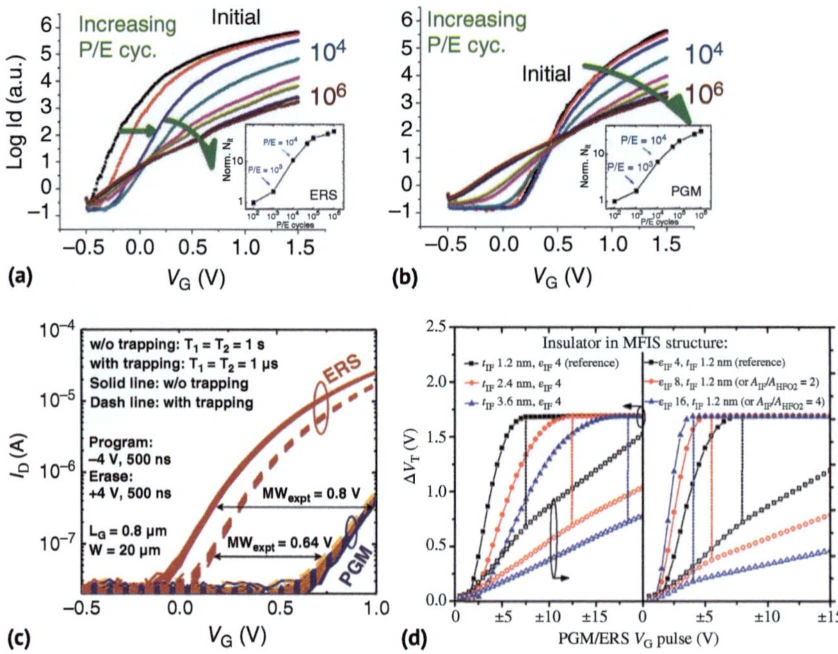

Figure 9.9 (a) and (b) $I_d - V_g$ curves. (a) ERS state (after "+" ERS pulse) shows parallel V_{th} shift up to 1000 P/E cycles, followed by slope degradation.[25] (b) PGM state (after "–" PGM pulses) exhibits slope degradation. *Source*: [25] / with permission of IEEE. (c) $I_D - V_G$ characteristics after program and erase pulses. A read delay of T1 $T_1 = T_2 = 1$ s allows trapped charge to release, approaching MW_{ideal}, while a delay of $T_1 = T_2 = 1\,\mu s$ narrows MW_{expt}. *Source*: [26] / with permission of IEEE. (d) Simulated ΔV_T (primary y-axis) and interfacial field stress E_{IF} (secondary y-axis) as a function of the PGM/ERS voltage. A variation of the thickness of the interfacial layer t_{IF} is on the left, and a variation of the dielectric constant ε_{IF} of the interfacial layer is on the right. *Source*: [27] / with permission of IEEE.

on the MW. They found that electrons and holes are released within 10 and 100 ms, respectively, significantly influencing MW performance. Additionally, they observed that charge trapping increases nearly in proportion to the thickness of HfO_2.[26]

Muller et al. investigated gate dielectric degradation using sub-loop polarization switching and found that reduced interfacial field strength under low-field conditions suppresses charge injection and defect generation, thereby improving endurance, as highlighted in Figure 9.9d.[27] Wang et al. analyzed defect generation and its impact on the MW by analyzing V_{th} variations during charge trapping and polarization switching. Moreover, they used changes in gate-induced drain leakage (GIDL) current to further assess the formation and evolution of interface defects.[28] In summary, these studies employed various techniques, monitoring both dynamic behaviors (e.g., rapid charge trapping) and long-term trends (e.g., interface trap generation). They provide a comprehensive analysis of the impact of charge trapping and trap generation on MW degradation, as well as critical data for memory endurance and reliability optimization.

The high dielectric constant and large ΔE of materials such as Al_2O_3 make them excellent candidates for the interfacial layer, which can reduce the voltage drop across the layer and suppress charge trapping. Wang et al. conducted multiple experiments on FeFET endurance and concluded that inserting an Al_2O_3 interfacial layer can mitigate the trend of MW degradation. They further proposed that the recovery pulse should be applied before trap generation rather than waiting until the MW decreases to the minimum sense margin. Figure 9.10a shows the endurance characteristics of FeFET with and without Al_2O_3 interfacial layer. With the insertion of the Al_2O_3 interfacial layer, the endurance of FeFETs is significantly improved, and the MW degradation is effectively slowed down. At the same time, the leakage current of the FeFETs with Al_2O_3 is maintained at a relatively low level, indicating that fewer defects are generated after cycling. Consequently, devices with Al_2O_3 will require recovery pulses later than those without, as shown in Figure 9.10b.

The endurance degradation mechanism consists of two main stages. First, electron detrapping in the HZO layer induces the domain pinning effect, leading to ferroelectric fatigue. This process is recoverable, and applying an appropriate long-time voltage can help restore the MW (Figure 9.10c). Second, electron detrapping

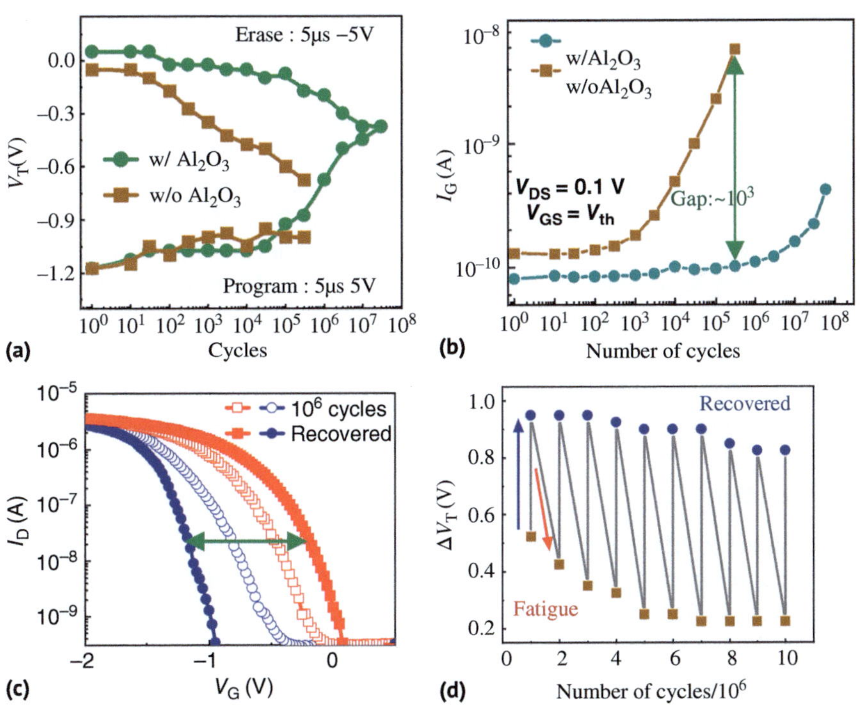

Figure 9.10 (a) Endurance of FeFETs with Al_2O_3 can be improved to 10^7 cycles, compared with those without Al_2O_3. (b) Evolution of gate leakage current of FeFET with and without Al_2O_3 with switching cycles. (c) Schematic of the recovery process and the experimental results. (d) MW evolution after each 10^6 cycles and recovered states. *Source*: [29] / with permission of IEEE.

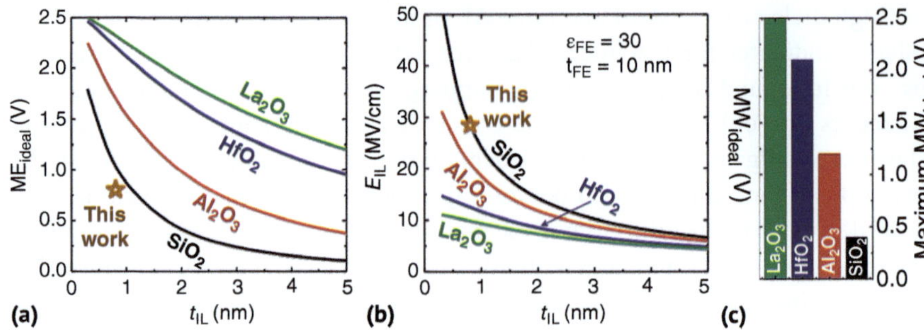

Figure 9.11 (a) Simulated MW and (b) maximum EIL as a function of IL composition/ thickness. (c) Maximum MW satisfying all the constraints for four different IL dielectrics. *Source*: [26] / with permission of IEEE.

causes a gradual accumulation of holes in the HZO layer, which affects the electric field distribution and increases the bias voltage across the Al_2O_3 layer. When the electric field reaches a critical level, irreversible interface traps are generated, further exacerbating device degradation. Therefore, improving endurance requires delaying the hole accumulation process and applying recovery pulses prior to the formation of interface traps. Figure 9.10d illustrates the MW trend after optimized recovery pulses. After 10^7 cycles with optimized recovery strategies, the MW is well-maintained, and the endurance improves by at least 100 times compared to devices without Al_2O_3.[29]

Ni et al. utilized high-dielectric-constant interfacial layers to optimize the electric field distribution in the gate stack. Figure 9.11a,b presents the calculated MW and IL electric field for different interfacial layer materials. The results indicate that increasing the dielectric constant of IL can improve the MW and reduce the electric field strength within the IL simultaneously, thereby achieving higher endurance. Additionally, MW was tested for different IL thicknesses, FE thicknesses, and IL compositions, as shown in Figure 9.11c. The results demonstrate that thinner ILs and higher dielectric constants can increase the FE thickness while expanding the MW. This further validates the feasibility of inserting a high-dielectric-constant interfacial layer (e.g., Al_2O_3) as an effective optimization method.[26]

9.3.2 Retention

While the depolarization field is a significant factor in retention degradation, its impact varies across different FeFET materials. In Hf-based FeFETs, for example, the depolarization field has a lesser impact compared to perovskite-based FeFETs. Gong et al. highlighted that the determinant of early retention decay in FeFETs is not the depolarization field (E_{dep}) itself but its ratio to the coercive field (E_{dep}/E_C). When E_{dep} significantly exceeds E_C, retention loss becomes substantial. However, polarization reversal is negligible if E_{dep} is comparable to or smaller than E_C. With $E_C \approx 1$ MV/cm in FE-HfO_2 compared to $E_C \approx 30 - 90$ kV/cm in perovskite materials, the impact of the depolarization field on retention loss in Hf-based FeFETs is relatively minor, as shown in Figure 9.12a.[30]

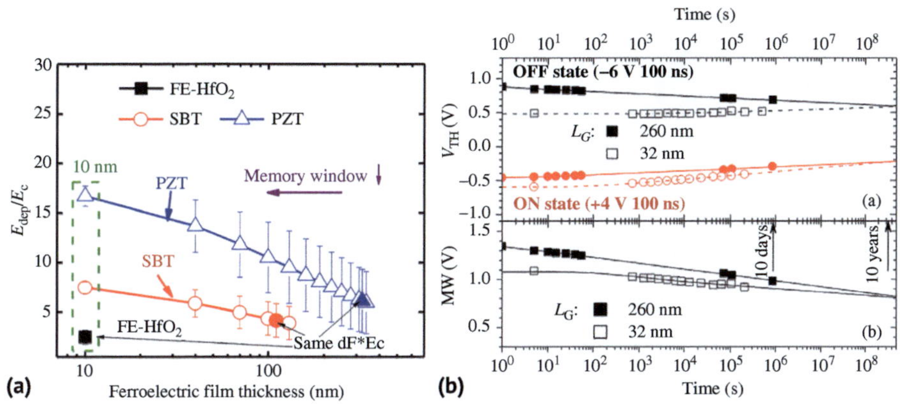

Figure 9.12 (a) The ratio of depolarization field (E_{dep}) to coercive field (E_C) in three cases: (1) FE-HfO_2 + 1.2 nm SiO_2 buffer layer, (2) SBT + 1.2 nm SiO_2, and (3) PZT + 1.2 nm SiO_2. E_{dep}/E_C in PZT and SBT FeFETs are significantly higher than in FE-HfO_2, leading to faster retention decay. *Source*: [30] / with permission of IEEE. (b) Impact of scaling on the retention properties of Si:HfO_2-based FeFET devices (4.4 mol% SiO_2): (1) Retention behavior of ON- and OFF-memory states. (2) Time dependence of MW for devices with gate widths and lengths of 2 μm : 260 nm and 80 : 32 nm. *Source*: [31] / with permission of IEEE.

In addition to the depolarization field, charge trapping also plays a significant role in retention degradation. Yurchuk et al. analyzed the ON/OFF state voltage drift in short and long channel devices and concluded that retention asymmetry is primarily caused by interfacial charge trapping rather than the depolarization field (Figure 9.12b).[31] These studies suggest that while the depolarization field contributes to retention loss, interfacial charge trapping is the dominant factor driving FeFET retention degradation.

Temperature is another critical factor that exacerbates retention loss. Ni et al. analyzed the retention performance of HfO_2-based FeFETs at different temperatures (Figure 9.13a) and observed a decrease in the MW as temperature increased.[26] This behavior can be attributed to the reduction in polarization strength at higher temperatures, which leads to a loss of memory retention (Figure 9.13b).

Similar results were obtained by Ali et al., who evaluated the retention characteristics of FeFETs at 0 and 40 °C (Figure 9.13c). They concluded that within a suitable temperature range, the polarization strength remains relatively high, but retention characteristics weaken as temperature increases. At high temperatures, enhanced charge injection effects and depolarization fields were observed. Moreover, the asymmetry in retention, where the low-V_{th} state remains stable while the high-V_{th} state degrades, suggests that charge trapping plays a more significant role in retention loss than depolarization. Consequently, cooling the device can recover memory characteristics affected by thermal depolarization, mitigating the retention loss at high temperatures.[32]

Improving the retention characteristics of FeFETs has become a critical area of research, driven not only by the need to address temperature-induced degradation

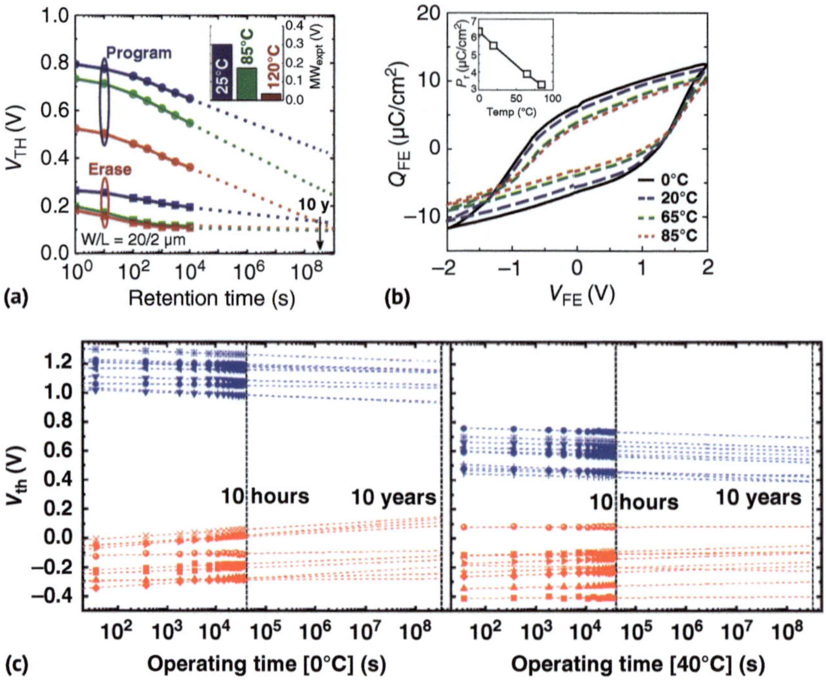

Figure 9.13 (a) V_{th} during retention after program/erase at different temperatures: Low-V_{th} state remains stable, while high-V_{th} state degrades over time. (b) Polarization decreases with increasing temperature, leading to reduced MW at higher temperatures. *Source*: [26] / with permission of IEEE. (c) Statistical FeFET retention as a function of operating temperature: 0 and 40 °C, measured for over 10 hours and extrapolated to meet the 10-year retention criterion. *Source*: [32] / with permission of IEEE.

but also by advancements in the ferroelectric HfO_2 layer and interface engineering. Muller et al. utilized sub-loop polarization operation to avoid high-voltage programming and erasing operations. This approach reduces the interface electric field (E_{IF}) and consequently minimizes charge injection and interface charge trapping effects, leading to improved retention characteristics. Additionally, Muller et al. employed a 10 nm thick Si-doped HfO_2 thin film that underwent thermal treatment to stabilize its ferroelectric properties. When combined with a TiN metal gate (TiN-MG), this structure exhibited a stable MW, with retention characteristics predicting a residual MW of 620 mV after 10 years, as shown in Figure 9.14a.[33]

In a subsequent study, Ali et al. focused on the interface layer to further improve retention. By introducing nitrogen into the conventional SiO_2 interface layer, they created a SiON interface layer, which significantly increased the dielectric constant of the interface. The nitrogen passivation in the SiON layer not only enhanced the interface strength but also reduced both the interface electric field and the depolarization field. As shown in Figure 9.14b, the SiON interface layer outperforms its SiO_2 counterpart in terms of retention characteristics, achieving a residual MW of 0.5 V after 10 years under 150 °C bake testing (Figure 9.14c).[34]

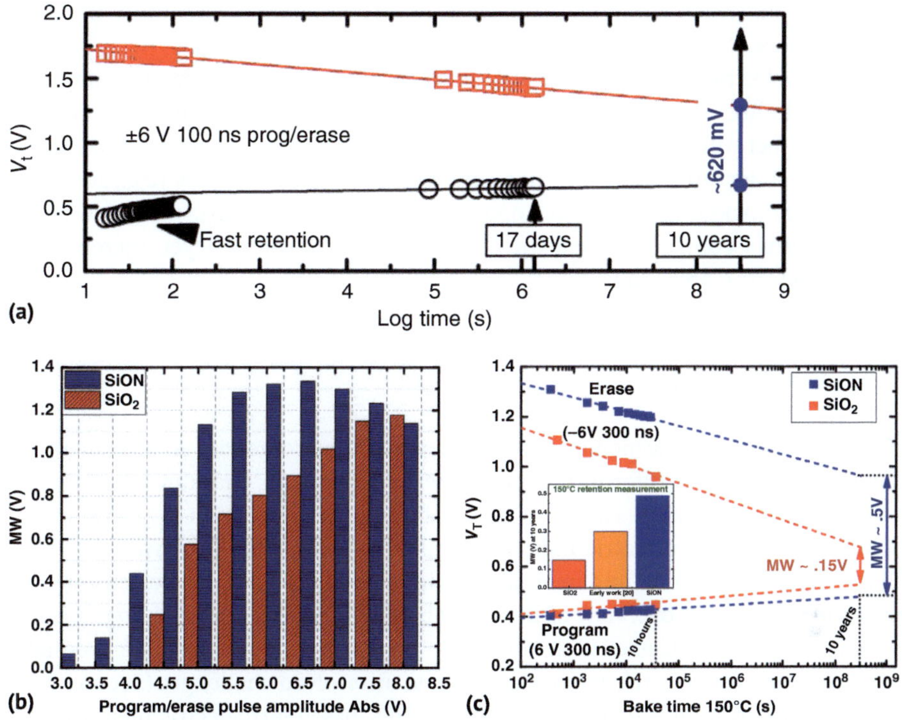

Figure 9.14 (a) Retention of a Si:HfO_2-based FeFET: despite an initial fast retention loss, extrapolation predicts an MW of 620 mV after 10 years. *Source*: [33] / with permission of IEEE. (b) Extracted MW for SiON versus SiO_2 demonstrating a maximized MW at lower voltages for the SiON-based FeFET. (c) FeFET retention characteristics for the SiON versus SiO_2 interfaces under a 150 °C aging test. A 10-year extrapolation indicates improved retention for the SiON interface with a residual MW of 0.5 V. *Source*: [34] / with permission of IEEE.

9.4 Reliability Challenges in HfO_2-based Ferroelectric Memories

Reliability has been a critical barrier to the widespread adoption of HfO_2-based ferroelectric memories. Despite significant progress made in recent years, comprehensive solutions remain underdeveloped.

In FeCAPs, HfO_2-based materials that rely on the $Pca2_1$ ferroelectric phase exhibit inherent reliability limitations. A critical aspect of addressing these constraints involves correlating the polarization switching mechanism and device reliability. Two main strategies are being explored: (1) developing new HfO_2-based ferroelectric materials or reducing the coercive field of the $Pca2_1$ phase while retaining CMOS compatibility. A lower coercive field would enable low-voltage operation, thereby improving reliability and (2) enhancing the breakdown electric field through techniques such as IL engineering, thickness reduction, or grain boundary minimization, which eliminates leakage paths. These approaches aim to improve the endurance of HfO_2 thin films.

For ferroelectric transistors, several approaches are being pursued to improve device reliability. One route focuses on improving the interface contacts and developing high-permittivity interfacial layers. Another strategy is to adopt non-silicon channel architectures, such as oxide semiconductor-based devices (e.g., indium–gallium–zinc oxide, IGZO). This approach eliminates the interfacial oxide layer, enabling direct contact between the channel and ferroelectric layer, thereby preventing dielectric breakdown issues.

9.5 Conclusion

HfO_2-based ferroelectric memory stands out as a leading contender among emerging memory technologies, offering exceptional write/read speed, endurance, scalability, and CMOS compatibility. However, its reliability – in comparison to conventional perovskite-based ferroelectric memory – still requires further improvement. This chapter begins with the fundamental principles of HfO_2-based ferroelectric materials and provides an overview of their reliability challenges. It then systematically reviews recent advancements in reliability, focusing on both capacitor-type and transistor-type ferroelectric memories. Despite unresolved challenges (Section 9.3), continued innovation is expected to solidify HfO_2-based ferroelectric memory as a transformative technology in the memory industry.

Acknowledgments

The author acknowledges the financial support in part from MOST of China (2022YFB3608400); the National Natural Science Foundation of China (NSFC) (62425407, 62304246, 62404246); and the China Postdoctoral Science Foundation (BX20230405).

References

1 Schroeder, U., Park, M. H., Mikolajick, T. et al. (2022). The fundamentals and applications of ferroelectric HfO_2. *Nature Reviews Materials* 7 (8): 653–669.

2 Boescke. T. S., Muller, J., Brauhaus, D. et al. (2011). Ferroelectricity in hafnium oxide thin films (in English). *Applied Physics Letters* 99 (10): 102903.

3 Wei, W., Zhang, W., Wang, F. et al. (2020). Deep insights into the failure mechanisms in field-cycled ferroelectric $Hf_{0.5}Zr_{0.5}O_2$ thin film: TDDB characterizations and first-principles calculations. In: *2020 IEEE International Electron Devices Meeting (IEDM)*. IEEE. pp. 39.6.1–39.6.4.

4 Li, X., Wei, W., Wu, J. et al. (2022). Experimental investigations on ferroelectric dielectric breakdown in sub-10 nm $Hf_{0.5}Zr_{0.5}O_2$ film through comprehensive TDDB characterizations. *Japanese Journal of Applied Physics* 61 (10): 101002.

5 Xu, Y., Yang, Y., Zhao, S. et al. (2021). Robust breakdown reliability and improved endurance in $Hf_{0.5}Zr_{0.5}O_2$ ferroelectric using grain boundary interruption. *IEEE Transactions on Electron Devices* 69 (1): 1–4.

6 Lee, H. J., Lee, M., Lee, K. et al. (2020). Scale-free ferroelectricity induced by flat phonon bands in HfO_2 (in English). *Science* 369 (6509): 1343–1347.

7 Wei, W., Zhao, G., Zhan, X. et al. (2022). Switching pathway-dependent strain-effects on the ferroelectric properties and structural deformations in orthorhombic HfO_2. *Journal of Applied Physics* 131 (15): 154101.

8 El Boutaybi, A., Maroutian, T., Largeau, L. et al. (2022). Stabilization of the epitaxial rhombohedral ferroelectric phase in ZrO_2 by surface energy. *Physical Review Materials* 6 (7): 074406.

9 Bégon-Lours, L., Mulder, M., Nukala, P. et al. (2020). Stabilization of phase-pure rhombohedral $HfZrO_4$ in pulsed laser deposited thin films. *Physical Review Materials* 4 (4): 043401.

10 Zhang, Y. K., Yang, Q., Tao, L. L. et al. (2020). Effects of strain and film thickness on the stability of the rhombohedral phase of HfO_2 (in English). *Physical Review Applied* 14 (1): 014068.

11 Wei, Y., Nukala, P., Salverda, M. et al. (2018). A rhombohedral ferroelectric phase in epitaxially strained $Hf_{0.5}Zr_{0.5}O_2$ thin films. *Nature Materials* 17 (12): 1095–1100.

12 Nukala, P., Wei, Y., de Haas, V. et al. (2020). Guidelines for the stabilization of a polar rhombohedral phase in epitaxial $Hf_{0.5}Zr_{0.5}O_2$ thin films. *Ferroelectrics* 569 (1): 148–163.

13 Guo, J., Tao, L., Xu, X. et al. (2024). Rhombohedral R3 phase of Mn-doped $Hf_{0.5}Zr_{0.5}O_2$ epitaxial films with robust ferroelectricity. *Advanced Materials* 36 (47): 2406038.

14 Kaiser, N., Song, Y.-J., Vogel, T. et al. (2023). Crystal and electronic structure of oxygen vacancy stabilized rhombohedral hafnium oxide. *ACS Applied Electronic Materials* 5 (2): 754–763.

15 Wang, Y., Tao, L., Guzman, R. et al. (2023). A stable rhombohedral phase in ferroelectric $Hf(Zr)_{1+x}O_2$ capacitor with ultralow coercive field. *Science* 381 (6657): 558–563.

16 Pešić, M., Fengler, F. P. G., Larcher, L. et al. (2016). Physical mechanisms behind the field-cycling behavior of HfO_2-based ferroelectric capacitors. *Advanced Functional Materials* 26 (25): 4601–4612.

17 Liao, P. J., Chang, Y. K., Lee, Y. H. et al. (2021). Characterization of fatigue and its recovery behavior in ferroelectric HfZrO. In: *2021 Symposium on VLSI Technology*. IEEE. pp. 1–2.

18 Gong, T., Tao, L., Li, J. et al. (2021). $10^5\times$ endurance improvement of FE-HZO by an innovative rejuvenation method for 1z node NV-DRAM applications. In: *2021 Symposium on VLSI Technology*. IEEE. pp. 1–2.

19 Li, X., Wu, J., Tai, L. et al. (2022). Temperature-dependent defect behaviors in ferroelectric $Hf_{0.5}Zr_{0.5}O_2$ thin film: Re-wakeup phenomenon and underlying mechanisms. In: *2022 International Electron Devices Meeting (IEDM)*. IEEE. pp. 32.3.1–32.3.4.

20 Müller, J., Böscke, T. S., Müller, S. et al. (2013). Ferroelectric hafnium oxide: A CMOS-compatible and highly scalable approach to future ferroelectric memories. In: *2013 IEEE International Electron Devices Meeting*. IEEE. pp. 10.8.1–10.8.4.

21 Vishnumurthy, P., Alcala, R., Mikolajick, T. et al. (2024). Ferroelectric HfO_2-based capacitors for FeRAM: Reliability from field cycling endurance to retention (invited). In: *2024 IEEE International Reliability Physics Symposium (IRPS)*. IEEE. pp. 1–10.

22 Yuan, P., Mao, G.-Q., Cheng, Y. et al. (2022). Microscopic mechanism of imprint in hafnium oxide-based ferroelectrics. *Nano Research* 15 (4): 3667–3674.

23 Lee, S. K., Kim, Y. T., Kim, S.-I. et al. (2002). Effects of coercive voltage and charge injection on memory windows of metal-ferroelectric-semiconductor and metal-ferroelectric-insulator-semiconductor gate structures. *Journal of Applied Physics* 91 (11): 9303–9307.

24 Yurchuk, E., Mueller, S., Martin, D. et al. (2014). Origin of the endurance degradation in the novel HfO2-based 1T ferroelectric non-volatile memories. In *2014 IEEE International Reliability Physics Symposium*. IEEE. pp. 2E.5.1–2E.5.5.

25 Gong, N. and Ma, T. P. (2018). A study of endurance issues in HfO_2-based ferroelectric field effect transistors: Charge trapping and trap generation. *IEEE Electron Device Letters* 39 (1): 15–18.

26 Ni, K., Sharma, P., Zhang, J. et al. (2018). Critical role of interlayer in $Hf_{0.5}Zr_{0.5}O_2$ ferroelectric FET nonvolatile memory performance. *IEEE Transactions on Electron Devices* 65 (6): 2461–2469.

27 Muller, J., Polakowski, P., Muller, S. et al. (2016). High endurance strategies for hafnium oxide based ferroelectric field effect transistor. In: *2016 16th Non-Volatile Memory Technology Symposium (NVMTS)*. IEEE. pp. 1–7.

28 Wang, Y., Yang, Y., Jiang, P. et al. (2023). HfO_2-based junctionless FeFET array on FDSOI platform for NAND memory applications. *IEEE Electron Device Letters* 44 (6): 943–946.

29 Wang, Y., Huang, Y., Yuan, H. et al. (2024). Endurance improvement of HfO_2-based FeFETs on FDSOI platform by Al_2O_3 insertion layer and leakage-awareness recover (LAR) strategies. *IEEE Transactions on Electron Devices* 71 (12): 8007–8010.

30 Gong, N. and Ma, T. P. (2016). Why is FE–HfO_2 more suitable than PZT or SBT for scaled nonvolatile 1-T memory cell? A retention perspective. *IEEE Electron Device Letters* 37 (9): 1123–1126.

31 Yurchuk, E., Müller, J., Paul, J. et al. (2014). Impact of scaling on the performance of HfO_2-based ferroelectric field effect transistors. *IEEE Transactions on Electron Devices* 61 (11): 3699–3706.

32 Ali, T., Kühnel, K., Czernohorsky, M. et al. (2020). A study on the temperature-dependent operation of fluorite-structure-based ferroelectric HfO_2 memory FeFET: Pyroelectricity and reliability. *IEEE Transactions on Electron Devices* 67 (7): 2981–2987.

33 Muller, J., Boscke, T. S., Schroder, U. et al. (2012). Nanosecond polarization switching and long retention in a novel MFIS-FET based on ferroelectric HfO_2. *IEEE Electron Device Letters* 33 (2): 185–187.

34 Ali, T., Polakowski, P., Riedel, S. et al. (2018). High endurance ferroelectric hafnium oxide-based FeFET memory without retention penalty. *IEEE Transactions on Electron Devices* 65 (9): 3769–3774.

10

Reliability of HfO_2-based Ferroelectric Thin Films and Field-effect Transistors

Min Liao[1], Binjian Zeng[2], Shuaibing Gao[1], and Yichun Zhou[1]*

[1]*School of Advanced Materials and Nanotechnology, Xidian University, Xi'an, China*
[2]*School of Materials Science and Engineering, Xiangtan University, Xiangtan, China*

*Corresponding author: mliao@xidian.edu.cn

10.1 Introduction

In the 1920s, Valasek discovered ferroelectricity while investigating the linear dielectric response, nonlinear dielectric properties, piezoelectric characteristics, pyroelectric phenomena, and other macroscopic physical properties of Rochelle salt single crystals. This seminal discovery initiated extensive explorations into ferroelectrics.[1–4] Today, ferroelectric materials play a crucial role in modern technology due to their unique properties. Their primary applications include piezoelectric transducers for sensors and actuators, nonvolatile ferroelectric memories, electro-optic devices, energy harvesting systems, and so on.[5–7] As a highly significant category of applications, ferroelectric memories, proposed as early as the 1950s, have garnered considerable research interest and widespread attention. Among these, ferroelectric random access memory (FeRAM) and ferroelectric field-effect transistor (FeFET) are the two most representative devices. The basic unit of FeRAM typically consists of one ferroelectric capacitor and one metal–oxide–semiconductor field-effect transistor (MOSFET). FeRAM, initially commercialized in the 1980s, has established itself as a prominent nonvolatile memory technology with broad applications, including the Internet of Things (IoT), automotive electronics, industrial control systems, and medical devices. However, the widespread adoption of FeRAM has been constrained by two significant limitations: its relatively high manufacturing costs and comparatively low storage density. By integrating ferroelectric thin films into the gate structure of conventional MOSFETs, FeFET achieves a compact one-transistor (1T) architecture, offering the potential for enhanced storage density and improved scalability. As shown in Figure 10.1, the working principle of FeFET relies on the switchable ferroelectric polarization to modulate the inversion or accumulation at the semiconductor surface, thereby controlling the on-state and off-state of the device and enabling the storage of binary data ("0" and "1").[8] In addition to inheriting the key features of FeRAM, such as non-volatility, low power consumption, and high-speed operation, FeFET also offers the advantage of non-destructive readout. However, unlike FeRAM, FeFET has yet to achieve industrialization. This is primarily due to the complex interfacial effects associated with ferroelectric thin films in FeFET. For instance, during the fabrication process, interfacial reactions occur between ferroelectric thin films such as lead zirconate titanate ($PbZr_xTi_{1-x}O_3$, PZT) or strontium bismuth tantalate ($SrBi_2Ta_2O_9$, SBT) and silicon substrate. These reactions lead to undesirable element diffusion (e.g., Pb or Bi), which significantly deteriorates the device's electrical performance. Additionally, the presence of interfacial layers induces a large depolarization field (E_{dep}). Given the relatively small coercive fields of PZT and SBT, this depolarization effect accelerates polarization degradation, ultimately impairing the retention characteristics of the device. Another critical limitation is the difficulty in scaling down the thickness of PZT and SBT films. Thicker ferroelectric layers are incompatible with transistor integration, particularly in modern semiconductor devices that demand ever-smaller feature sizes.

In 2011, Böscke et al. first reported that Si-doped HfO_2 thin films exhibited ferroelectricity. Prior to this discovery, HfO_2 thin films had already been widely adopted as high-k dielectric layers in MOSFETs for advanced chip manufacturing. The revelation of their ferroelectric properties sparked significant excitement, positioning HfO_2-based ferroelectric (FE-HfO_2) thin films as a highly promising material for FeFET fabrication and underscoring its potential to overcome the technical challenges associated with FeFETs based on PZT and SBT.[9,10] Beyond the fabrication advantages, FE-HfO_2 thin films also surpass PZT and SBT in several critical material properties, making them more suitable for FeFET applications. These advantages include a larger band gap, a higher coercive field, a relatively lower dielectric constant, excellent thermal stability, and a low crystallization temperature.[11–16]

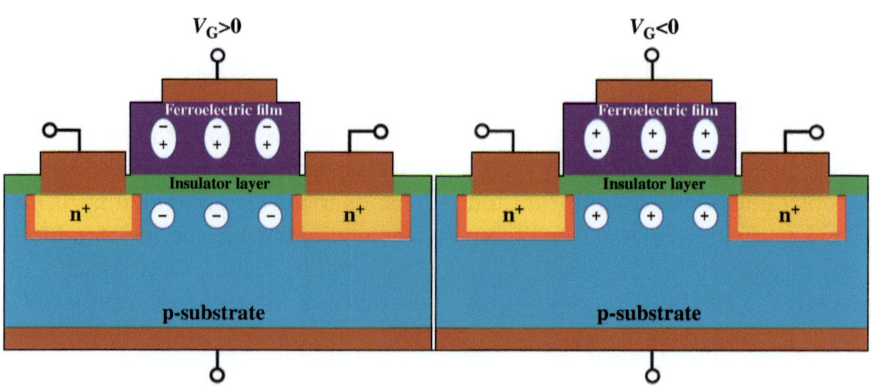

Figure 10.1 The basic structure and operation principle of FeFET.

More importantly, FE-HfO_2 thin films retain robust ferroelectricity and switchable polarization even when scaled down to a few nanometers in thickness.[17–19] These characteristics make the fabrication of HfO_2-based FeFETs relatively straightforward, as their manufacturing process is highly compatible with the start-of-the-art complementary metal–oxide–semiconductor (CMOS) technologies. As a result, they have rapidly enabled the successful tape-out of high-density memory arrays. Despite these significant breakthroughs in FeFET manufacturing using FE-HfO_2 thin films, the device reliability remains a critical challenge that must be addressed before commercialization.[20,21]

In this chapter, we will provide a critical review of reliability aspects, from fundamental material properties of FE-HfO_2 thin films to their operational performance in FeFET devices. We will explore the factors contributing to reliability issues, the underlying mechanisms, potential solutions, and provide an outlook on the future of HfO_2-based FeFETs. The chapter is organized as follows: the first section discusses the fundamental material properties of FE-HfO_2 thin films; the second section thoroughly examines the reliability of FE-HfO_2 thin films; the third section focuses on the operational reliability of HfO_2-based FeFETs; and the fourth section offers a forward-looking perspective on the future of HfO_2-based FeFETs.

10.2 Fundamentals of Ferroelectric HfO_2

10.2.1 Ferroelectricity in HfO_2

Ferroelectricity is a property of certain dielectric materials that exhibit spontaneous electric polarization, which can be reversed by an applied electric field. The polarization-electric field (*P-E*) hysteresis loop serves as a crucial indicator for verifying ferroelectric properties in thin films.[9,22,23] Figure 10.2a displays a typical *P-E* ferroelectric hysteresis loop, demonstrating several key features: (1) nonlinear polarization evolution: the polarization increases nonlinearly with electric field until reaching saturation (P_s), beyond which further electric field increase yields a linear polarization increase dominated by dielectric response, (2) remanent polarization (P_r): when the electric field returns to zero, the material retains a nonzero polarization ($\pm P_r$), measured at the *y*-axis intercepts, and (3) coercive field (E_c): the *x*-axis intercepts represent the critical field ($\pm E_c$) required to switch polarization direction. P_r and E_c are the key parameters for information storage, determining the discrimination of states "0" and "1" and the operating voltage, respectively. The phenomenon of ferroelectricity was successfully induced in HfO_2 by employing a variety of elemental doping techniques. This has led to the emergence and rapid expansion of a diverse family of HfO_2-based ferroelectric materials. Although the P_r of FE-HfO_2 thin films is smaller than that of PZT thin films, it remains adequate for the efficient operation of FeFET devices. Meanwhile, the relatively large E_c in FE-HfO_2 thin films, typically exceeding 1 MV/cm, significantly enhances the retention performance of FeFET devices. Importantly, this large E_c does not lead to an increase in the operating voltage, thanks to the nanoscale thickness.

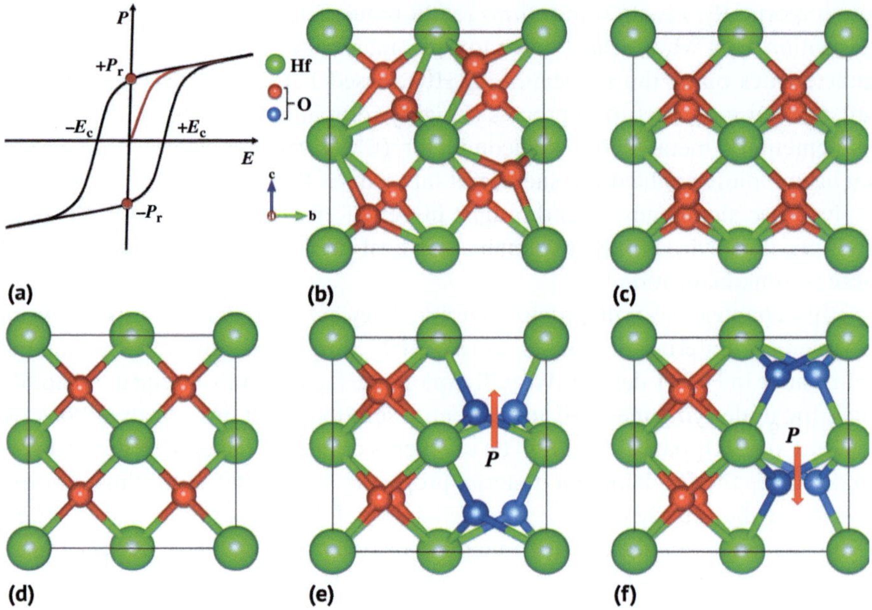

Figure 10.2 (a) Typical *P-E* loop of a ferroelectric material. The crystal structures of HfO_2 with different phases: (b) monoclinic, (c) tetragonal, (d) cubic, (e) and (f) the non-centrosymmetric orthorhombic phase, the red arrow indicates the direction of polarization.

10.2.2 Origin of Ferroelectricity in HfO_2

In general, there are three thermodynamically stable phase structures in crystalline HfO_2, namely the monoclinic phase (M-phase, space group $P2_1/c$), the tetragonal phase (T-phase, space group $P4_2/nmc$), and the cubic phase (C-phase, space group $Fm\bar{3}m$), as shown in Figure 10.2b–d.[24–26] However, all these phases are centrosymmetric and consequently lack spontaneous polarization or ferroelectricity. The ferroelectricity in HfO_2-based thin films is primarily attributed to the presence of the non-centrosymmetric orthorhombic phase (O-phase, space group $Pca2_1$, Figure 10.2e,f).[27,28] This phase is known to exhibit ferroelectric properties due to the movement of tricoordinate oxygen atoms along the *c*-axis, which results in changes in the polarization state. This ferroelectric orthorhombic phase exists as a metastable intermediate state. Park et al. demonstrated that the direct phase transition from tetragonal/cubic to monoclinic phases requires substantial energy, leading to the formation of the orthorhombic phase as a transitional structure.[24] However, the orthorhombic phase remains energetically less favorable than the monoclinic phase, resulting in its metastable nature.[25] Additionally, other phases like the rhombohedral phase (R-phase, space group R3m) have also been identified as contributing to ferroelectric behavior. But the R3m phase is also metastable and requires specific conditions to be stabilized.

10.3 Reliability Aspects of Ferroelectric HfO_2

Given the reliability concerns associated with HfO_2-based FeFETs, this section will provide an overview of the key reliability challenges observed in FE-HfO_2 thin films, including wake-up effect, polarization fatigue, imprint effect, leakage current, and dielectric breakdown.

10.3.1 Wake-up Effect

Wake-up effect refers to the phenomenon where the remanent polarization of FE-HfO_2 thin films increases with the number of electric field cycles before the onset of polarization fatigue.[12,29] This phenomenon indicates that a critical number of switching cycles must be applied to activate optimal ferroelectric performance. In HfO_2-based FeFETs, this manifests as a small memory window (MW) before wake-up cycling due to the pinched *P-E* hysteresis loop (Figure 10.3). Following wake-up treatment, the initially pinched *P-E* hysteresis loop exhibits significant opening, accompanied by the merging of the two distinct transient current peaks in the current-voltage (*I-V*) characteristic into a single dominant peak.[29] The wake-up phenomenon has attracted significant research attention, with extensive studies conducted to elucidate the fundamental physical mechanisms underlying this characteristic behavior. Current understanding suggests the effect arises from complex interactions between defect dynamics, phase transformations, domain structures, and doping effects.

1. Defect-mediated mechanism: Pešić et al. proposed that the wake-up effect involves redistribution of existing defects rather than generation of new ones, with oxygen vacancy diffusion identified as the primary driver for enhanced

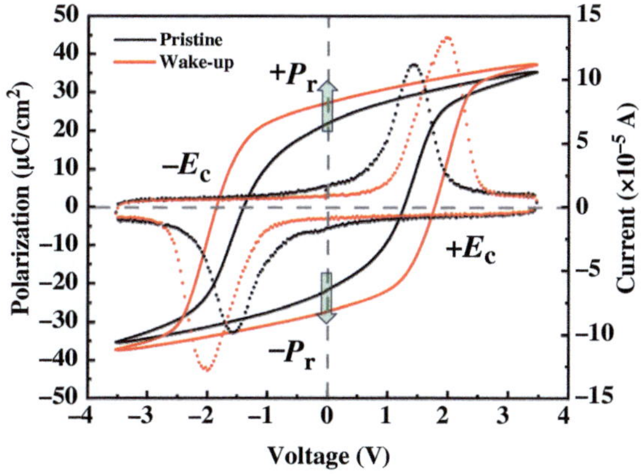

Figure 10.3 *P-V* and *I-V* loops before and after wake-up cycling.

phase transition and remanent polarization.[29] Numerous studies have corroborated that the redistribution of defects (e.g., oxygen vacancies) can result in the depinning of domains, which are the fundamental origin of the wake-up effect.[30–37] This view is also supported by Mehmood et al., who identified two distinct wake-up stages: initial charge redistribution and ferroelectric domain depinning, followed by interface-related phase transformation or degradation after 1×10^4 cycles.[38] Lee et al. reported that oxygen vacancies and trapped charges predominantly originate from dipole defects or domain pinning sites at ferroelectric/electrode interfaces. Their study revealed that the pinning and depinning processes of domains are strongly governed by the local distribution of point defects, which directly hinder domain wall propagation.[39]

2. Phase transformation mechanism: Pešić et al. and Grimley et al. directly observed through transmission electron microscopy the field-cycling-induced transition from tetragonal/monoclinic to orthorhombic phases.[29,40] These structural changes, occurring predominantly during the initial cycling stage as confirmed by *P-V* and piezoresponse force microscopy (PFM) measurements, correlate strongly with the emergence of ferroelectric properties.[41]
3. Domain dynamics: advanced characterization by Zhang et al. using integrated differential phase contrast scanning transmission electron microscopy (iDPC-STEM) revealed atomic-scale domain evolution during wake-up (Figure 10.4a–d), showing reorganization from 90° to 180° domain walls in $Hf_{0.5}Zr_{0.5}O_2$ (HZO) films.[42]
4. Doping effects: Yu et al. systematically explored the suppression mechanism of the wake-up effect in La-doped HZO ferroelectric thin films using density functional theory (DFT) calculations (Figure 10.4e–g). Their findings revealed that La doping markedly elevates the activation energy (ΔE) of V_O^{2+} migration, thereby substantially impeding oxygen vacancy diffusion and consequently inhibiting the wake-up phenomenon.[43] In contrast, Liu et al. studied the possible mechanism of switching current peak splitting in the pristine state of Al-doped HZO films, and considered that wake-up is more likely to occur in Al-doped HZO films.[34]

Extensive research efforts have been devoted to suppressing the wake-up phenomenon in FE-HfO_2. The majority of studies have employed oxygen-scavenging electrodes composed of oxidation-resistant metals (e.g., TaN, RuO_2, W, and IrO_2) to mitigate the wake-up effect. These oxide electrodes serve dual functions: providing supplemental oxygen to FE-HfO_2, minimizing oxygen vacancy concentrations at interfaces, and enhancing reliability by reducing defect-mediated degradation.[44–47] Besides, several innovative synthesis methods have demonstrated success in producing wake-up-free FE-HfO_2. For example, Kashir et al. found that prolonging the duration of the ozone pulse can effectively reduce oxygen vacancies and carbon concentrations in FE-HfO_2. This treatment promotes the formation of the tetragonal phase and renders a stable polarization during the electric field cycle,

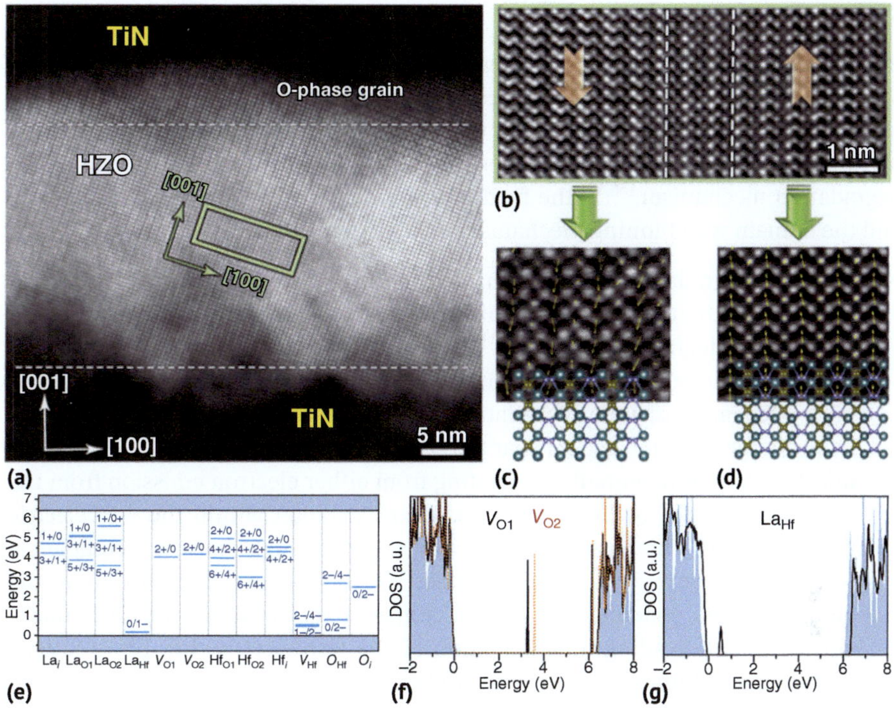

Figure 10.4 (a) The HAADF-STEM image of the woken-up HZO film. (b) iDPC-STEM image acquired from the green square area in (a). The polarization mapping of (c) the left region and (d) the right region in (b). *Source*: [42] / American Association for the Advancement of Science / CC BY 4.0. (e) The calculated charge-transition levels of intrinsic defects and La-related defects. Calculated densities of states (DOSs) for defect models of (f) V_{O1}, V_{O2}, and (g) La_{Hf} in the *f*-HfO_2. *Source*: [43] / with permission of American Physical Society.

ultimately yielding the negligible wake-up effects in the films.[48,49] Xu et al. synthesized wake-up-free yttrium-doped HfO_2 (Y:HfO_2) crystals via a state-of-the-art laser-diode-heated floating zone (LDFZ) technique. The resultant crystals exhibited no interfacial defects capable of inducing phase transitions or domain wall pinning, ensuring intrinsic structural homogeneity.[50] Silva et al. successfully prepared R3m epitaxial strained ZrO_2 films on (111)-oriented Nb:$SrTiO_3$ substrates by ion beam sputtering deposition technology, which did not require any wake-up cycle.[51] Kim et al. reported the adoption of cyclopentadiene-based precursors to obtain the wake-up free HZO ferroelectric films by atomic layer deposition.[52] Complementary approaches have also shown effectiveness in wake-up suppression: Pt interlayer insertion at metal/insulator interfaces, oxygen-ambient annealing processes, and adjusting the annealing process.[53–55] Overall, reducing the formation of oxygen vacancies and promoting the distribution uniformity of orthorhombic phases contribute to the mitigation of wake-up effects in FE-HfO_2.

10.3.2 Polarization Fatigue

Polarization fatigue refers to the phenomenon where the switchable polarization of woken-up FE-HfO_2 progressively decreases with electric field cycling, ultimately leading to functional degradation of the device. Several mechanisms have been revealed to be the causes of polarization fatigue in FE-HfO_2, including the interface degradation mechanism,[56,57] the field-induced phase transition mechanism,[29,58] and the domain wall pinning mechanism,[59,60] among others.

1. Interface degradation mechanism: Lou et al. proposed a significant perspective on the fatigue failure mechanism in ferroelectric thin films by developing an interface degradation model.[61] Their work highlights that charge injection from electrodes and subsequent interface degradation constitute one of the primary mechanisms underlying fatigue failure. When subjected to an electric field, charge injection occurs near the electrode-ferroelectric interface, with the injected charges potentially originating from either electron emission from the electrode material or field-induced carrier excitation within the ferroelectric thin film itself. These accumulated charges form a high-density, incompletely screened charge region at the interface. During polarization switching, the bound charges generate a E_{dep} that enhances the local electric field intensity, ultimately leading to degradation of the ferroelectric thin film's fatigue performance. Building upon this fundamental interface degradation model, numerous researchers have investigated strategies to mitigate or prevent fatigue failure in FE-HfO_2. For instance, Cao et al. demonstrated that optimizing electrode materials could effectively reduce charge injection and consequently enhance the fatigue characteristics of HZO.[62] Other researchers have proposed alternative approaches, such as introducing engineered interfacial layers between the electrodes and ferroelectric thin films.[63–65] These interfacial modifications aimed to improve interface quality by lowering defect density and optimizing electric field distribution, thereby improving fatigue performance.
2. Field-induced phase transition mechanism: as seen in the discussion of wake-up effects in FE-HfO_2, the repeated electrical stimuli could induce the phase transitions from non-ferroelectric phases to ferroelectric phases. Conversely, polarization fatigue in FE-HfO_2 has been attributed to the reverse process – the transformation from ferroelectric phases to non-ferroelectric phases under prolonged electrical stress. For example, Fields et al. employed synchrotron X-ray diffraction (XRD) to conduct an in-depth analysis of phase evolution in HZO films with different electrode materials during wake-up and fatigue processes.[58] Their study revealed that the HZO films with Pt electrodes exhibited significant phase transitions during fatigue, with an increase in monoclinic phase content. Besides, Cheng et al. examined the crystal structures of pristine, woken-up, fatigued, and rejuvenated O-phase grains in atomic layer-deposited ferroelectric HZO films in detail using the high-angle annular dark

field (HAADF)-STEM and annular bright-field (ABF)-STEM techniques.[66] They found that the interfacial T-phase grew as the number of switching cycles increased to a degree that was also dependent on the applied voltage. This interfacial T-phase generated a depolarizing field, which reduced the effective field applied to the remaining O-phase film. As the interfacial T-phase evolves in tandem with the increasing numbers of cycles, the effective field then becomes lower than the critical field, even at the given applied voltage; this is accompanied by the transition from wake-up to fatigue. Because the fatigue in this material was not induced by a reversible phase transition between the polar and antipolar phases, the fatigued sample could then be rejuvenated by applying a voltage that increased the effective field to exceed the critical field.

3. Domain wall pinning mechanism: oxygen vacancies, as the main defects in FE-HfO_2, play a pivotal role in polarization fatigue through domain wall pinning mechanisms. These mobile defects preferentially accumulate at critical locations, including film-electrode interfaces and grain boundaries during electric field cycling, significantly impacting device reliability. Several studies have provided compelling evidence for this mechanism. Huang et al. systematically investigated charge trapping phenomena, demonstrating that injected charges captured at shallow defect potential wells significantly promote domain wall pinning.[60] Their temperature-dependent studies revealed that thermal activation facilitates the release of trapped charges, thereby mitigating pinning effects and providing direct evidence for the reversibility of this fatigue mechanism. Furthermore, Fengler et al. conducted comprehensive defect characterization through impedance spectroscopy and thermally stimulated depolarization current measurements, revealing a substantial increase in both singly-and doubly-charged oxygen vacancy concentrations after electrical cycling.[67] This defect accumulation directly enhanced domain wall pinning, leading to progressive polarization degradation. In contrast, Kim et al. demonstrated an alternative approach using cyclopentadienyl precursors in atomic layer deposition, achieving fatigue-resistant HZO films. This is because the TiN bottom electrode will be oxidized during high-temperature heat treatment, and a large number of oxygen vacancies will be generated at the interface between the TiN electrode and the HZO film.[52]

Significant progress has been made in improving the fatigue resistance of FE-HfO_2 through various material engineering approaches (Figure 10.5a–d). The use of conductive oxide electrodes, such as indium-tin-oxide (ITO), RuO_2, IrO_x, and $SrRuO_3$, has demonstrated remarkable fatigue resistance up to 10^4–10^8 cycles by effectively suppressing oxygen vacancy generation during field cycling.[45,47,56,68] By changing the exposure time of H_2O and oxygen plasma during the deposition process to adjust the concentration of oxygen vacancies, Bao et al. obtained HZO films with good fatigue resistance.[69] Chen et al. used NH_3 plasma treatment to effectively

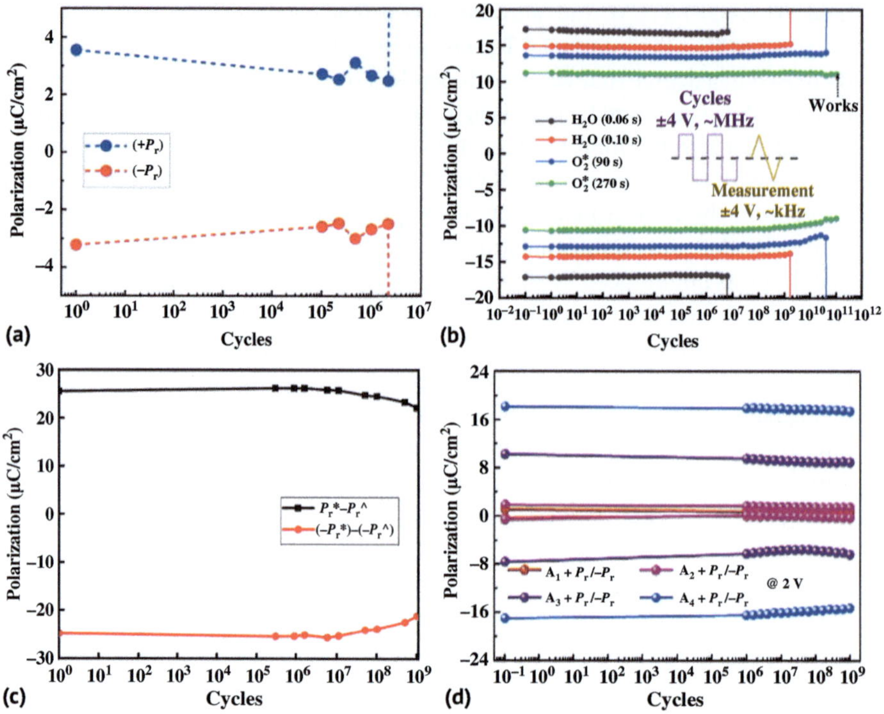

Figure 10.5 (a) The fatigue characteristics of the ITO/HZO structure. *Source*: [68] / with permission of IOP Publishing. (b) The fatigue performance of the HZO samples with different H_2O dose times and O_2 plasma times. *Source*: [69] / with permission of American Chemical Society. (c) The fatigue characteristics of $Hf_{0.85}Ce_{0.15}O_2$ films measured under 2.9 MV/cm with a frequency of 100 kHz. *Source*: [71] / with permission of AIP Publishing LLC. (d) The fatigue properties of HZO films with different Hf/Zr atomic ratios. *Source*: [75] / with permission of Elsevier.

suppress the generation of oxygen vacancies in HZO, which significantly reduced the fatigue effect.[70] Some studies have greatly improved the fatigue performance of FE-HfO_2 by adjusting the doping content of Pr, Sr, La, Ce, Y, etc.[12,15,71–73] Zheng et al. proved that suitable annealing temperature and atmosphere could improve the fatigue performance of CeO_2-HfO_2 films.[74] Yan et al. effectively enhanced the fatigue resistance of HZO thin films through systematic optimization of the Hf/Zr atomic ratio and post-deposition annealing temperature.[75] Hao et al. demonstrated that the composition-graded HZO ferroelectric thin film exhibited improved endurance characteristics compared to the conventional composition-uniform one.[14] Liang et al. demonstrated the highly robust endurance characteristics of HZO by improving the interface quality between TiN electrodes and HZO layer through "in-situ" like atomic layer deposition.[76] Chen et al. reported that an ultra-thin Al_2O_3 layer was inserted into the HZO film by atomic layer deposition to modify the interface, and good polarization performance and stability were obtained.[77]

Furthermore, lowering the operation voltage also contributed to a high fatigue resistance, such as decreasing the E_c.[78] The collective findings demonstrate that optimizing the fatigue endurance of FE-HfO_2 requires comprehensive control over oxygen vacancy dynamics through four key strategies: (i) suppressing vacancy generation, (ii) impeding their field-driven migration, (iii) preventing interfacial accumulation, and (iv) mitigating associated charge trapping effects.

10.3.3 Imprint Effect

Imprint effect refers to a phenomenon where ferroelectric thin films, after being maintained in a particular polarization state for an extended period, require a higher applied reverse electric field to switch to the opposite polarization state. Macroscopically, this manifests as a horizontal shift in the *P-V* hysteresis curve, accompanied by a corresponding displacement of the coercive voltage (V_c). When the imprint effect becomes sufficiently pronounced, it can lead to read/write errors in memory devices. Imprint can be quantitatively defined as the shift in V_c, namely $V_{shift} = (V_c + V_{c-})/2$, where V_{c+} and V_{c-} represent positive and negative coercive voltages, respectively. In FE-HfO_2, imprint effect can be categorized into two types based on its formation conditions: (1) electric field cycling-induced imprint and (2) time-dependent imprint. For electric field cycling-induced imprint, which is primarily observed during wake-up and fatigue studies, researchers have proposed that similarities exist between wake-up/fatigue characteristics and imprint behavior. During repeated electric field cycling, the ferroelectric domains gradually align preferentially in one direction, leading to the development of imprint characteristics. In contrast, the time-dependent imprint usually occurs during polarization retention. Most in-depth research on imprint phenomena in FE-HfO_2 has focused on its time-dependent characteristics. Several mechanisms have been revealed to be responsible for the imprint effect, including charge trapping and de-trapping in the interfacial layer, the redistribution of mobile charges in the dead layer by E_{deps}, and domain wall pinning caused by oxygen vacancies.

1. Interfacial layer effects: Abe et al. proposed the existence of a non-switchable interfacial layer with irreversible spontaneous polarization between the ferroelectric layer and the bottom electrode, which significantly influences imprint characteristics.[79] Tagantsev et al. attributed imprint phenomena to charge injection and subsequent accumulation in the electrode's passive layer adjacent to the ferroelectric film.[80] Zacharaki et al. employed plasma-assisted atomic oxygen deposition to fabricate HZO films on germanium (Ge) substrates. The resulting HZO displayed remarkably weak imprint effects. Additionally, the impacts of annealing processes, specifically flash laser annealing (FLA) and rapid thermal annealing (RTA), on the imprint effects were thoroughly investigated. The drift of V_c over time and temperature suggested that the capture and de-trapping of interfacial oxygen vacancy defects, which

are controlled by the TiN/HZO electrodes, are the primary contributors to the imprint phenomenon.[81] Furthermore, Chen et al. found that the imprint effect in HZO films was tightly related to the thickness of the interfacial layer.[77]

2. Redistribution of mobile charges: Li et al. studied the relationship between imprint effects and remanent polarization by varying the temperature (−23 °C to 127 °C) and concluded that imprint was caused by E_{dep} effects.[82] Specifically, the E_{dep} drove the migration of oxygen vacancies along this field direction towards the interface. This, in turn, led to the pinning effect of ferroelectric domains, thereby forming the imprint phenomenon. Besides, Takada et al. conducted the time-domain measurements of imprint, distinguishing the charge redistribution processes from the origin. In addition, there are correlations between the amount of redistributed charge and the dielectric relaxation of the HZO film deposited by atomic layer deposition, indicating that the mobile charge in the HZO film contributes to the time-dependent imprint.[83]
3. Domain wall pinning: Fengler et al. performed the imprint test on HZO films after wake-up to analyze the charge diffusion without application of an external electrical field. The observed imprint caused by charge alignment at domain walls followed an Arrhenius behavior with an activation energy of around 1 eV.[84] Besides, Higashi et al. performed the switching time measurements on the HZO films with and without imprint effects, and corresponding fitting results with the nucleation-limited switching (NLS) model strongly indicated that imprint was caused by domain-seeds-pinning.[85]

Currently, various strategies have been developed to suppress or mitigate the imprint effects in FE-HfO_2. They could be sorted into two categories: (i) decreasing oxygen vacancy concentration[69,70,86] and (ii) improving interface quality, such as reducing the interface trap density.[87]

10.3.4 Leakage Current

Leakage current serves as a crucial performance metric for ferroelectric thin films, with significant implications for device reliability. Excessive leakage current critically compromises device performance through three primary failure mechanisms: (i) retention degradation; (ii) functional failure; (iii) breakdown. In FE-HfO_2, a specific type of polarization fatigue, typically manifested through breakdown or failure, is closely associated with leakage current. As the number of field cycles increases, the remanent polarization of FE-HfO_2 gradually diminishes, while the leakage current surges by more than 50 times.[40,59,66,88] Meanwhile, driven by the demand for scaling memory devices, FE-HfO_2 is often integrated into devices with ultrathin thicknesses.[89] However, the polarization switching of these films requires a significantly high E_c, often exceeding 1 MV/cm, which operates precariously close to the breakdown limit. Under these circumstances, leakage current emerges as an inevitable issue, severely impacting the stability, endurance, and overall performance of memory devices. Therefore, understanding the leakage current mechanisms of FE-HfO_2 is crucial for their practical application in memory devices.

Several competing models have been proposed to explain the leakage mechanisms, including the Poole–Frenkel emission mechanism,[90] tunneling mechanisms,[91,92] and the Schottky emission mechanism.[93] Overall, the leakage mechanism can be influenced by various factors such as the phase, crystalline quality, defect type and concentration, and film thickness. Correspondingly, the strategies of reducing oxygen vacancies, suppressing defect generation, and inserting the interlayers have been developed to reduce leakage current.[69,94,95] Besides, interface engineering has been proved to effectively reduce leakage current. For example, Kashir et al. demonstrated significant leakage current suppression (~20× reduction) by strategically inserting a 10 nm Pt interlayer at the W/HZO interface. This improvement originated from two synergistic mechanisms: (i) enhanced Schottky barrier height at the Pt/HZO interface ($\Delta\Phi_B \approx 0.3 - 0.5$ eV increase vs. W/HZO) and (ii) effective prevention of interfacial oxide (WO_x) formation at the electrode interface, thereby eliminating a major leakage pathway.[49]

10.3.5 Dielectric Breakdown

Dielectric breakdown represents a critical failure mode in insulating materials, characterized by an abrupt transition from high resistivity to conductive behavior when the applied electric field exceeds a threshold value. This phenomenon primarily manifests through two distinct mechanisms: hard breakdown and soft breakdown. Hard breakdown is an abrupt, irreversible failure that typically results in permanent damage, such as burning or complete device malfunction. In contrast, soft breakdown exhibits a progressive deterioration in dielectric performance, typically marked by gradual leakage current increase, parametric drift, or localized conduction paths, without immediate catastrophic consequences.

Considering the structural similarity between HfO_2-based FeFETs and conventional MOSFETs, the breakdown mechanisms of the high-k dielectric HfO_2 layer in MOSFETs should serve as the primary reference for understanding breakdown phenomena in FE-HfO_2 in FeFETs. Kim et al. provided a comprehensive review of breakdown mechanisms in high-k dielectrics, systematically differentiating between hard breakdown and soft breakdown. Their analysis identified key factors influencing breakdown, including defect density, defect size, current type (AC and DC), voltage frequency, film thickness, and interface effect.[96] While soft breakdown behavior in high-k dielectrics resembled that observed in ultra-thin SiO_2 gate dielectrics, a critical distinction lays in the post-breakdown leakage current magnitude.[96,97] Since the high-k dielectric was generally a bilayer structure (comprising an interface layer and a bulk high-k layer), the radius of the soft breakdown path or the origin of the soft breakdown may be very different. Furthermore, the HfO_2 gate layer exhibited polarity-dependent tunneling conductance and defect generation dynamics. The interface layer played a pivotal role in soft breakdown initiation, which may arise from an enhanced defect generation rate due to high local stress fields or a reduced critical defect density attributable to its thinner dimension.[96,98–101] In contrast, the hard breakdown was mainly affected by the defect generation and charge trapping in bulk HfO_2. The difference between the soft and hard breakdown of HfO_2 can also be traced to the interface effects. The interface layer experienced significantly greater charge accumulation and electric field intensity compared to the bulk HfO_2 layer under substrate injection. Conversely, gate injection induced distinct defect formation modes and charge effects between the bulk HfO_2 and the interface layer. These differences depended on the composition and thickness of the interface layer. The disparate physicochemical properties of the interface layer and the bulk layer of HfO_2 ultimately dictated their respective breakdown behaviors.[96,102–105]

A critical challenge in FE-HfO_2 stems from the relatively small margin between the E_c (~1 MV/cm) and dielectric breakdown field (~5–6 MV/cm).[106–108] A notable manifestation of this limitation was demonstrated by Pešić et al., who observed that cycling-induced oxygen vacancy defects formed conductive leakage paths in Sr-doped HfO_2 ferroelectric thin films. This defect-mediated conduction mechanism ultimately precipitated dielectric breakdown prior to the onset of polarization fatigue.[109]

10.4 Operational Reliability of HfO_2-Based FeFETs

This section first elucidates the fundamental operational principles of HfO_2-based FeFETs. Subsequently, it provides a systematic examination of their primary reliability challenges, with particular emphasis on program/erase (P/E) endurance characteristics and data retention performance. Furthermore, a dedicated analysis of their radiation tolerance will be presented, evaluating their potential suitability for mission-critical applications in aerospace, military, and nuclear environments.

10.4.1 Principle of HfO_2-Based FeFETs

Figure 10.6a illustrates the operational mechanism of the FeFET memory device. During the erase operation, application of a positive gate voltage ($+V_G$) pulse induces downward polarization, resulting in a leftward shift of the transfer characteristic ($I_D - V_G$) curve. This corresponds to the low threshold voltage (V_{TH1}) state. Conversely, a negative programming pulse ($-V_G$) generates upward polarization, producing a rightward curve shift to the high threshold state (V_{TH2}). The voltage difference between these two states constitutes the MW, which serves as the fundamental basis for binary data storage ("0" and "1"). The MW represents a crucial performance metric for FeFET devices, with its theoretical maximum value being twice the coercive voltage ($2V_c$) of the ferroelectric thin film.[110] The read operation is accomplished through application of small bias voltages to both gate and drain terminals, enabling nondestructive state detection via drain current measurement. Compared to FeRAMs, the FeFET offers several distinct advantages: (1) simplified device structure, (2) inherent nondestructive readout capability, and (3) compatibility with 3D NAND Flash fabrication processes. These characteristics enable FeFET technology to achieve high-density nonvolatile memory integration while maintaining fast read/write speeds and low power consumption, making it particularly promising for next-generation memory applications.[111]

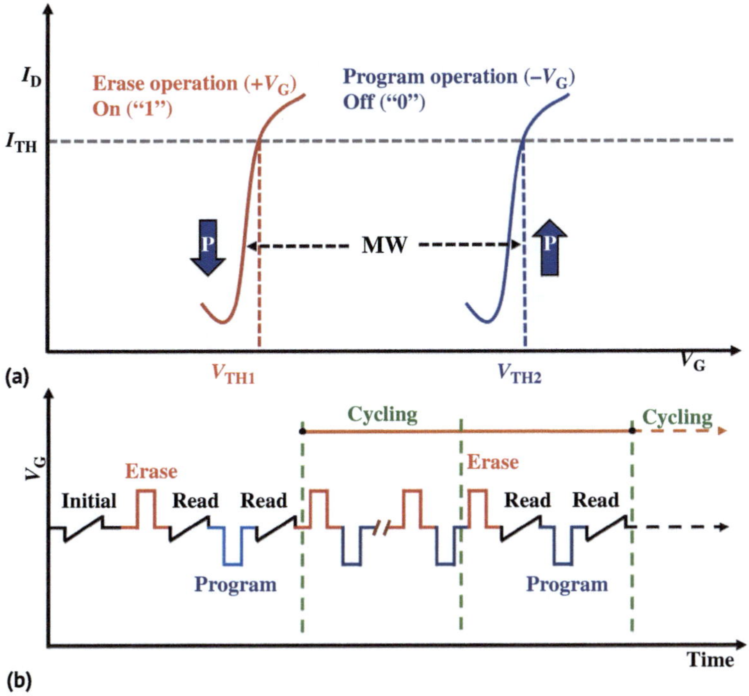

Figure 10.6 (a) Schematic diagram of the MW of the FeFET. (b) Test sequences used for MW and endurance measurements.

10.4.2 Endurance

A potential problem with HfO_2-based FeFET memory is its limited endurance, manifested through progressive degradation of the MW during repetitive P/E cycling (the test sequence used for endurance measurements is shown in Figure 10.6b).[112] Current experimental results demonstrated that the endurance performance of HfO_2-based FeFETs still fell significantly short of the International Roadmap for Devices and Systems (IRDS) target of 10^{14} cycles. The most optimistic reported endurance characteristics remain in the range of 10^{10} to 10^{12} cycles, indicating substantial room for improvement to meet the expectations.[113,114] Gong et al. proposed that endurance failure in HfO_2-based FeFETs occurred through two distinct mechanisms: trapping and generation of interface traps.[20] In a complementary study, Yurchuk et al. systematically investigated the cycling degradation mechanisms in HfO_2-based FeFETs. Their research revealed that the observed reduction in MW during endurance cycling primarily originated from degradation in the interface layer, rather than from FE-HfO_2 degradation. Significantly, the researchers established a direct correlation between: (1) the increasing interface trap density (as evidenced by rising charge pump current) and (2) the escalating gate leakage current during cycling. These findings strongly supported the conclusion that interface layer degradation constituted the dominant factor in device endurance limitation.[115] Zeng et al. systematically investigated the P/E cycling degradation

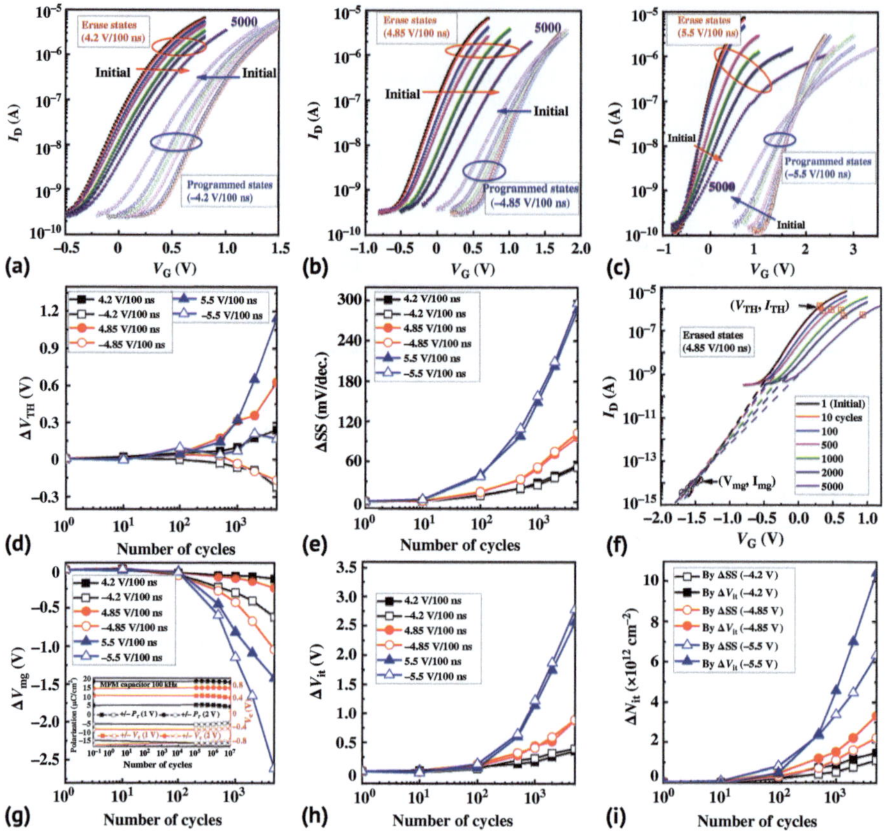

Figure 10.7 Evolutions of I_D-V_G characteristics with (a) +/−4.2 V/100 ns, (b) +/−4.85 V/100 ns, and (c) +/−5.5 V/100 ns cycling. The corresponding (d) ΔV_{TH} and (e) ΔSS values for (a)–(c). (f) Subthreshold-current curves for the erased states (−4.85 V/100 ns). (g) ΔV_{mg} evolutions extracted from (a)–(c); inset in (g): Evolutions of $\pm P_r$ and $\pm V_c$ with cycling number under 1 V/100 kHz and 2 V/100 kHz. (h) ΔV_{it} and (i) ΔN_{it} evolutions extracted from (a)–(c). (I_D: drain current, V_G: gate voltage, ΔV_{TH}: threshold-voltage shift, ΔSS: subthreshold-swing change, ΔV_{mg}: midgap-voltage shift, ΔV_{it}: voltage shift due to interface-trap generation, and ΔN_{it}: interface-trap density change relative to the initial state). *Source*: [116] / with permission from IEEE.

mechanism in HZO-based FeFETs using the mid-gap voltage method (Figure 10.7). It was revealed that charge trapping dominated the mid-gap voltage shift in the gate stack, and the asymmetric threshold voltage variations between programmed and erased states originated from differential mid-gap voltage shift magnitudes. Notably, the MW degradation was found to correlate directly with this shift disparity, which was fundamentally influenced by interface trap generation.[116] Complementing these findings, Yurchuk et al. employed a single-pulse I_D-V_G characterization to elucidate the interplay between charge trapping and ferroelectric polarization switching. Their work demonstrated that the parasitic traps produced charge trapping during device operation, while the ferroelectric polarization in Si-HfO_2 layer substantially enhanced trapping rates (the characteristic trap time in the ferroelectric gate stack is shorter than that in the standard high-k metal gate stack). This charge trapping/de-trapping process was identified as the primary mechanism for interface layer degradation, establishing parasitic charge trapping as the fundamental limitation for the HfO_2-FeFET endurance characteristics.[117] Further mechanistic studies by Ichihara et al. involved real-time current measurements during storage operations to quantify spontaneous polarization and interface trap charge. Their results indicated that endurance limitations arise from both polarization reduction during cycling and increased polarization-shielding trap sites at the ferroelectric/interface layer boundary.[118] These findings underscored the critical importance of suppressing charge injection through interfacial SiO_2 layers for achieving enhanced endurance.[119] Theoretical modeling approaches have also contributed to understanding these phenomena. Zagni et al. developed an analytical MW expression for metal–ferroelectric–insulator–semiconductor (MFIS) FeFETs that explicitly incorporated interface and oxide trap contributions, providing a quantitative framework for endurance evaluation.[120] Concurrently, Hao et al. revealed the crucial role of spatial trap distribution at the Si/SiO_2 interface in governing polarization switching dynamics (Figure 10.8). Their work established that the coupling effects of interface charge trapping and polarization switching fundamentally determined the evolution of $I_D - V_G$ characteristics.[121]

Multiple effective approaches have been developed to improve the P/E cycling degradation in HfO_2-based FeFETs: (1) optimizing operational conditions: employing lower-voltage operations with unsaturated polarization and the smaller E_c maintained

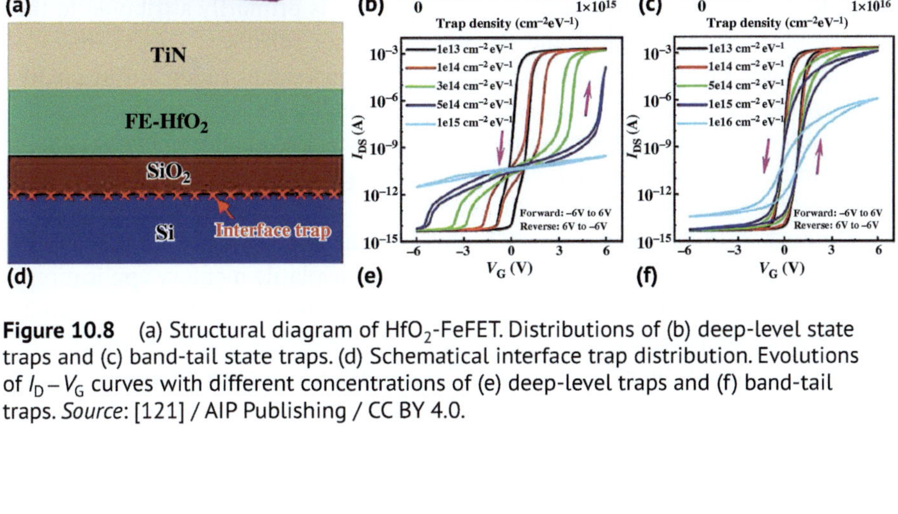

Figure 10.8 (a) Structural diagram of HfO_2-FeFET. Distributions of (b) deep-level state traps and (c) band-tail state traps. (d) Schematical interface trap distribution. Evolutions of $I_D - V_G$ curves with different concentrations of (e) deep-level traps and (f) band-tail traps. *Source*: [121] / AIP Publishing / CC BY 4.0.

a smaller MW while substantially reducing the charge trapping. This approach minimized electric field stress across the interface layer,[122] (2) interface engineering: implementation of high-k oxides as interface layers or seed layers leveraged their large valence band offset (ΔE_v) to reduce voltage drop and effectively suppress hole trapping on the interface layers.[123] More importantly, specific ZrO_2 seed layers enhanced the thermodynamic stability of the orthorhombic ferroelectric phase in HfO_2 thin films.[124–127] Besides, NH_3 plasma treatment effectively passivated oxygen vacancies while simultaneously inhibiting Hf/Zr interdiffusion, thereby significantly improving the interface layer quality,[70,128] (3) material optimization: strategic element doping served to enhance ferroelectric properties, reduce E_c, and enable lower operational voltages while maintaining high polarization.[72,74] Appropriate gate metal selection reduced oxygen vacancy formation,[129] and (4) device structure design: the design of metal–ferroelectric–metal–insulator–semiconductor (MFMIS) structure could effectively improve the endurance.[130,131] Kim et al. proposed a new approach of a spherical recess channel FeFET to improve endurance performance.[132] These strategies collectively addressed the fundamental mechanisms of degradation while optimizing device performance parameters. The approaches ranged from material-level modifications to architectural innovations, providing a comprehensive toolkit for endurance enhancement in HfO_2-FeFET memory technologies.

10.4.3 Retention

Retention in FeFETs is a critical metric that reflects the ability to maintain memory states (stored data) over time without requiring power refreshment. This nonvolatile memory characteristic is essential for the operation of FeFETs and is quantified by measuring how long the device can reliably preserve its threshold voltage states under specific environmental conditions. The retention performance is primarily governed by two critical factors: leakage current and E_{dep}. The leakage current induces local compensation charges through the charge trapping processes, which gradually screen the ferroelectric polarization. Meanwhile, the E_{dep} emerges as a consequence of incomplete charge screening, which is primarily attributed to the presence of the interface layer and the semiconductor substrate. This incomplete screening leads to domain back-switching and a consequent reduction in polarization within the ferroelectric layer.[20,126] Additionally, the operational temperature significantly impacts the retention performance. Elevated temperatures tend to accelerate retention loss by thermally activating depolarization processes and enhancing trap-assisted leakage pathways.[113]

Traditional FeFETs employing PZT and SBT have long faced challenges in achieving the 10-year retention time required for nonvolatile memory applications. However, the advent of HfO_2-based FeFETs has breathed new life into this technology, with demonstrations of extrapolated 10-year retention time in 28 nm FeFETs. Gong et al. explored the reasons behind the significantly longer retention time of HfO_2-based FeFETs compared to their PZT and SBT counterparts and identified two key contributing factors: (1) high E_c and comparable E_{dep}: FE-HfO_2 exhibits a very high E_c, more than one order of magnitude higher than those of PZT and SBT. Additionally, the E_{dep} in FE-HfO_2 is comparable to its E_c, which substantially reduces the rate of the fast decay component of memory retention. In contrast, both SBT and PZT have much higher E_{dep}/E_c ratios than FE-HfO_2, leading to significantly larger retention loss due to the E_{dep} and (2) lower trap concentration: FE-HfO_2 has a much lower trap concentration compared to SBT and PZT. This lower trap density results in significantly less trapping-related retention loss, thereby contributing to the superior retention performance of HfO_2-based FeFETs.[133] Nevertheless, several new retention loss phenomena have been observed in HfO_2-based FeFETs, as detailed below. Notably, the observed asymmetric retention behaviors between high and low V_{TH} states suggested that the E_{dep} plays a secondary role in HfO_2-based FeFETs.[119] Experimental measurements revealed that interface traps at the FE-HfO_2/interface layer boundary compensate approximately 90% of the polarization charge during retention tests, with the primary retention loss mechanism being polarization reduction induced by E_{dep}.[118] Additional findings included: gate leakage-induced charge trapping as the dominant factor in long-term retention degradation;[111] progressive polarization loss due to imprint effects;[134] superior retention characteristics of memory states with unsaturated polarization compared to those with saturated polarization, attributed to a smaller E_{dep}/E_c and less charge injection (Figure 10.9).[135] In summary, while HfO_2-based FeFETs demonstrate relatively superior retention characteristics, practical applications require minimization of both E_{dep} effects and leakage current to achieve optimal performance.

10.4.4 Radiation Tolerance

Radiation tolerance is a crucial figure of merit for nonvolatile memory devices, particularly for applications in radiation-harsh environments such as space and nuclear systems. The primary radiation-induced degradation mechanisms consist of total ionizing dose (TID) effects, single-event effects (SEE), and displacement damage.[136–138]

Extensive studies have established the exceptional radiation hardness of FE-HfO_2 under various radiation conditions.[139–146] For example, the HZO ferroelectric thin film exhibited remarkable radiation stability, demonstrating essentially unchanged surface morphology, leakage current density, remanent polarization, relative dielectric constant, dielectric loss, endurance, and retention characteristics following ^{60}Co γ-rays irradiation at an ultra-high total dose of 10 Mrad(Si).[138,139] Huang et al. also found that γ-ray irradiation had little effect on the electrical properties of Y-doped HfO_2 ferroelectric thin films.[138] Wang et al. further proved that proton radiation had no effect on the properties of Y-doped HfO_2 ferroelectric thin films.[141] While FE-HfO_2 demonstrated excellent intrinsic radiation tolerance, the degradation mechanisms of HfO_2-based FeFETs under ionizing radiation exhibited fundamentally different characteristics. Chen et al. systematically investigated

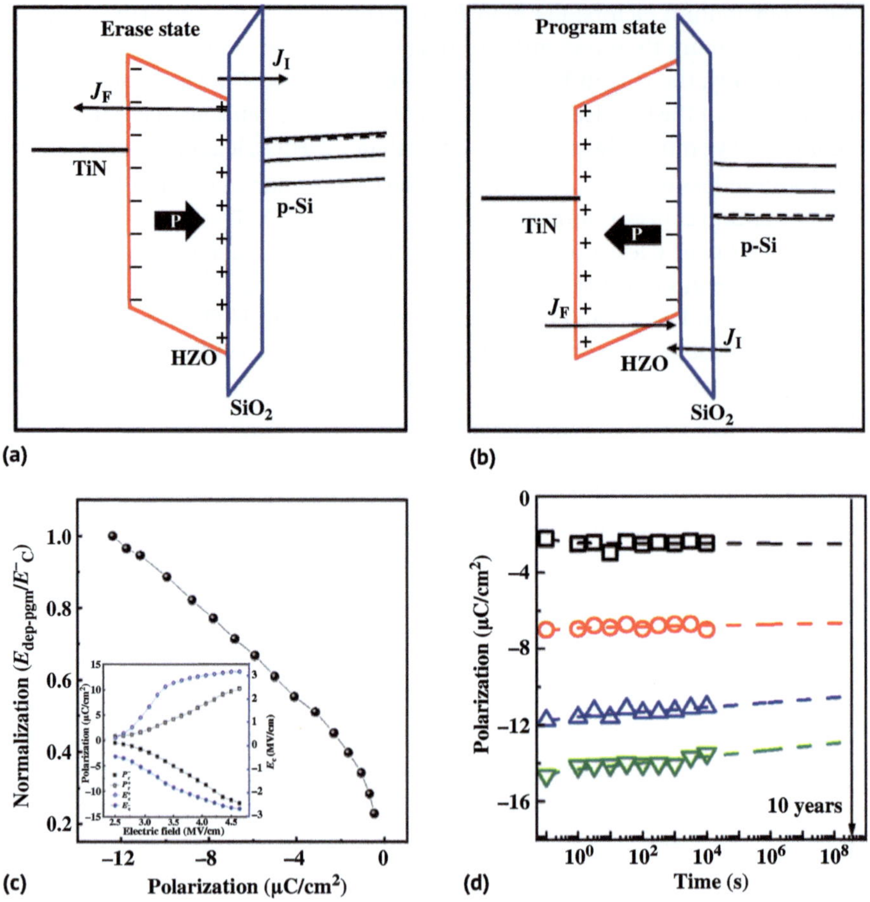

Figure 10.9 Schematic band diagrams of the HZO-based FeFETs for the (a) erased and (b) programmed states. (c) Normalized E_{dep}/E_c vs. ($-P_r$) of the FeFETs. The inset in (c) shows the polarization and E_c characteristics of the HZO ferroelectric thin film vs. applied electric field. (b) Retention properties of various polarization states of the HZO ferroelectric thin films. *Source*: [135] / Elsevier / CC BY 4.0.

the impact of γ-ray radiation on the performance of HZO-based FeFET devices. They found that the remanent polarization decreased with radiation dose due to the radiation-induced oxygen vacancy and lattice distortion. However, for devices with 300 krad dose, the MW still maintained as in the nonirradiated devices and showed stable 10-year retention characteristics. The most significant radiation-induced degradation was observed in endurance performance, primarily caused by hole trap accumulation from radiation-generated defects.[139] Dai et al. performed a comprehensive study of the effects of TID on the erased/programmed states of HfO_2-based FeFETs, specifically analyzing the evolutions of erased and programmed states with irradiation, as illustrated in Figure 10.10a–h. Through the analysis of V_{TH} drift, subthreshold swing evolution, and ferroelectric polarization

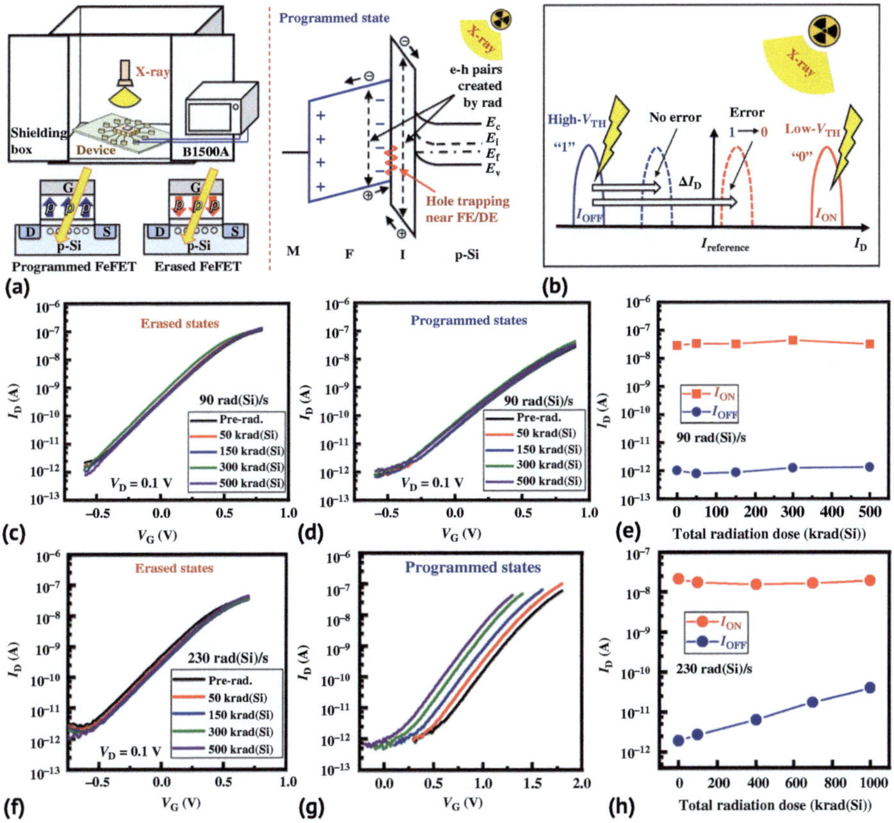

Figure 10.10 (a) The schematic diagram of FeFETs programming and erasing under irradiation; (b) Schematic radiation-induced bit error (i.e., I_D evolution) in the FeFET memory cell. Evolutions of $I_D - V_G$ characteristics of HZO-based FeFETs under irradiation with a dose rate of (c)–(d) 90 rad(Si)/s and (f)–(g) 230 rad(Si)/s. (e)–(h) Corresponding I_{ON} and I_{OFF} values at different radiation doses. (I_{ON} and I_{OFF} denote the drain current in the erased and programmed states, respectively, measured at the read voltage). *Source*: [146] / with permission of American Chemical Society

stability, they revealed that the degradation of the erased state was caused by oxide-trapped charges rather than interface degradation or polarization switching.[146] Furthermore, interface engineering has been demonstrated as a crucial approach for enhancing MW, endurance, and radiation tolerance in HfO_2-based FeFETs. For instance, Xiao et al. reported that MW and reliability with respect to endurance could be improved by employing ZrO_2 seed layers.[125] Liu et al. demonstrated that inserting an HfO_2 seed layer between the HZO ferroelectric layer and the SiO_2 interface layer significantly improved the immunity of the FeFETs to γ-ray radiation.[147] Peng et al. demonstrated the HZO-based FeFETs with AlON interface layers exhibiting stable MW and desirable retention performance independent of radiation dose in the range between 300 krad and 1 Mrad (Si).[148]

Overall, extensive research efforts to date have primarily focused on the TID effects on HfO_2-based FeFETs, with consistent findings demonstrating their superior radiation tolerance compared to conventional Flash memory technologies. However, several critical aspects of radiation tolerance remain underexplored, such as single-event effects.

10.5 Summary and Perspective

Extensive research has confirmed the great potential of HfO_2-FeFETs for next-generation memory applications, yet reliability remains a critical concern. This chapter comprehensively reviews key reliability concerns, including P/E endurance, retention degradation, and radiation tolerance, along with their respective optimization strategies. The study highlights that further advancements in P/E endurance are essential and can be achieved through interdisciplinary approaches combining materials engineering, device physics, and operational optimization. While HfO_2-based FeFETs exhibit relatively robust retention characteristics, minimizing depolarization field effects and leakage currents remain crucial for optimal performance in practical applications. Additionally, despite progress in radiation hardness, further in-depth investigations – particularly on single-event effects – are necessary to ensure device stability in harsh radiation environments. A critical research priority should be to systematically establish the reliability transfer mechanisms from FE-HfO_2 materials to FeFET devices, employing in-situ diagnostic tools and multiscale computational modeling to connect microscopic material behavior with macroscopic device reliability.

Acknowledgment

The authors acknowledge the financial support from the National Natural Science Foundation of China (No. 52122205).

References

1 Chen, H. Y., Zhou, X. F., Tang, L. et al. (2022). HfO_2-based ferroelectrics: From enhancing performance, material design, to applications. *Applied Physics Reviews* 9 (1): 011307.
2 Valasek, J. (1921). Piezo-electric and allied phenomena in Rochelle Salt. *Physical Review* 17 (4): 475–481.
3 Valasek, J. (1971). The early history of ferroelectricity. *Ferroelectrics* 2 (1): 239–244.
4 Busch, G. (1987). Early history of ferroelectricity. *Ferroelectrics* 74 (1): 267–284.
5 Lines, M. E. and Glass, A. M. (2002). Principles and applications of ferroelectrics and related materials. *Materials Research Bulletin* 37 (11): 1893.

6 Uchino, K. (2018). *Ferroelectric Devices*. CRC Press.

7 Liao, J. J., Dai, S. W., Peng, R. C. et al. (2023). HfO_2-based ferroelectric thin film and memory device applications in the post-Moore era: A review. *Fundamental Research* 3 (3): 332–345.

8 Park, B.-E., Ishiwara, H., Okuyama, M. et al. (2016). Features, principles and development of ferroelectric–gate field-effect transistors. In: *Ferroelectric-Gate Field Effect Transistor Memories: Device Physics and Applications*. Dordrecht: Springer Netherlands, pp. 3–20.

9 Böscke, T. S., Müller, J., Bräuhaus, D. et al. (2011). Ferroelectricity in hafnium oxide thin films. *Applied Physics Letters* 99 (10): 102903.

10 Böscke, T. S., Müller, J., Bräuhaus, D. et al. (2011). Ferroelectricity in hafnium oxide: CMOS compatible ferroelectric field effect transistors. In: *2011 IEEE International Electron Devices Meeting*. IEEE. pp. 24.5. 1–24.5. 4.

11 Xiao, W. W., Peng, Y., Zheng, S. Z. et al. (2018). Integration and electrical properties of ferroelectric $Hf_{0.5}Zr_{0.5}O_2$ thin film on bulk β-Ga_2O_3(−201) substrate for memory applications. *IEEE Electron Device Letters* 39 (10): 1504–1507.

12 Liu, H., Zheng, S. Z., Chen, Q. et al. (2019). Structural and ferroelectric properties of Pr doped HfO_2 thin films fabricated by chemical solution method. *Journal of Materials Science: Materials in Electronics* 30: 5771–5779.

13 Zhang, X. Y., Hsu, C. H., Cho, Y. S. et al. (2018). Rapid thermal processing of hafnium dioxide thin films by remote plasma atomic layer deposition as high-k dielectrics. *Thin Solid Films* 660: 797–801.

14 Hao, P. Q., Zheng, S. Z., Zeng, B. J. et al. (2023). Highly enhanced polarization switching speed in HfO_2-based ferroelectric thin films via a composition gradient strategy. *Advanced Functional Materials* 33 (313): 2301746.

15 Yin, L., Gong, S. Q., Li, X. Y. et al. (2022). Improvement of ferroelectricity and endurance in Sr doped $Hf_{0.5}Zr_{0.5}O_2$ films. *Journal of Alloys and Compounds* 914: 165301.

16 Liao, J. J., Zeng, B. J., Sun, Q. et al. (2019). Grain size engineering of ferroelectric Zr-doped HfO_2 for the highly scaled devices applications. *IEEE Electron Device Letters* 40 (11): 1868–1871.

17 Hao, P. Q., Li, H. S., Zeng, B. J. et al. (2023). Effects of thickness scaling on the dielectric properties of $Hf_{0.5}Zr_{0.5}O_2$ ferroelectric thin films. *Journal of Materials Science: Materials in Electronics* 34 (13): 1089.

18 Cheema, S. S., Kwon, D., Shanker, N. et al. (2020). Enhanced ferroelectricity in ultrathin films grown directly on silicon. *Nature* 580 (7804): 478–482.

19 Lee, H. J., Lee, M., Lee, K. et al. (2020). Scale-free ferroelectricity induced by flat phonon bands in HfO_2. *Science* 369 (6509): 1343–1347.

20 Gong, N. and Ma, T. P. et al. (2018). A study of endurance issues in HfO_2-based ferroelectric field effect transistors: Charge trapping and trap generation. *IEEE Electron Device Letters* 39 (1): 15–18.

21 Mikolajick, T., Schroeder, U., Slesazeck, S. (2020). The past, the present, and the future of ferroelectric memories. *IEEE Transactions on Electron Devices* 67 (4): 1434–1443.

22 Jiang, A. Q., Wang, C., Jin, K. J. et al. (2011). A resistive memory in semiconducting $BiFeO_3$ thin-film capacitors. *Advanced Materials* 23 (10): 1277–1281.

23 Wang, C., Jin, K. J., Xu, Z. T. et al. (2011). Switchable diode effect and ferroelectric resistive switching in epitaxial $BiFeO_3$ thin films. *Applied Physics Letters* 98 (19): 192901.

24 Park, M. H., Lee, Y. H., Kim, H. J. et al. (2015). Ferroelectricity and antiferroelectricity of doped thin HfO_2-based films. *Advanced Materials* 27 (11): 1811–1831.

25 Huan, T. D., Sharma, V., Rossetti, G. A. et al. (2014). Pathways towards ferroelectricity in hafnia. *Physical Review B* 90 (6): 064111.

26 Yang, J. H., Liao, J. J., Huang, J. et al. (2024). Kinetical phase transition paths and phase stability in ferroelectric HfO_2. *Scripta Materialia* 242: 115953.

27 Li, G., Yan, S. A., Liu, Y. L. et al. (2025). Unraveling the origins of ferroelectricity in doped hafnia through carrier-mediated phase transitions. *npj Computational Materials* 11 (1): 34.

28 Sang, X. H., Grimley, E. D., Schenk, T. et al. (2015). On the structural origins of ferroelectricity in HfO_2 thin films. *Applied Physics Letters* 106 (16): 162905.

29 Pešić, M., Fengler, F. P. G., Larcher, L. et al. (2016). Physical mechanisms behind the field-cycling behavior of HfO_2-based ferroelectric capacitors. *Advanced Functional Materials* 26 (25): 4601–4612.

30 Starschich, S., Menzel, S., Böttger, U. J. (2016). Evidence for oxygen vacancies movement during wake-up in ferroelectric hafnium oxide. *Applied Physics Letters* 108 (3): 032903.

31 Park, M. H., Kim, H. J., Kim, Y. J. et al. (2015). Study on the internal field and conduction mechanism of atomic layer deposited ferroelectric $Hf_{0.5}Zr_{0.5}O_2$ thin films. *Journal of Materials Chemistry C* 3 (24): 6291–6300.

32 Zhou, D., Xu, J., Li, Q. et al. (2013). Wake-up effects in Si-doped hafnium oxide ferroelectric thin films. *Applied Physics Letters* 103 (19): 192904.

33 Park, M. H., Kim, H. J., Kim, Y. J. et al. (2016). Effect of Zr content on the wake-up effect in $Hf_{1-x}Zr_xO_2$ films. *ACS Applied Materials & Interfaces* 8 (24): 15466–15475.

34 Liu, X., Zhao, W. D., Wang, J. W. et al. (2024). Contribution of the interfacial oxygen vacancies on the asymmetric switching behaviors of the Al: $Hf_{0.5}Zr_{0.5}O_2$ ferroelectric films. *Journal of Applied Physics* 136 (20): 205302.

35 Lee, J., Song, M. S., Jang, W. S. et al. (2022). Modulating the ferroelectricity of hafnium zirconium oxide ultrathin films via interface engineering to control the oxygen vacancy distribution. *Advanced Materials Interfaces* 9 (7): 2101647.

36 Kim, B. S., Hyun, S. D., Moon, T. et al. (2020). A comparative study on the ferroelectric performances in atomic layer deposited $Hf_{0.5}Zr_{0.5}O_2$ thin films using tetrakis (ethylmethylamino) and tetrakis (dimethylamino) precursors. *Nanoscale Research Letters* 15 (1): 72.

37 Xu, Z., Zhu, X., Zhao, G. D. et al. (2024). Oxygen vacancies stabilized 180° charged domain walls in ferroelectric hafnium oxide. *Applied Physics Letters* 124 (1): 012902.

38 Mehmood, F., Mikolajick, T., Schroeder, U. et al. (2020). Wake-Up mechanisms in ferroelectric lanthanum-doped $Hf_{0.5}Zr_{0.5}O_2$ thin films. *Physica Status Solidi A* 217 (22): 2000281.

39 Jan, A., Fraser, S. A., Moon, T. et al. (2024). Resetting the drift of oxygen vacancies in ultrathin HZO ferroelectric memories by electrical pulse engineering. *Small Science* 4 (11): 2400223.

40 Grimley, E. D., Schenk, T., Sang, X. H. et al. (2016). Structural changes underlying field-cycling phenomena in ferroelectric HfO_2 thin films. *Advanced Electronic Materials* 2 (9): 1600173.

41 Chouprik, A. A., Spiridonov, M., Zarubin, S. et al. (2019). Wake-up in a $Hf_{0.5}Zr_{0.5}O_2$ film: A cycle-by-cycle emergence of the remnant polarization via the domain depinning and the vanishing of the anomalous polarization switching. *ACS Applied Electronic Materials* 1 (3): 275–287.

42 Zhang, S. R., Zhang, Q. H., Meng, F. Q. et al. (2023). Domain wall evolution in $Hf_{0.5}Zr_{0.5}O_2$ ferroelectrics under field-cycling behavior. *Research* 6: 0093.

43 Yu, T., Geng, S. N., Zeng, B. J. et al. (2023). Origin of la doping-induced endurance improvement and wake-up effect reduction in ferroelectric HfO_2 thin films. *Physical Review Applied* 20 (5): 054052.

44 Hoffmann, M., Schroeder, U., Schenk, T. et al. (2015). Stabilizing the ferroelectric phase in doped hafnium oxide. *Journal of Applied Physics* 118 (7): 072006.

45 Goh, Y., Cho, S. H., Park, S. H. K. et al. (2020). Oxygen vacancy control as a strategy to achieve highly reliable hafnia ferroelectrics using oxide electrode. *Nanoscale* 12 (16): 9024–9031.

46 Wang, D., Zhang, Y., Wang, J. L. et al. (2022). Enhanced ferroelectric polarization with less wake-up effect and improved endurance of $Hf_{0.5}Zr_{0.5}O_2$ thin films by implementing W electrode. *Journal of Materials Science & Technology* 104: 1–7.

47 Yadav, M., Kashir, A., Oh, S. et al. (2022). High polarization and wake-up free ferroelectric characteristics in ultrathin $Hf_{0.5}Zr_{0.5}O_2$ devices by control of oxygen-deficient layer. *Nanotechnology* 33 (8): 085206.

48 Kashir, A., Oh, S., Hwang, H. et al. (2021). Defect engineering to achieve wake-up free HfO_2-based ferroelectrics. *Advanced Engineering Materials* 23 (1): 2000791.

49 Kashir, A., Kim H., Oh, S. et al. (2021). Large remnant polarization in a wake-up free $Hf_{0.5}Zr_{0.5}O_2$ ferroelectric film through bulk and interface engineering. *ACS Applied Electronic Materials* 3 (2): 629–638.

50 Xu, X. H., Huang, F. T., Qi, Y. B. et al. (2021) Kinetically stabilized ferroelectricity in bulk single-crystalline HfO_2: Y. *Nature materials* 20 (6): 826–832.

51 Silva, J. P. B., Negrea, R. F., Istrate, M. C. et al. (2021). Wake-up free ferroelectric rhombohedral phase in epitaxially strained ZrO_2 thin films. *ACS Applied Materials & Interfaces* 13 (43): 51383–51392.

52 Kim, H. B., Jung, M., Oh, Y. et al. (2021). Superior and stable ferroelectric properties of hafnium-zirconium-oxide thin films deposited via atomic layer deposition using cyclopentadienyl-based precursors without annealing. *Nanoscale* 13 (18): 8524–8530.

53 Samanta, S., Anoop, G., Joh, H. J. et al. (2021). Multi-step chemical solution deposition annealing process toward wake-up free ferroelectricity in Y:HfO_2 films. *Advanced Materials Interfaces* 8 (18): 2100907.

54 Kim, H., Kashir, A., Oh, S. et al. (2021). Effects of high pressure oxygen annealing on $Hf_{0.5}Zr_{0.5}O_2$ ferroelectric device. *Nanotechnology* 32 (31): 315712.

55 Mittmann, T., Fengler, F. P. G., Richter, C. et al. (2017). Optimizing process conditions for improved $Hf_{1-x}Zr_xO_2$ ferroelectric capacitor performance. *Microelectronic Engineering* 178: 48–51.

56 Do, M. T., Gauquelin, N., Nguyen, M. D. et al. (2021). Interface degradation and field screening mechanism behind bipolar-cycling fatigue in ferroelectric capacitors. *APL Materials* 9 (2): 021113.

57 Zhang, F., Luo, Z. D., Yang, Q. Y. et al. (2022). Evolution of the interfacial layer and its impact on electric-field-cycling behaviors in ferroelectric $Hf_{1-x}Zr_xO_2$. *ACS Applied Materials & Interfaces* 14 (8): 11028–11037.

58 Fields, S. S., Smith, S. W., Ryan, P. J. et al. (2020). Phase-exchange-driven wake-up and fatigue in ferroelectric hafnium zirconium oxide films. *ACS Applied Materials & Interfaces* 12 (23): 26577–26585.

59 Li, S. D., Zhou, D. Y., Shi Z. X. et al. (2020). Involvement of unsaturated switching in the endurance cycling of Si-doped HfO_2 ferroelectric thin films. *Advanced Electronic Materials* 6 (8): 2000264.

60 Huang, F., Chen, X., Liang, X. et al. (2017). Fatigue mechanism of yttrium-doped hafnium oxide ferroelectric thin films fabricated by pulsed laser deposition. *Physical Chemistry Chemical Physics* 19 (5): 3486–3497.

61 Lou, X. J., Zhang, M., Redfern, S. A. T. et al. (2006). Local phase decomposition as a cause of polarization fatigue in ferroelectric thin films. *Physical Review Letters* 97 (17): 177601.

62 Cao, R. R., Song, B., Shang, D. S. et al. (2019). Improvement of endurance in HZO-based ferroelectric capacitor using Ru electrode. *IEEE Electron Device Letters* 40 (11): 1744–1747.

63 Kim, B. Y., Park, H. W., Hyun, S. D. et al. (2022). Enhanced ferroelectric properties in $Hf_{0.5}Zr_{0.5}O_2$ films using a $HfO_{0.61}N_{0.72}$ interfacial layer. *Advanced Electronic Materials* 8 (6): 2100042.

64 Li, Y. C., Li, X. X., Wu, M. K. et al. (2022). Dual $Al_2O_3/Hf_{0.5}Zr_{0.5}O_2$ stack thin films for improved ferroelectricity and reliability. *IEEE Electron Device Letters* 43 (8): 1235–1238.

65 Liu, Y. L., Cao, Y. Y., Zhu, H. et al. (2021). $HfZrO_x$-based ferroelectric tunnel junction with crested symmetric band structure engineering. *IEEE Electron Device Letters* 42 (9): 1311–1314.

66 Cheng, Y., Gao, Z. M., Ye, K. H. et al. (2022). Reversible transition between the polar and antipolar phases and its implications for wake-up and fatigue in HfO_2-based ferroelectric thin film. *Nature Communications* 13 (1): 645.

67 Fengler, F. P. G., Nigon, R., Muralt, P. et al. (2018). Analysis of performance instabilities of hafnia-based ferroelectrics using modulus spectroscopy and thermally stimulated depolarization currents. *Advanced Electronic Materials* 4 (3): 1700547.

68 Liao, J. J., Yang, T. Y., Ju, C. F. et al. (2023). Improved ferroelectric and endurance properties of $Hf_{0.5}Zr_{0.5}O_2$ thin films by a replacement indium-tin-oxide top electrode process. *Journal of Physics D: Applied Physics* 56 (38): 385301.

69 Bao, K. Y., Liao J. J., Yan, F. et al. (2023). Enhanced endurance and imprint properties in $Hf_{0.5}Zr_{0.5}O_{2-\delta}$ ferroelectric capacitors by tailoring the oxygen vacancy. *ACS Applied Electronic Materials* 5 (8): 4615–4623.

70 Chen, K. Y., Chen, P. H., Wu, Y. H. et al. (2017). Excellent reliability of ferroelectric $HfZrO_x$ free from wake-up and fatigue effects by NH_3 plasma treatment. In: *2017 Symposium on VLSI Circuits*. IEEE, pp. T84–T85.

71 Zheng, S. Z., Zhao, Z. D., Liu, Z. T. et al. (2020). Improvement of remanent polarization of CeO_2-HfO_2 solid solution thin films on Si substrates by chemical solution deposition. *Applied Physics Letters* 117 (21): 212904.

72 Li, X. F., Li, C., Xu, Z. Y. et al. (2021). Ferroelectric properties and polarization fatigue of La:HfO_2 thin-film capacitors. *Physica Status Solidi (RRL)–Rapid Research Letters* 15 (4): 2000481.

73 Müller, J., Schröder, U., Böscke, T. S. et al. (2011). Ferroelectricity in yttrium-doped hafnium oxide. *Journal of Applied Physics* 110 (11): 114113.

74 Zheng, S. Z., Jing, C., Liu, Z. T. et al. (2023). Effects of annealing conditions on structural and ferroelectric properties of CeO_2-HfO_2 solid solution thin films on InAs substrates fabricated by chemical solution deposition. *Journal of Alloys and Compounds* 967: 171541.

75 Yan, F., Liao, J. J., Cao, K. et al. (2023). Achieving excellent ferroelectric and dielectric performance of $HfO_2/ZrO_2/HfO_2$ thin films under low operating voltage. *Journal of Alloys and Compounds* 968: 172267.

76 Liang, Y. K., Wu, J. S., Teng, C. Y. et al. (2021). Demonstration of highly robust 5 nm $Hf_{0.5}Zr_{0.5}O_2$ ultra-thin ferroelectric capacitor by improving interface quality. *IEEE Electron Device Letters* 42 (9): 1299–1302.

77 Chen, H. Y., Tang, L., Liu, L. Y. (2021). Significant improvement of ferroelectricity and reliability in $Hf_{0.5}Zr_{0.5}O_2$ films by inserting an ultrathin Al_2O_3 buffer layer. *Applied Surface Science* 542: 148737.

78 Wang, Y., Tao, L., Guzman, R. et al. (2023). A stable rhombohedral phase in ferroelectric $Hf(Zr)_{1+x}O_2$ capacitor with ultralow coercive field. *Science* 381 (6657): 558–563.

79 Abe, K., Yanase, N., Yasumoto, T. et al. (2002). Nonswitching layer model for voltage shift phenomena in heteroepitaxial barium titanate thin films. *Japanese Journal of Applied Physics* 41 (10R): 6065–6071.

80 Tagantsev, A. K., Stolichnov, I., Setter, N. et al. (2004). Nature of nonlinear imprint in ferroelectric films and long-term prediction of polarization loss in ferroelectric memories. *Journal of Applied Physics* 96 (11): 6616–6623.

81 Zacharaki, C., Tsipas, P., Chaitoglou, S. et al. (2020). Reliability aspects of ferroelectric $TiN/Hf_{0.5}Zr_{0.5}O_2/Ge$ capacitors grown by plasma assisted atomic oxygen deposition. *Applied Physics Letters* 117 (21): 212905.

82 Li, X. P., Tai, L., Sang, P. P. et al. (2024). Imprint-correlated retention loss in $Hf_{0.5}Zr_{0.5}O_2$ ferroelectric thin film through wide-temperature characterizations. *IEEE Transactions on Electron Devices* 71 (9): 5361–5366.

83 Takada, K., Takarae, S., Shimamoto, K. et al. (2021). Time-dependent imprint in $Hf_{0.5}Zr_{0.5}O_2$ ferroelectric thin films. *Advanced Electronic Materials* 7 (8): 2100151.

84 Fengler, F. P. G., Pešić, M., Starschich, S. et al. (2017). Domain pinning: Comparison of hafnia and PZT based ferroelectrics. *Advanced Electronic Materials* 3 (4): 1600505.

85 Higashi, Y., Kaczer, B., Verhulst, A. S. et al. (2020). Investigation of imprint in FE-HfO_2 and its recovery. *IEEE Transactions on Electron Devices* 67 (11): 4911–4917.

86 Mehmood, F., Hoffmann, M., Lomenzo, P. D. et al. (2019). Bulk depolarization fields as a major contributor to the ferroelectric reliability performance in lanthanum doped $Hf_{0.5}Zr_{0.5}O_2$ capacitors. *Advanced Materials Interfaces* 6 (21): 1901180.

87 Yuan, P., Mao, G. Q., Cheng, Y. et al. (2022). Microscopic mechanism of imprint in hafnium oxide-based ferroelectrics. *Nano Research* 15 (4): 3667–3674.

88 Fields, S. S., Smith, S. W., Jaszewski, S. T. et al. (2021). Wake-up and fatigue mechanisms in ferroelectric $Hf_{0.5}Zr_{0.5}O_2$ films with symmetric RuO_2 electrodes. *Journal of Applied Physics* 130 (13):134101.

89 Park, J. Y., Choe, D. H., Lee, D. H. et al. (2023). Revival of ferroelectric memories based on emerging fluorite-structured ferroelectrics. *Advanced Materials* 35 (43): 2204904.

90 Tan, T., Liu, Z., Tian, H. et al. (2010). Low voltage stress-induced leakage current in HfO_2 dielectric films. *Materials Science and Engineering: B* 171 (1-3): 159–161.

91 Islamov, D. R., Perevalov, T. V., Gritsenko, V. A. et al. (2015). Charge transport in amorphous $Hf_{0.5}Zr_{0.5}O_2$. *Applied Physics Letters* 106 (10): 102906.

92 González, Y., Youssef, A. H., Dörfler, A. et al. (2021). Competing tunneling conduction mechanisms in oxygen deficient $Hf_{0.5}Zr_{0.5}O_2$. *Applied Physics Letters* 119 (13): 132906.

93 Gonzalez, Y., Hadj, Y. A., Katoch, R. et al. (2022). Mechanism of oxygen vacancy filament-based resistive switching in non-polar $Hf_{0.5}Zr_{0.5}O_{2-\delta}$ thin films. *ACS Applied Electronic Materials* 4 (3): 1311–1317.

94 Kim, S., Lee, S. H., Jo, I. H. et al. (2024). Reduced leakage current in atomic-layer-deposited HfO_2 thin films deposited at low temperature by in-situ defect passivation. *Applied Surface Science* 645: 158790.

95 Umezawa, N., Shiraishi, K., Ohno, T. et al. (2005). First-principles studies of the intrinsic effect of nitrogen atoms on reduction in gate leakage current through Hf-based high-k dielectrics. *Applied Physics Letters* 86 (14): 143507.

96 Kim, Y. H. and Lee, J. C. (2004). Reliability characteristics of high-k dielectrics. *Microelectronics Reliability* 44 (2): 183–193.

97 Tao, F., Yang, H., Tang, B. et al. (2014). TDDB characteristic and breakdown mechanism of ultra-thin SiO_2/HfO_2 bilayer gate dielectrics. *Journal of Semiconductors* 35 (6): 064003.

98 Degraeve, R., Groeseneken, G., Bellens, R. et al. (1995). A consistent model for the thickness dependence of intrinsic breakdown in ultra-thin oxides. In: *1995 Proceedings of International Electron Devices Meeting*. IEEE, pp. 863–866.

99 Houssa, M., Nigam, T., Mertens, P. W. et al. (1998). Model for the current-voltage characteristics of ultra-thin gate oxide after soft breakdown. *Journal of Applied Physics* 84 (8): 4351–4355.

100 Stathis, J. H. and DiMaria, D. J. et al. (1998). Reliability projection for ultrathin oxides at low voltage. In: *1998 IEEE International Electron Devices Meeting (IEDM)*. IEEE, pp. 167–170.

101 Stathis, J. H. (2001). Physical and predictive models of ultrathin oxide reliability in CMOS devices and circuits. *IEEE Transactions on Device and Materials Reliability* 1 (1): 43–59.

102 Kirsch, P. D., Kang, C. S., Lozano, J. et al. (2002). Electrical and spectroscopic comparison of HfO_2/Si interfaces on nitrided and un-nitrided Si(100). *Journal of Applied Physics* 91 (7): 4353–4363.

103 Kim, Y. H., Onishi, K., Kang, C. S. et al. (2002). Area dependence of TDDB characteristics for HfO_2 gate dielectrics. *IEEE Electron Device Letters* 23 (10): 594–596.

104 DiMaria, D. J. and Stathis, J. H. et al. (1997). Explanation for the oxide thickness dependence of breakdown characteristics of metal-oxide-semiconductor structures. *Microelectronic Engineering* 36 (1–4): 317–320.

105 Stathis, J. H. (1999). Percolation models for gate oxide breakdown. *Journal of Applied Physics* 86 (10): 5757–5766.

106 Park, J. Y., Yang, K., Lee D. H. et al. (2020). A perspective on semiconductor devices based on fluorite-structured ferroelectrics from the materials-device integration perspective. *Journal of Applied Physics* 128 (24): 240904.

107 Pešić, M., Schroeder, U., Slesazeck, S. et al. (2018). Comparative study of reliability of ferroelectric and anti-ferroelectric memories. *IEEE Transactions on Device and Materials Reliability* 18 (2): 154–162.

108 Shao, M. H., Zhao, R. T., Liu, H. et al. (2024). Challenges and recent advances in HfO_2-based ferroelectric films for non-volatile memory applications. *Chip* 3: 100101.

109 Pešić, M., Fengler, F. P. G., Slesazeck S. et al. (2016). Root cause of degradation in novel HfO_2-based ferroelectric memories. In: *2016 IEEE International Reliability Physics Symposium (IRPS)*. IEEE, MY–3–1–MY–3–5.

110 Ali, T., Polakowski P., Riedel S. et al. (2018). Silicon doped hafnium oxide (HSO) and hafnium zirconium oxide (HZO) based FeFET: A material relation to device physics. *Applied Physics Letters* 112 (22): 222903.

111 Ma, T. P. and Han, J. P. (2002). Why is nonvolatile ferroelectric memory field-effect transistor still elusive? *IEEE Electron Device Letters* 23 (7): 386–388.

112 Zeng, B. J., Liao, M., Peng, Q. X. et al. (2018). 2-bit/cell operation of $Hf_{0.5}Zr_{0.5}O_2$ based FeFET memory devices for NAND applications. *IEEE Journal of the Electron Devices Society* 7: 551–556.

113 Cheng, C. H. and Chin, A. (2014). Low-leakage-current DRAM-like memory using a one-transistor ferroelectric MOSFET with a Hf-based gate dielectric. *IEEE Electron Device Letters* 35 (1):138–140.

114 Tan, A. J., Liao, Y. H., Wang, L. C. et al. (2021). Ferroelectric HfO_2 memory transistors with high-k interfacial layer and write endurance exceeding 10^{10} cycles. *IEEE Electron Device Letters* 42 (7): 994–997.

115 Yurchuk, E., Mueller, S., Martin, D. et al. (2014). Origin of the endurance degradation in the novel HfO_2-based 1T ferroelectric non-volatile memories. In: *2014 IEEE International Reliability Physics Symposium*. IEEE, 2E.5.1–2E.5.5.

116 Zeng, B. J., Liao, M., Liao, J. J. et al. (2019). Program/erase cycling degradation mechanism of HfO_2-based FeFET memory devices. *IEEE Electron Device Letters* 40 (5): 710–713.

117 Yurchuk, E., Müller, J., Müller, S. et al. (2016). Charge-trapping phenomena in HfO_2-based FeFET-type nonvolatile memories. *IEEE Transactions on Electron Devices* 63 (9): 3501–3507.

118 Ichihara, R., Suzuki, K., Kusai, H. et al. (2020). Re-examination of V_{th} window and reliability in HfO_2 FeFET based on the direct extraction of spontaneous polarization and trap charge during memory operation. In: 2020 *IEEE Symposium on VLSI Technology*. IEEE, pp. 1–2.

119 Yurchuk, E., Müller, J., Paul, J. et al. (2014). Impact of scaling on the performance of HfO_2-based ferroelectric field effect transistors. *IEEE Transactions on Electron Devices* 61 (11): 3699–3706.

120 Zagni, N., Pavan, P., Alam, M. A. et al. (2020). A memory window expression to evaluate the endurance of ferroelectric FETs. *Applied Physics Letters* 117 (15): 152901.

121 Hao, T. Q., Zeng, B. J., Sun, Z. J. et al. (2024). Coupling effects of interface charge trapping and polarization switching in HfO_2-based ferroelectric field effect transistors. *APL Materials* 12 (1): 011108.

122 Muller, J., Polakowski, P., Muller, S. et al. (2016). High endurance strategies for hafnium oxide based ferroelectric field effect transistor. In: *2016 16th Non-Volatile Memory Technology Symposium (NVMTS)*. IEEE, pp. 1–7.

123 Chan, C. Y., Chen, K. Y., Peng, H. K. et al. (2020). FeFET memory featuring large memory window and robust endurance of long-pulse cycling by interface engineering using high-k AlON. In: *IEEE Symposium on VLSI Technology*. IEEE, pp. 1–2.

124 Onaya, T., Nabatame, T., Sawamoto, N. et al. (2019). Improvement in ferroelectricity of $Hf_xZr_{1-x}O_2$ thin films using top-and bottom-ZrO_2 nucleation layers. *APL Materials* 7 (6): 061107.

125 Xiao, W. W., Liu, C., Peng, Y. et al. (2019). Performance improvement of $Hf_{0.5}Zr_{0.5}O_2$-based ferroelectric-field-effect transistors with ZrO_2 seed layers. *IEEE Electron Device Letters* 40 (5): 714–717.

126 Xiao, W. W., Liu, C., Peng, Y. et al. (2019). Memory window and endurance improvement of $Hf_{0.5}Zr_{0.5}O_2$-based FeFETs with ZrO_2 seed layers characterized by fast voltage pulse measurements. *Nanoscale Research Letters* 14 (1): 254.

127 Huang, C., Zhang, Y. K., Zheng, S. Z. et al. (2021). Interface effects induced by a ZrO_2 seed layer on the phase stability and orientation of HfO_2 ferroelectric thin films: A first-principles study. *Physical Review Applied* 16 (4): 044048.

128 Chen, Y. H., Su, C. J., Yang, T. H. et al. (2020). Improved TDDB reliability and interface states in 5-nm $Hf_{0.5}Zr_{0.5}O_2$ ferroelectric technologies using NH_3 plasma and microwave annealing. *IEEE Transactions on Electron Devices* 67 (4): 1581–1585.

129 Zeng, B. J., Xiao, W. W., Liao, J. J. et al. (2018). Compatibility of HfN metal gate electrodes with $Hf_{0.5}Zr_{0.5}O_2$ ferroelectric thin films for ferroelectric field-effect transistors. *IEEE Electron Device Letters* 39 (10): 1508–1511.

130 Choi, S. N., Moon, S. E., Yoon, S. M. (2020). Impact of oxide gate electrode for ferroelectric field-effect transistors with metal-ferroelectric-metal-insulator-semiconductor gate stack using undoped HfO_2 thin films prepared by atomic layer deposition. *Nanotechnology* 32 (8): 085709.

131 Yoon, S. J., Min, D. H., Moon, S. E. et al. (2020). Improvement in long-term and high-temperature retention stability of ferroelectric field-effect memory transistors with metal–ferroelectric–metal–insulator–semiconductor gate-stacks using Al-doped HfO_2 thin films. *IEEE Transactions on Electron Devices* 67 (2): 499–504.

132 Kim, T., Hwang, J., Kim, G. et al. (2021). High-performance and high-endurance HfO_2-based ferroelectric field-effect transistor memory with a spherical recess channel. *Physica Status Solidi (RRL)–Rapid Research Letters* 15 (5): 2100018.

133 Gong, N. and Ma, T. P. et al. (2016). Why is FE-HfO_2 more suitable than PZT or SBT for scaled nonvolatile 1-T memory cell? a retention perspective. *IEEE Electron Device Letters* 37 (9): 1123–1126.

134 Florent, K., Lavizzari, S., Piazza, L. D. et al. (2017). Reliability study of ferroelectric Al:HfO_2 thin films for DRAM and NAND applications. *IEEE Transactions on Electron Devices* 64 (10): 4091–4098.

135 Liu, C., Zeng, B. J., Dai, S. W. et al. (2022). Robustly stable intermediate memory states in HfO_2 based ferroelectric field-effect transistors. *Journal of Materiomics* 8 (3): 685–692.

136 Gao, K. X., Chen Y. Q., Zheng S. Z. et al. (2021). Degradation behavior and mechanism of SiC power MOSFETs by total ionizing dose irradiation under different gate voltages. In: *IEEE Workshop on Wide Bandgap Power Devices and Applications in Asia (WiPDA Asia)*. IEEE, pp. 46–50.

137 Xiao, W. W., Liu, C., Peng, Y. et al. (2019). Thermally stable and radiation hard ferroelectric $Hf_{0.5}Zr_{0.5}O_2$ thin films on muscovite mica for flexible nonvolatile memory applications. *ACS Applied Electronic Materials* 1 (6): 919–927.

138 Huang, F., Wang, Y., Liang, X. et al. (2017). HfO_2-based highly stable radiation immune ferroelectric memory. *IEEE Electron Device Letters* 38 (3): 330–333.

139 Chen, K. Y., Tsai, Y. S., Wu, Y. H. et al. (2019). Ionizing radiation effect on memory characteristics for HfO_2-based ferroelectric field-effect transistors. *IEEE Electron Device Letters* 40 (9): 1370–1373.

140 Kämpfe, T., Vogel, T., Olivo, R. et al. (2020). Heavy ion irradiation effects on structural and ferroelectric properties of HfO_2 films. In *2020 Joint Conference of the IEEE International Frequency Control Symposium and International Symposium on Applications of Ferroelectrics (IFCS-ISAF)*. IEEE, pp. 1–3.

141 Wang, Y., Huang, F., Hu, Y. et al. (2018). Proton radiation effects on Y-doped HfO_2-based ferroelectric memory. *IEEE Electron Device Letters* 39 (6): 823–826.

142 Zhang, W. L., Li, G., Long, X. J. et al. (2020). A comparative study of the γ-ray radiation effect on Zr-doped and al al-doped HfO_2-based ferroelectric memory. *Physica Status Solidi (B)* 257 (5): 1900736.

143 Liu, W. Y., Liao, J. J., Jiang, J. et al. (2020). Highly stable performance of flexible $Hf_{0.6}Zr_{0.4}O_2$ ferroelectric thin films under multi service conditions. *Journal of Materials Chemistry C* 8 (11): 3878–3886.

144 Sun, Q., Liao, J. J., Peng, Q. X. et al. (2020). Total ionizing dose effects of ^{60}co γ-rays radiation on $Hf_xZr_{1-x}O_2$ ferroelectric thin film capacitors. *Journal of Materials Science: Materials in Electronics* 31: 2049–2056.

145 Liao, N., Lin, X., Zhu, B. et al. (2024). Optimization of ferroelectric properties of HfO_2-based thin films by ion irradiation. *IEEE Transactions on Nuclear Science* 71: 1208–1217.

146 Dai, S. W., Yang, Q. J., Zeng, B. J. et al. (2022). Robustly stable ferroelectric polarization states enable long-term nonvolatile storage against radiation in HfO_2-based ferroelectric field-effect transistors. *ACS Applied Materials & Interfaces* 14 (45): 51459–51467.

147 Liu, C., Xiao W. W., Peng Y. et al. (2021). $Hf_{0.5}Zr_{0.5}O_2$-based ferroelectric field-effect transistors with HfO_2 seed layers for radiation-hard nonvolatile memory applications. *IEEE Transactions on Electron Devices* 68 (9): 4368–4372.

148 Peng, H. K., Lai, T. C., Kao, T. H. et al. (2022). Improved reliability and read latency under radiation observed in $HfZrO_x$ based p-FeFETs with AlON interfacial layer. *IEEE Electron Device Letters* 43 (3): 494–497.

11

Hafnia-based Materials for Neuromorphic Devices

Hai Zhong[1,2]*, Kui-juan Jin*[2,3*]*, and Chen Ge*[2,3*]

[1] *School of Physics and Optoelectronic Engineering, Ludong University, Yantai, China*
[2] *Beijing National Laboratory for Condensed Matter Physics, Institute of Physics, Chinese Academy of Sciences, Beijing, China*
[3] *University of Chinese Academy of Sciences, Beijing, China*

*Corresponding authors: gechen@iphy.ac.cn; kjjin@iphy.ac.cn

11.1 Introduction

The emergence of artificial intelligence (AI) and the proliferation of big data are causing exponential data growth, increasing the demand for data storage and necessitating new computing architectures to manage data loads and real-world challenges. These trends require the development of novel memory devices. Current memories have limitations: dynamic random-access memory (DRAM) and static random-access memory (SRAM) offer fast access (~10 ns) but are volatile; hard disk drives (HDDs) are nonvolatile (NV), have high storage densities, and are cost-effective but lack random-access functionality and have slow access speeds; flash memory (solid-state disks and USB drives) is NV and of high density but limited by slower programming/erasing speeds. Consequently, new memory devices are needed to meet modern data-intensive demands.[1]

An ideal memory should concurrently fulfill three key criteria: rapid access speed, high storage density, and non-volatility. Figure 11.1 illustrates several types of NV memories that are currently under active development, among which ferroelectric (FE) memories, such as ferroelectric field-effect transistors (FeFETs), ferroelectric random-access memories (FeRAMs), and ferroelectric tunnel junctions (FTJs), are regarded as ideal candidates for low-power NV memories.

Unlike other emerging memory technologies, including magnetoresistive random-access memories (MRAMs), resistive random-access memories (RRAMs), and phase-change memories (PCMs), the writing operation in FE memories is purely field-driven, thus avoiding the problem of inefficiency.[2] However,

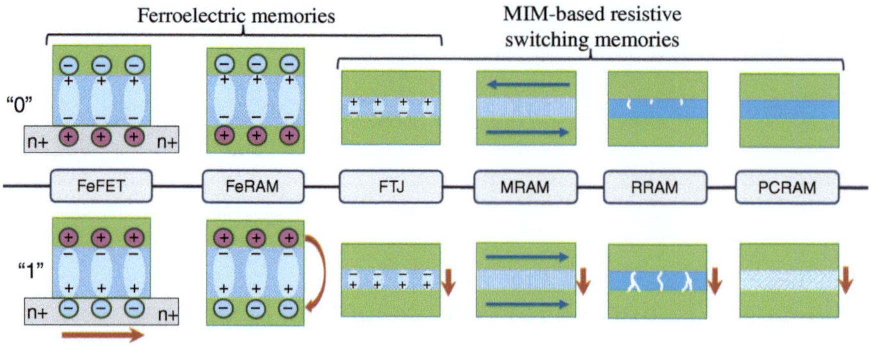

Figure 11.1 NV memories during development. From left to right, the sequence includes the FeFET, FeRAM, FTJ, MRAM, RRAM, and PCM.

conventional perovskite FE materials have encountered a technological bottleneck at the hundred-nanometer node, and their complex fabrication processes are not compatible with complementary metal–oxide–semiconductor (CMOS) processes. The emergence of hafnia-based FEs perfectly circumvents both of these problems, as they demonstrate excellent compatibility with CMOS and preserve stable FE properties at the nanoscale.[3–5] This has established them as the most promising FE materials in the field of nanoelectronics, particularly for applications involving low-power NV memories.[6]

Hafnia-based materials have extensive applications beyond their initial scope. Primarily, they serve as excellent dielectrics in advanced semiconductor manufacturing: HfO_2 is used as the gate dielectric in CPUs, and ZrO_2 is employed in DRAM capacitors.[7] Additionally, hafnia-based FE materials exhibit piezoelectric properties akin to AlN, indicating their potential in electromechanical systems such as sensors, resonators, and transducers.[8] They also display significant pyroelectricity.[9] Although their pyroelectric coefficients are lower than those of perovskite single crystals,[10] their compatibility with CMOS technology, environmental friendliness, and ease of fabrication make them promising for infrared detectors.[8] The piezoelectric and pyroelectric properties of hafnia-based materials can be adjusted by factors like film thickness, manufacturing methods, doping, and electric field cycling, allowing for customization to specific applications. Moreover, the negative capacitance in hafnia-based FEs is a promising approach for creating ultra-low-power nanoscale transistors.[11] Despite debates on the origin of the negative capacitance effect,[12] transistors with hafnia-based FE layers have shown the potential to exceed the subthreshold swing limit of 60 mV/dec at room temperature.[13] Finally, due to their high dielectric constant (κ) and breakdown voltage (E_{BD}), hafnia-based materials are promising for supercapacitors, making them attractive for energy storage applications.[14]

Beyond the applications already discussed, hafnia-based materials have extended their utility to the field of neuromorphic devices. These devices, which rely on memory devices, constitute the focus of this chapter. Whether in RRAM-type synaptic devices characterized by ionic migration and valence change or in FeRAM-, FTJ-, and FeFET-type neuromorphic devices characterized by controllable FE domain switching, research on hafnia-based materials for neuromorphic devices has achieved significant advancements. In the second part of this chapter, we begin with a basic introduction to neuromorphic computing and the functions of neuromorphic devices. The third section provides a detailed exploration of the fundamental material properties of hafnia-based materials, in conjunction with the essential requirements of neuromorphic devices. The fourth part will elaborate on the application of hafnia-based materials in different device functions and configurations. Finally, we will also address the current prospects and challenges in this field and propose possible solutions.

11.2 Neuromorphic Computing and Neuromorphic Devices

In October 2024, Nobel Prizes in Physics and Chemistry were unveiled, igniting widespread public discourse. The Physics award was conferred upon John Hopfield and Geoffrey Hinton for their seminal contributions and inventions in machine learning through artificial neural networks. The Prize in Chemistry was granted to David Baker, Demis Hassabis, and John Jumper for their contributions to computational protein design and the application of AI methodologies for predicting protein structures. These accolades underscore the recognition that AI has transcended the boundaries of computer science and become deeply embedded in various scientific disciplines. AI has evolved into an extremely broad field of study, encompassing multiple research branches, such as machine learning, deep learning, natural language recognition, and computer vision. Among these, neuromorphic computing, which aims to emulate the functions of the biological brain to efficiently address complex practical challenges, emerges as a significant and prominent research domain within the AI landscape.[15–17]

11.2.1 Neuromorphic Architecture

In classical computing systems, data processing and storage are distinct operations that are executed by CPUs and RAMs, respectively. These two components exhibit significant differences in terms of working speed, operational modes, and manufacturing technologies. The von Neumann tackled compatibility discrepancies with a constructive approach, thereby advancing the evolution of computational technology. However, due to the considerable time and energy consumption required to transfer data between the memory and processing units, the performance and efficiency of data centers are constrained by circuit delays and power issues, a phenomenon commonly referred to as the von Neumann bottleneck.[18]

One might presume that, despite this constraint, contemporary computers can execute fundamental arithmetic operations at speeds reaching billions of operations

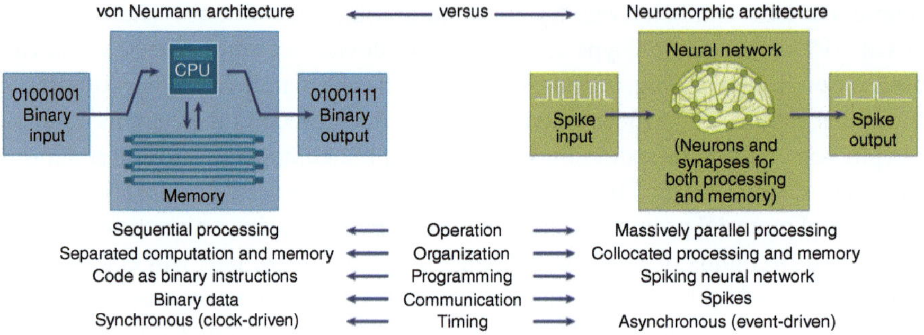

Figure 11.2 Comparison of von Neumann and neuromorphic architectures. *Source*: [18] / with permission of Springer Nature.

per second, vastly outpacing the computational velocity of biological entities such as the human brain. Undoubtedly, in specific domains, such as numerical computation, von Neumann architectures not only eclipse the human brain in speed but also uphold superior precision. Nevertheless, when addressing intricate real-world challenges, such as image recognition, natural language processing, and autonomous driving – data-intensive tasks that are common in our daily lives – von Neumann computers are not only sluggish but also consume a significant amount of energy. For instance, quickly identifying various modes of transportation or pedestrians on a congested urban road, or the simple act of extending an arm to steadily lift a glass of water to one's lips, are feats that remain beyond the reach of even the most sophisticated personal computers available today. However, these tasks are managed effortlessly by the human brain.

The human brain, a high-density neural network, comprises approximately 10^{12} neurons and 10^{15} synapses. It occupies less than 2L of space and consumes about 20W of power, yet it can outperform supercomputers in many tasks.[19,20] With the idea and hypothesis of constructing a brain-like computing machine, the concept of neuromorphic computing was proposed by Dr. Mead.[21] The significance of neuromorphic computing is not only in building a more powerful computer but also in potentially revealing the fundamental operating mechanisms of the human brain. Neuromorphic computers store and process data in a completely different mode than von Neumann computers, as shown in Figure 11.2. The main features include:

1. Highly parallel operation: In contrast to the sequential processing in von Neumann computers, neuromorphic computers are inherently parallel, with all neurons and synapses potentially operating simultaneously. This parallelism allows neuromorphic systems to process complex data patterns more efficiently and rapidly than traditional computers, particularly in tasks that involve real-time learning and adaptation.[18]
2. Collocated processing and memory: In neuromorphic hardware, there is no distinction between processing and memory. Typically, both neurons and synapses can process and store values. This co-location of processing and memory can effectively circumvent the von Neumann bottleneck problem.

3. Efficient signal encoding schemes: Unlike computing in CPUs, which performs binary code-based computations, neurons and synapses within the brain represent data in sequences of spikes to generate signals akin to those in the brain. The amplitude and morphology of these spikes can be harnessed to encode numerical information. Although it is possible to convert binary values into spike patterns and back, the exact methods for this conversion are still a subject of ongoing research in the field of neuromorphic computing.
4. Event-driven computation: Neuromorphic computers leverage event-driven computation and temporally sparse activity to achieve exceptionally efficient processing.[22] Neurons and synapses only engage in computation when there are spikes to handle, and typically, spikes are relatively sparse within the operation of the network.
5. New logical algorithms: Given the unique non-von Neumann architecture and spiking encoding scheme of neuromorphic computing systems, traditional binary algebra is no longer applicable. Machine learning algorithms based on neural networks are widely considered as the ideal candidate for executing on a neuromorphic computing system.[23]
6. Inherent scalability: Neuromorphic computers are designed to be inherently scalable because incorporating additional neuromorphic chips entails increasing the number of neurons and synapses available. It is feasible to aggregate numerous physical neuromorphic chips and operate them as a unified, expansive neuromorphic system, capable of running increasingly larger networks.[18]
7. Stochasticity: Neuromorphic computers can incorporate a notion of randomness, such as neuronal firing, to accommodate noise.[18]

Neuromorphic computers are distinguished by their remarkable energy efficiency and typically consume significantly less energy than conventional computing systems. This efficiency is due to their event-driven and massively parallel architecture, which ensures that only a fraction of the system is active at any given time, while the rest is dormant. Currently, by mimicking the functions of neurons and synapses, numerous physical implementations of neuromorphic hardware are either under development or are already available for use by the research community. For instance, the European Union's Human Brain Project funded projects such as SpiNNaker[24] and BrainScaleS[25] to facilitate large-scale neural science simulations. The Tianjic chip,[26] developed by Tsinghua University, serves as a neuromorphic platform that supports both spiking neural networks (SNNs) and traditional artificial neural networks capable of tackling a diverse range of problems. IBM's TrueNorth chip,[27] which has 540 million transistors and 4,096 cores, integrates 1 million neurons and 256 million synapses. It achieves supercomputer-level performance, while each core consumes only 70mW of power. The chip efficiently executes cognitive tasks, such as image and speech recognition, surpassing traditional computers in terms of both energy efficiency and speed. Intel's Loihi[28] is arguably the most advanced neuromorphic chip available to date. The latest Loihi 2 chip, fabricated using a 7nm process, integrates one million neurons and over 120 million synapses on each chip. It implements SNNs with programmable dynamics, modular connectivity, and optimized scale, speed, and efficiency.

11.2.2 Functions of Neuromorphic Devices

Current neuroscience indicates that biological brain functions are primarily executed by neurons and synapses. As shown in Figure 11.3, the structure of a complete neural unit typically comprises four components.[30]

1. Dendrite: This receptive organ receives spiking signals from other neurons.
2. Soma (neuron body): It integrates incoming spike signals, once the aggregated signal intensity surpasses a certain threshold voltage, and generates and transmits action potentials to the axon.
3. Axon: This component is responsible for propagating spiking signals to other neurons.
4. Synapse: This is a junction between two neurons, enabling the transfer of signals from the axon of one neuron to the dendrites of another. It facilitates communication by releasing neurotransmitters that bind to the receptors on the receiving neuron. It is widely considered a memory organ in the brain, as it encodes memory information through the strength of its connectivity.

The construction of neuromorphic computers involves the design of artificial devices that emulate the functions of biological neurons and synapses. Within the neuroscience domain, research continues to explore the intricacies of neurons and synapses. The current understanding of the nervous system's operations can be concisely articulated as follows: dendrites collect signals and convey them to the soma. The soma serves as a central processing unit, in which nonlinear transformations occur. Upon surpassing a certain threshold, the input signals trigger the generation of an output signal. This output signal propagates along the axon and is conveyed to other neurons via the synapses. Throughout this process, the roles of neurons and synapses will be elucidated individually.

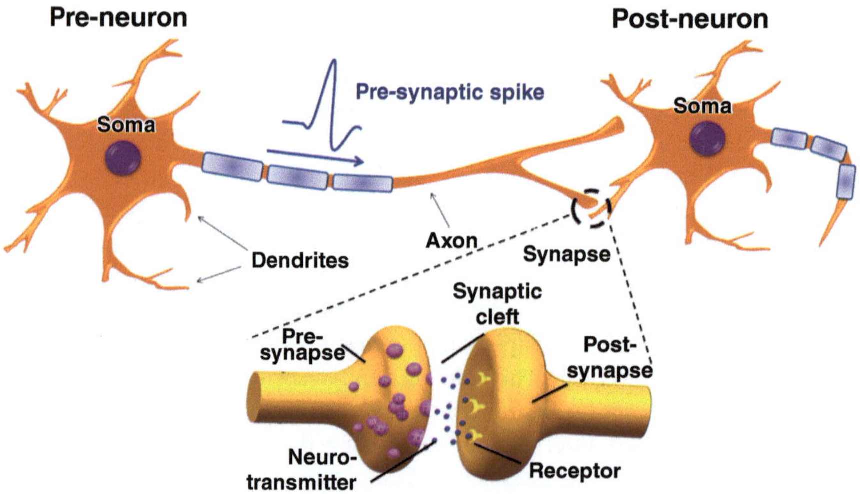

Figure 11.3 Schematic representation of correlated neurons and synapses. *Source*: [29] / with permission of IOP Publishing.

11.2.2.1 Neuron Models

A signal, also referred to as a stimulus, travels from dendrites to the soma. If the stimulus is weak and fails to attain the critical threshold level, the membrane potential dissipates. Conversely, when the stimulus is sufficiently robust, an action potential is generated. After the firing process, the neuron undergoes a refractory period, rendering it unlikely to activate once more and eventually reset to its initial state. The oscillation between activation and repose is known as the activity of biological neurons, as shown in Figure 11.4. Several representative neuron models have been proposed by researchers, including the integrate-and-fire (IF) model,[31] the Fitzhugh–Nagumo (FN) model,[32] the Hodgkin–Huxley (HH) model,[33] and the leaky integrate-and-fire (LIF) model.[34]

11.2.2.2 Synapse Models

Upon reaching the presynaptic terminal, a neuron's signal prompts exocytosis of neurotransmitters into the synaptic cleft. Among these neurotransmitters, some attach to receptors on the postsynaptic terminal, which may encompass ion channels that facilitate the movement of positively charged ions into or out of the cell. At excitatory synapses, the influx of sodium ions is common, creating an excitatory postsynaptic current/potential (EPSC/EPSP), thereby increasing the likelihood that the postsynaptic neuron will discharge an action potential.[35] In contrast, inhibitory synapses generate an inhibitory postsynaptic current/potential (IPSC/IPSP), which hinders the postsynaptic neuron from reaching the threshold required to trigger an action potential, as shown in Figure 11.5a. This modulation of the connection between neurons changes the synapse's weight through synaptic plasticity, which is crucial for learning and memory.[36,37]

Synaptic plasticity is categorized into short- and long-term based on its duration. Short-term plasticity occurring within milliseconds to seconds includes short-term

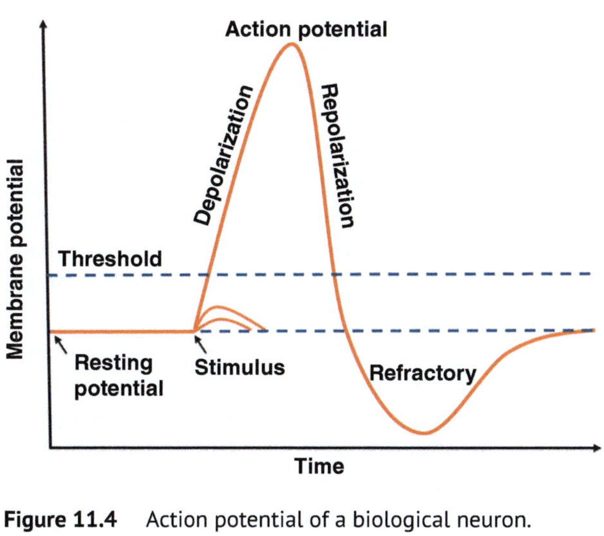

Figure 11.4 Action potential of a biological neuron.

potentiation (STP) and short-term depression (STD). Paired-pulse facilitation (PPF) is a phenomenon in which the EPSC evoked by a second pulse is larger than that of the first when the pulses are closely spaced.[37,38] The PPF ratio is calculated as PPF = $(A_2/A_1) \times 100\%$, where A_1 and A_2 are the peak EPSCs of the first and second pulses, respectively. This ratio decreases with the interval between pulses and can be modeled by a double exponential decay (Figure 11.5b).[36] Paired-pulse depression (PPD) is the opposite phenomenon, where the EPSC of the second pulse is smaller than the first, and its ratio increases with the interval between pulses.[39–41] PTP is a temporary increase in synaptic weight caused by intense, rapid synaptic activity, similar to an extension of PPF.[42] An example of PTP can be seen in Figure 11.5a–c, where 10 repeated presynaptic spikes at 50 Hz trigger the phenomenon.[43]

Long-term plasticity, which lasts for hours or more, is divided into long-term potentiation (LTP) and depression (LTD) (Figure 11.5d). LTP strengthens synaptic connections permanently, while LTD weakens them, both through repeated stimulation, as shown in Figure 11.5d.[46,47] LTP is considered crucial for memory formation in the brain.[48,49] The learning capabilities of the biological brain are fundamentally rooted in Hebbian learning principles,[50] which are mainly reflected in two types of synaptic plasticity: spike-timing-dependent plasticity (STDP) and spike rate-dependent plasticity (SRDP).[51,52] STDP denotes that synaptic weight alterations are a function of the temporal discrepancy between presynaptic and postsynaptic spikes (Figure 11.5e), whereas SRDP signifies that synaptic weight changes are contingent upon the frequency of presynaptic spikes (Figure 11.5f).

It can be confidently stated that the integrate-fire function of neurons and synaptic plasticity constitute the biological foundations of our brain's learning and memory processes. Our understanding of brain mechanisms is still limited. Nonetheless, we can still endeavor to construct neural networks and implement neuromorphic computing by simulating the structures and functions of neurons and synapses. Large-scale neuromorphic systems such as TrueNorth and Loihi, as previously mentioned, are based on traditional CMOS technology. They leveraged transistors, capacitors, inductors, and other circuit elements to emulate the functionalities of neurons and synapses for neuromorphic computing. Undoubtedly, this kind of neuromorphic device simulation is inefficient; for instance, IBM's TrueNorth chip employs 5.4 billion transistors to simulate one million programmable neurons and 256 million programmable synapses, averaging about 20 transistors per synapse. The demand for innovative devices that mimic the functionalities of neurons and synapses, known as neuromorphic devices, is increasing. Indeed, the neuromorphic community is actively engaged in extensive research aimed at creating novel materials and devices tailored for neuromorphic applications, such as memristors, PCM, electrochemical transistors, ferromagnetic devices, FE devices, etc.

Neuromorphic devices, which are capable of simultaneously storing and processing data, frequently originate from memory technologies. Among the NV memory technologies, FE memories stand out for their low power consumption, rapid operation, and robust resistance to fatigue, rendering them a subject of intense research. The spontaneous polarization of FE thin films, which is NV, can be reversed by applying an electric field. By controlling the intensity and duration of this electric field, the

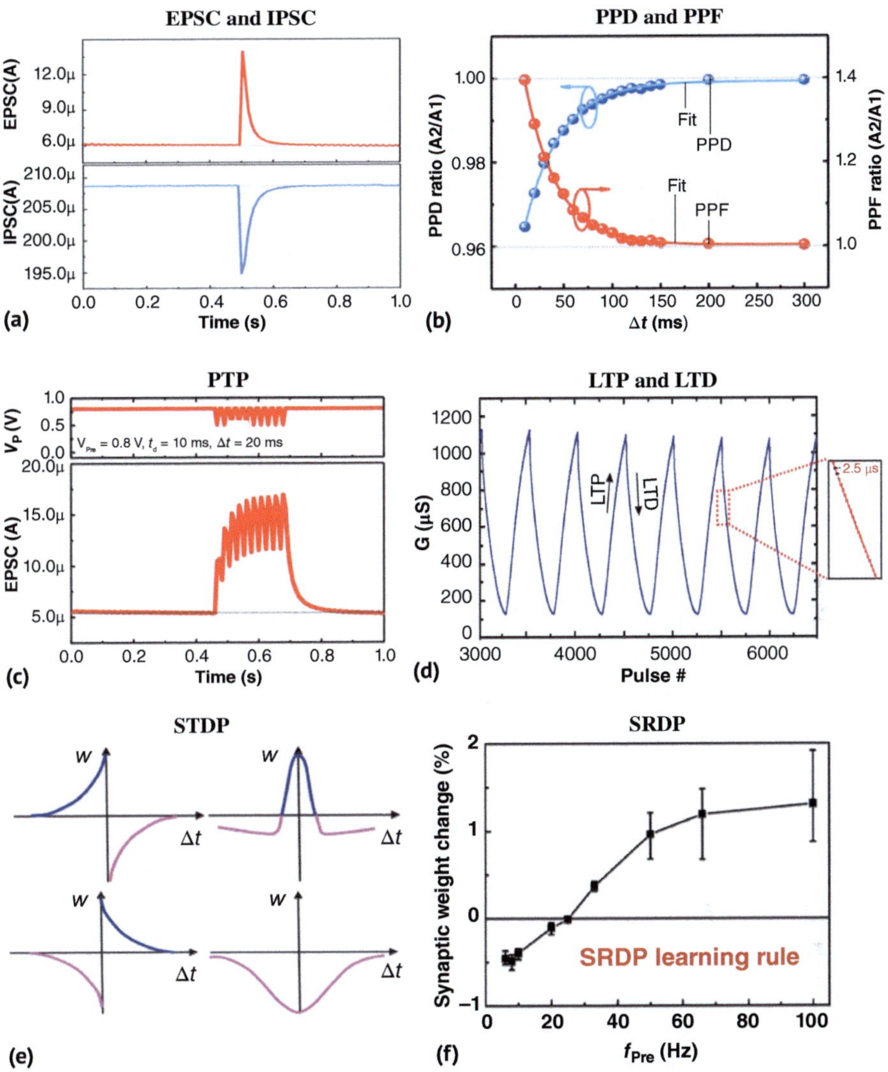

Figure 11.5 Synaptic plasticity and Hebb learning rules. (a) Typical postsynaptic current behaviors: EPSC & IPSC; (b) PPD & PPF; (c) PTP. *Source*: [43] / with permission of John Wiley & Sons. (d) Demonstration of LTP & LTD. (e) Typical STDP behaviors. *Source*: [44] / with permission of John Wiley & Sons. (f) SRDP learning rule. *Source*: [45] / with permission of American Chemical Society.

state of the FE domain can be continuously altered, emulating the synaptic plasticity observed in biological synapses. Moreover, the accumulative switching behavior of the FE domains resembles the integrate-fire characteristics of the neurons. Consequently, FE devices are well-positioned to serve as exceptional candidates for neuromorphic applications. Later in this chapter, we will delve into the properties of HfO_2-based materials and their utilization in neuromorphic devices, with a particular emphasis on the role of HfO_2-based FE devices in these applications.

11.3 Properties Related to Neuromorphic Devices

Hafnia-based materials possess a plethora of physicochemical properties, which have paved the way for their widespread application in the fields of electricity, optics, mechanics, and thermal protection.[53] However, in this section, we will concentrate primarily on material properties pertinent to the application of neuromorphic devices, including their FE and antiferroelectric (AFE) properties, dielectric and valence-change characteristics, and the ultrahigh oxygen-ion conductivity associated with the FE phase. A detailed discussion is provided on the correlation between the FE properties and functionality of neuromorphic devices.

11.3.1 Ferroelectricity and Antiferroelectricity

The most notable characteristic of hafnia-based materials is their exceptional FE properties. Since the discovery of ferroelectricity in Si-doped HfO_2 in 2011,[54] hafnia-based FE materials have garnered widespread interest from researchers in both academia and industry.[4,5,55] Hafnia-based materials exhibit a highly intricate and sensitive crystal structure,[56] and their structure and FE performance are influenced and regulated by factors such as dopants, oxygen vacancies, stress, electrode materials, and field cycling. Taking the most extensively studied $Hf_xZr_{1-x}O_2$ (HZO) films as an example, it was observed that pure HfO_2 films generally display paraelectric behavior, whereas pure ZrO_2 films usually exhibit AFE characteristics. However, when doped at the correct ratios, they display pronounced FE properties, as shown in Figure 11.6a.[57] This essentially corresponds to the changes in the structural phases of the films as a function of the doping concentration: HfO_2, which tends to stabilize in the monoclinic (m) phase, and ZrO_2, which tends to stabilize in the tetragonal (t) phase (space group: $P4_2/nmc$), HZO stabilizes in the $Pca2_1$ (o-I) phase within a suitable doping range. Currently, this o-I phase is widely considered to be the primary source of ferroelectricity in hafnia-based FE materials.

Hafnia-based FEs exhibit several appealing characteristics, notably their compatibility with CMOS technology and ability to maintain stable FE properties even at extremely thin thicknesses. These attributes strongly indicate their considerable potential for future applications in the micro- and nano-electronics industries.[4] Moreover, hafnia-based FEs exhibit a distinct characteristic that distinguishes them from traditional FE materials. Density functional theory (DFT) calculations revealed the presence of flat polar phonon bands in hafnia-based FEs, leading to extreme localization of electric dipoles within the confines of their irreducible half-unit cells.[55] This is markedly different from the dispersed dipoles typically observed in conventional FEs. The capacity to individually control these localized dipoles offers unprecedented opportunities for the development of dense, cell-by-cell FE switching devices. However, achieving unit-cell-level polarization control is not yet feasible. Polarization switching in FE materials is a predominantly domain-driven process. The FE domains are closely related to the grains, and the dimensions, shapes, and orientations of these grains influence the morphology, migration, and evolution of the domains, thereby affecting the macroscopic electrical properties

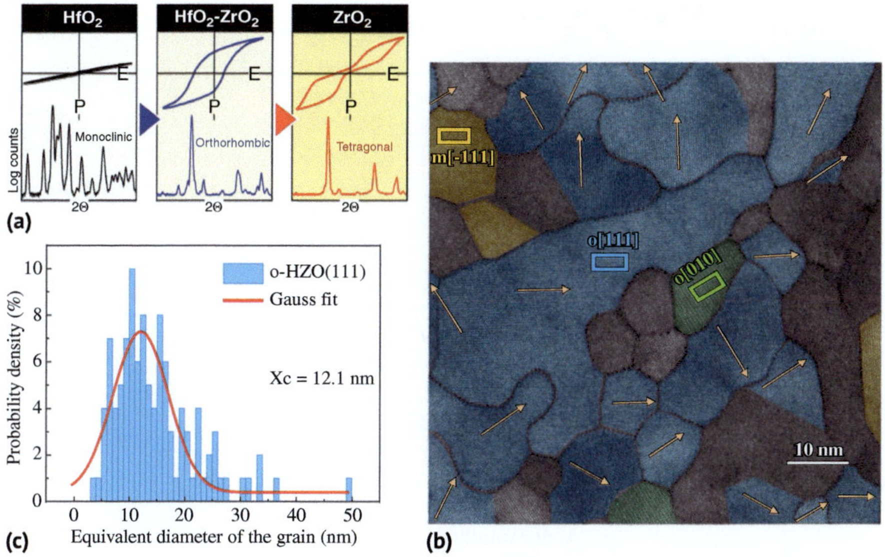

Figure 11.6 The AFE properties of hafnia-based materials. (a) The ferroelectricity and antiferroelectricity of $Hf_xZr_{1-x}O_2$. *Source*: [57] / with permission of American Chemical Society. (b) Atomically resolved plan view of the 10nm HZO film and (c) Grain-size distribution of the FE o-(111) phase in the 10nm HZO film. *Source*: [58] / with permission of John Wiley & Sons.

of FE materials. Enhancing the performance of FE materials and gaining microscopic insights into novel FE devices are essential for meticulously characterizing and controlling the microstructure of hafnia-based FEs. In our recent work,[58] we employed scanning transmission electron microscopy (STEM) to conduct an atomic-level analysis of HZO films, enabling us to accurately ascertain the distribution, proportion, and orientation of various phases within the thin films, as well as the dimensions and grain boundary structure of the domains and grains. We discovered that the average equivalent diameter of the grains is approximately 12.1nm, with a standard deviation of approximately 10nm, and a diameter ranging from 4 to 50nm, as shown in Figure 11.6b, c.

In FE devices, the resistance or capacitance state is determined by the polarization state of the FE domains. The writing operation for all types of FE devices involves the reversal of the FE domains, which is typically executed by applying an electric field that surpasses the coercive field (E_c) of the FE layer. For hafnia-based FEs, their E_c ranges widely from 0.65 to 5.0 MV/cm, influenced by factors such as dopants, strain, and electrode materials.[6] Depending on the device dimensions and the configuration of the material stacks, three types of switching dynamics can be observed, as illustrated in Figure 11.7. In large-area devices, an ensemble of multiple domains dictates the switching kinetics. When an electric field significantly exceeding the E_c threshold is applied, all FE domains undergo uniform polarization reversal, resulting in digital switching of the device, which facilitates the storage of binary data. In small-area devices with a single domain or very few

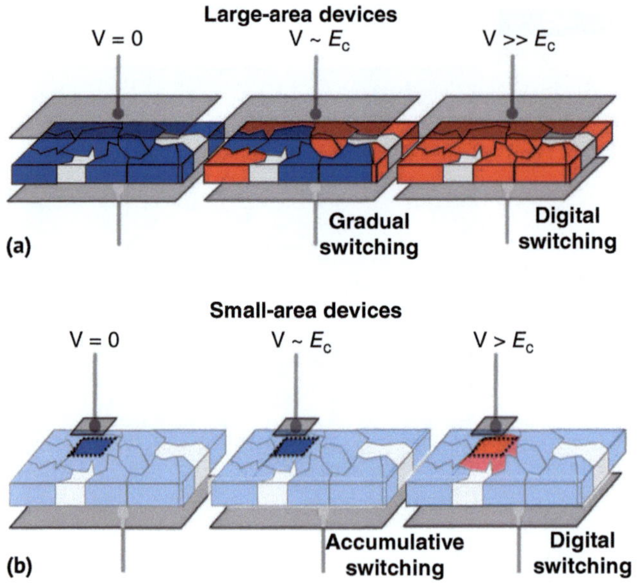

Figure 11.7 Switching kinetics of FE devices. (a) In large-area devices, different electric field stimuli produce gradual and digital switching. (b) In small-area devices, different electric field stimuli produce accumulative and digital switching. *Source*: [60] / John Wiley & Sons / CC BY 4.0.

domains, the application of a single electric field pulse that is slightly less than the E_c value does not result in polarization switching. However, the cumulative effect of multiple electric field pulses leads to domain reversal, thus endowing the device with accumulative switching characteristics,[59] which can be used to simulate the integrate-fire function of the neurons. Owing to the internal gain of the FeFET and its nondestructive indirect readout mechanism, this is the only device where this effect can be observed. When the electric field exceeded E_c, digital switching was also observed. Therefore, for FE synaptic devices to achieve multiple resistance or capacitance states, larger device dimensions are required. When developing FE synaptic devices based on thin films with a grain feature size of approximately 10nm, as observed earlier, the minimum size threshold should be in the order of tens of nanometers. By contrast, FE neuron devices typically require smaller device dimensions to perform their functions. To optimize device performance, crystallization engineering is essential for precisely controlling the orientation and size of the FE film grains.

Different device configurations and functionalities typically require different properties of FEs. For instance, the remanent polarization (P_r) value reflects the amount of charge retained in FE capacitor structures, and a high P_r value is crucial for FeRAMs. However, in FeFETs, an excessively high P_r can lead to a substantial depolarization field (E_{dep}) and an increased concentration of trapped charges, which would compromise device reliability. Therefore, from a materials science perspective, to achieve superior device performance or to cater the diverse requirements of different devices, it is necessary to control and optimize multiple parameters of hafnia-based FE thin films, including P_r, E_c, polarization switching speed, retention, endurance, and spatial uniformity. For an in-depth discussion on these subjects, Ref. [60] can be consulted.

Although current technological capabilities do not permit polarization control at the single-cell level, this concept is theoretically feasible for hafnia-based FEs (Figure 11.8). A study employing integrated differential phase contrast (iDPC)-STEM technology for atomic-scale imaging of ZrO_2 thin films demonstrated that the local electric field generated by electron irradiation can induce polarization reversal at the single-cell level, as shown in Figure 11.8c,d.[61] Although the structure in Figure 11.8d could be explained by the *Pbca* (o-II) phase, the reversible polarization switching at the single-cell level in the experiment does not strictly adhere to the antipolar structure of the *Pbca* phase, and still retains macroscopic FE properties. This result also suggests that with the right manipulation techniques, it may be possible to achieve multilevel memory or synaptic plasticity in FE devices of the several-nanometer scale.

Compared with ferroelectricity, the AFE characteristics of hafnia-based materials have garnered relatively less attention. Increasing evidence suggests that hafnia-based AFEs possess significant potential for a variety of applications.[62] From a microscopic perspective, in traditional AFEs, the electric dipoles in adjacent unit

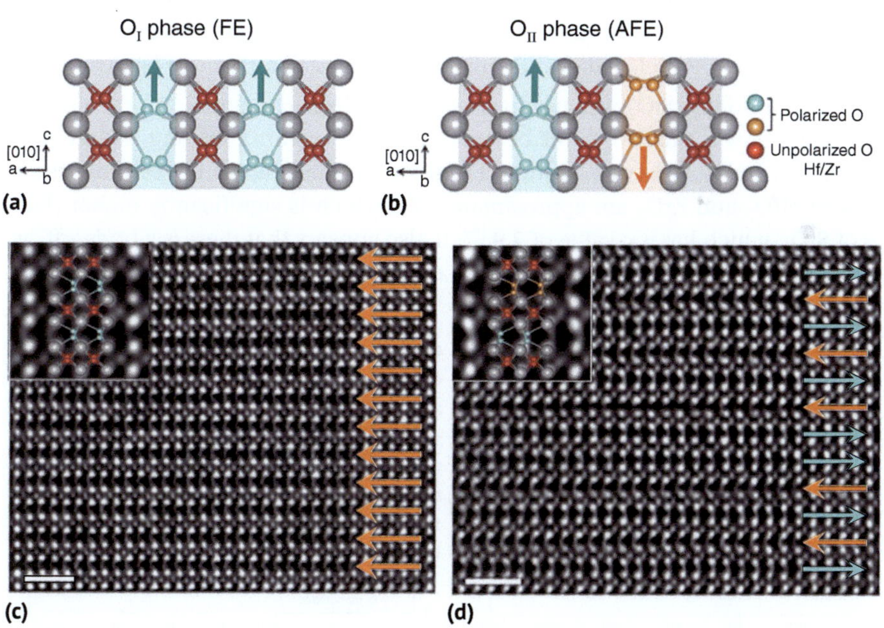

Figure 11.8 Single-cell domain switching in hafnia-based materials. (a) and (b) show lattice structure projection diagrams of the *Pca*2_1-FE and *Pbca*-AFE phases, respectively. (c) and (d) are iDPC-STEM images of the same grain at different moments under electron irradiation, with arrows indicating the polarization orientation. The insets show magnified views of local structures. *Source*: [61] / with permission of John Wiley & Sons.

cells are arranged antiparallel, resulting in the cancellation of macroscopic polarization, which renders them unobservable in the absence of an E_{ext}. As illustrated in Figure 11.8a, b, the o-II phase exhibits typical AFE characteristics. Interestingly, in addition to the traditional AFE o-II phase, hafnia-based materials also display typical AFE properties in their t-phase, which is not an inherent AFE structure. It is widely recognized that hafnia-based films rich in ZrO_2 typically stabilize in a nonpolar t-phase at zero or low E_{ext}. When E_{ext} exceeds the critical field, a phase transition occurs from the nonpolar t-phase to the polar o-phase. Upon removal of E_{ext}, the material reverts to the nonpolar t-phase, with any residual polarization dissipating under zero-field conditions. Consequently, in P-E measurement, the nonpolar t-phase exhibits double hysteresis loops, which are characteristic of AFE properties.[57,63] This feature indicates that hafnia-based AFEs are suitable for simulating the integrate-fire function of neurons. In summary, the inherent FE and AFE properties of hafnia-based materials determine their potential in neuromorphic devices.

11.3.2 Dielectric and Valence-change Properties

Dielectric insulators play a crucial role in microelectronic devices by enhancing the capacitive coupling between adjacent metals and semiconductors. It is essential to curtail the leakage current between electrodes to prevent negative impacts on energy consumption and device reliability.[64] Actually, prior to its recognition of FE properties, it had already been utilized in CMOS technology as a gate dielectric material owing to its high κ. An ideal dielectric material possesses both a large bandgap (E_g) and high κ. Initially, SiO_2 was used as the gate dielectric material. As the devices were scaled down, the SiO_2 layer became very thin (<2nm), leading to excessive gate leakage and power consumption. From Figure 11.9a, it can be seen that the κ values of HfO_2 and ZrO_2 are approximately 25, which is significantly higher than that of SiO_2, which has a κ-value of 3.9.[64] It also appears that there is a trade-off, as no single material can have both large E_g and high κ. As a compromise, HfO_2 and

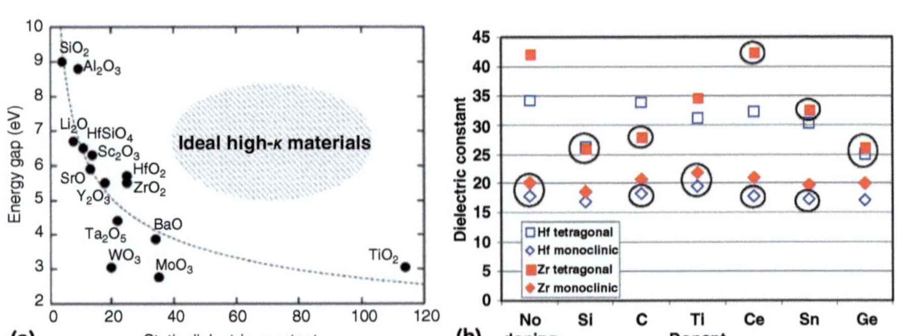

Figure 11.9 (a) Band gap and dielectric constants of well-known oxides. *Source*: [66] / with permission of Springer Nature. (b) Dielectric constants of HfO_2 and ZrO_2 for undoped oxides and doped systems at a doping level of 12.5%. *Source*: [67] / with permission of AIP Publishing.

ZrO_2 have become favored dielectrics due to their relatively high κ values, thermal stability, excellent silicon interface quality, and low density of bulk defects.[65]

It should be noted that hafnia-based materials may exhibit different crystal structures under various conditions, such as dopants, stress, and temperature, each associated with distinct dielectric constants. The cubic and tetragonal phases of these materials have a higher κ value than the thermodynamically most stable m-phase.[68] Dominik Fischera and Alfred Kersch conducted a detailed study on the impact of different dopants on the dielectric properties of HfO_2 and ZrO_2.[67] They found that dopants can stabilize the t-phase, which in turn increases the dielectric constant of the material, with Si being the most effective stabilizing atom. ZrO_2 tends to be more readily stabilized in the t-phase than HfO_2, and under identical dopants and crystal phases, it typically exhibits higher κ values. As shown in Figure 11.9b, the highest κ-value exceeding 40 was achieved in the t-phase of ZrO_2. Crystallization engineering is an essential method for enhancing the dielectric performance of these materials. Sabina Spiga et al. have demonstrated that doping HfO_2 with 4.5% Al can lead to a higher proportion of cubic/tetragonal phases after thermal annealing, thereby attaining a κ-value of approximately 32.[69]

In addition to their high κ and large E_g, hafnia-based materials exhibit valence-change properties, which are characterized by local electrochemical oxidation-reduction reactions under the influence of an electric field. When exposed to an electric field, these materials can drive oxygen ions to vacate their original lattice positions, effectively transforming HfO_2 into $HfO_{2\text{-}x}$. These unique characteristics have led to the widespread use of hafnia-based materials in RRAM-type memristors, which are among the earliest electronic devices proposed for simulating synaptic functions. The typical structure of HfO_2-based RRAM devices consists of a HfO_2 layer serving as the switching medium, with thicknesses ranging from 5 to 30 nm, and electrodes fabricated from metals or nitrides such as Pt, Au, W, Ni, TiN, and TaN at both the bottom and top.[70] The switching mechanism is often described as filamentary switching, with resistance changes attributed to ionic migration and local redox reactions.[71] The forming process in these devices is recognized as a time-dependent dielectric breakdown. Under a strong electric field, oxygen atoms are displaced from the lattice of the insulating layer and migrate toward the anode. Simultaneously, oxygen vacancies, which are considered a type of defect, were generated and accumulated near the cathode. The localized concentration of oxygen vacancies leads to the formation of a conductive filament in the sub-oxide phase bridging the anode and cathode, resulting in resistive switching from a high-resistance state (HRS) to a low-resistance state (LRS). The transition from the LRS back to the HRS is due to the rupture of the conductive filament. Applying a negative voltage pulse to the anode triggers reverse ionic migration, neutralizing the oxygen vacancies and converting the sub-stoichiometric region to a stoichiometric composition. In the latest research on HfO_2-based RRAM devices, over 2,048 conductance levels were achieved with memristors in fully integrated chips featuring 256 × 256 device arrays monolithically integrated on CMOS circuits,[72] demonstrating significant potential in a neuromorphic computing system.

11.3.3 Ultrahigh Oxygen-ion Conductivity

As previously mentioned, the motion of oxygen ions is a key factor in both resistive switching of hafnia-based RRAMs and polarization switching of hafnia-based FEs. The difference lies in that the former is characterized by the thermally activated, long-range movement of oxygen ions, whereas the latter involves the collective and coordinated local motion of oxygen ions propelled by an external electric field. In hafnia-based FEs, oxygen migration and polarization switching appear to be intertwined, and discerning whether their effects are synergistic or independent is crucial for device manipulation. In 2021, Nukala et al. conducted in-situ measurements on $La_{0.67}Sr_{0.33}MnO_3$ (LSMO)/HZO capacitors using atomic resolution microscopy with direct oxygen imaging and in-situ electrical bias.[73] Their experimental findings indicated that oxygen ions are capable of reversible and hysteretic migration through the HZO layer from the bottom electrode to the top electrode. When reactive top electrodes such as TiN and LSMO are used, the HZO layer merely acts as a conduit for the reversible migration of oxygen between the electrodes, where topological changes occur at the electrodes. However, when noble metal top electrodes such as Au are used, HZO acts both as a sink and source of oxygen ions. This migration can occur within milliseconds. This evidence indicates a strong correlation between the oxygen-ion migration and polarization switching in hafnia-based FEs.

Extensive theoretical work has delved into the relationship between polarization switching and migration pathways of oxygen ions in hafnia-based FEs.[74–77] There are currently two prevailing viewpoints: one posits that oxygen ions are more inclined to engage in shift-across motion,[78] while the other contends that shift-inside polarization switching presents a lower energy barrier.[75] However, there is a consensus that under the influence of a substantial electric field, both shift-across and shift-inside mechanisms could be active concurrently. Recently, Liyang Ma et al. explored three polarization switching pathways in FE-HfO_2 through large-scale molecular dynamics simulations, as illustrated in Figure 11.10a.[79] Their results

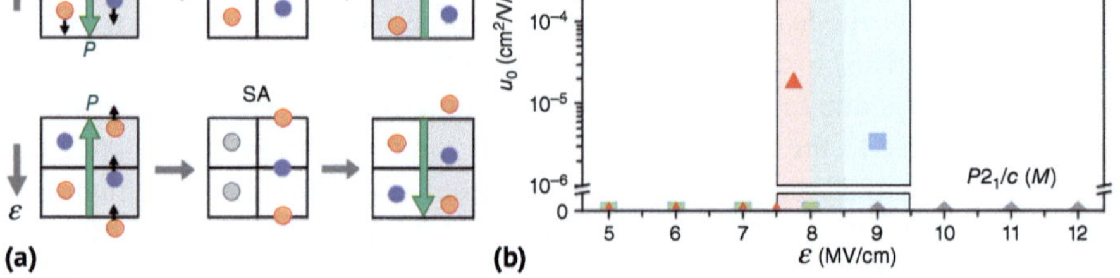

Figure 11.10 (a) Polarization switching pathways in FE HfO_2. (b) Ferroelectricity-promoted oxygen ion mobility in *Pca*2$_1$ HfO_2. *Source*: [79] / with permission of American Physical Society.

indicated that when subjected to an E_{ext} of a magnitude pertinent to the operational conditions of the devices, all three oxygen motion pathways can be triggered at room temperature. They quantitatively estimated the mobility of oxygen ions in FE-HfO_2 across a specific temperature and E_{ext} range, as depicted in Figure 11.10b. Close to the temperature-dependent critical electric field, the mobility of oxygen ions in FE-HfO_2 experiences a sharp increase from near zero to a significant value of approximately ~$10^{-3}cm^2/V/s$. However, in m-HfO_2 at 600 K, even with a larger E_{ext}, the mobility remains near zero. This bias-driven continuous FE switching in FE-HfO_2 fosters exceptionally high mobility of oxygen ions at moderate temperatures, highlighting the potential of combining ferroelectricity and ionic conductivity for the development of innovative materials and technologies.

Considering that doped ZrO_2 and doped CeO_2, both of which possess a fluorite structure, are extensively utilized in solid oxide fuel cells because of their excellent oxygen-ion conductivity at elevated temperatures,[80] it is reasonable to anticipate that hafnia-based materials would also demonstrate exceptional oxygen-ion conductivity. However, it is important to highlight that the ultrahigh oxygen-ion conductivity mentioned in this context originates from the sustained switching of FE polarization, which is unique to hafnia-based FE materials. The synergistic integration of ferroelectricity and ionic conductivity paves the way for the development of innovative device architectures such as FE chemical ionic synapses.

11.4 Neuromorphic Devices based on Hafnia-based Materials

Based on our previous discussion, hafnia-based materials with FE and valence change properties are prime candidates for applications in FE memory devices and RRAM. The device structure and operating principles are shown in Figure 11.11.

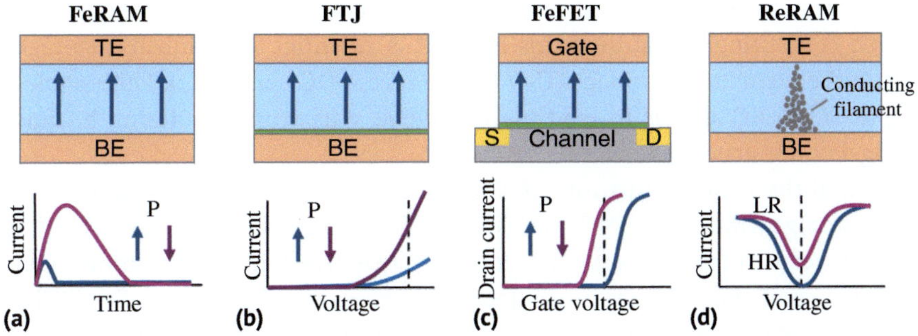

Figure 11.11 Schematic of memory devices based on FE materials. (a) In a FeRAM, if a field larger than the E_c is applied to the FE, a large or small current flows through the FE, depending on whether it switches. (b) In an FTJ, the current flowing through the structure depends on the polarization direction. (c) In a FeFET, the threshold voltage of the FET depends on the FE film polarization. (d) In RRAM, the conductance of the device depends on the state of the conducting filaments in the insulator.

These devices often show significant variations in material stack design, thermal budgets during the manufacturing process, integration strategies, programming conditions, and operational mechanisms.[70] This diversity suggests that they can be tailored for specific applications or functions within different areas of the same chip. In the subsequent sections of this chapter, we provide an in-depth examination and analysis of these materials, focusing on their roles in artificial synapse devices and artificial neuron devices.

11.4.1 Artificial Synapses based on Hafnia-based Materials

11.4.1.1 FeRAM-type Synapses

The concept of the FeRAM was first proposed by Buck in 1952. In its traditional form, FeRAM mirrors the structure of DRAM memory cells, consisting of a transistor and an FE capacitor (denoted 1T-1C). The information is stored in the form of a polarization state and is read by switching the polarization in the capacitor, as shown in Figure 11.11a. FeRAM, which uses traditional perovskite FE materials, encounters limitations owing to the requirement of a significant thickness for maintaining FE states and the extensive area required for charge detection, making it difficult to achieve high-density integration.[81] Fortunately, the emergence of hafnia-based FE materials, along with back-end-of-line (BEOL) compatibility, low-voltage operation,[82] and nanosecond-level switching speeds,[83] has successfully addressed these challenges.

The write operation of the FeRAM involves applying a voltage between the electrodes to realign the dipoles in the FE interlayer. By applying an appropriate voltage, it is possible to achieve partial polarization in the multidomain interlayer, which facilitates multilevel data writing. Once the writing is complete, the FE interlayer retains its polarization state, ensuring NV memory retention. The read operation in the FeRAM is executed by applying a read voltage to ascertain its polarization state and measuring the charge that passes through the capacitor. This typically requires a sufficiently large voltage to realign the dipoles of the FE material. Consequently, the read operation of the FeRAM is destructive, rendering it unsuitable for neuromorphic computing. To address this issue, it is essential to establish a storage window in the capacitance–voltage (C–V) curve at zero DC bias (V_{bias}). This enables the use of small AC signals to read the capacitive state without altering it, thus achieving a nondestructive readout.

Yu et al., at the Georgia Institute of Technology, successfully achieved a stable capacitive storage window in TiN/HZO/TiN capacitors, as shown in Figure 11.12a.[84] Here, the dielectric constant (ε_r) represents the capacitive state. The asymmetry in the capacitance of the device is attributed to an excess of oxygen vacancies at the bottom electrode interface, which induces domain wall (DW) pinning effects (even after positive scanning, it remains polarized downward), leading to a higher number of DWs and, consequently, a greater charge in the high-capacitance state (HCS), as depicted in Figure 11.12b. Subsequently, they constructed an NV capacitive crossbar array based on this asymmetric HZO capacitor.[85] This capacitive array leveraged the small-signal capacitance state as weights, consuming only dynamic power.

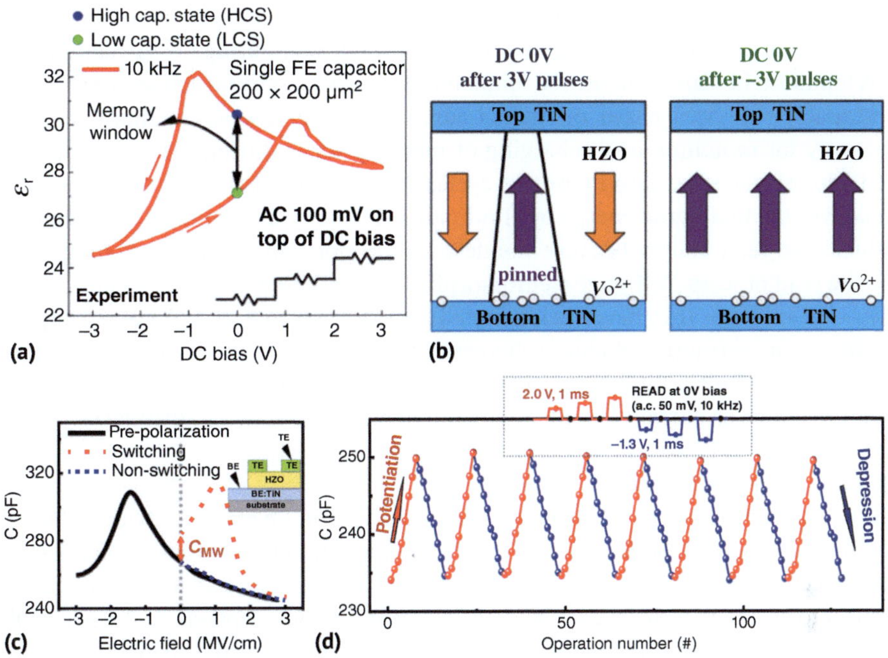

Figure 11.12 FeRAM for synaptic devices. (a) Asymmetric C–V characteristics of TiN/HZO/TiN capacitors. (b) Physical illustration of asymmetric C–V at a DC of 0V. The positively charged oxygen vacancies at the BE pinned some domains up-polarized. *Source*: [85] / John Wiley & Sons / CC BY 4.0. (c) C–V curves of the TiN/HZO/Pt capacitor under different historical polarization states and (d) potentiation and depression characteristics of the capacitance states under incremental amplitude pulses. *Source*: [86] / with permission of AIP Publishing.

It effectively prevents the quiescent path current and IR drop, demonstrating faster speeds and higher energy efficiency in in-memory computation.

Recently, following a similar approach, Wu et al. also effectively modulated a stable capacitive storage window at zero V_{bias} by introducing electrodes with different work functions, as shown in Figure 11.12c.[86] The polarization states of the capacitors were written with voltage pulses of varying amplitudes and read using an AC signal (50 mV, 10 kHz) at zero bias. The devices were able to sustain 8 distinct analog states over 100 cycles of enhancement and suppression, demonstrating excellent cycle-to-cycle consistency, as shown in Figure 11.12d. These results demonstrate the potential application of FeRAM devices in emulating synaptic functions. Additionally, Parkin et al. demonstrated a scheme for neuromorphic computing using FE capacitor devices in 2021.[87] They utilized an FE capacitive device with a structure similar to that of a field-effect transistor, named memcapacitor devices. Their method of using memcapacitor devices for neuromorphic computing showed remarkable energy efficiency, with 29.6 trillion operations per second per watt while ensuring high precision (6–8 bits). This technology has the potential to be scaled down to a lateral dimension of approximately 45 nm.

Lee et al. recently discussed the advantages and disadvantages of resistive and capacitive approaches in achieving neuromorphic computing.[88] A significant development in this field is the implementation of nondestructive readout in FE-capacitor devices based on hafnia-based FE materials, as it eliminates the necessity for switching or discharging of memory states during the access operation. This obviates the need for refresh cycles. This characteristic of FeRAM not only conserves the total write energy by eliminating refresh cycles, but also significantly reduces the read energy because the device does not undergo FE switching during the read process. Although neuromorphic devices based on FeRAMs are still in their nascent stage, they have demonstrated considerable potential. However, implementing nondestructive readout schemes for device capacitance may increase the complexity of the circuit design. In contrast, readout schemes focused on electrical quantities such as current or voltage could potentially offer more advantages.

11.4.1.2 FTJ-type Synapses

The typical structure of an FTJ closely resembles that of an FE capacitor, with the main difference being that the FE layer in the FTJ is much thinner (1–5nm), which facilitates effective electron tunneling. Additionally, asymmetric electrodes are essential in the FTJ to ensure that the barrier height at electron injection depends on the polarization direction, as shown in Figure 11.13a. It is also common practice to incorporate an intermediate layer at the FE/electrode interface to improve device performance.[89,90] The polarization state of the FE layer, which corresponds to the resistance state of the device, is used for information storage.[91] For an FTJ device with a multi-domain structure, assuming that all domain orientations are perpendicular to the electrodes, the FTJ can be considered as a bundle of parallel resistors, as shown in Figure 11.13b. Multiple polarization states can be achieved by applying an appropriate programming voltage. Clearly, the FTJ devices can achieve multilevel memory and simulate synaptic learning. More importantly, the read operations in FTJs are nondestructive by applying an electric field that is less than E_c.

Weight updates in the FTJ-type synapses are accomplished by applying voltage pulses. Obviously, the number of accessible storage states is intrinsically linked to the absolute number of FE domains within the device, and the dimensions of the domains depend on the material properties of the thin film. It is evident that analog functionality deteriorates when the device size becomes comparable to the FE grain size. Moreover, weight updates require robust linearity, and superior linearity contributes to increased computational precision. Therefore, the pulses utilized for weight updates can be optimally designed based on the exponential relationship between pulse duration and the fraction of switched domains.[95,96] Similarly, higher linearity can be attained by programming voltage pulses post-amplitude adjustment, as shown in Figure 11.13c.[92] For the convenience of implementation with CMOS technology, it is advisable to employ the same pulse sequence to progressively change the device state. From a circuit-design perspective, this method is simpler and more efficient. One solution is to harness the cumulative switching effect of FE materials by integrating the effects of multiple pulses with amplitudes below E_c to achieve gradual changes in the device state.[97] However, this approach carries the risk of unintended data manipulation in suppressed units during iterative write operations in FTJ arrays. We propose a complementary solution here, which is to utilize the relationship between E_c and grains for reverse grain control, optimize the grain size distribution and stress environment, and disrupt the Lorentz distribution characteristics of the polarization reversal time, making it possible to achieve relatively linear weight updates with the same pulse.

FTJ-type synapses are capable of implementing STDP, where the synaptic strength varies with the temporal sequence and duration between the presynaptic and postsynaptic action potentials, as shown in Figure 11.13d.[92] Recently, algorithms such as STDP learning schemes have been leveraged for unsupervised training within SNNs, indicating that the learning outcomes significantly depend on the STDP profile of the electrical synapses.[98] Additionally, to tailor the synaptic behavior to different requirements, the plasticity behavior of FTJ-type synapses can be adjusted by changing the pulse amplitude and pulse width of the presynaptic and postsynaptic action potentials, consequently modifying the shape of the STDP curve.[99]

The tunneling electro-resistance ratio (TER) is a key metric for FTJs because a superior TER is indispensable for reducing read/write errors and energy expenditure. However, HfO_2-based FTJs typically exhibit much lower TERs, only in the range of a few to tens.[100,101] Researchers usually explore electrode engineering to

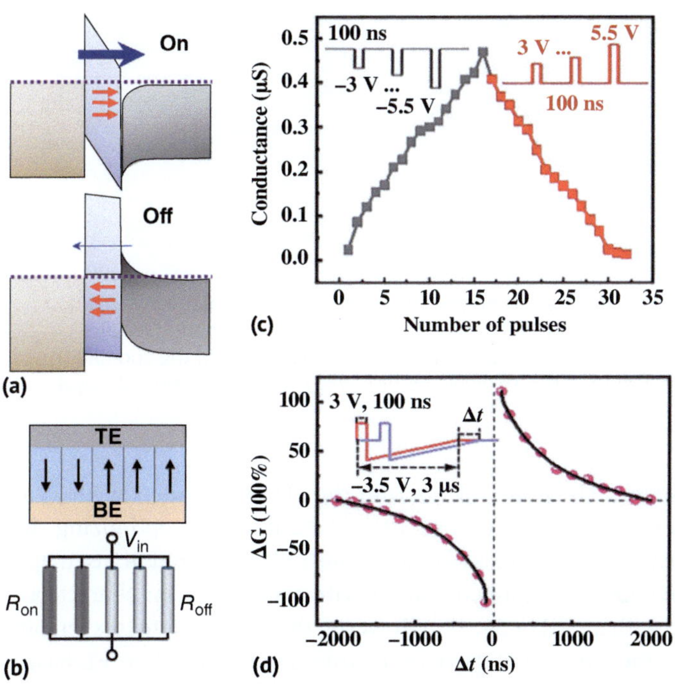

Figure 11.13 FTJs for synaptic devices. (a) Schematic representation of the potential energy profile of an FTJ. (b) Schematic domain configuration for the FTJ and the corresponding parallel resistor model. (c) The LTP, LTD and (d) STDP curves of an FTJ-type synapse. *Source*: [92] / with permission of John Wiley & Sons.

(Continued)

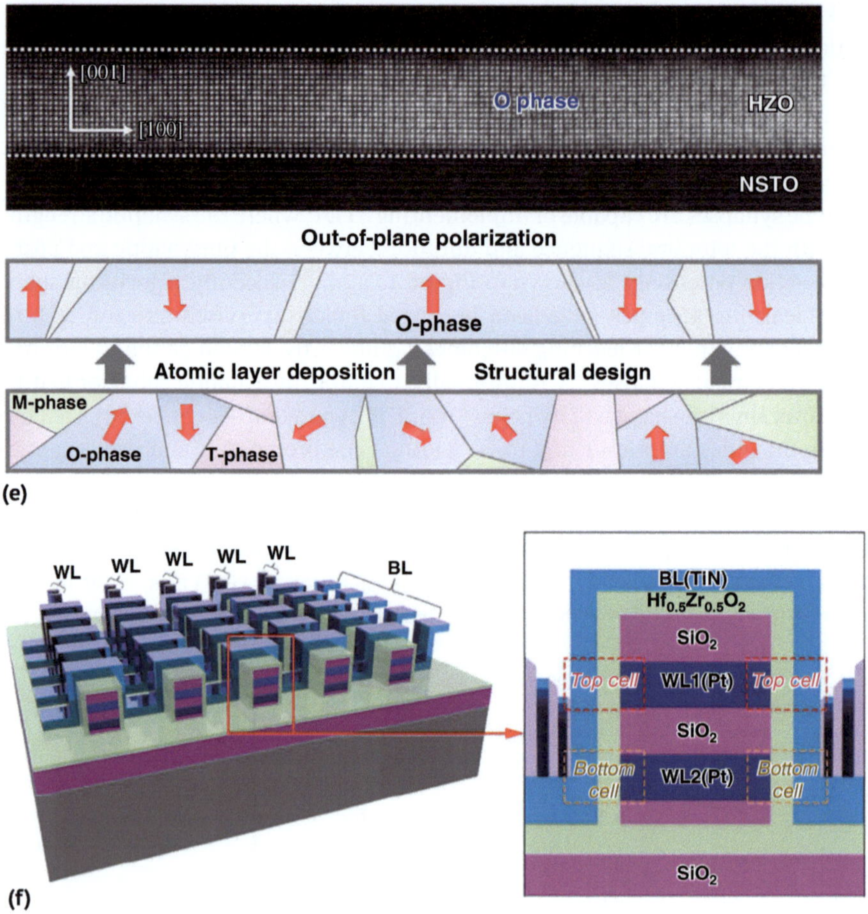

Figure 11.13 (Cont'd) (e) Crystallization-engineering-induced out-of-plane polarization and crystal structure control in HZO films. *Source*: [93] / with permission of Elsevier. (f) Schematic diagram of high-density 3D vertical HZO-based FTJs array and the zoomed-in schematic of the devices located at the sidewall. *Source*: [94] / with permission of Royal Society of Chemistry.

enhance TER. For instance, HZO-based FTJs with the 2D semiconductor MoS_2 as electrodes exhibited a commendable TER of 10^3.[102] Additionally, optimizing the properties of the tunneling layer can also enhance the TER. Müller et al. used epitaxial Y-doped HfO_2 to endow the tunneling layer with semiconductor characteristics, rendering a device with a TER of 540.[103] However, these measurements remain considerably less than the functional capacity of traditional perovskite FE-based FTJs.[104] Gao et al. recently unveiled a noteworthy breakthrough; they employed crystalline engineering to fabricate [001] textured HZO thin films on Nb-doped $SrTiO_3$ substrates (Figure 11.13e), achieving an extremely high TER of up to 2×10^7.[93] This result demonstrates the tremendous potential of crystalline engineering in enhancing the performance of FE devices.

As two-terminal devices, an often-overlooked advantage of hafnia-based FTJs is their immense potential for high-density 3D integration, which plays an instrumental role in enabling large-scale parallel and dense neuromorphic systems. Chen et al. fabricated a 3D vertical FTJ array based on HZO, as illustrated in Figure 11.13f.[94] This is a promising method for high-density integration of two-terminal devices, as more FTJ layers can be incorporated in 3D by cyclically depositing metal and insulator layers. However, the typical current density of hafnia-based FTJs is less than 1 pA/μm^2. While this low current enables ultralow-power synaptic operation, it also poses significant challenges for readout. Therefore, novel approaches should be investigated to strike a balance between power consumption and readout capability. Nonetheless, compared to other FEs, FTJs stand to benefit the most from the advantages of hafnia-based FEs.[8] FTJs offer considerable promise for the development of neuromorphic computing applications.

11.4.1.3 FeFET-type Synapses

As shown in Figure 11.11c, the FeFET has a three-terminal structure similar to that of the MOSFET. The incorporation of FE materials into the gate stack ensures non-volatility.[105,106] Compared to two-terminal devices, FeFETs demonstrate distinct advantages in mimicking synaptic functions. First, the write and read paths are separated spatially. Input pulses from the presynaptic neuron are applied to the gate electrode, whereas electrical signals from the postsynaptic neurons originate from the drain electrode. This allows for the complete decoupling of signal transmission and weight updates during readout, thereby supporting intricate neuromorphic systems.[107] Second, gate contact allows deselecting specific devices and applying specific suppressive voltages in array configurations for selective programming. Furthermore, it mitigates the sneak current in crossbar arrangements.[108] Third, owing to their similarity to high-κ metal-gate transistors, they can effortlessly integrate into highly scalable technology nodes.[109] Fourth, owing to the capacitive coupling between the polarization state and the channel surface potential, a high on/off ratio can be achieved.[108] This is beneficial for increasing the number of multilevel states. Fifth, the polarization state of the FE layer in the FeFETs determines the conductance state of the channel. For large-scale FeFET devices with a multi-domain structure, the polarization state can be modulated by gate pulses, allowing for the gradual alteration of their threshold voltage and channel conductance under the influence of an external electric field, as shown in Figure 11.14a,b.[110] By applying multiple voltage pulses with appropriate amplitude and duration, the weight values in FeFET-type synapses can be precisely controlled.[111]

The application of FeFETs presents certain obstacles in mimicking neuromorphic devices. The use of an additional terminal increases the structural complexity and space occupation of the device.[108] Similar to FTJs, they grapple with a compromise between the device dimensions and the capacity of stored states. For relatively large-scale FeFETs, the channel conductance can be adjusted gradually. However, highly scalable synaptic transistors are indispensable for the realization of dense neuromorphic computing systems. As the gate size shrinks, multilevel states become constrained by the number of FE domains. When the size drops to less than approximately 30nm, FeFETs exhibit abrupt and accumulative switching behavior, which is more suitable for constructing artificial neurons.[115] Fortunately, compared with FTJs, FeFETs might potentially surmount this hurdle with relative ease. For instance, technologies such as BEOL FeFETs and 3D FeFETs can mitigate the device size requirements by providing a sufficient number of switchable domains within the gate stack.[116] Additionally, Zeng et al. proposed a method to adjust the electric field gradient on the HZO FE film along the FeFET channel direction, which is capable of efficiently controlling various NV storage states in FeFETs with a channel length of less than 20nm.[117] Considering the unique properties of hafnia-based FE materials, the critical domain size can reach the scale of a single cell, and meticulous engineering could afford robust multilevel control in nanometer-scale FeFET devices.

There are various device configurations in FeFET; depending upon their gate stack structure, they can be categorized into metal–ferroelectric–semiconductor (MFS), metal–ferroelectric–insulator–semiconductor (MFIS),[118] and metal–ferroelectric–metal–insulator–semiconductor (MFMIS).[119] Kim et al. fabricated an MFS-FeFET device with an HZO FE gate and an IGZO oxide semiconductor to emulate synaptic

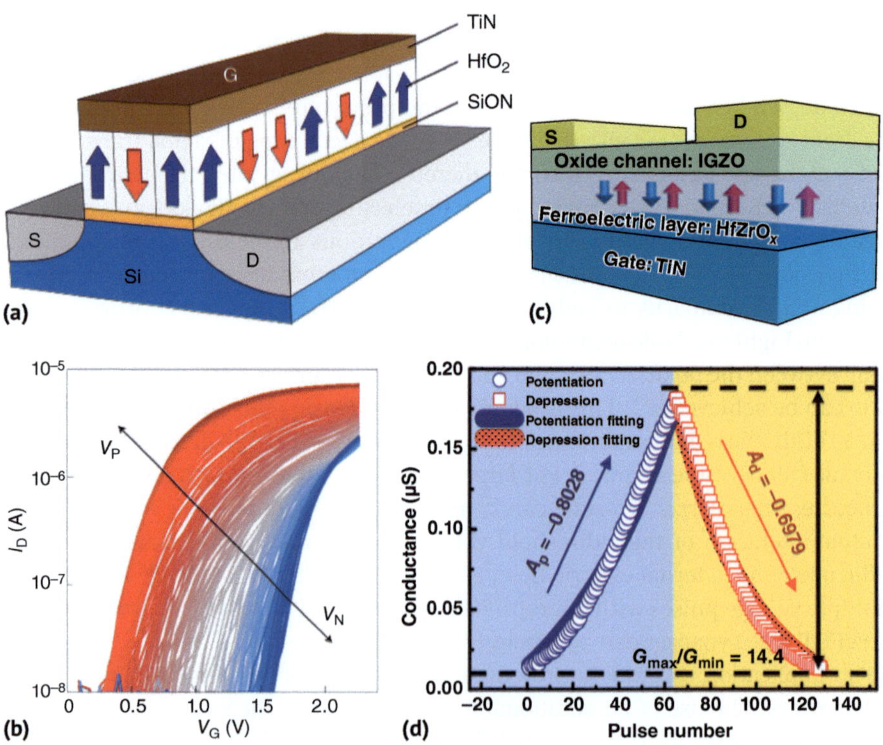

Figure 11.14 FeFETs for synaptic devices. (a) Schematic of partial polarization switching in FeFETs. The arrows represent the polarization directions of the FE domains. (b) Transfer characteristics of the FeFET devices. *Source*: [110] / with permission of Springer Nature. (c) Schematic structure of HZO-IGZO MFS-FeFET. (d) Potentiation and depression properties of the MFS-FeFET. *Source*: [112] / with permission of American Chemical Society.

(Continued)

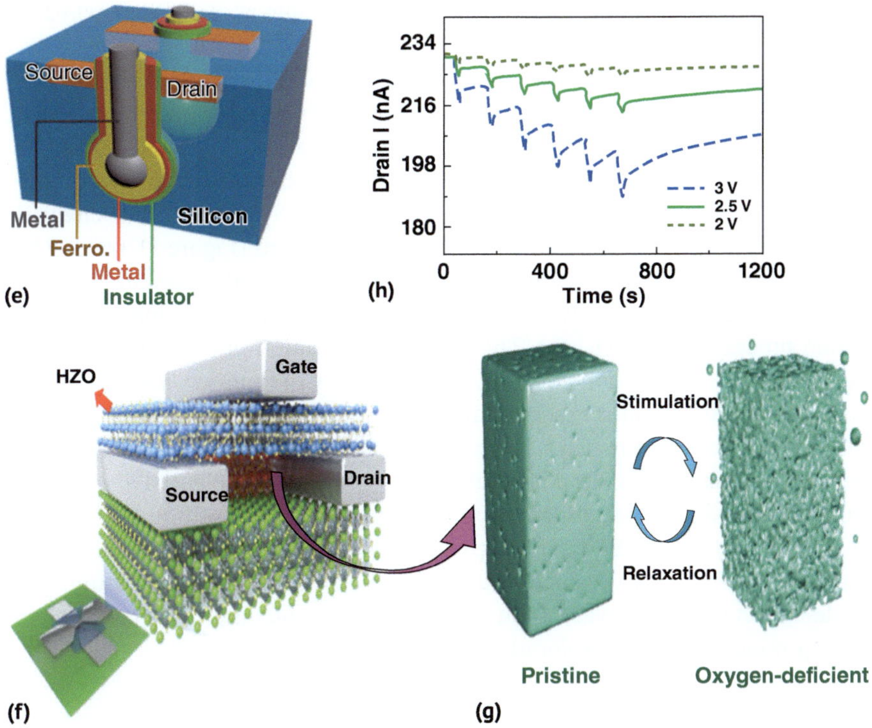

Figure 11.14 (Cont'd) (e) Schematic structure of the spherical recess channel MFMIS-FeFET. *Source*: [113] / with permission of John Wiley & Sons. (f) Structure of an FE-HZO-based FET device for reservoir computing and (g) schematic representation of the structural transition between the pristine and oxygen-deficient phases. (h) Evolution of drain current as a function of time during different positive voltage stimulations. *Source*: [114] / with permission of Springer Nature.

behavior.[112] The devices exhibited characteristics such as potentiation and depression properties, including high linearity, multiple operation, and minimal cycle-to-cycle/device-to-device fluctuations, as shown in Figure 11.14c,d. In simulations with measured properties, a neuromorphic system with these devices displayed an astounding 91.1% handwritten digit recognition rate. The MFS gate stack represents the optimal scenario; in the majority of practical configurations, it is challenging to completely suppress the formation of interlayers between the FE layer and the semiconductor. The presence of interlayers leads to a reduced voltage drop across the FE layer, and additional interlayer capacitance can hinder the accurate compensation of the polarization charge, subsequently causing E_{dep}, which degrades the retention of polarization in the FE film.[120] The MFMIS gate stack may provide a more suitable solution. By independently modifying the area proportion of the FE and dielectric capacitors to reduce the voltage drop across the dielectric, E_{dep} can be substantially alleviated, thereby enhancing retention.[119] As depicted in Figure 11.14e, an MFMIS FeFET employing a spherical recess channel transistor was designed to address the challenges posed by device size limitations in high-density integration.[113] The device demonstrated superior memory window (MW) and reliability properties compared to the MFIS configuration.

Interestingly, the flexible configuration of FETs enables them to possess diverse functionalities. Our current investigation involving hafnia-based FETs demonstrated that we effectively modulated the channel conductivity of the Mott material LSMO by leveraging the unique coupling characteristics of polarization and oxygen migration in HZO.[114] In this FeFET-like device, the conductivity of the LSMO channel was influenced by the extraction and insertion of oxygen ions (Figure 11.14f,g). The disparity in oxygen affinities within the HZO film prompts the accumulation of oxygen ions, which subsequently migrate towards the channel, triggering a dynamic relaxation process (Figure 11.14h). This inherent relaxation property is employed judiciously in physical memory computing, particularly in time-sensitive tasks. For instance, it has demonstrated exceptional learning efficiency and classification precision in speech-recognition tasks. In fact, given the ultrahigh oxygen conductivity exhibited by hafnia-based materials with FE properties, their potential application in electrolyte-gated neuromorphic computing systems could be substantial.

The functionality of FeFET-type synapses is influenced by several factors. Within the FeFET community, issues such as depolarization effects, charge trapping, and the trade-off between the polarization switching speed and endurance continue to be fertile grounds for future exploration.

11.4.1.4 RRAM-type Synapses

RRAM, also known as a memristor, features a simple metal–insulator–metal structure, as shown in Figure 11.11d. Because of the substantial resemblance between the ions' dynamic state within the device and the transmission of neurotransmitters across synaptic clefts, it is regarded as an artificial synapse.[121] The remarkably scalable nature of RRAM devices, encompassing both device size and 3D integration capacities,[122] enables the construction of dense neural networks crucial for advanced deep learning architectures and brain-inspired circuits. Hafnia-based RRAM has emerged as a hot research topic owing to its simple structure, fast switching speed (<10 ns), low power consumption (<10 pJ), and compatibility with standard CMOS fabrication techniques. They have been extensively studied for simulating synaptic plasticity phenomena.[123–125]

Both the 1 resistor (1R) and 1 transistor-1 resistor (1T-1R) configurations, as depicted in Figure 11.15a, were utilized for characterization and implementation in memory or computing systems. The 1T-1R configuration has great advantages for various applications. First, this configuration facilitates large array integration, as the transistor can selectively access memory cells, effectively mitigating the sneak path current issues in large-scale arrays, which is not possible with the 1R structure.[126] Second, in the 1T-1R structure, the transistor provides a more reliable and accurate current limitation through the gate voltage, significantly enhancing the device performance. In contrast, the 1R structure may suffer from cycle-to-cycle variability and device failure owing to current overshoot. Third, the transistor in the 1T-1R configuration not only supports multiple selections of memory cells but also assists in programming for memory computing applications, which is challenging to achieve with the 1R structure. Moreover, owing to the constraints imposed by parasitic capacitance effects,[127] using the 1T-1R configuration can more precisely manage the current compliance during forming and setting operations, thereby facilitating energy-efficient operation. Therefore, for a hafnia-based RRAM, employing the 1T-1R configuration yields optimal device performance.

Optimization through material engineering of the stack was employed to enhance the electrical performance of hafnia-based RRAM devices. One common optimization strategy is to insert Ti (or Hf) between the HfO_2 layer and the electrode layer.[130–132] Ti expedites the transfer of oxygen from HfO_2, forming a TiO_x layer that acts as an oxygen interchange layer during switching operations. In addition, the formation

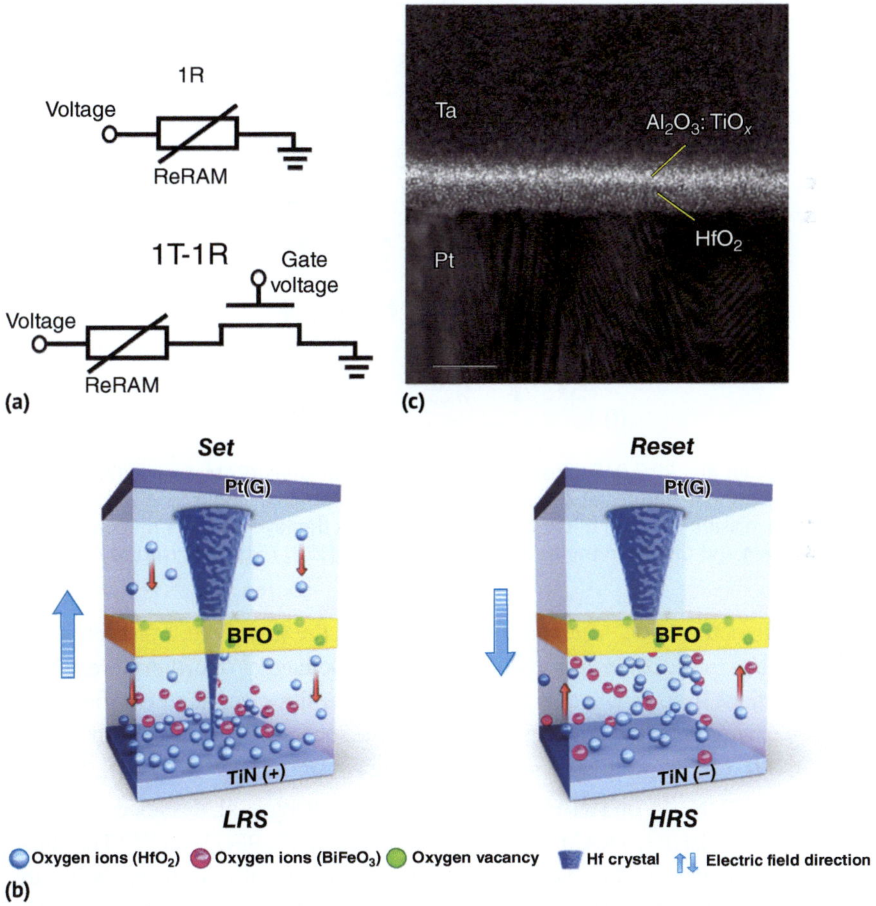

Figure 11.15 FeFETs for synaptic devices. (a) Schematic of the 1R and 1T1R configurations. (b) Schematic of the resistive switching mechanism for the Pt/HfO_2/BFO/HfO_2/TiN device. *Source*: [128] / with permission of John Wiley & Sons. (c) Magnification of the material stack of a thousand-level RRAM device. Scale bar, 5 nm.

(Continued)

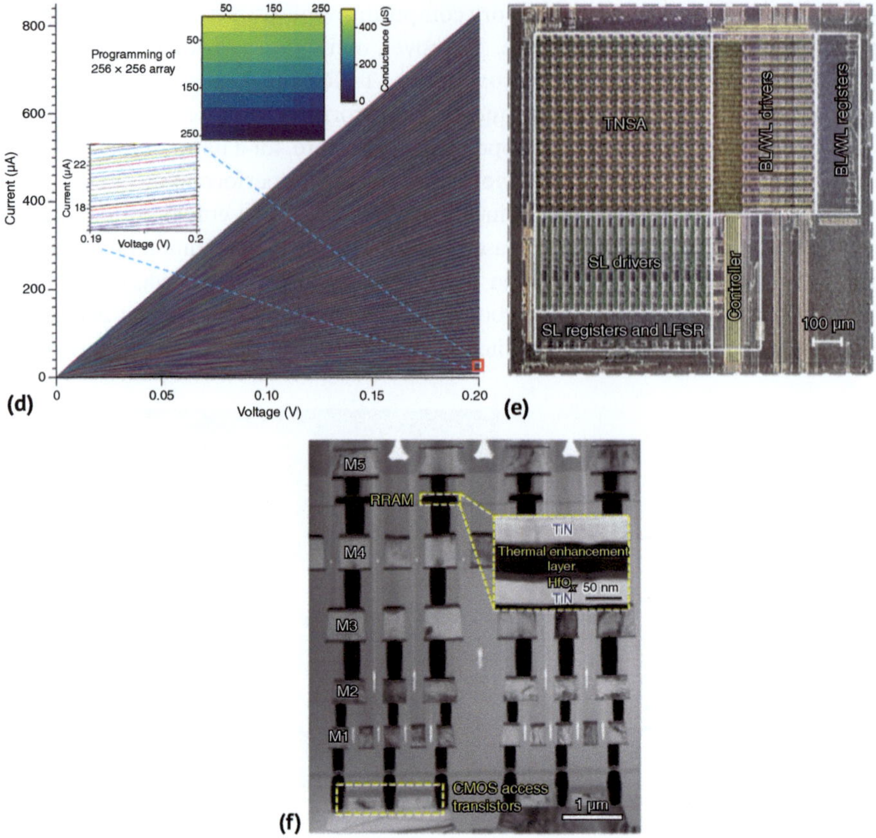

Figure 11.15 (Cont'd) (d) 2,048 resistance levels of an individual RRAM. *Source*: [72] / with permission of Springer Nature. (e) Zoomed-in chip micrograph on a single core and (f) cross-sectional transmission electron microscopy image showing the layer stack of monolithically integrated RRAM and CMOS on the NeuRRAM chip. *Source*: [128] / Springer Nature / CC BY 4.0.

of sub-stoichiometric HfO_x layers at the interface and the asymmetric distribution of oxygen vacancies in the oxide layer reduce the forming voltage.[133] Furthermore, numerous investigations have shown that the use of multilayer films can augment the analog modulation of conductivity.[134–136] In particular, some successful instances of multilevel analog RRAMs utilize multilayer stacks, such as Pt/HfO_x/TiO_x/HfO_x/TiO_x/TiN,[134] TiN/TaO_x/HfO_x/TiN,[135] or TiN/$HfAlO_x$/TaO_x/TiN.[136] Peng et al. designed a tri-layer HfO_2/$BiFeO_3$ (BFO)/HfO_2 RRAM by inserting a traditional FE BFO layer.[128] The abundant oxygen vacancies in the inserted BFO film play a crucial role in regulating the morphology and growth of Hf conductive filaments, as shown in Figure 11.15b. The device exhibited excellent resistive switching performance, such as a storage window of 10^4 and multilevel storage ability. Pattern recognition simulation based on neuromorphic networks attained a commendable recognition accuracy of 91.2%. Recently, Rao et al. designed a Pt/HfO_2/Al_2O_3 : TiO_x/Ta stack structure and realized 2,048 conductance levels, equivalent to an 11-bit resolution, on a fully integrated chip produced by a commercial foundry through a 256 × 256 device array, as shown in Figure 11.15c,d.[72] This signifies a significant stride towards the commercialization of RRAM technology as hardware accelerators of machine learning and AI for neuromorphic computing.

RRAMs, as one of the most comprehensively researched and technologically mature artificial synapses, have witnessed their progression and research extend beyond the device level, with various system-level computing chips leveraging RRAM. In 2018, Panasonic ingeniously engineered a chip with 2 Mb analogue RRAM to implement a perceptron neural network.[137] In 2019, the University of Massachusetts developed a 128 × 64 RRAM crossbar system capable of tackling tangible issues, including the regression of the number of airline passengers and the classification of individual humans by their gaits, using long short-term memory, a variant of recurrent neural networks.[138] Lately, Wan et al. developed a compute-in-memory (CIM) chip called NeuRRAM based on RRAM.[129] This chip includes 48 CIM cores (Figure 11.15e), and can execute calculations concurrently, enhancing the processing capabilities for AI tasks. It is adept at handling diverse AI tasks, including image classification, speech recognition, and image recovery. The chip maintains ultrahigh energy efficiency while maintaining an inference accuracy comparable to that of the software models. Its core element is predicated on HfO_2 RRAM, as shown in Figure 11.15f. HfO_2 has emerged as one of the most important RRAM materials for the demonstration and prototyping of neuromorphic circuits.

11.4.2 Artificial Neurons based on Hafnia-based Materials

Compared with synaptic devices, research on neuronal devices is relatively sparse. This scarcity is partly attributed to the fact that the number of neurons is significantly lower than that of synapses in neuromorphic systems, which diminishes the urgency to find a more compact solution than CMOS-based neurons. However, materials capable of effectively mimicking the characteristic behavior of biological neurons are rare. However, simulating the function of a neuron using CMOS technology typically requires multiple capacitors and transistors (Figure 11.16a). Such circuit designs are not only structurally complex, but also highly inefficient, making them unsuitable for large-scale neural networks.[139] There is a need for simpler solutions. In this field, neuron devices based on hafnia materials have garnered extensive attention.[139–142]

Figure 11.7b shows that the polarization of small-area FE devices exhibits accumulative switching behavior, a characteristic that can be harnessed to simulate the IF function of the neurons. To this end, Mulaosmanovic et al. fabricated nanoscale FeFETs with channel lengths and widths of 30 and 80 nm, respectively, to emulate the characteristics of IF neurons.[141] Accumulation and random switching in the Si:HfO_2-FeFET occur abruptly, leading to a sudden surge in channel current, which can mimic the spikes of biological neurons. Figure 11.16b shows the microstructure of the device. Firstly, a long pulse with an amplitude of −3.25 V was applied on the gate to erase the current state (ERS), followed by a series of +2.2 V, 1 μs wide positive pulses (V_p). 2.2 V is lower than the coercive voltage, which means that a single V_p pulse is not sufficient for programming the transition; however, after a

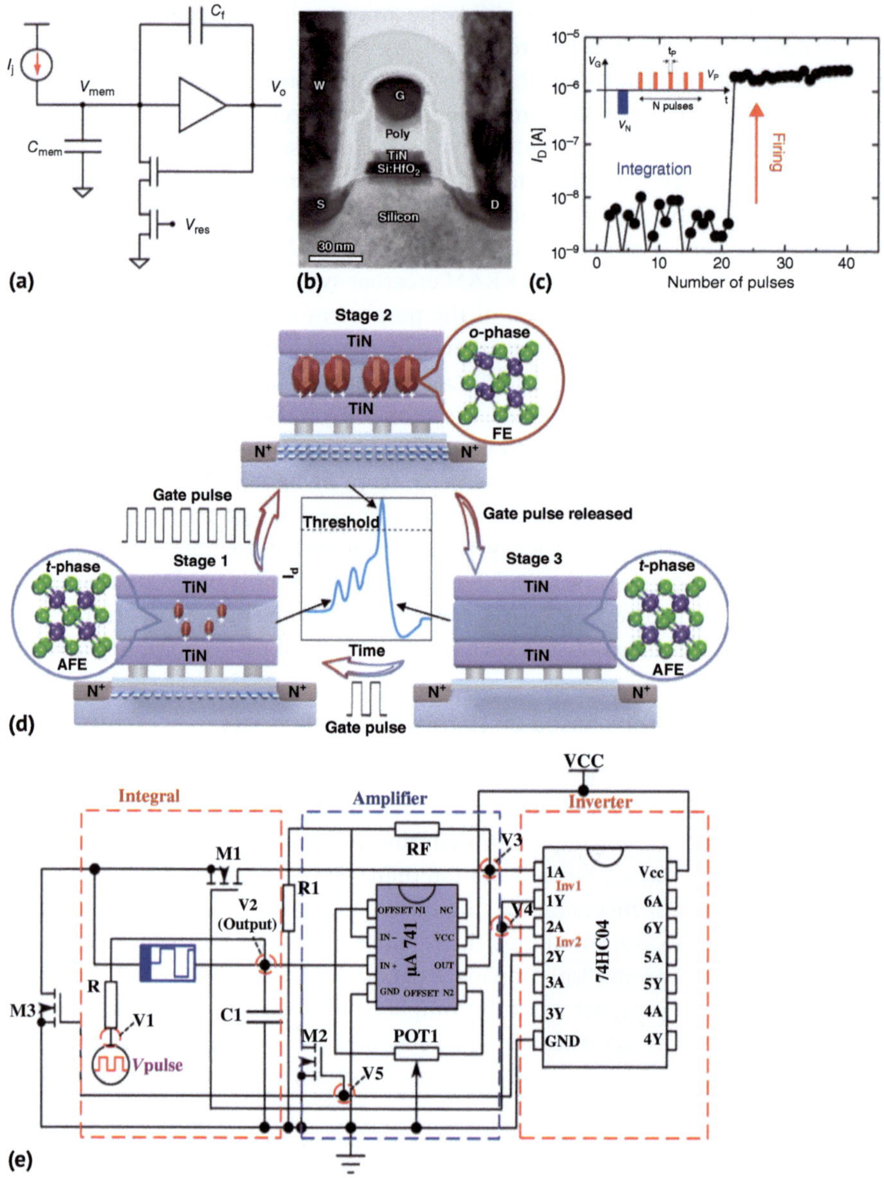

Figure 11.16 Hafnia-based artificial neurons. (a) Schematic of Axon-Hillock CMOS neuron circuit. (b) Cross-sectional TEM image of the FeFET-type neuron with L = 30nm and W = 80nm for channel length and width, respectively. (c) Accumulative switching in FeFET-type neurons. *Source*: [141] / The Royal Society of Chemistry. (d) The IF dynamics of the AFeFET-type neuron. *Source*: [143] / Springer Nature / CC BY 4.0. (e) Circuit based on HfO_2-based RRAM-CMOS LIF neuron model. *Source*: [140] / with permission of AIP Publishing.

certain number of V_p pulses, the drain current I_D experiences a sudden increase, as shown in Figure 11.16c. This phenomenon is due to the abrupt polarization reversal of the entire grain, which is triggered by a critical number of nanodomains nucleated after a sufficient number of pulses. At a lower number of pulses, the dispersed nanodomains failed to induce substantial charge alterations within the channel, resulting in the absence of conduction paths (OFF state). After the device was triggered, a reset voltage was immediately applied to reset the device to the ERS state, thus achieving repeated triggering. Moreover, the firing rate depends on the amplitude and duration of the V_p. It should be noted that this phenomenon does not occur in FTJs, where the tunneling current traverses the FE layer and each nucleated domain contributes to the total current, making it difficult to observe the sudden increase in conductivity.

Interestingly, by employing an AFE layer as a substitute for the FE layer in the FeFET to form an AFeFET device, similar neuronal functions can be achieved. Gao et al. recently demonstrated this phenomenon.[143] As shown in Figure 11.16d, the transistor consisted of a gate dielectric material of $Hf_{0.2}Zr_{0.8}O_2$. The excess Zr induces the film to adopt a t-phase structure in its ground state, exhibiting AFE characteristics. Upon exposure to a sequence of subthreshold gate pulses, the AFE t-phase experienced a comparable cumulative transition behavior, transforming into the FE o-phase. When these gate pulses are subsequently withdrawn, the HZO layer inherently shifts back from the o-phase to the t-phase, and the channel resumes its corresponding low-conductance state. This dynamic manifests as an IF behavior in channel conductance.

In addition to utilizing FeFET devices for simulating neurons, there have also been demonstrations of utilizing RRAM-CMOS hybrid circuits to mimic neuronal functions.[140,142] The resistance state change in RRAM is contingent on whether the applied voltage exceeds its threshold voltage, and the resistance variation can be used to generate action potentials. Therefore, the threshold voltage characteristics of RRAM can be employed to construct artificial neurons. Based on this fundamental principle, Zhang et al. designed an artificial neuron device, as shown in Figure 11.16e, where an RC integrator circuit was utilized to simulate the excitation and firing processes, with an HfO_2-based RRAM-CMOS hybrid implementation to demonstrate LIF functionality.[140] The model charges a capacitor to trigger an RRAM's state change, mimicking the neuron's threshold-driven firing, followed by a leaky discharge process, effectively replicating the dynamics of the LIF neuron model. Nonetheless, it is evident that this design is relatively intricate and poses challenges when implemented in large-scale neuromorphic computing systems.

11.5 Summary and Outlooks

In this chapter, we discuss the fundamental attributes of hafnia-based materials relevant to neuromorphic devices, the fundamental traits of neuromorphic computing, and neuromorphic devices developed based on the properties of hafnia-based materials. From a materials science standpoint, hafnia-based materials possess numerous physical and chemical properties. Notably, their unexpected ferroelectricity has garnered widespread attention owing to their general advantages, such as CMOS compatibility, scale-free behavior,[55] scalability, and well-established deposition procedures. Traditional perovskite FE materials, exemplified by $Pb(Zr,Ti)O_3$, have been applied in various devices such as FeRAM owing to their mature FE performance and cyclical stability, and their restricted scalability and incompatibility with CMOS restrict their commercial applications to niche markets.[144] In 2019, FE $Al_xSc_{1-x}N$ was first reported. Relative to hafnia-based FEs, it boasts a higher P_r and a smaller κ, and it also maintains CMOS compatibility, seemingly a formidable contender against hafnia-based FEs.[145] However, its relatively higher E_c results in poorer endurance, significantly curtailing its application in commercial devices.[146] At present, hafnia-based materials continue to serve as the most valuable FE materials within the realm of nanoelectronics. However, there are numerous challenges associated with the use of hafnia-based materials in research and applications. Firstly, the competition among different crystal phases at the nanoscale engenders intricate polymorphism, which poses a challenge for manufacturing uniform integrated devices. Moreover, the growth of hafnia-based FE thin films currently does not typically follow the anticipated optimal crystal orientation, and it is difficult to control the proportion of their crystal phases. This may necessitate a profound comprehension of the physics of the material and the application of crystalline engineering for optimization. Second, their relatively subpar endurance, especially for devices with FE/Si interfaces, is another challenge. Interface engineering, such as the use of high-κ interlayers and annealing processes, might rectify this predicament. Moreover, when annealing hafnia-based films to crystallize the FE phase, it is necessary to carefully control the thermal budget to prevent excessive thermal diffusion that could impact the performance and dependability of the devices.

From a device perspective, it is essential to optimize the material properties to attain exceptional device performance. For instance, in research on FTJs, the TER of hafnia-based FTJs has invariably been below those of FTJs predicated on conventional perovskite FE materials. Ultimately, through the application of crystalline engineering techniques, we successfully produced HZO films with optimal orientation, leading to high TER values that are comparable to those of conventional FTJs.[93] Crystalline engineering can fundamentally optimize the performance of FE devices by efficiently controlling the phase composition, orientation, and grain size in hafnia-based FEs. On the other hand, device design also exercises a pivotal function in performance. For example, the MFS, MFIS, and MFMIS configurations developed in FeFETs illustrate distinct performance characteristics to cater to diverse prerequisites. The insertion of a Ti interlayer in RRAM devices can substantially bolster device stability, and to harness or mitigate the depolarization field, different interfacial contact materials must be chosen according to the device specifications.[116,147] Overall, research on the application of hafnia-based materials, especially hafnia-based FEs, in neuromorphic devices is experiencing a surge in growth. We believe that the multifaceted material properties of hafnia, particularly its unique FE characteristics and related ultrahigh oxygen-ion conductivity, may inspire the development of more innovative neuromorphic devices and attain superior success in forthcoming bioinspired computing paradigms.

Acknowledgments

We gratefully acknowledge funding from the National Key R&D Program of China (No. 2024YFA1409500), the National Natural Science Foundation of China (Grants Nos. 12304123, 12222414, and 12074416), the Natural Science Foundation of Shandong Province (Grant No. ZR2023QA085), the Youth Innovation Technology Project of Higher School in Shandong Province (No. 2023KJ211), and the Youth Innovation Promotion Association of CAS (No. Y2022003).

References

1 Meena, J. S., Sze, S. M., Chand, U. et al. (2014). Overview of emerging nonvolatile memory technologies. *Nanoscale Research Letters* 9: 526. https://doi.org/10.1186/1556-276x-9-526.

2 Mikolajick, T., Schroeder, U., and Slesazeck, S. (2020). The past, the present, and the future of ferroelectric memories. *IEEE Transactions on Electron Devices* 67: 1434–1443. https://doi.org/10.1109/TED.2020.2976148.

3 Fan, Z., Chen, J., and Wang, J. (2016). Ferroelectric HfO_2-based materials for next-generation ferroelectric memories. *Journal of Advanced Dielectrics* 06: 1630003. https://doi.org/10.1142/S2010135X16300036.

4 Cheema, S. S., Kwon, D., Shanker, N. et al. (2020). Enhanced ferroelectricity in ultrathin films grown directly on silicon. *Nature* 580: 478–482. https://doi.org/10.1038/s41586-020-2208-x.

5 Cheema, S. S., Shanker, N., Hsu, S. L. et al. (2022). Emergent ferroelectricity in subnanometer binary oxide films on silicon. *Science* 376: 648–652. https://doi.org/10.1126/science.abm8642.

6 Schroeder, U., Park, M. H., Mikolajick, T. et al. (2022). The fundamentals and applications of ferroelectric HfO_2. *Nature Reviews Materials* 7: 653–669. https://doi.org/10.1038/s41578-022-00431-2.

7 Kim, S. K., Lee, S. W., Han, J. H. et al. (2010). Capacitors with an equivalent oxide thickness of <0.5 nm for nanoscale electronic semiconductor memory. *Advanced Functional Materials* 20: 2989–3003. https://doi.org/10.1002/adfm.201000599.

8 Silva, J. P. B., Alcala, R., Avci, U. E. et al. (2023). Roadmap on ferroelectric hafnia- and zirconia-based materials and devices. *APL Materials* 11: 089201. https://doi.org/10.1063/5.0148068.

9 Jachalke, S., Schenk, T., Park, M. H. et al. (2018). Pyroelectricity of silicon-doped hafnium oxide thin films. *Applied Physics Letters* 112: 142901. https://doi.org/10.1063/1.5023390.

10 Zhang, Q. and Whatmore, R. W. (2003). Improved ferroelectric and pyroelectric properties in Mn-doped lead zirconate titanate thin films. *Journal of Applied Physics* 94: 5228–5233. https://doi.org/10.1063/1.1613370.

11 Jo, S., Lee, H., Choe, D. H. et al. (2023). Negative differential capacitance in ultrathin ferroelectric hafnia. *Nature Electronics* 6: 390–397. https://doi.org/10.1038/s41928-023-00959-3.

12 Hoffmann, M., Slesazeck, S., Schroeder, U. et al. (2020). What's next for negative capacitance electronics? *Nature Electronics* 3: 504–506. https://doi.org/10.1038/s41928-020-00474-9.

13 Kwon, D., Cheema, S., Shanker, N. et al. (2019). Negative capacitance FET with 1.8-nm-thick Zr-doped HfO_2 oxide. *IEEE Electron Device Letters* 40: 993–996. https://doi.org/10.1109/LED.2019.2912413.

14 Hoffmann, M., Fengler, F. P. G., Max, B. et al. (2019). Negative capacitance for electrostatic supercapacitors. *Advanced Energy Materials* 9: 1901154. https://doi.org/10.1002/aenm.201901154.

15 Marković, D., Mizrahi, A., Querlioz, D. et al. (2020). Physics for neuromorphic computing. *Nature Reviews Physics* 2: 499–510. https://doi.org/10.1038/s42254-020-0208-2.

16 Burr, G. W., Shelby, R. M., Sebastian, A. et al. (2017). Neuromorphic computing using non-volatile memory. *Advances in Physics: X* 2: 89–124. https://doi.org/10.1080/23746149.2016.1259585.

17 Furber, S. (2016). Large-scale neuromorphic computing systems. *Journal of Neural Engineering* 13: 051001. https://doi.org/10.1088/1741-2560/13/5/051001.

18 Schuman, C. D., Kulkarni, S. R., Parsa, M. et al. (2022). Opportunities for neuromorphic computing algorithms and applications. *Nature Computational Science* 2: 10–19. https://doi.org/10.1038/s43588-021-00184-y.

19 Ho, V. M., Lee, J.-A., and Martin, K. C. (2011). The cell biology of synaptic plasticity. *Science* 334: 623–628. https://doi.org/10.1126/science.1209236.

20 Abbott, L. F. and Regehr, W. G. (2004). Synaptic computation. *Nature* 431: 796–803. https://doi.org/10.1038/nature03010.

21 Mead, C. (1990). Neuromorphic electronic systems. *Proceedings of the IEEE* 78: 1629–1636. https://doi.org/10.1109/5.58356.

22 Mostafa, H., Müller, L. K., and Indiveri, G. (2015). An event-based architecture for solving constraint satisfaction problems. *Nature Communications* 6: 8941. https://doi.org/10.1038/ncomms9941.

23 Nandakumar, S. R., Kulkarni, S. R., Babu, A. V. et al. (2018). Building brain-inspired computing systems: Examining the role of nanoscale devices. *IEEE Nanotechnology Magazine* 12: 19–35. https://doi.org/10.1109/MNANO.2018.2845078.

24 Furber, S. B., Galluppi, F., Temple, S. et al. (2014). The SpiNNaker project. *Proceedings of the IEEE* 102: 652–665. https://doi.org/10.1109/JPROC.2014.2304638.

25 Pehle, C., Billaudelle, S., Cramer, B. et al. (2022). The BrainScaleS-2 accelerated neuromorphic system with hybrid plasticity. *Frontiers in Neuroscience* 16: 795876. https://doi.org/10.3389/fnins.2022.795876.

26 Pei, J., Deng, L., Song, S. et al. (2019). Towards artificial general intelligence with hybrid Tianjic chip architecture. *Nature* 572: 106–111. https://doi.org/10.1038/s41586-019-1424-8.

27 Merolla, P. A., Arthur, J. V., Alvarez-Icaza, R. et al. (2014). A million spiking-neuron integrated circuit with a scalable communication network and interface. *Science* 345: 668–673. https://doi.org/10.1126/science.1254642.

28 Davies, M., Srinivasa, N., Lin, T. H. et al. (2018). Loihi: A neuromorphic manycore processor with on-chip learning. *IEEE Micro* 38: 82–99. https://doi.org/10.1109/MM.2018.112130359.

29 Xu, W., Cho, H., Kim, Y. H. et al. (2016). Organometal halide perovskite artificial synapses. *Advanced Materials* 28: 5916–5922. https://doi.org/10.1002/adma.201506363.

30 Hongyu, A., Kangjun, B., and Yang, Y. (2018). *Advances in Memristor Neural Networks* (ed Ciufudean Calin) Ch. 2 (IntechOpen).

31 Burkitt, A. N. (2006). A review of the integrate-and-fire neuron model: I. Homogeneous synaptic input. *Biological cybernetics* 95: 1–19. https://doi.org/10.1007/s00422-006-0068-6.

32 Rocsoreanu, C., Georgescu, A., and Giurgiteanu, N. (2012). *The FitzHugh-Nagumo Model: Bifurcation and Dynamics*. Vol. 10. Springer Science & Business Media.

33 Häusser, M. (2000). The Hodgkin-Huxley theory of the action potential. *Nature Neuroscience* 3: 1165–1165. https://doi.org/10.1038/81426.

34 Teeter, C., Iyer, R., Menon, V. et al. (2018). Generalized leaky integrate-and-fire models classify multiple neuron types. *Nature Communications* 9: 709. https://doi.org/10.1038/s41467-017-02717-4.

35 Takagi, H. (2000). Roles of ion channels in EPSP integration at neuronal dendrites. *Neuroscience Research* 37: 167–171. https://doi.org/10.1016/S0168-0102(00)00120-6

36 Zucker, R. S. and Regehr, W. G. (2002). Short-term synaptic plasticity. *Annual Review of Physiology* 64: 355–405. https://doi.org/10.1146/annurev.physiol.64.092501.114547.

37 Fioravante, D. and Regehr, W. G. (2011). Short-term forms of presynaptic plasticity. *Current Opinion in Neurobiology* 21: 269–274. https://doi.org/10.1016/j.conb.2011.02.003.

38 Jackman, S. L., Turecek, J., Belinsky, J. E. et al. (2016). The calcium sensor synaptotagmin 7 is required for synaptic facilitation. *Nature* 529: 88–91. https://doi.org/10.1038/nature16507.

39 Debanne, D., Guérineau, N. C., Gähwiler, B. H. et al. (1996). Paired-pulse facilitation and depression at unitary synapses in rat hippocampus: Quantal fluctuation affects subsequent release. *The Journal of Physiology* 491: 163–176. https://doi.org/10.1113/jphysiol.1996.sp021204.

40 Sullivan, J. M. (2007). A Simple depletion model of the readily releasable pool of synaptic vesicles cannot account for paired-pulse depression. *Journal of Neurophysiology* 97: 948–950. https://doi.org/10.1152/jn.00554.2006.

41 Waldeck, R. F., Pereda, A., and Faber, D. S. (2000). Properties and plasticity of paired-pulse depression at a central synapse. *The Journal of Neuroscience* 20: 5312–5320. https://doi.org/10.1523/jneurosci.20-14-05312.2000.

42 Thompson, R. (1986). The neurobiology of learning and memory. *Science* 233: 941–947. https://doi.org/10.1126/science.3738519.

43 Ling, H., Wang, N., Yang, A. et al. (2019). Dynamically reconfigurable short-term synapse with millivolt stimulus resolution based on organic electrochemical transistors. *Advanced Materials Technologies* 4: 1900471. https://doi.org/10.1002/admt.201900471.

44 Sharbati, M. T., Du, Y., Torres, J. et al. (2018). Low-power, electrochemically tunable graphene synapses for neuromorphic computing. *Advanced Materials* 30: 1802353. https://doi.org/10.1002/adma.201802353.

45 Yang, Y., Wen, J., Guo, L. et al. (2016). Long-term synaptic plasticity emulated in modified graphene oxide electrolyte gated IZO-based thin-film transistors. *ACS Applied Materials & Interfaces* 8: 30281–30286. https://doi.org/10.1021/acsami.6b08515.

46 Bear, M. F. and Abraham, W. C. (1996). Long-term depression in hippocampus. *Annual Review of Neuroscience* 19: 437–462. https://doi.org/10.1146/annurev.ne.19.030196.002253.

47 Kullmann, D. M. and Lamsa, K. P. (2007). Long-term synaptic plasticity in hippocampal interneurons. *Nature Reviews Neuroscience* 8: 687–699. https://doi.org/10.1038/nrn2207.

48 Bear, M. F. and Malenka, R. C. (1994). Synaptic plasticity: LTP and LTD. *Current Opinion in Neurobiology* 4: 389–399. https://doi.org/10.1016/0959-4388(94)90101-5.

49 Bliss, T. V. P. and Collingridge, G. L. (1993). A synaptic model of memory: Long-term potentiation in the hippocampus. *Nature* 361: 31–39. https://doi.org/10.1038/361031a0.

50 Bi, G.-q. and Poo, M.-m. (2001). Synaptic modification by correlated activity: Hebb's postulate revisited. *Annual Review of Neuroscience* 24: 139–166. https://doi.org/10.1146/annurev.neuro.24.1.139.

51 He, Y., Yang, Y., Nie, S. et al. (2018). Electric-double-layer transistors for synaptic devices and neuromorphic systems. *Journal of Materials Chemistry C* 6: 5336–5352. https://doi.org/10.1039/C8TC00530C.

52 Tang, J., Yuan, F., Shen, X. et al. (2019). Bridging biological and artificial neural networks with emerging neuromorphic devices: Fundamentals, progress, and challenges. *Advanced Materials* 31: 1902761. https://doi.org/10.1002/adma.201902761.

53 Johnson, B. and Jones, J. L. (2019). *Ferroelectricity in Doped Hafnium Oxide: Materials, Properties and Devices*. Schroeder, U., Hwang, C. S., and Funakubo, H. (eds.). Woodhead Publishing. pp. 25–45.

54 Böscke, T. S., Müller, J., Bräuhaus, D. et al. (2011). Ferroelectricity in hafnium oxide thin films. *Applied Physics Letters* 99: 102903. https://doi.org/10.1063/1.3634052.

55 Lee, H.-J., Lee, M., Lee, K. et al. (2020). Scale-free ferroelectricity induced by flat phonon bands in HfO_2. *Science* 369: 1343–1347. https://doi.org/10.1126/science.aba0067.

56 Huan, T. D., Sharma, V., Rossetti, G. A. et al. (2014). Pathways towards ferroelectricity in hafnia. *Physical Review B* 90: 064111. https://doi.org/10.1103/PhysRevB.90.064111.

57 Müller, J., Boscke, T. S., Schroder, U. et al. (2012). Ferroelectricity in Simple Binary ZrO_2 and HfO_2. *Nano Letters* 12: 4318–4323. https://doi.org/10.1021/nl302049k.

58 Zhong, H., Li, M., Zhang, Q. et al. (2022). Large-scale $Hf_{0.5}Zr_{0.5}O_2$ membranes with robust ferroelectricity. *Advanced Materials* 34: 2109889. https://doi.org/10.1002/adma.202109889.

59 Mulaosmanovic, H., Dünkel, S., Trentzsch, M. et al. (2020). Investigation of accumulative switching in ferroelectric FETs: Enabling universal modeling of the switching behavior. *IEEE Transactions on Electron Devices* 67: 5804–5809. https://doi.org/10.1109/TED.2020.3031249.

60 Mikolajick, T., Park, M. H., Begon-Lours, L. et al. (2023). From ferroelectric material optimization to neuromorphic devices. *Advanced Materials* 35: 2206042. https://doi.org/10.1002/adma.202206042.

61 Li, X., Zhong, H., Lin, T. et al. (2023). Polarization switching and correlated phase transitions in fluorite-structure ZrO_2 nanocrystals. *Advanced Materials* 35: 2207736. https://doi.org/10.1002/adma.202207736.

62 Xu, K., Wang, T., Yu, J. et al. (2024). Fluorite-structured antiferroelectric hafnium-zirconium oxide for emerging nonvolatile memory and neuromorphic-computing applications. *Applied Physics Reviews* 11: 021303. https://doi.org/10.1063/5.0175439.

63 Wang, C.-Y., Wang, C. I., Yi, S. H. et al. (2020). Paraelectric/antiferroelectric/ferroelectric phase transformation in As-deposited ZrO_2 thin films by the TiN capping engineering. *Materials & Design* 195: 109020. https://doi.org/10.1016/j.matdes.2020.109020.

64 Yim, K., Yong, Y., Lee, J. et al. (2015). Novel high-κ dielectrics for next-generation electronic devices screened by automated ab initio calculations. *NPG Asia Materials* 7: e190–e190. https://doi.org/10.1038/am.2015.57.

65 Robertson, J. (2004). High dielectric constant oxides. *European Physical Journal-Applied Physics* 28: 265–291.

66 Jeon, W. (2020). Recent advances in the understanding of high-k dielectric materials deposited by atomic layer deposition for dynamic random-access memory capacitor applications. *Journal of Materials Research* 35: 775–794. https://doi.org/10.1557/jmr.2019.335.

67 Fischer, D. and Kersch, A. (2008). The effect of dopants on the dielectric constant of HfO_2 and ZrO_2 from first principles. *Applied Physics Letters* 92: 012908. https://doi.org/10.1063/1.2828696.

68 Wang, Y., Zahid, F., Wang, J. et al. (2012). Structure and dielectric properties of amorphous high-k oxides: HfO_2, ZrO_2, and their alloys. *Physical Review B* 85: 224110. https://doi.org/10.1103/PhysRevB.85.224110.

69 Spiga, S., Driussi, F., Congedo, G. et al. (2018). Sub-1 nm equivalent oxide thickness Al-HfO_2 trapping layer with excellent thermal stability and retention for nonvolatile memory. *ACS Applied Nano Materials* 1: 4633–4641. https://doi.org/10.1021/acsanm.8b00918.

70 Brivio, S., Spiga, S., and Ielmini, D. (2022). HfO_2-based resistive switching memory devices for neuromorphic computing. *Neuromorphic Computing and Engineering* 2: 042001. https://doi.org/10.1088/2634-4386/ac9012.

71 Zhu, J., Zhang, T., Yang, Y. et al. (2020). A comprehensive review on emerging artificial neuromorphic devices. *Applied Physics Reviews* 7: 011312. https://doi.org/10.1063/1.5118217.

72 Rao, M., Tang, H., Wu, J. et al. (2023). Thousands of conductance levels in memristors integrated on CMOS. *Nature* 615: 823–829. https://doi.org/10.1038/s41586-023-05759-5.

73 Nukala, P., Ahmadi, M., Wei, Y. et al. (2021). Reversible oxygen migration and phase transitions in hafnia-based ferroelectric devices. *Science* 372: 630–635. https://doi.org/10.1126/science.abf3789.

74 Maeda, T., Magyari-Kope, B., and Nishi, Y. (2017). Identifying Ferroelectric Switching Pathways in HfO2: First Principles Calculations under Electric Fields. In: *2017 IEEE International Memory Workshop (IMW)*. Monterey, CA, USA. pp. 1–4. https://doi.org/10.1109/IMW.2017.7939087.

75 Dou, X., Wei, W., Sang, P. et al. (2024). Polarization switching pathways of ferroelectric Zr-doped HfO_2 based on the first-principles calculation. *Applied Physics Letters* 124: 092901. https://doi.org/10.1063/5.0194409.

76 Wei, W., Zhao, G., Zhan, X. et al. (2022). Switching pathway-dependent strain-effects on the ferroelectric properties and structural deformations in orthorhombic HfO_2. *Journal of Applied Physics* 131: 154101. https://doi.org/10.1063/5.0084660.

77 Silva, A., Fina, I., Sanchez, F. et al. (2023). Unraveling the ferroelectric switching mechanisms in ferroelectric pure and La doped HfO_2 epitaxial thin films. *Materials Today Physics* 34: 101064. https://doi.org/10.1016/j.mtphys.2023.101064.

78 Wu, Y., Zhang, Y., Jiang, J. et al. (2023). Unconventional polarization-switching mechanism in $(Hf, Zr)O_2$ ferroelectrics and its implications. *Physical Review Letters* 131: 226802. https://doi.org/10.1103/PhysRevLett.131.226802.

79 Ma, L., Wu, J., Zhu, T. et al. (2023). Ultrahigh oxygen ion mobility in ferroelectric hafnia. *Physical Review Letters* 131: 256801. https://doi.org/10.1103/PhysRevLett.131.256801.

80 Qiu, P., Li, C., Liu, B. et al. (2023). Materials of solid oxide electrolysis cells for H_2O and CO_2 electrolysis: A review. *Journal of Advanced Ceramics* 12: 1463–1510. https://doi.org/10.26599/JAC.2023.9220767.

81 Majumdar, S. (2022). Back-end CMOS compatible and flexible ferroelectric memories for neuromorphic computing and adaptive sensing. *Advanced Intelligent Systems* 4: 2100175. https://doi.org/10.1002/aisy.202100175.

82 Pešić, M., Li, T., Di Lecce, V. et al. (2018). Built-in bias generation in anti-ferroelectric stacks: Methods and device applications. *IEEE Journal of the Electron Devices Society* 6: 1019–1025. https://doi.org/10.1109/JEDS.2018.2825360.

83 Chang, S.C., Haratipour, N., Shivaraman, S. et al. (2021). FeRAM using Anti-ferroelectric Capacitors for High-speed and High-density Embedded Memory. In: *2021 IEEE International Electron Devices Meeting (IEDM)*. San Francisco, CA, USA. pp. 33.32.31–33.32.34. https://doi.org/10.1109/IEDM19574.2021.9720510.

84 Luo, Y.-C., Hur, J., Wang, P. et al. (2020). Non-volatile, small-signal capacitance in ferroelectric capacitors. *Applied Physics Letters* 117: 073501. https://doi.org/10.1063/5.0018937.

85 Hur, J., Luo, Y. C., Lu, A. et al. (2022). Nonvolatile capacitive crossbar array for in-memory computing. *Advanced Intelligent Systems* 4: 2100258. https://doi.org/10.1002/aisy.202100258.

86 Wu, S., Zhang, X., Cao, R. et al. (2024). Multi-state nonvolatile capacitances in HfO_2-based ferroelectric capacitor for neuromorphic computing. *Applied Physics Letters* 124: 102902. https://doi.org/10.1063/5.0180088

87 Demasius, K.-U., Kirschen, A., and Parkin, S. (2021). Energy-efficient memcapacitor devices for neuromorphic computing. *Nature Electronics* 4: 748–756. https://doi.org/10.1038/s41928-021-00649-y.

88 Lee, M., Narayan, D. M., Kim, J. H. et al. (2024). Hafnium oxide-based ferroelectric devices for in-memory computing: Resistive and capacitive approaches. *ACS Applied Electronic Materials* 6: 5391–5401. https://doi.org/10.1021/acsaelm.4c01071

89 Caffrey, N. M., Archer, T., Rungger, I. et al. (2012). Coexistance of giant tunneling electroresistance and magnetoresistance in an all-oxide composite magnetic tunnel junction. *Physical Review Letters* 109: 226803. https://doi.org/10.1103/PhysRevLett.109.226803.

90 Yin, Y., Burton, J. D., Kim, Y. M. et al. (2013). Enhanced tunnelling electroresistance effect due to a ferroelectrically induced phase transition at a magnetic complex oxide interface. *Nature Materials* 12: 397–402.

91 Tan, H., Quintana, A., Dix, N. et al. (2024). Photovoltaic-driven dual optical writing and non-destructive voltage-less reading of polarization in ferroelectric $Hf_{0.5}Zr_{0.5}O_2$ for energy efficient memory devices. *Nano Energy* 123: 109384. https://doi.org/10.1016/j.nanoen.2024.109384.

92 Cao, Y., Liu, Y., Yang, Y. et al. (2023). Structural engineering of $H_{0.5}Z_{0.5}O_2$-based ferroelectric tunneling junction for fast-speed and low-power artificial synapses. *Advanced Electronic Materials* 9: 2201247. https://doi.org/10.1002/aelm.202201247.

93 Gao, Z., Zhang, W., Zhong, Q. et al. (2023). Giant electroresistance in hafnia-based ferroelectric tunnel junctions via enhanced polarization. *Device* 1: 100004. https://doi.org/10.1016/j.device.2023.100004.

94 Chen, L., Wang, T. Y., Dai, Y. W. et al. (2018). Ultra-low power $Hf_{0.5}Zr_{0.5}O_2$ based ferroelectric tunnel junction synapses for hardware neural network applications. *Nanoscale* 10: 15826–15833. https://doi.org/10.1039/C8NR04734K.

95 Bégon-Lours, L., Halter, M., Puglisi, F. M. et al. (2022). Scaled, ferroelectric memristive synapse for back-end-of-line integration with neuromorphic hardware. *Advanced Electronic Materials* 8: 2101395. https://doi.org/10.1002/aelm.202101395.

96 Song, T., Sánchez, F., and Fina, I. (2022). Impact of non-ferroelectric phases on switching dynamics in epitaxial ferroelectric Hf0.5Zr0.5O2 films. *APL Materials* 10: 031108. https://doi.org/10.1063/5.0083661.

97 Boyn, S., Grollier, J., Lecerf, G. et al. (2017). Learning through ferroelectric domain dynamics in solid-state synapses. *Nature Communications* 8: 14736. https://doi.org/10.1038/ncomms14736.

98 Yang, S.-T., Li, X. Y., Yu, T. L. et al. (2022). High-performance neuromorphic computing based on ferroelectric synapses with excellent conductance linearity and symmetry. *Advanced Functional Materials* 32: 2202366. https://doi.org/10.1002/adfm.202202366.

99 Max, B., Hoffmann, M., Mulaosmanovic, H. et al. (2020). Hafnia-based double-layer ferroelectric tunnel junctions as artificial synapses for neuromorphic computing. *ACS Applied Electronic Materials* 2: 4023–4033. https://doi.org/10.1021/acsaelm.0c00832.

100 Cervo Sulzbach, M., Tan, H., Estandía, S. et al. (2021). Polarization and resistive switching in epitaxial 2 nm $Hf_{0.5}Zr_{0.5}O_2$ tunnel junctions. *ACS Applied Electronic Materials* 3: 3657–3666. https://doi.org/10.1021/acsaelm.1c00604.

101 Long, X., Tan, H., Estandía, S. et al. (2022). Enhanced electroresistance endurance of capped Hf0.5Zr0.5O2 ultrathin epitaxial tunnel barriers. *APL Materials* 10: 031114. https://doi.org/10.1063/5.0076865.

102 Chaudhary, P., Buragohain, P., Kozodaev, M. et al. (2021). Electroresistance effect in MoS2-Hf0.5Zr0.5O2 heterojunctions. *Applied Physics Letters* 118: 083106. https://doi.org/10.1063/5.0035306.

103 Müller, M. L., Becker, M. T., Strkalj, N. et al. (2022). Schottky-to-Ohmic switching in ferroelectric memristors based on semiconducting $Hf_{0.93}Y_{0.07}O_2$ thin films. *Applied Physics Letters* 121: 093501. https://doi.org/10.1063/5.0095762.

104 Li, J., Li, N., Ge, C. et al. (2019). Giant electroresistance in ferroionic tunnel junctions. *iScience* 16: 368–377. https://doi.org/10.1016/j.isci.2019.05.043.

105 Mulaosmanovic, H., Breyer, E. T., Dünkel, S. et al. (2021). Ferroelectric field-effect transistors based on HfO_2: A review. *Nanotechnology* 32: 502002. https://doi.org/10.1088/1361-6528/ac189f.

106 Lee, D. H., Park, G. H., Kim, S. H. et al. (2022). Neuromorphic devices based on fluorite-structured ferroelectrics. *InfoMat* 4: e12380. https://doi.org/10.1002/inf2.12380.

107 Breyer, E. T., Mulaosmanovic, H., Mikolajick, T. et al. (2021). Perspective on ferroelectric, hafnium oxide based transistors for digital beyond von-Neumann computing. *Applied Physics Letters* 118: 050501. https://doi.org/10.1063/5.0035281.

108 Halter, M., Bégon-Lours, L., Bragaglia, V. et al. (2020). Back-end, CMOS-compatible ferroelectric field-effect transistor for synaptic weights. *ACS Applied Materials & Interfaces* 12: 17725–17732. https://doi.org/10.1021/acsami.0c00877.

109 Lederer, M., Kämpfe, T., Ali, T. et al. (2021). Ferroelectric field effect transistors as a synapse for neuromorphic application. *IEEE Transactions on Electron Devices* 68: 2295–2300. https://doi.org/10.1109/TED.2021.3068716.

110 Mulaosmanovic, H., Breyer, E. T., Mikolajick, T. et al. (2020). Reconfigurable frequency multiplication with a ferroelectric transistor. *Nature Electronics* 3: 391–397. https://doi.org/10.1038/s41928-020-0413-0.

111 Oh, S., Kim, T., Kwak, M. et al. (2017). HfZrOx-based ferroelectric synapse device with 32 levels of conductance states for neuromorphic applications. *IEEE Electron Device Letters* 38: 732–735. https://doi.org/10.1109/LED.2017.2698083.

112 Kim, M.-K. and Lee, J.-S. (2019). Ferroelectric analog synaptic transistors. *Nano Letters* 19: 2044–2050. https://doi.org/10.1021/acs.nanolett.9b00180.

113 Kim, T., Hwang, J., Kim, G. et al. (2021). High-performance and high-endurance HfO_2-based ferroelectric field-effect transistor memory with a spherical recess channel. *Physica Status Solidi (RRL) – Rapid Research Letters* 15: 2100018. https://doi.org/10.1002/pssr.202100018.

114 Liu, Z., Zhang, Q., Xie, D. et al. (2023). Interface-type tunable oxygen ion dynamics for physical reservoir computing. *Nature Communications* 14: 7176. https://doi.org/10.1038/s41467-023-42993-x.

115 Mulaosmanovic, H., Ocker, J., Müller, S. et al. (2017). Switching kinetics in nanoscale hafnium oxide based ferroelectric field-effect transistors. *ACS Applied Materials & Interfaces* 9: 3792–3798. https://doi.org/10.1021/acsami.6b13866.

116 Covi, E., Mulaosmanovic, H., Max, B. et al. (2022). Ferroelectric-based synapses and neurons for neuromorphic computing. *Neuromorphic Computing and Engineering* 2: 012002. https://doi.org/10.1088/2634-4386/ac4918.

117 Zeng, B., Liu, C., Dai, S. et al. (2021). Electric field gradient-controlled domain switching for size effect-resistant multilevel operations in HfO_2-based ferroelectric field-effect transistor. *Advanced Functional Materials* 31: 2011077. https://doi.org/10.1002/adfm.202011077.

118 Khan, A. I., Keshavarzi, A., and Datta, S. (2020). The future of ferroelectric field-effect transistor technology. *Nature Electronics* 3: 588–597. https://doi.org/10.1038/s41928-020-00492-7.

119 Tokumitsu, E., Fujii, G., and Ishiwara, H. (2000). Electrical properties of metal-ferroelectric-insulator-semiconductor (MFIS)- and metal-ferroelectric-metal-insulator-semiconductor (MFMIS)-FETs using ferroelectric $SrBi_2Ta_2O_9$ film and $SrTa_2O_6$/SiON buffer layer. *Japanese Journal of Applied Physics* 39: 2125. https://doi.org/10.1143/JJAP.39.2125.

120 Muller, J., Polakowski, P., Muller, S. et al. (2016). High endurance strategies for hafnium oxide based ferroelectric field effect transistor. In: *2016 16th Non-Volatile Memory Technology Symposium (NVMTS)*. Pittsburgh, PA, USA. pp. 1–7. https://doi.org/10.1109/NVMTS.2016.7781517.

121 Yan, X., Pei, Y., Chen, H. et al. (2019). Self-assembled networked PbS distribution quantum dots for resistive switching and artificial synapse performance boost of memristors. *Advanced Materials* 31: 1805284. https://doi.org/10.1002/adma.201805284.

122 Seok, J. Y., Song, S. J., Yoon, J. H. et al. (2014). A review of three-dimensional resistive switching cross-bar array memories from the integration and materials property points of view. *Advanced Functional Materials* 24: 5316–5339. https://doi.org/10.1002/adfm.201303520.

123 Banerjee, W., Zhang, X., Luo, Q. et al. (2018). Design of CMOS compatible, high-speed, highly-stable complementary switching with multilevel operation in 3D vertically stacked novel HfO_2/Al_2O_3/TiO (HAT) RRAM. *Advanced Electronic Materials* 4: 1700561. https://doi.org/10.1002/aelm.201700561.

124 Stecconi, T., Guido, R., Berchialla, L. et al. (2022). Filamentary TaO/HfO_2 ReRAM devices for neural networks training with analog in-memory computing. *Advanced Electronic Materials* 8: 2200448. https://doi.org/10.1002/aelm.202200448.

125 Li, X., Zhang, L., Guo, R. et al. (2020). A flexible transient biomemristor based on hybrid structure HfO_2/BSA:Au double layers. *Advanced Materials Technologies* 5: 2000191. https://doi.org/10.1002/admt.202000191.

126 Chen, H.-Y., Brivio, S., Chang, C. C. et al. (2017). Resistive random access memory (RRAM) technology: From material, device, selector, 3D integration to bottom-up fabrication. *Journal of Electroceramics* 39: 21–38. https://doi.org/10.1007/s10832-017-0095-9.

127 Ielmini, D. (2011). Modeling the universal set/reset characteristics of bipolar RRAM by field- and temperature-driven filament growth. *IEEE Transactions on Electron Devices* 58: 4309–4317. https://doi.org/10.1109/TED.2011.2167513.

128 Peng, Z., Wu, F., Jiang, L. et al. (2021). HfO_2-based memristor as an artificial synapse for neuromorphic computing with tri-layer HfO_2/$BiFeO_3$/HfO_2 design. *Advanced Functional Materials* 31: 2107131. https://doi.org/10.1002/adfm.202107131.

129 Wan, W., Kubendran, R., Schaefer, C. et al. (2022). A compute-in-memory chip based on resistive random-access memory. *Nature* 608: 504–512. https://doi.org/10.1038/s41586-022-04992-8.

130 Chen, Y. Y., Goux, L., Clima, S. et al. (2013). Endurance/retention trade-off on HfO_2/metal cap 1T1R bipolar RRAM. *IEEE Transactions on Electron Devices* 60: 1114–1121. https://doi.org/10.1109/TED.2013.2241064.

131 Calka, P., Sowinska, M., Bertaud, T. et al. (2014). Engineering of the chemical reactivity of the Ti/HfO_2 interface for RRAM: Experiment and theory. *ACS Applied Materials & Interfaces* 6: 5056–5060. https://doi.org/10.1021/am500137y.

132 Clima, S., Sankaran, K., Chen, Y. Y. et al. (2014). RRAMs based on anionic and cationic switching: A short overview. *Physica Status Solidi (RRL) – Rapid Research Letters* 8: 501–511. https://doi.org/10.1002/pssr.201409054.

133 Young-Fisher, K. G., Bersuker, G., Butcher, B. et al. (2013). Leakage current-forming voltage relation and oxygen gettering in HfOx RRAM devices. *IEEE Electron Device Letters* 34: 750–752. https://doi.org/10.1109/LED.2013.2256101.

134 Yu, S., Gao, B., Fang, Z. et al. (2013). A low energy oxide-based electronic synaptic device for neuromorphic visual systems with tolerance to device variation. *Advanced Materials* 25: 1774–1779. https://doi.org/10.1002/adma.201203680.

135 Yao, P., Wu, H., Gao, B. et al. (2020). Fully hardware-implemented memristor convolutional neural network. *Nature* 577: 641–646. https://doi.org/10.1038/s41586-020-1942-4.

136 Yao, P., Wu, H., Gao, B. et al. (2017). Face classification using electronic synapses. *Nature Communications* 8: 15199. https://doi.org/10.1038/ncomms15199.

137 Mochida, R., Kouno, K., Hayata, Y. et al. (2018). A 4M Synapses integrated Analog ReRAM based 66.5 TOPS/W Neural-Network Processor with Cell Current Controlled Writing and Flexible Network Architecture. In: *2018 IEEE Symposium on VLSI Technology*. Honolulu, HI, USA. pp. 175–176. https://doi.org/10.1109/VLSIT.2018.8510676.

138 Li, C., Wang, Z., Rao, M. et al. (2019). Long short-term memory networks in memristor crossbar arrays. *Nature Machine Intelligence* 1: 49–57. https://doi.org/10.1038/s42256-018-0001-4.

139 Chen, C., Yang, M., Liu, S. et al. (2019). Bio-inspired neurons based on novel leaky-FeFET with ultra-low hardware cost and advanced functionality for all-ferroelectric neural network. In: *2019 Symposium on VLSI Technology*. Kyoto, Japan. pp. T136–T137. https://doi.org/10.23919/VLSIT.2019.8776495.

140 Zhang, Y., Fang, Z., and Yan, X. (2022). HfO_2-based memristor-CMOS hybrid implementation of artificial neuron model. *Applied Physics Letters* 120: 213502. https://doi.org/10.1063/5.0091286.

141 Mulaosmanovic, H., Chicca, E., Bertele, M. et al. (2018). Mimicking biological neurons with a nanoscale ferroelectric transistor. *Nanoscale* 10: 21755–21763. https://doi.org/10.1039/C8NR07135G.

142 Woo, J., Lee, D., Koo, Y. et al. (2017). Dual functionality of threshold and multilevel resistive switching characteristics in nanoscale HfO_2-based RRAM devices for artificial neuron and synapse elements. *Microelectronic Engineering* 182: 42–45. https://doi.org/10.1016/j.mee.2017.09.001.

143 Cao, R., Zhang, X., Liu, S. et al. (2022). Compact artificial neuron based on anti-ferroelectric transistor. *Nature Communications* 13: 7018. https://doi.org/10.1038/s41467-022-34774-9.

144 Mikolajick, T., Slesazeck, S., Mulaosmanovic, H. et al. (2021). Next generation ferroelectric materials for semiconductor process integration and their applications. *Journal of Applied Physics* 129: 100901. https://doi.org/10.1063/5.0037617.

145 Fichtner, S., Wolff, N., Lofink, F. et al. (2019). AlScN: A III-V semiconductor based ferroelectric. *Journal of Applied Physics* 125: 114103. https://doi.org/10.1063/1.5084945.

146 Zhang, Y., Zhu, Q., Tian, B. et al. (2024). New-generation ferroelectric AlScN materials. *Nano-Micro Letters* 16: 227. https://doi.org/10.1007/s40820-024-01441-1.

147 Toprasertpong, K., Nako, E., Wang, Z. et al. (2022). Reservoir computing on a silicon platform with a ferroelectric field-effect transistor. *Communications Engineering* 1: 21. https://doi.org/10.1038/s44172-022-00021-8.

12

HfO_2-based Ferroelectric Tunnel Junctions

*Zheng Wen**

College of Electronics and Information, Shandong Key Laboratory of Micro-Nano Packaging and System Integration, Qingdao University, Qingdao, China

*Corresponding author: zwen@qdu.edu.cn

Ferroelectric tunnel junctions (FTJs) have great potential for applications in nonvolatile memories, memristive devices, and neuromorphic networks, owing to the advantages of high data storage density, high writing/erasure speed, low power consumption, and high tunability in multilevel resistance states. The study of FTJs starts from perovskite oxides since most of the ferroelectrics that have superior properties are in ABO_3 structures, which can also be epitaxially grown on commercial single-crystalline substrates, such as $SrTiO_3$ (STO), $DyScO_3$ (DSO), and $LaAlO_3$ (LAO), buffered with metallic-oxide bottom electrodes, such as (La, Sr) MnO_3 (LSMO), $SrRuO_3$ (SRO), and $LaNiO_3$ (LNO). Pioneering works can be found in $BaTiO_3$ (BTO)/LSMO and BTO/SRO ultrathin heterostructures by Garcia et al. and Gruverman et al., respectively, using conductive AFM tips as top electrodes (TEs) in 2009,[1,2] as well as in FTJ devices with the Au/Co/BTO/LSMO architecture by Chanthbouala et al. in 2010.[3] However, in the perovskite ferroelectric materials, there exists a long-standing compatibility problem with current CMOS process due to their high annealing temperature for crystallization and structural mismatching with the silicon wafers. These limit the broad applications and further industrialization of FTJs in the next-generation information technologies.

Most recently, the discovery of ferroelectricity in ionic-doped HfO_2, such as the Si:HfO_2, provides ideal solutions since the HfO_2 itself is a kind of high-k gate oxide of the field-effect transistors in the state-of-the-art integrated circuits.[4] The ferroelectricity comes from the polar orthorhombic (o) phase in HfO_2 thin films, which is a metastable phase and can be stabilized by ionic substitution, applied strain, interfacial configuration, and film thickness, as discussed in chapter 1. In addition, there is another advantage in the HfO_2-based thin films that, differing from the perovskite oxides, the polarizations have been observed to be increased with decreasing thickness, which makes the ferroelectric HfO_2 suitable to serve as the tunneling barrier. In this chapter, we address the key scientific challenges and technological advancements in the FTJs with HfO_2-based ferroelectric barriers. We first introduce the fundamental principles of FTJs and then discuss the HfO_2-based FTJ devices with diverse structures fabricated on single-crystalline perovskite substrates and CMOS-compatible wafers. We also discuss reliability issues and memristive properties of the HfO_2-based FTJs. Finally, we give an outlook for the development of the devices.

12.1 Fundamental Principles of FTJs

12.1.1 Device Structure and Tunneling Electroresistance

FTJs are nanoelectronic devices that regulate quantum tunneling effects via polarization switching in ferroelectric thin-film heterostructures. The device structure in general consists of an ultrathin ferroelectric barrier of several nanometers in thickness sandwiched between two metallic electrodes.[5] Figure 12.1a depicts a typical metal/ferroelectric/metal (MFM, assuming that the electrostatic screening length (δ) of M_1 is smaller than that of M_2) FTJ, in which the polarization switching modulates barrier profile of the ferroelectric layer. Specifically, as demonstrated in Figures 12.2a and b, when the polarization is pointing to the metal electrode (M_2) that has a larger δ, the depolarization field $\left(E_{\text{dep}}\right)$, due to the incomplete electrostatic screening of the ferroelectric bounded charges at the two-side electrode surfaces, lowers average height of the barrier and induces the low resistance state (the ON state, I_{ON}). However, when the polarization is switched against the M_2, the E_{dep} increases the barrier height and yields the high resistance state (the OFF state, I_{OFF}). Therefore, the so-called tunneling electroresistance (TER) is achieved in the FTJ device with the bistable and reversible nonvolatile states. The TER can be defined as ON/OFF current (resistance) ratio or the TER current (resistance) ratio, which corresponds to $\frac{I_{\text{ON}}}{I_{\text{OFF}}}\left(\frac{R_{\text{OFF}}}{R_{\text{ON}}}\right)$ or $\frac{I_{\text{ON}}-I_{\text{OFF}}}{I_{\text{OFF}}}\left(\frac{R_{\text{OFF}}-R_{\text{ON}}}{R_{\text{ON}}}\right)$, mainly stemming from the modulation of the transmittance (T) in quantum tunneling by the barrier height (Φ) in the MFM FTJ. In 2005, Zhuravlev et al. formulated the TER effect associated with the Thomas–Fermi electrostatic screening model (Figure 12.2a,b) and shown that the ON/OFF ratio increases with increasing difference between the screening lengths of the two metallic electrodes (Figure 12.2c).[6] Certainly, the ON/OFF ratio is also observed to increase with increasing polarization and thickness in the regime of quantum tunneling of the ferroelectric barrier (Figure 12.2d).

In addition to the barrier height, the barrier width (d) is also important to the tunneling transmittance since the T depends exponentially on d as well, following the formular $T \propto \exp(d\sqrt{\Phi})$. For a ferroelectric, the piezoelectric effect inherently exists, which yields a butterfly-type strain loop and hence gives rise to the change of barrier width upon polarization switching of the ferroelectric layer. However, such a width change is small in the nanometer-thick barrier and the induced TER ratio is in general not obvious in practical FTJ devices. In Figure 12.2a,b, one may find that the barrier width can also be modulated if there is a large difference in

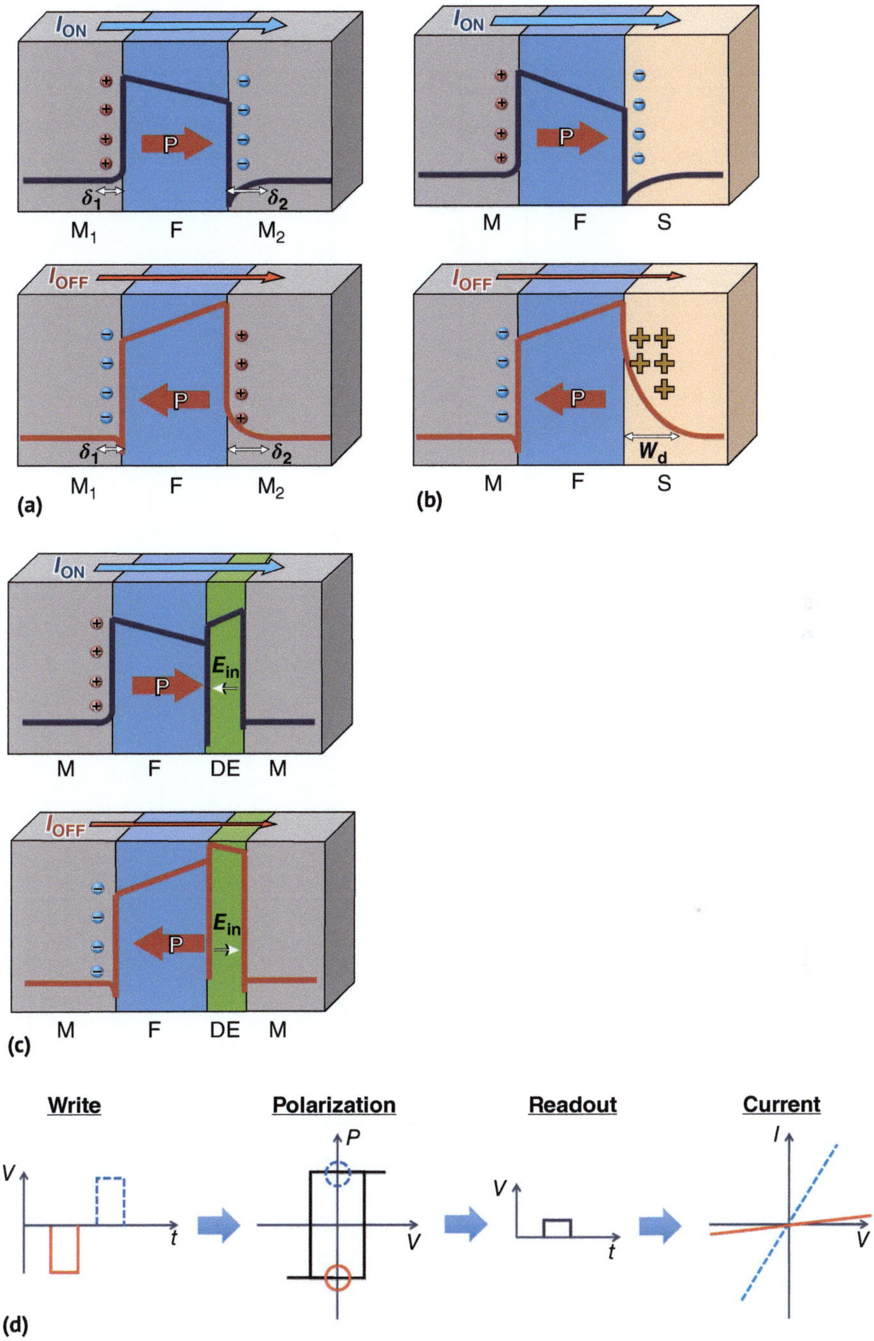

Figure 12.1 Barrier profiles and polarization switching of the (a) MFM-, (b) MFS-, and (c) MFIM(S)-type FTJs for the ON (I_{ON}) and OFF (I_{OFF}) states. (d) Write/readout operations of the FTJ devices. *Source*: [5] / with permission of John Wiley & Sons.

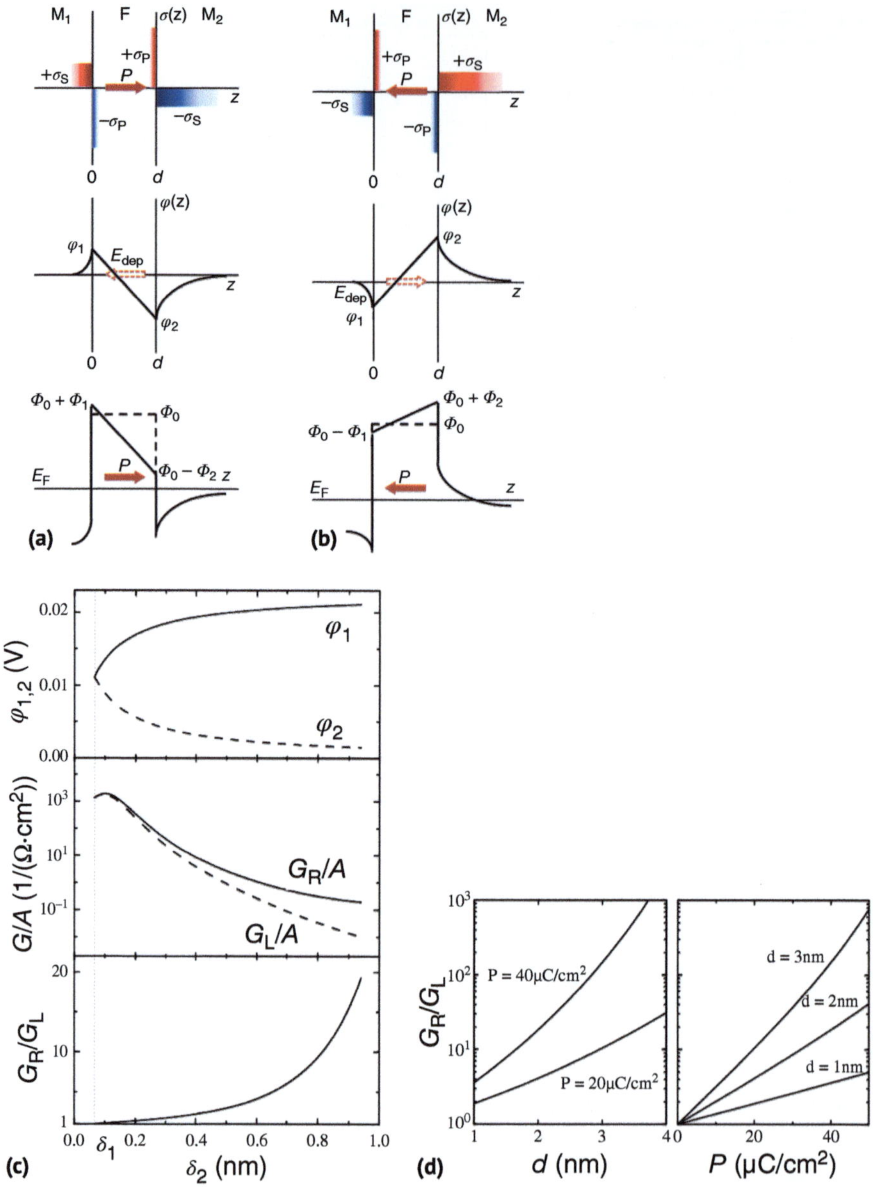

Figure 12.2 Fundamental mechanisms of the TER in MFM-type FTJs with dissimilar top (M_1) and bottom (M_2) electrodes. Charge distributions (top panels), electrostatic potential profiles (middle panels), and barrier profiles (bottom panels) for the (a) ON and (b) OFF states with polarization pointing toward/against the metallic electrode with larger screening length (M_2). (c) Calculated potentials, conductance (G/A), and TER ratio (G_R/G_L) as a function of the difference in screening length between M_1 and M_2 electrodes. (d) TER ratio as functions of polarization and thickness of the ferroelectric barrier, in which the screening length difference is fixed. *Source*: [6] / with permission of American Physical Society.

the screening length between the two-side electrodes.[6] However, it is limited by the MFM structure since the metals always exhibit similar values of δ in the range of a few nanometers. Employing semiconducting electrodes has been proved as an effective approach for realizing substantial width changes of the overall barriers in FTJ devices, as depicted in Figure 12.1b.[7] Taking n-type semiconductor as an example, when the polarization is pointing to the semiconductor, the positive bounded charges are screened by the major carriers, electrons, at the ferroelectric/semiconductor interface. The aggregation of electrons makes the semiconductor surface accumulated. It can be regarded as a metallic electrode with a screening length much larger than that of common metals, such as Pt, Au, and $SrRuO_3$. As a result, the average height of the trapezoidal barrier is well lowered and the MFS FTJ is switched to the ON state with a large I_{ON} (Figure 12.1b). When the polarization is switched against the semiconductor, the negative bounded charges sweep the electrons away from the ferroelectric/semiconductor interface and the semiconductor surface is depleted. The bounded charges are compensated by the ionized donors in the space charge region with a finite depletion width (W_d). The W_d can be as wide as more than 100 nm, which depends upon the doping concentration (N_d) and the dielectric permittivity of the semiconductor, following $W_d = \sqrt{\frac{2\phi_b \varepsilon_0 \varepsilon_s}{qN_d}}$, where ϕ_b is the barrier height, ε_0 the vacuum permittivity, ε_s the relative dielectric constant of the semiconductor, and q the elementary charge.[8] As a result, a remarkable Schottky barrier rises in the space charge region, which serves as an extra barrier and effectively increases the overall barrier width. It in turn substantially suppresses the junction current, accompanying with the increased barrier height of the ferroelectric layer itself, yielding the OFF state. Although the MFS FTJs are first realized in perovskite-oxide heterostructures, such as the Pt/BTO/Nb:STO, such a semiconductor-electrode tunnel junctions could be suitable for the HfO_2-based FTJs owing to the well-developed CMOS process on Si, Ge, and GaAs semiconducting wafers.

In addition to the MFM and MFS structures, there are also MFIM(S) structures in the HfO_2-based FTJs, in which the insulating layer can be deliberately inserted and/or unavoidably formed at the ferroelectric/metal (ferroelectric/semiconductor) interfaces (Figure 12.1c). The dielectric layer (DE) can serve as a switch to help the ferroelectric barrier to open/shut the junction current, making the polarization modulation become more effective.

Figure 12.1d demonstrates the write/readout operations of FTJ devices. The polarization of the ferroelectric barrier is switched by a pulse with the voltage larger than the coercivity. Then the resultant resistance state is read by a small voltage. During the readout process, the stored resistance states aren't disturbed owing to the ferroelectric hysteretic character. Therefore, the resistive-type FTJ devices are nondestructive-readout memory, overcoming a main drawback of the capacitive-type ferroelectric random-access memory (FeRAM).

12.1.2 Conduction Mechanisms

In ideal FTJs with the MFM structure, the junction conduction is predominantly regulated by quantum tunneling in the direct tunneling (DT) mechanism, as depicted in Figure 12.3a. The DT current density (J_{DT}) is given by:[2,9]

$$J_{DT} = C\frac{\exp\left\{\alpha(V)\left[\left(\Phi_2 - \frac{qV}{2}\right)^{3/2} - \left(\Phi_1 + \frac{qV}{2}\right)^{3/2}\right]\right\}}{\alpha^2(V)\left[\left(\Phi_2 - \frac{qV}{2}\right)^{1/2} - \left(\Phi_1 + \frac{qV}{2}\right)^{1/2}\right]^2} \times \sinh\left\{\frac{3}{2}\alpha(V)\left[\left(\Phi_2 - \frac{qV}{2}\right)^{1/2} - \left(\Phi_1 + \frac{qV}{2}\right)^{1/2}\right]\frac{qV}{2}\right\} \quad (12.1)$$

where $C = -(4qm_n^*)/(9\pi^2\hbar^3)$ and $\alpha(V) = [4d(2m_n^*)^{1/2}]/[3\hbar(\Phi_1 + qV - \Phi_2)]$, $\Phi_1(\Phi_2)$ correspond to the barrier heights at the two interfaces, m_n^* the effective electron mass, and $\hbar$ the reduced Planck constant. The DT transport behavior occurs near the Fermi level (E_F) of the FTJ device, in which the barrier should not be too thick and the applied voltage cannot be too high to keep the barrier in a trapezoidal shape.

Gruverman et al.[2] have simplified the DT formula and the J_{DT} is written as.

$$J_{DT} \approx \frac{q^2}{h}\frac{1}{2\pi d}\frac{\sqrt{2m_0\overline{\Phi}}}{\hbar}\exp\left[-2\frac{\sqrt{2m_0\overline{\Phi}}}{\hbar}d\right]V \quad (12.2)$$

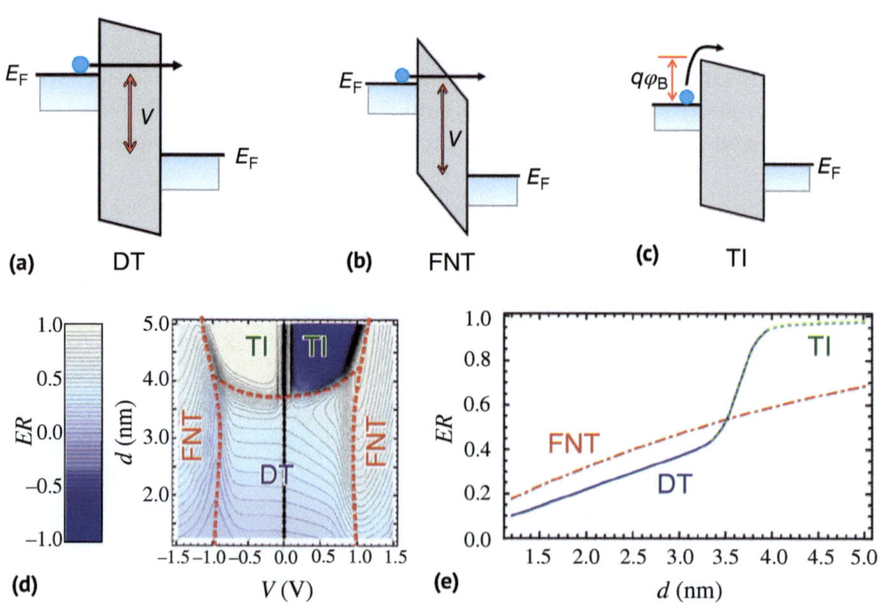

Figure 12.3 Conduction mechanisms in FTJ devices. (a) DT, (b) FNT, (c) TI, as well as the change of conduction mechanism and corresponding TER as functions of (d) applied voltage and (e) barrier thickness. *Source*: [10] / with permission of John Wiley & Sons.

where the $\overline{\Phi}$ is average barrier height, which is calculated by $\frac{\Phi_1 + \Phi_2}{2}$, and m_0 is the electron mass. The TER ratio can be obtained as $\frac{J_{\mathrm{DT,ON}}}{J_{\mathrm{DT,OFF}}} \approx \exp\left(\frac{2\sqrt{2m_0}}{\hbar} \cdot \delta\sqrt{\overline{\Phi}} \cdot d\right)$, where $\delta\sqrt{\overline{\Phi}} = \sqrt{\overline{\Phi}_{\mathrm{OFF}}} - \sqrt{\overline{\Phi}_{\mathrm{ON}}}$ and the barrier width is considered as the same for the ON and OFF states.[2] Therefore, the TER ratio can be very large in the FTJs since there is an exponential dependence upon the barrier height. In addition, if the barrier width can also be changed upon the polarization switching, dramatically improved TER ratio can be achieved in the FTJ device, just like the MFS structure. The DT transport in general takes place when the barrier isn't too thick and the applied voltage isn't too high. If the applied voltage is high, the band shift of the electrode makes the trapezoidal barrier becomes a triangle barrier and the transport evolves to the Fowler–Nordheim tunneling (FNT, Figure 12.3b), which is given by.[9,10]

$$J_{\mathrm{FNT}} = \frac{q^3 m_0}{8\pi h m_n^* \Phi_{\mathrm{B}}} E^2 \exp\left[-\frac{8\pi\sqrt{2m_n^*}}{3hq}\frac{\Phi_{\mathrm{B}}^{\frac{3}{2}}}{E}\right] \tag{12.3}$$

As depicted in Figure 12.3b, unlike the DT where electrons traverse the entire barrier, the FNT describes the electron tunneling through only a partial width of the barrier. Besides, the FNT is also observed in the FTJ with thicker barrier under a high bias.

When the barrier is too thick, such as the OFF state of the MFS and MFIS devices, the electron tunneling is forbidden and the transport is in general dominated by the Thermionic injection (TI) across a Schottky barrier (Figure 12.3c). The J_{TI} is written as.[9,11]

$$J_{\mathrm{TI}} = A^{**}T^2 \exp\left[-\frac{1}{k_{\mathrm{B}}T}\left(\Phi_{\mathrm{B}} - \sqrt{\frac{q^3 E}{4\pi\varepsilon_0 \varepsilon_{ifj}}}\right)\right] \tag{12.4}$$

where A^{**} is the effective Richardson constant, T the temperature, k_{B} the Boltzmann constant, and ε_{ifj} the relative dielectric constant of the interfacial layer. The TI describes the classical transport mechanism where thermally excited electrons surmount the potential barrier at metal-semiconductor or metal-insulator interfaces.

Pantel et al. have calculated the transport of FTJ devices with varying applied bias and barrier thickness and given a TER diagram associated with the DT, FNT, and TI mechanisms, as demonstrated in Figure 12.3d,e.[9] In addition, Dong et al. have calculated tunneling current in HfO_2-based FTJs by using the nonequilibrium Green's function formalism with the parameterized effective Hamiltonian, involving the MFM and MFS structures.[12]

12.2 HfO_2-based FTJ Devices

12.2.1 Fabricated on Single-crystalline Perovskite Substrates

Following the idea of perovskite tunnel junctions, lots of HfO_2-based FTJs have been fabricated on single-crystalline perovskite substrates, such as the STO, buffering with an epitaxial metallic-oxide electrode. In this regime, $Hf_{0.5}Zr_{0.5}O_2$ (HZO) and LSMO have been usually adopted as the ferroelectric barrier and the bottom electrode, respectively. For example, Figure 12.4a exhibits a typical MFM FTJ device in Pt/HZO/LSMO structure (the inset (i)), in which the HZO layer, grown by means of pulsed laser deposition (PLD), is stabilized in rhombohedral (*r*) polar phase on the epitaxial LSMO, perhaps due to the large compressive strain (the inset (iii)).[13] Along the $(111)_r$ orientation, robust ferroelectricity can persist down to 1.0 nm-thick HZO layer, which is about two unit cells (the inset (ii)). The existence of considerable polarization at nanoscale is an advantage for the utilization of HfO_2-based thin films as tunneling barriers in FTJs. Therefore, the polarization switching modulates junction conduction, as shown in current-voltage (*I-V*) curves of the ON and OFF states, and gives rise to the ON/OFF ratio of ~135 for ~1.3 nm-thick HZO barrier and of ~10^5 when the barrier is increased to ~2.5 nm. Based on the tight-binding model, the pronounced TER effects have been explained as the polarization-modulation on the barrier height and also the existence of *n*-type semiconducting interfacial layer, associated with the field-migration of oxygen vacancies, between the HZO and LSMO.[13] The change of transport and spin properties of the LSMO electrode associated oxygen vacancies migration has also observed by Wei et al. in Co/HZO/LSMO multiferroic tunnel junctions, which yields a large TER ratio of 10^4 and strong magnetoelectric coupling with both the sign and magnitude changed in the magnetoresistance effect.[14]

The field-controlling of transport properties of the electrode has also shown in the MFS-type FTJs. As shown in Figure 12.4b, a typical device comprises the Pt/HZO/Nb:STO fabricated by Jiao et al., in which the 5.8 nm-thick HZO exhibits the mixture of nonpolar monoclinic (m) and ferroelectric o phases along the (111) orientation (the inset (i)).[15] The polarization of the HZO barrier switches the *n*-type Nb:STO surface between accumulated and depleted states of electrons through the ferroelectric-field effect. The Schottky barrier yielded in the depletion region can serve as an extra barrier and improve the TER effect due mainly to the change of barrier width and hence the junction conduction (the inset (iii)). The ON (accumulated) state is dominated by the DT transport with a large current, while the electron tunneling is suppressed in the OFF state, which is regulated by the TI mechanism and exhibits dramatically lowered junction current. As a result, a large ON/OFF ratio of ~800 is achieved in the Pt/HZO/Nb:STO FTJs.[15] In Figure 12.4c, Du et al. have improved the TER effect of Pt/HZO/Nb:STO device by buffering a one-unit-cell LSMO layer (the insets (i) and (ii)), which facilitates the stabilization of o-phase because of the reduction of the interface energy between the Nb:STO and the HZO.[16] The authors not only demonstrated the increased ON/OFF ratio but also achieved high-speed switching of ~20 ns with multiple resistance levels with

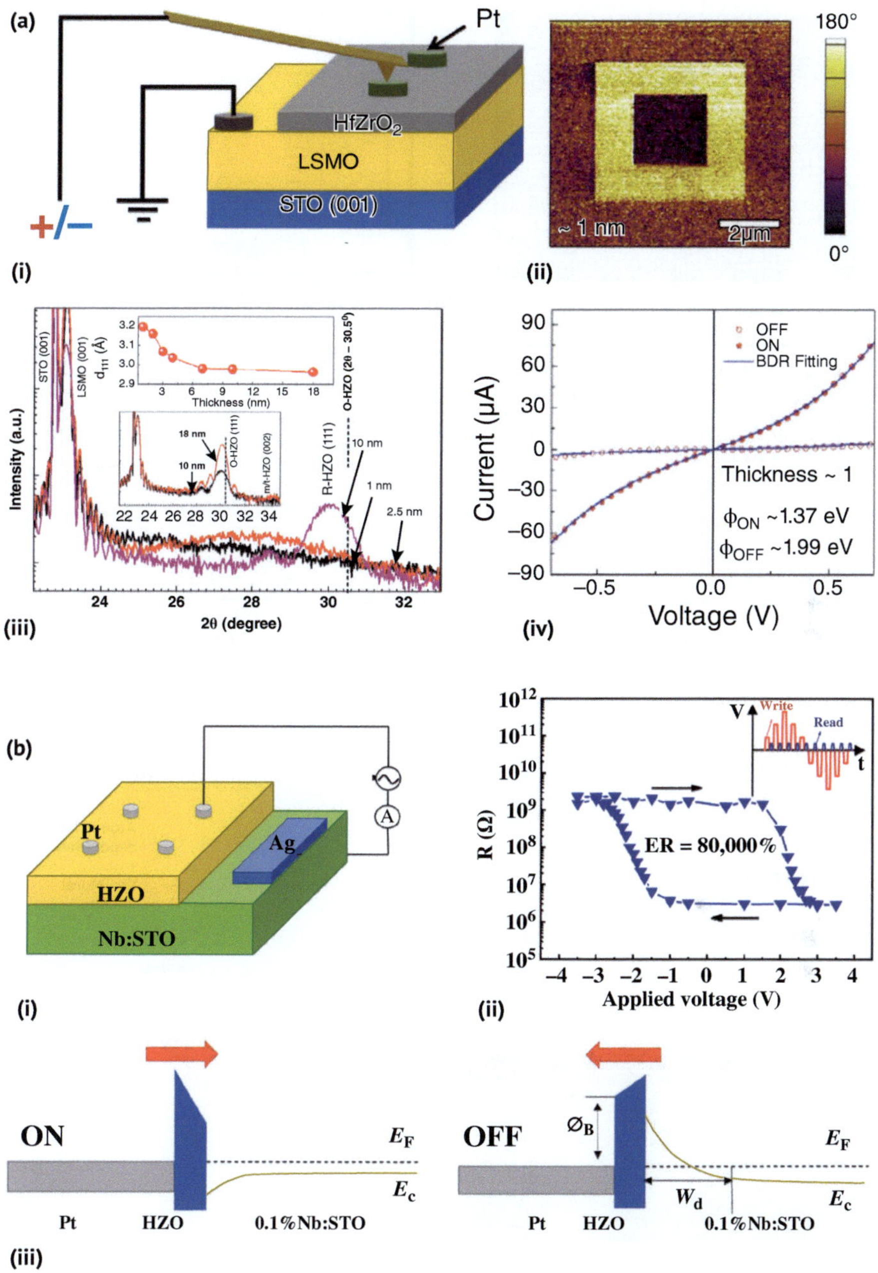

Figure 12.4 TER properties of the HfO_2-based FTJs on single-crystalline perovskite substrates. (a) Pt/HZO/LSMO FTJs: (i) device sketch, (ii) PFM phase image, (iii) XRD θ-2θ scans, and (iv) *I*-*V* curves of the ON and OFF. *Source*: [13] / with permission of John Wiley & Sons. (b) Pt/HZO/Nb:STO FTJs: (i) device sketch, (ii) *R*-*V* hysteresis loop, and (iii) barrier profiles of the ON and OFF states. *Source*: [15] / with permission of AIP Publishing.

(*Continued*)

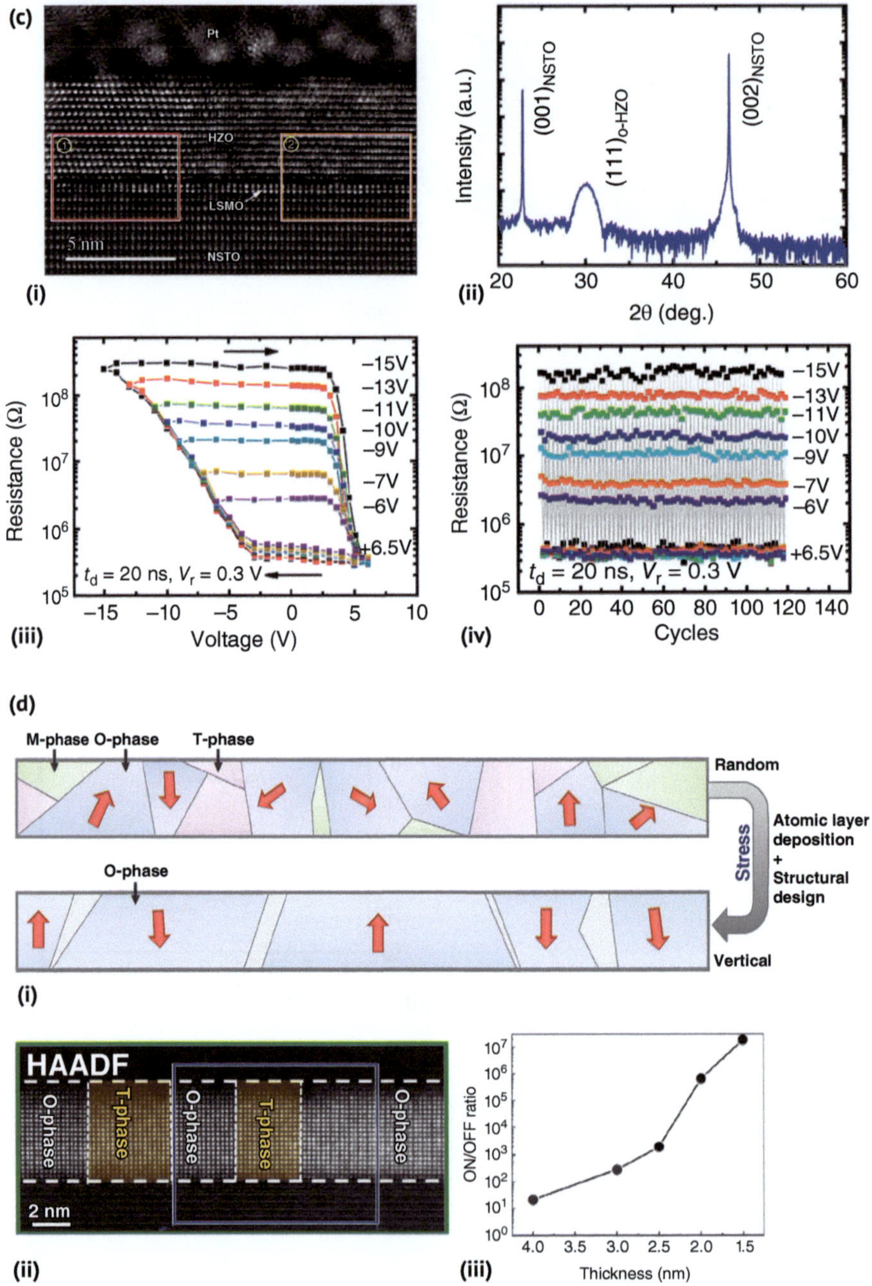

Figure 12.4 (Cont'd) (c) Pt/HZO/LSMO/Nb:STO FTJs: (i) HAADF image, (ii) XRD θ-2θ scan, (iii) *R-V* hysteresis loops at various switching voltages, and (iv) the corresponding multiple resistance levels. *Source*: [16] / with permission of American Chemical Society. (d) Pt/HZO/Nb:STO FTJs: (i) structural design for [001] texture of the HZO layer, (ii) HAADF image of the HZO, and (iii) ON/OFF ratio as a function of HZO thickness. *Source*: [17] / with permission of Elsevier.

good retention between the ON and OFF states (the inset (iii) and (iv)).[16] The other improvement of the TER effect in the Pt/HZO/Nb:STO device has been reported by Gao et al. by modulating crystalline texture of the HZO layer with stress control in rapid thermal annealing of the atomic layer deposition (ALD) process, as shown in the inset (i) in Figure 12.4d.[17] The optimized HZO exhibits the (001) orientation of o-phase directly grown on the Nb:STO stabilized by the coexistence of tetragonal (t) phase. Since the polar axis of the HZO is along the $(001)_o$ direction, the effective polarization along the thickness direction is improved, which makes the polarization-modulation of barrier profiles, including the barrier height and the interfacial Schottky barrier, become more remarkable. As a result, a giant ON/OFF ratio of ~2×10^7 is achieved in the Pt/HZO/Nb:STO FTJ with the barrier thickness of HZO of ~1.5 nm, which is a record-high value in 2023 and before (Table 12.1) and also a promising performance parameter comparable with that of FLASH memories (the inset (iii)).[17]

Table 12.1 TER properties of the HfO_2-based FTJ devices.

Device structure	Thickness (nm)	ON/ OFF ratio	Retention (Second)	Endurance (Cycling number)	References
Fabricated on single-crystalline perovskite substrates					
$Pt/Hf_{0.5}Zr_{0.5}O_2$/Nb:STO	1.5	2×10^7	5×10^5	3×10^5	[17]
$Pt/Hf_{0.5}Zr_{0.5}O_2$/Nb:STO	5.8	800	10^4	1500	[15]
$Pt/Hf_{0.5}Zr_{0.5}O_2$/LSMO	2.5	10^5	/	/	[13]
$Au/Hf_{0.5}Zr_{0.5}O_2$/LSMO/ Nb:STO	4.3	834	10^4	1000	[16]
$Pt/Hf_{0.5}Zr_{0.5}O_2$/LSMO	4.6	4	/	/	[59]
$Pt/Hf_{0.5}Zr_{0.5}O_2$/LSMO	2	2	10^3	10^4	[60]
$Pt/STO/Hf_{0.5}Zr_{0.5}O_2$/LSMO	4.6	14	/	/	[61]
Fabricated on CMOS-compatible substrates					
$TiN/Hf_{0.5}Zr_{0.5}O_2/p$-Si	4.5	80	/	/	[62]
TiN/HfAlO/GaAs	10	15	10^5	10^8	[20]
$TiN/Hf_{0.5}Zr_{0.5}O_2$/Ge	6	30	10^8	10^7	[19]
$TiN/Hf_{0.5}Zr_{0.5}O_2$/W	4.5	16	10^8	10^8	[63]
$TiN/Hf_{0.5}Zr_{0.5}O_2$/Pt	2.8	15	10^4	1000	[21]
$Au/Ti/HfAlO_x$/Pt/Ti/Si	10	10	/	/	[64]
$RuO_2/Hf_{0.5}Zr_{0.5}O_2$/ITO/ SiO_2/Si	2.6	75	10^4	/	[25]
$Al/Hf_{0.5}Zr_{0.5}O_2/n$-Si	4	30	10^8	1000	[65]
$W/Hf_xZr_{1-x}O_2$/TiN	4.5	3	10^4	10^6	[66]
$Au/Ti/Hf_{0.5}Zr_{0.5}O_2$/Pt/Ti/Si	10	50	10^4	10^4	[67]
Ag/Y-HfO_2/TiN	3	100	/	/	

(Continued)

Table 12.1 (Continued)

Device structure	Thickness (nm)	ON/ OFF ratio	Retention (Second)	Endurance (Cycling number)	References
Cr/Ni/$Hf_{1-x}Zr_xO_2$/TiN/Si	4.5	8	10^8	10^8	[22]
Pt/$Hf_{0.5}Zr_{0.5}O_2$/Pt	2.8	20	/	/	[68]
TiN/$Hf_{0.5}Zr_{0.5}O_2$/*p*-Ge	8	20	/	/	[69]
Ti/Au/Al_2O_3/$Hf_{0.5}Zr_{0.5}O_2$/ ZrO_2/*p*-Si	10	30	10^4	10^4	[35]
TiN/$Hf_{0.5}Zr_{0.5}O_2$/Ta_2O_5/TiN	4	10^8	/	/	[43]
TiN/Al_2O_3/HZO/TaN	10	100	10^4	10^8	[34]
Al/TiN/Al_2O_3/$Hf_{0.5}Zr_{0.5}O_2$/ TiN/*p*-Si	10	20	10^8		[35]
TiN/Al_2O_3/$Hf_{0.5}Zr_{0.5}O_2$/Ge	8.4	14	10^9	10^6	[36]
Ni/HfO_2/MgS/Ni	4.75 u.c.	4580	/	/	[42]
Ti/Au/Al_2O_3/ $Hf_{0.5}Zr_{0.5}O_2$/*p*-Si	10-12	25	10^8	10^7	[56]
TiN/HfO_x/SiO_x/*n*-Si	6	100	10^4	10^4	[32]
TiN/Y-HfO_2/SiO_x/*n*-Si	1	10^6	/	/	[29]
Au/Ti/ZrO_2/HZO/Al_2O_3/ HZO/ZrO_2/Pt	12	13.8	10^8	10^4	[70]
W/HfAlO/TiN/SiO_2/Si	9	1.86	10^4	10^6	[71]
Au/Ti/La:HfO_2/ZrO_2/ La:HfO_2/TiN	10	15.1	/	10^7	[72]
W/Zr:HfO_2/SiO_2/*p*-Si	1	200	10^6	1000	[30]
TiN/Si:HfO_2/SiON/*n*-Si	8	7.9	10^8	2000	[33]
Pt/SiO_2/$Hf_{0.5}Zr_{0.5}O_2$/TiN/Si	2	850	10^5	10^7	[31]
Ni/$Hf_{0.5}Zr_{0.5}O_2$/TiN	9	100	10^8	10^7	[55]
TiN/$Hf_{0.5}Zr_{0.5}O_2$/Pt/	9	5	/	1000	[73]
TiN/$Hf_{0.5}Zr_{0.5}O_2$/TaN/W	5.5	100	10^8	10^8	[74]
TiN/TiO_2/HZO/TiN	6	11.6	10^8	10^7	[75]
TiN/HZO/SiO_2/Si	8	11	10^8	10^4	[76]
TiN/HZO/Al_2O_3/Ge	12	78	/	/	[77]
MoS_2/HZO/W/Si	4.5	1000	/	/	[78]
TiN/Al_2O_3/HZO/TiN	6.2	2470	/	/	[79]

12.2.2 Fabricated on CMOS-compatible Substrates

As high-*k* gate oxides of commercial field-effect transistors, the HfO_2-based thin films have well-studied fabrication processes on CMOS-compatible substrates, such as Si, Ge, and GaAs wafers, by means of ALD. Based on the oxide/semiconductor heterostructures, the HfO_2-based FTJs have been investigated, in which the semiconducting wafers are directly used as the bottom electrodes, forming the MFS architectures. Growing oxides on Si and similar alloying semiconductors, it is possible to form dead layers due to the interdiffusion at the interfaces, which may degrade and even kill ferroelectricity of the HfO_2-based thin films, especially that the films are only several nanometers in thickness. In 2016, Chernikova et al. have grown 2.5 nm-thick HZO films on *n*-type Si substrates, which are polycrystalline with the mixed m- and o-phases shown in plan-view transmission electron microscopy (TEM) image, and demonstrated switchable ferroelectric polarizations by piezoresponse force microscopy (PFM) technique and positive-up-negative-down (PUND) method with TiN TEs (Figure 12.5a).[18] Goh et al. have fabricated HZO FTJs with different bottom electrodes (the inset (i)), in which robust ferroelectricity with remanent polarizations higher that 12 $\mu C/cm^2$ have been achieved on the TiN, SiGe, and Ge wafers.[19] However, considerable TER effects are only observed in the TiN/HZO/SiGe and TiN/HZO/Ge devices due to the *p*-type bottom electrodes improving the polarization-modulation on the junction conductance, associated with the Schottky barriers (the inset (iii)). The ON/OFF ratio is around one order with the 8.0 nm-thick HZO barrier layer (Figure 12.5b).[19] These results also show that the o-phase stability is dramatically affected by the bottom electrodes. As shown in Figure 12.5c, Li et al. have studied the phase stability of ultrathin Al-doped HfO_2 ($HfAlO_x$) films in experimental in TiN/HfAlO/Si, TiN/HfAlO/Pt, and TiN/HfAlO/GaAs FTJs and in theoretical by the first-principles calculations (the inset (i)).[20] High volume fraction of o-phase has been observed on the GaAs bottom electrode, yielding the TER ratio larger than 10 and better retention character of resistance states (the insets (ii) and (iii)).[20]

In addition to the direct growth of HfO_2-based films on semiconductor wafers, TiN has been frequently used as the bottom electrode because of its good compatibility with the ALD process, facilitating the fabrication of MFM-type FTJs on compatible substrates. Figure 12.6a shows a typical device with Pt/HZO/TiN structure on Si wafer, in which the 2.8 nm-thick HZO is dominated by o-phase along the $(100)_o$ orientation (the inset (i)).[21] Polarization switching leads to distinguishable ON and OFF states with the electron transport regulated by the DT mechanism (the inset (ii)). The authors also give further analysis on the barrier profiles combining the X-ray photoelectron spectroscopy (XPS) measurements (the inset (iii)) and the DT fitting (the inset (iv)). The resultant barrier height changes are 0.16 and 0.89 eV at Pt/HZO and HZO/TiN interfaces, respectively, yielding the TER ratio of ~15.[21] Such an improved TER effect may be ascribed to the stabilization of the o-phase on the TiN bottom electrode, even though there is no Schottky barrier at the ferroelectric/metal interface. Athle et al. have shown the effects of TEs, such as Ni, Cr, and W, on the TER effects of the metal/HZO/TiN FTJs, in which the work functions (WF) and electron densities (n_e) of the TE metals have been separately studied.[22] The WF gives rise to the different band alignments and hence the built-in electric fields of the metal/HZO/TiN devices (the inset (ii)). The n_e results in the difference in the polarization-modulation of barrier height in the theoretical

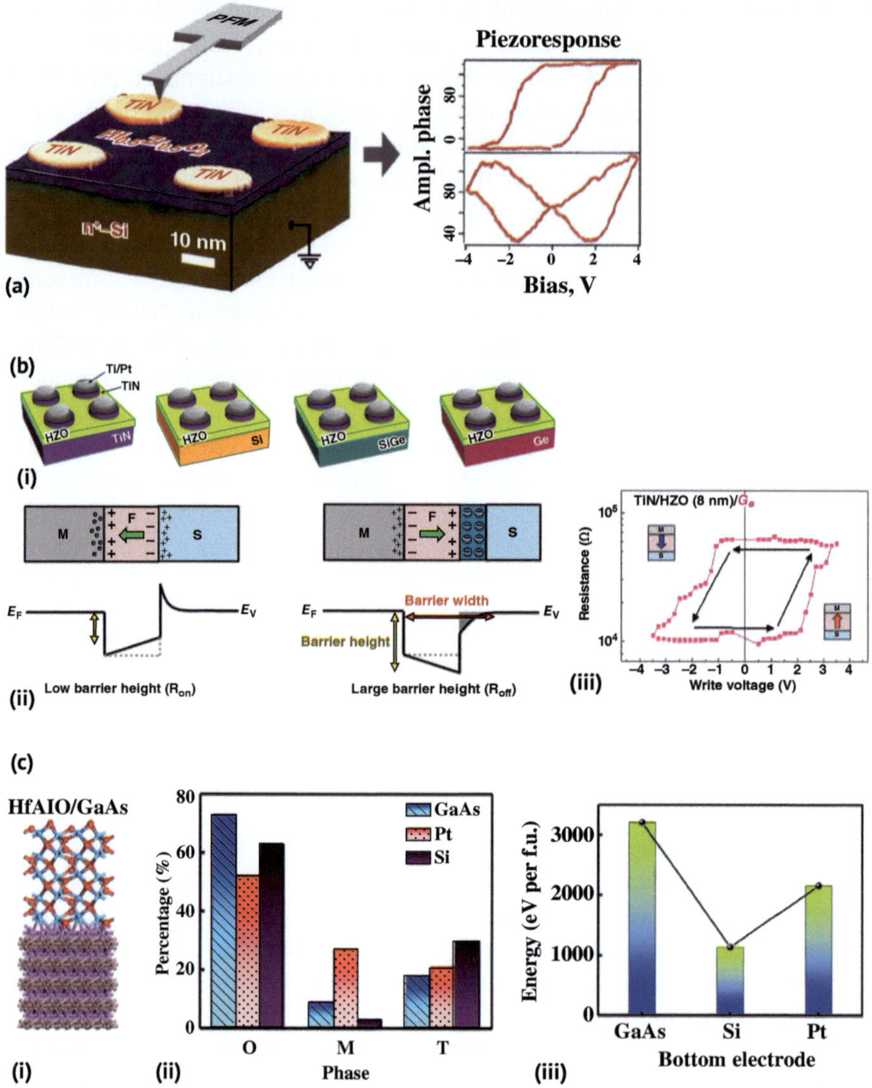

Figure 12.5 MFS-type HfO_2-based FTJs. (a) PFM characterization of a 2.5 nm-thick HZO film directly grown on *n*-type Si wafer. *Source*: [18] / with permission of American Chemical Society. (b) Effects of bottom electrode on TER: (i) device sketches, (ii) barrier profiles for the ON and OFF states, and (iii) *R-V* hysteresis loop for the TiN/HZO/Ge device. *Source*: [19] / with permission of IOP Publishing. (c) Effects of bottom electrode on the stabilization of *o* phase: (i) atomic structure model for the calculation, (ii) phase volume fractions, and (iii) energies of the $HfAlO_x$ on different bottom electrodes. *Source*: [20] / with permission of IEEE.

framework of Thomas–Fermi screening, since there is an inverse relation between the screening length and the n_e, following $\delta = \sqrt{\frac{\varepsilon_0 \hbar^2}{2m_n^* q^2 n_e}}$.[23] The largest TER ratio is observed in the Ni/HZO/TiN device due to the largest difference of n_e between Ni ($n_e = 9.1 \times 10^{28}$ m^{-3}) and TiN ($n_e = 5.5 \times 10^{28}$ m^{-3}) (the inset (iii)).[22] However, the improved TER ratio is at the expense of resistance retention since there exists a large built-in electric field, as indicative of the shift of the TER curve along the voltage axis that origin from the imprinted polarization switching, due to the large difference in WF between Ni (5.1 eV) and TiN (4.4 eV). Then, a trade-off approach by adopting a Cr/Ni bilayer TE is shown for designing and optimizing the TER properties (the inset (iv)) (Figure 12.6b).[22]

In fact, even though employing the same material in both top and bottom electrodes, the asymmetric atomistic structure at barrier/electrode interfaces can also give rise to considerable TER effect, which has been calculated by Seo et al. by the density functional theory (DFT) method in TiN/HZO/TiN FTJs (the inset (i)), as shown in Figure 12.6c.[24] The asymmetrical interface dipoles, associated

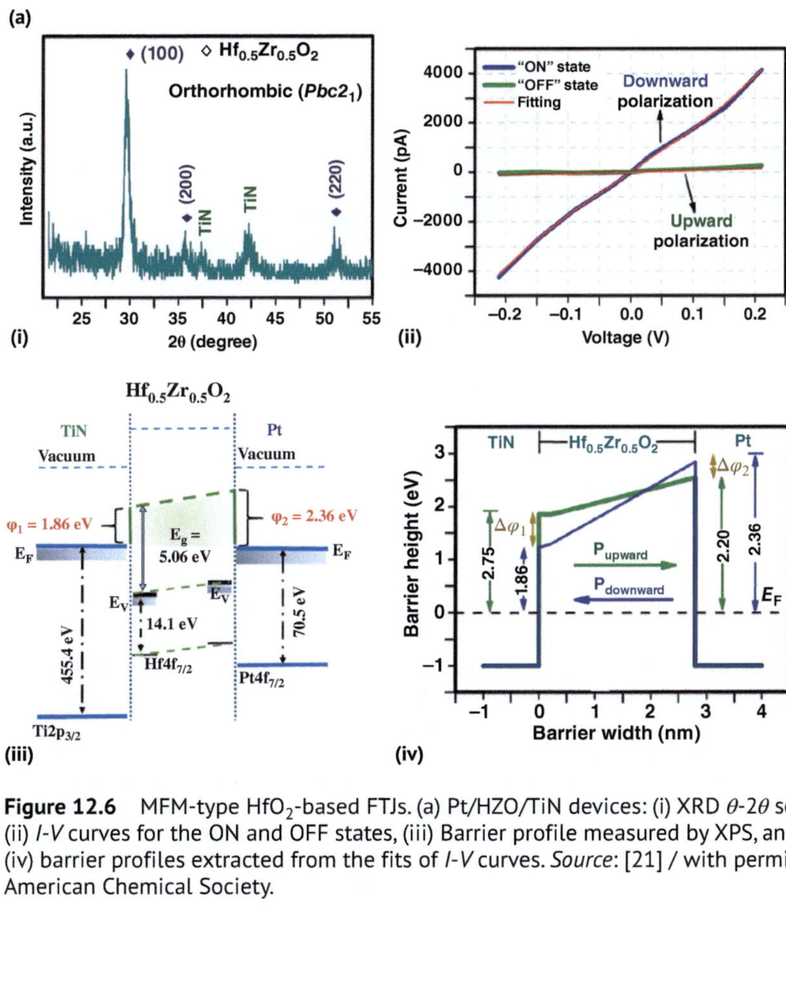

Figure 12.6 MFM-type HfO_2-based FTJs. (a) Pt/HZO/TiN devices: (i) XRD θ-2θ scan, (ii) *I-V* curves for the ON and OFF states, (iii) Barrier profile measured by XPS, and (iv) barrier profiles extracted from the fits of *I-V* curves. *Source*: [21] / with permission of American Chemical Society.

(Continued)

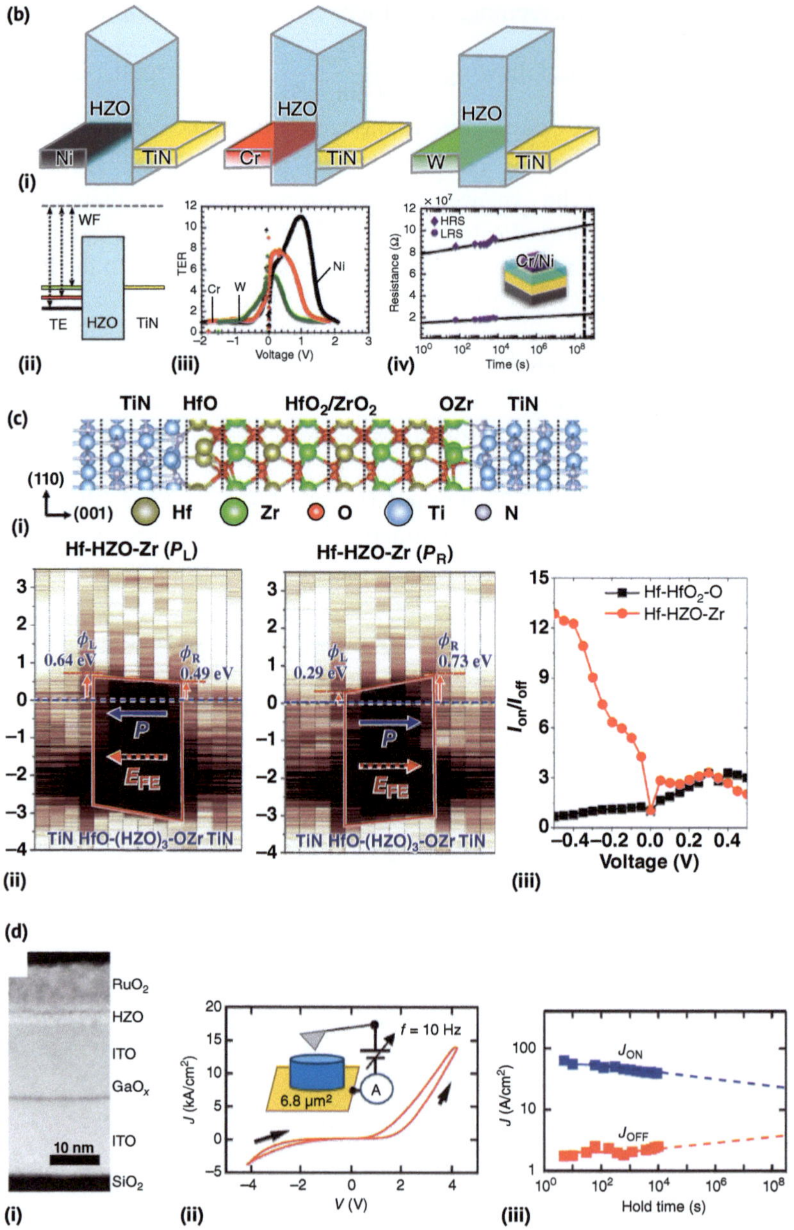

Figure 12.6 (Cont'd) (b) Effects of TEs on TER of the metal/HZO/TiN devices: (i) device structures, (ii) sketches for band alignments, (iii) TER ratios, and (iv) retention behaviors with Cr/Ni TE. *Source*: [22] / with permission of American Chemical Society. (c) DFT calculations of atomistic asymmetric effect on TER: (i) atomistic model of TiN/HZO/TiN device, (ii) LPDOS for barrier profiles of the ON and OFF states, and (iii) calculated ON/OFF ratios for asymmetric Hf-HZO-Zr and Hf-HfO_2-O models. *Source*: [24] / with permission of American Physical Society. (d) All-oxide RuO_2/HZO/ITO FTJs: (i) Cross-sectional TEM image of the device, (ii) Hysteretic *I-V* curve, and (iii) retention behaviors for the ON and OFF states. *Source*: [25] / with permission of American Chemical Society.

with specific terminations of the HZO layer, establish an effective built-in field, which facilitates the polarization-modulation of barrier profile (the inset (ii)) and hence the improvement of ON/OFF ratio (the inset (iii)).[24] However, in experiments, such a delicate microstructure with high-controllable interfacial atomic configurations isn't easy to achieve in the HfO_2-based FTJs since the ferroelectric HfO_2-based barriers are always polycrystalline and/or mixed by more than one phase on TiN, metal (e.g., Pt and W), and semiconductor (e.g., Si, Ge, and GaAs) bottom electrodes. The remarkable atomistic effect is in general averaged at the non-sharp interfaces. The growth of HfO_2-based thin-film heterostructures in epitaxial layer-by-layer mode, like what the perovskite FTJs done, may be a possible approach to utilize the interface effect to improve the TER effect, in which a bottom electrode that is lattice-match with the HfO_2 should be a prerequisite. Indium tin oxide (ITO) may be a good choice, and there are all-epitaxial ionic-doped HfO_2/ITO/YSZ (an acronym for yttria-stabilized zirconia) heterostructures for the study of ferroelectric properties.[26] Figure 12.6d shows an example of the ITO-electroded FTJ in RuO_2/HZO (2.9 nm)/ITO architecture integrated on the Si wafer (the inset (i)), in which the *I*-*V* loop exhibits clear hysteresis (the inset (ii)) and an improved TER ratio of ~75 with stable ON and OFF states is achieved (the inset (iii)), suggesting promising performances.[25] Unfortunately, atomic-scale structure and epitaxial character of the all-oxide HZO devices haven't been shown in this publication.

Another approach to improve the TER effects is the insertion of DE between the top and bottom electrodes, which combines with the ferroelectric layer to form composited barriers and hence the MFIS(M)-type FTJs. The effectiveness of ferroelectric/dielectric barriers has been manifested in perovskite FTJs, such as the epitaxial BTO/STO bilayers, in which the dielectric STO barrier that is shifted in response to the polarization-modulation of the ferroelectric BTO barrier can serve as a transport switch, showing the improvement in both the electroresistance and magnetoresistance.[27,28] In the HfO_2-based FTJs, the industrialized SiO_x, could be inherently utilized as the dielectric barrier, which can be achieved on the Si wafer by thermal oxidation and chemical/physical vapor processes. Figure 12.7a shows an example of the MFIS-type device with TiN/Y:HfO_2/SiO_x/*n*-Si structure (the inset (i)), in which the 1.0 nm-thick Y:HfO_2 is orthorhombic and along the $(001)_o$ orientation on the Si (001) substrate (the inset (ii)).[29] The polarization switching not only modulates the Y:HfO_2/SiO_x composited barrier but also results in accumulation/depletion of the *n*-type semiconducting Si electrode (the inset (iii)). As a result, the hysteretic *R*-*V* character with a giant ON/OFF ratio of $\sim 10^6$ is achieved (the inset (iv)). The electron transports across the Y:HfO_2/SiO_x barrier are found in DT (low bias) and FNT (high bias) mechanisms for the bistable resistance states (the inset (v)).[29] The Schottky barrier at the dielectric/semiconductor interface is measured by capacitance (*C*) method in C^{-2}-V plots, showing the W_d of ~0.5 and ~5.4 nm in the ON and OFF states, respectively (the inset (vi)).[29] Similar SiO_x-based composited barriers have also been reported in W/HZO/SiO_2/*p*-Si,[30] Pt/SiO_2/HZO/TiN,[31] TiN/HfO_x/SiO_x/Si,[32] as well as TiN/Si:HfO_2/SiON/Si devices.[33]

In addition to the SiO_x, Al_2O_3 is also a kind of widely adopted DE in CMOS-compatible electronic devices. By buffering 1.0 nm-thick Al_2O_3, an ON/OFF ratio of ~10 with good resistance retention has been reported in TiN/HZO/Al_2O_3/Ge FTJs.[36] Hur et al. have fabricated TiN/Al_2O_3/HZO/TiN FTJs and characterized the TER properties at cryogenic temperature, in which an ON/OFF ratio of ~50 after the wake-up and switching endurance more than 1,000 cycles without obvious fatigue are observed.[37] In Figure 12.7b, Hsing et al. have shown the insertion of an ultrathin Al_2O_3 layer into the $Hf_{1-x}Zr_xO_2$/TiN interface can also stabilize the ferroelectricity of the $Hf_{1-x}Zr_xO_2$ layers (the inset (i)) and the antiferroelectric-to-ferroelectric transition takes place even in the 10 at % Hf-doped ZrO_2 (the insets (ii) and (iii)).[34] The ON/OFF is observed to increase with increasing Zr concentration, and it reaches the maximum value of ~100 at $x = 0.9$ in the FTJ device with the $Hf_{0.1}Zr_{0.9}O_2$ (5.0 nm)/Al_2O_3 (1.0 nm) composited barrier (the inset (iv)).[34]

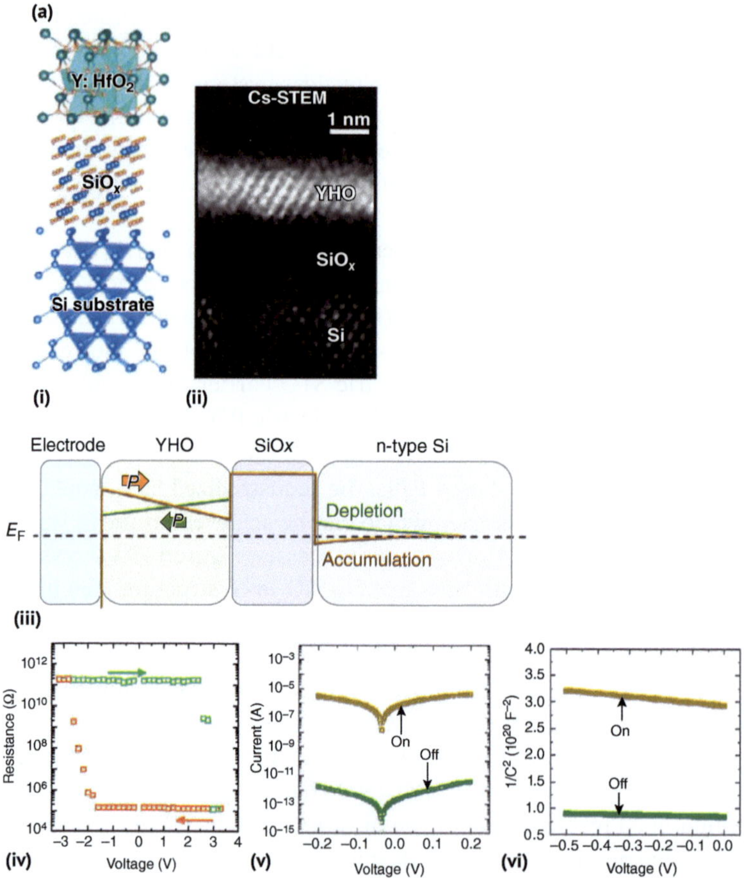

Figure 12.7 MFIS(M)-type HfO_2-based FTJs. (a) TiN/Y:HfO_2/SiO_x/*n*-Si devices: (i) structure sketch and (ii) cross-sectional HAADF image for the Y:HfO_2/SiO_x/Si heterostructure, as well as (iii) barrier profiles upon polarization switching, (iv) *R-V* hysteresis loops, (v) *I-V* curves for the ON and OFF states, and (vi) corresponding C^{-2}-*V* plots for deducing depleted regions of the semiconductor electrodes. *Source*: [29] / with permission of Elsevier.

The composited barrier can also be extended into three layers and even more complicated multiple layers. As shown in Figure 12.7c, Cao et al. have reported that, by sandwiching the ferroelectric HZO layer with ZrO_2 and Al_2O_3 layers (the insets (i), (ii), and (iii)), the polarization is increased and depends on the ZrO_2 thickness (the inset (iv)).[35] As a result, the TER effects are improved in the Ti/Au/Al_2O_3/HZO/ZrO_2/*p*-Si FTJs in comparison with that of the device with only the single HZO barrier.[35] The inset (v) shows *R-V* hysteresis loops of the optimized composite-barried devices and, compared with that of the Al_2O_3/HZO, the ON/OFF ratio is increased by about three times in the three-layer Al_2O_3/HZO/ZrO_2 barrier because the larger polarization makes the modulations of over-barrier profiles, including the width and height of the composited barrier, as well as the interfacial Schottky barriers, become more effective.[35]

Besides the barrier profiles, the flexible architecture of composited barrier with diverse ferroelectric and DEs can also modulate the electron tunneling behaviors, such as the resonant tunneling, which occurs in the quantum well when the energy of incoming electrons coincides with the quantized energy levels, generating the

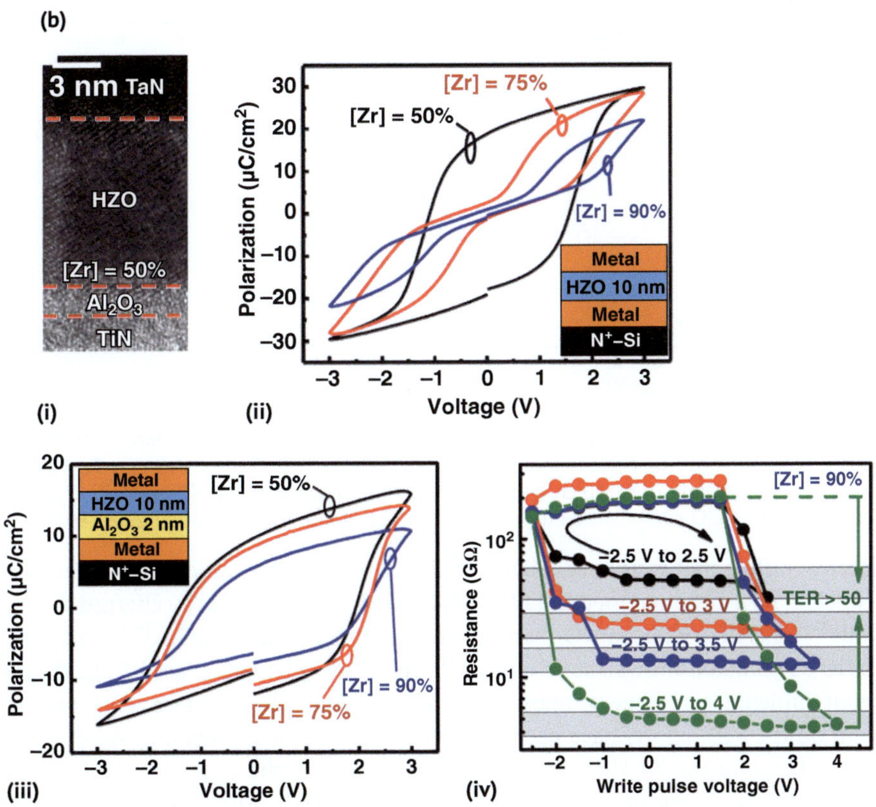

Figure 12.7 (Cont'd) (b) TaN/$Hf_{1-x}Zr_xO_2$/Al_2O_3/TiN devices: (i) cross-sectional TEM image, *P-V* hysteresis loops for the $Hf_{1-x}Zr_xO_2$ layers with various ZrO_2 concentrations in MFM structure, (ii) without and (iii) with the Al_2O_3 inserting layers, and (iv) *R-V* hysteresis loops. *Source*: [34] / with permission of IEEE.

(c)

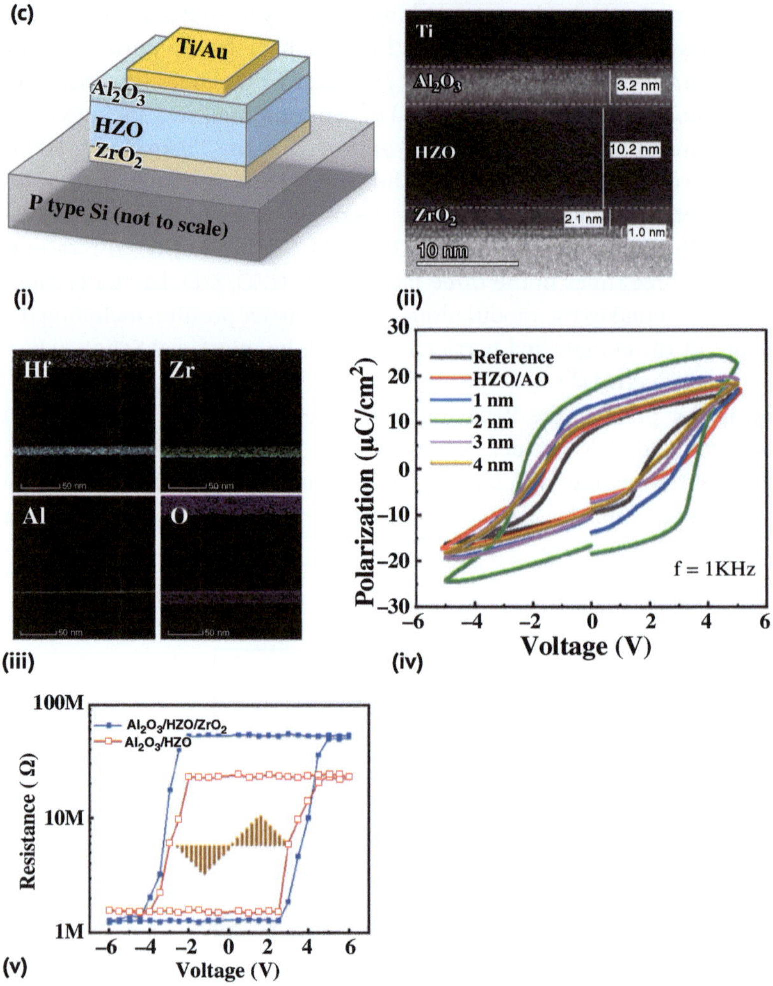

Figure 12.7 (Cont'd) (c) Ti/Au/Al_2O_3/HZO/ZrO_2/*p*-Si devices: (i) device sketch, (ii) cross-sectional TEM image, (iii) corresponding EDX mappings, (iv) *P*-*V* hysteresis loops for various ZrO_2 thicknesses (Reference: HZO), and (v) *R*-*V* hysteresis loops of the FTJs with Al_2O_3/HZO/ZrO_2 and Al_2O_3/HZO composited barriers. *Source*: [35] / John Wiley & Sons / CC BY 4.0.

negative differential resistance (NDR).[38] Experimental observations of the resonant tunneling have been reported in LSMO/BTO/LSMO FTJs assisted by charged domain walls,[39] $LaAlO_3$/BTO/$LaAlO_3$-barried resonant tunneling diodes,[40] and STO/$LaTiO_3$/STO quantum-well superlattices.[41] In MFIM-type HfO_2-based FTJs, the resonant tunneling across the composited ferroelectric/dielectric barrier and its improvement on the TER effects have been studied by theoretical simulations. As shown in Figure 12.8a, in the Ni/HfO_2 (o-phase)/MgS/Ni device (the inset (i)), the quantum well, which is electrically-erasable, is formed at the HfO_2/MgS interface due to the generation of two-dimensional electron gas (2DEG) when the polarization is pointing to the MgS (the inset (ii)).[42] The accumulation of electrons reduces the effective width of the barrier and causes the resonant tunneling across the FTJ in the ON state, which increases the transmission coefficient by more than two orders of magnitude (the inset (iii)).[42] As a result, the TER ratio is improved to as high as 4.58×10^3 since there is no electron accumulation at the HfO_2/MgS interface in the OFF state after the polarization is switched (the inset (ii)).[42] Figure 12.8b shows the other model of the resonant tunneling-enhanced TER by modulating the relative barrier height of the inserting DE in the TiN/HZO/SiO_2/TiN and TiN/HZO/Ta_2O_5/TiN FTJs.[43] For the high-barrier SiO_2, the quantum well is located in the HZO layer in the HZO/SiO_2 stack, which has only one quasi-bound state and the corresponding confined energy is far away from the Fermi energy (the inset (i)). In contrast, there are four quasi-bound states in HZO/Ta_2O_5 stack and the confined energies are much closer to the Fermi level due to the low barrier height of Ta_2O_5 (the inset (ii)), which makes the resonant tunneling easy to take place at the low applied bias for readout. As a result, the switching between the resonant tunneling in the ON state, as indicated by the NDR behavior in the *J-V* curve, and the DT in the OFF state dramatically improves the TER effect, leading to a giant ON/OFF ratio of ~10^8 in the TiN/HZO/Ta_2O_5/TiN device.[43] These resonant tunneling diagrams may guide the design of high-performance HfO_2-based FTJs by engineering band alignments of the composited barriers.

12.2.3 Device Reliability of the HfO_2-based FTJs

As a kind of ferroelectric memories, time retention and switching endurance of the bistable resistance states are regarded as two main reliability issues of the FTJ devices. The resistance retention character is mainly related to the reduction of effective polarization in the ON or OFF state, corresponding to the P_{up} or P_{down} or vice versa, of the ferroelectric HfO_2-based barrier by the E_{dep}. At the metal/ferroelectric interface, the E_{dep} is due to the incomplete screening of polarisation-bound charges in the metal electrode because of its finite δ (see Figure 12.2 for the electrostatic screening of the MFM structure), which can be written as

$$E_{dep} = \frac{P}{\varepsilon_0 \varepsilon_{FE}} \left[\frac{\frac{d_{FE}}{2\varepsilon_{FE}}}{\frac{d_{FE}}{2\varepsilon_{FE}} + \frac{\delta}{\varepsilon_M}} - 1 \right] \qquad (12.5)$$

where ε_M is the relative dielectric constant of the metal.[44] One can find that, for a certain ferroelectric barrier, the E_{dep} increases with the increase of δ. As discussed in Figure 12.2c, the improvement of the TER ratio always requires a large difference in the δ between the top and bottom electrodes, that is, one of the electrodes should have a larger δ. Therefore, retention is an issue that should be concerned in the FTJ devices, especially that the FTJ has an ultrathin ferroelectric layer in comparison with that of the FeRAMs and ferroelectric-field-effect transistors (FeFETs).

The retention problem may become more serious if there is a DE between the ferroelectric and metal, such as the MFIM structure in the HfO_2-based FTJs.[22,36,45,46] An example is shown in Figure 12.9 for the TiN/HZO/Al_2O_3/TiN device

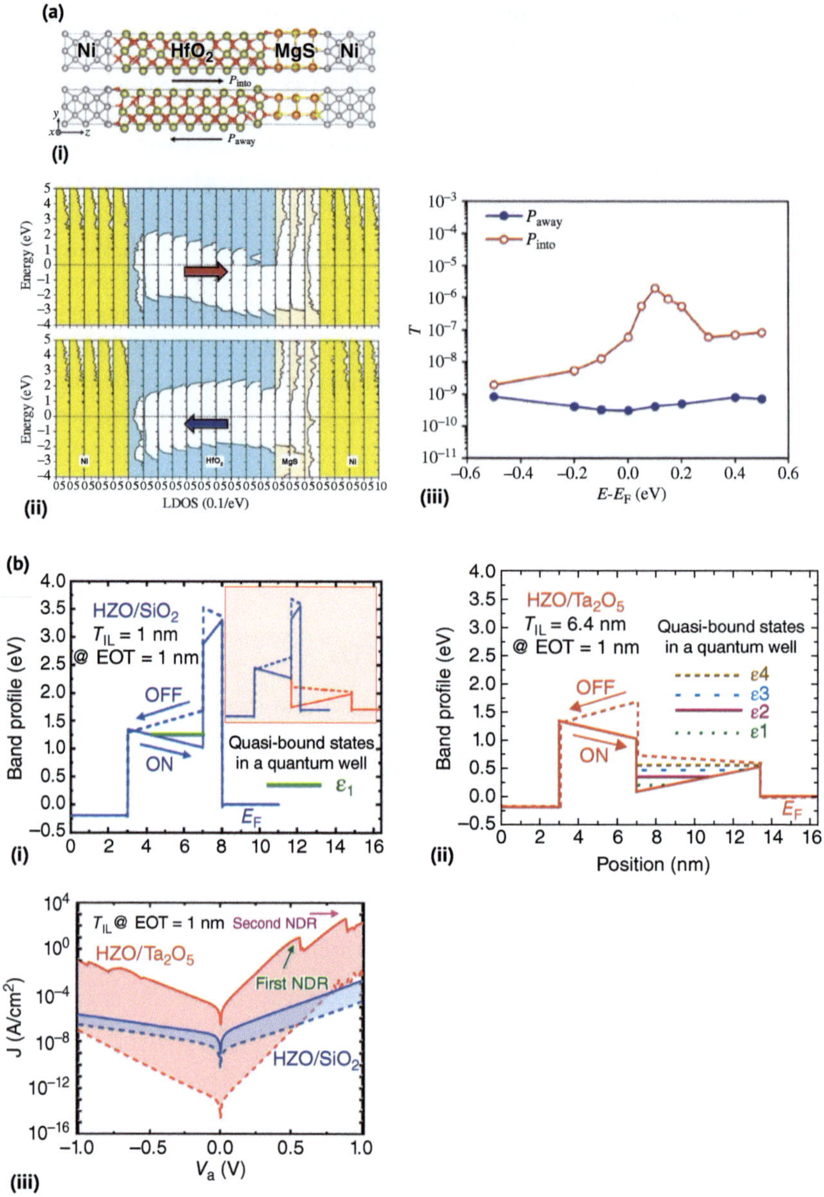

Figure 12.8 Theoretical calculations of resonant tunneling in MFIM-type HfO_2-based FTJs. (a) Ni/HfO_2/MgS/Ni devices: (i) optimized atomic structures and (ii) LDOS of the FTJ for the polarization pointing into (P_{into}, top panel) and away from (P_{away}, bottom panel) the MgS DE, (iii) the resultant transport characters for the P_{into} and P_{away} states. *Source*: [42] / with permission of AIP Publishing. (b) TiN/HZO/SiO_2/TiN and TiN/HZO/Ta_2O_5/TiN devices: (i, ii) band diagrams and (iii) *J-V* curves for the ON and OFF states of the HZO/SiO_2 and the HZO/Ta_2O_5 composited barriers. *Source*: [43] / International Conference on Simulation of Semiconductor Processes and Devices (SISPAD).

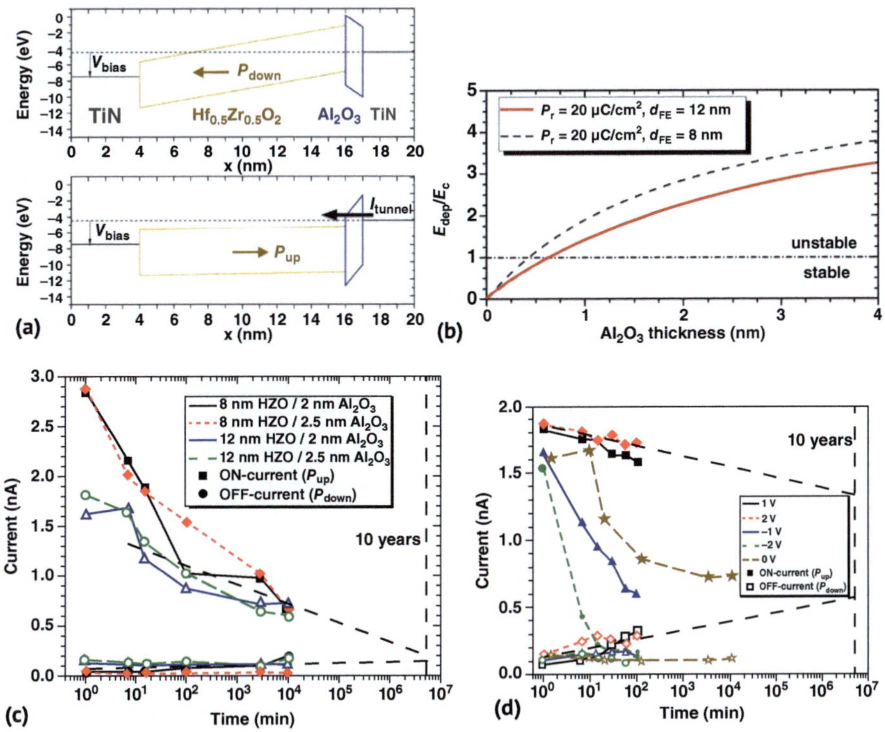

Figure 12.9 Retention study for the TiN/HZO/Al_2O_3/TiN FTJs. (a) Band alignments for the P_{down} and P_{up} states. (b) Calculated E_{dep} normalized to the E_c as a function of Al_2O_3 thickness for different HZO thicknesses, in which the dashed line indicates the E_c. (c) Retention characteristics and (d) their improvements under DC bias. *Source*: [45] / with permission of IEEE.

(Figure 12.9a), in which the E_{dep} follows $E_{dep} = \frac{P}{\varepsilon_0 \varepsilon_{FE}} \times \frac{\varepsilon_{FE} d_{DE}}{\varepsilon_{FE} d_{DE} + \varepsilon_{DE} d_{FE}}$.[45] The calculated E_{dep} normalized to the E_c of the 8.0 and 12 nm-thick HZO barriers as a function of Al_2O_3 thickness are plotted in Figure 12.9b. The E_{dep} is easy to be higher than the E_c even for the Al_2O_3 less than 1.0 nm in thickness, resulting in unstable polarizations of the HZO barrier. As shown in Figure 12.9c, the resistance relaxation is more pronounced in the ON state, which may also be ascribed to the asymmetric band alignment of the FTJ (Figure 12.9a). By applying the positive DC bias of 1–2 V to override the E_{dep}, the relaxation behaviors can be relieved, allowing the extrapolation of the ON and OFF states to 10 years. However, if the DC bias is changed to negative, the resistance relaxation accelerates. In addition to the E_{dep}, the existence of built-in field (E_{in}) result from the difference in WF between the top and bottom electrodes can also give rise to the retention issue.[45] Athle et al. have observed in the metal/HZO/TiN FTJs that the Ni TE can improve the TER ratio but leads to obvious resistance relaxation.[22]

In comparison with retention, the switching endurance, which is a dynamic process of the FTJ device under cycling electric field, may be more complicated. It is well known that the polarization fatigue is a common reliability issue in ferroelectric memories, which has been ascribed to several mainly scenarios, including the interfacial nucleation limitation, the domain wall pinning, and the phase transform/decomposition, based on numerous studies in the perovskite ferroelectric materials.[47,48] In the ferroelectric HfO_2-based thin films, similar fatigue phenomena and studies have also been reported.[49–53] Specifically, the structural transitions from the o- to m-phase and/or from the o- to t-phase under electrical stress have been considered as important mechanisms resulting in the polarization fatigue in ferroelectric HfO_2-based thin films.[53,54]

In the HfO_2-based FTJ devices, endurance degradation that the TER ratio decreases with increasing cycling number between the ON and OFF states have been frequently observed.[36,37,45,55] A kind of representative behavior of the resistance fatigue is shown in Figure 12.10a, in which the reduction of the TER ratios results from the decrease and increase in the resistance of the OFF and ON states. It may be understood that the effective polarizations (or switchable polarization) are decreased in both the ON and OFF states during the electric field cycling,

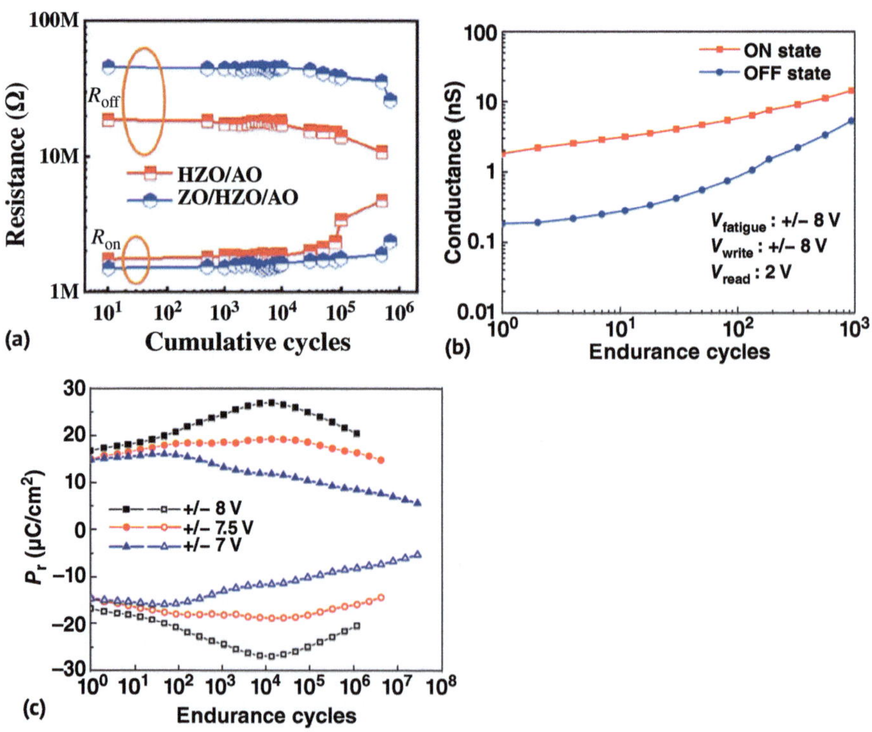

Figure 12.10 Switching endurance issues for (a) the Ti/Au/Al_2O_3/HZO/*p*-Si (HZO/AO) and Ti/Au/Al_2O_3/HZO/ZrO_2/*p*-Si (ZO/HZO/AO). *Source*: [35] / John Wiley & Sons / CC BY 4.0. (b) the Ti/Au/Al_2O_3/HZO/*p*-Si FTJs, and (c) polarization fatigue of the Ti/Au/Al_2O_3/HZO/*p*-Si device. *Source*: [56] / Springer Nature / CC BY 4.0.

which corresponds to the polarization fatigue of the HZO barriers.[35] Figure 12.10b shows the other typical resistance fatigue of the FTJs, in which both the ON and OFF become more and more conductive and the additional leakage currents are not modulated by the polarization, resulting in the reduction of the TER ratio.[56] Such endurance issue has been attributed to the formation of interface and bulk traps in the barriers during the high voltage pulses since, in the endurance range of 10^3 cycles, the P_r aren't observed to be decreased yet (Figure 12.10c).[56] Although a number of works have studied the endurance issues, there still lacks a deep understanding of the resistance fatigue, especially direct observations and proof measurements on the fatigued HfO_2-based devices.[57,58]

12.3 HfO_2-based FTJs as Memristors

12.3.1 Switching Dynamics

As a kind of resistive-switching memories, the capability of memristive properties is important for wide applications of the FTJs in neuromorphic computational circuits for artificial intelligence.[80] Chapter 11 has demonstrated detailed information for the neuromorphic computing. In this section, we only discussed the memristive behaviors, that is, the continuous and nonvolatile modulation of junction conductance based on the history of applied electric stimulus, of the HfO_2-based FTJs, which derive from switching dynamics of the ferroelectric barriers.[81–83] Figure 12.11a,b show the continuous changes of domain configurations of a 27 nm-thick Y:HfO_2 thin-film capacitor polarized by increasing pulse width and voltage, respectively.[84] One can find multiple nonvolatile states of the effective polarizations with high tunability. In the thin-film ferroelectric perovskites, the polarization switching has been described by the nucleation-limited-switching (NLS) model, which is given by

$$\Delta P = 2P_r \int_{-\infty}^{+\infty} \left\{ 1 - \exp\left[-\left(\frac{t_p}{t_0}\right)^n \right] \right\} F(\log t_0) d(\log t_0) \quad (12.6)$$

where ΔP is the switched polarization by the applied pulse, n the effective dimension, equaling 2 for thin films, and t_0 is the mean switching time. The $F(\log t_0)$ can be described as the Lorentzian distribution, that is $F(\log t_0) = \frac{A}{\pi}\left[\frac{W}{(\log t_p - \log t_0)^2 + W^2}\right]$ (A is a normalization constant and W is the half-width of half-maximum).[86] The NLS model has also been extended to quantitatively analyze the switching behaviors of FTJ devices, but adopting the resistance to denote the switched polarization.[87] Recently, a number of works have demonstrated that the NLS model is also available for the ferroelectric HfO_2-based thin films.[84,85,88] Figure 12.11c–e shows an example for ultrathin La and Ta co-doped HfO_2 films, in which the effective polarization, denoted by the $\Delta P/P_s$

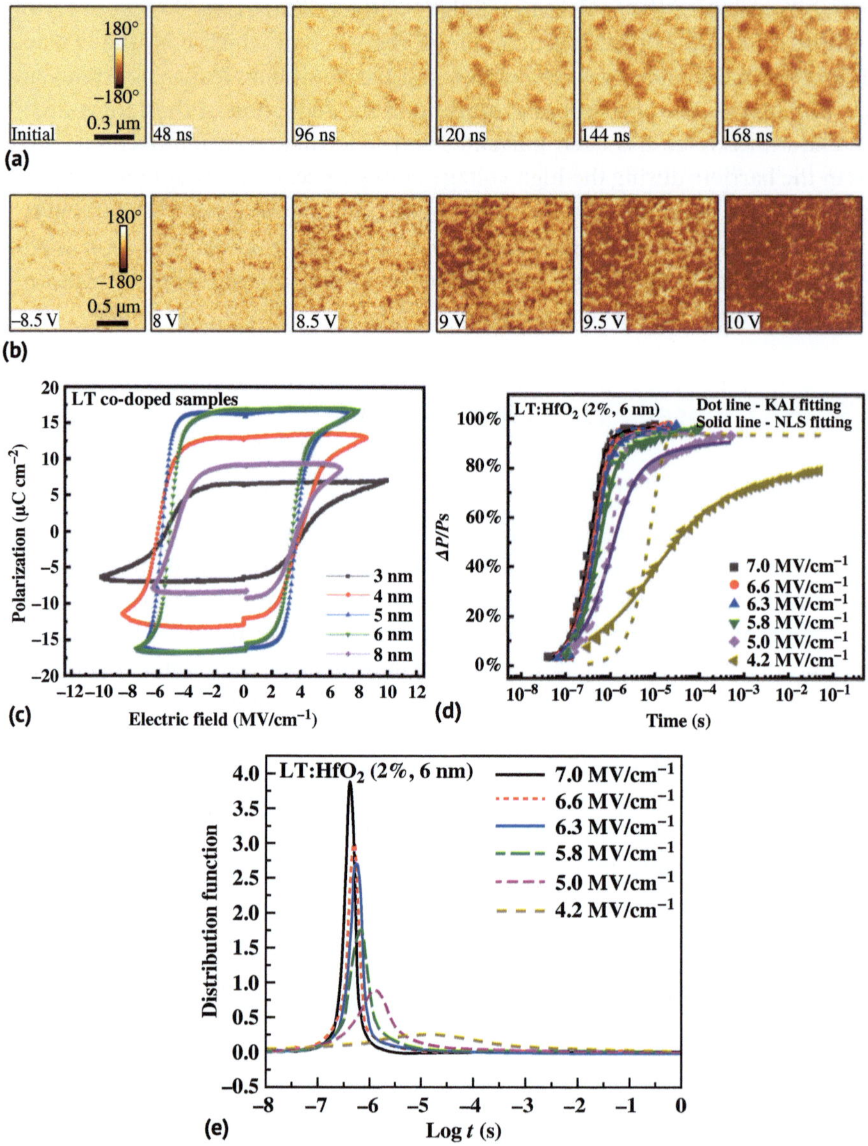

Figure 12.11 Switching dynamics. PFM phase images of a 27 nm-thick Y:HfO_2 thin-film capacitor with increasing (a) poling pulse width and (b) voltage. *Source*: [84] / with permission of John Wiley & Sons. (c) PUND hysteresis loops, (d) switchable polarization as a function of poling time at various electric fields, and (e) Lorentzian distribution of the mean switching time of the La and Ta co-doped HfO_2 (LT:HfO_2) thin films. *Source*: [85] / Springer Nature/CC BY 4.0.

(P_s, the spontaneous polarization), can be elaborately modulated by pulse width and voltage (Figure 12.11d), yielding the continuous tunability following the NLS mechanism.[85] The NLS fitting also revealed the distribution of mean switching time under various applied voltages (Figure 12.11e).[85] These NLS diagrams of polarization switching dynamics may guide the realization of memristive properties in the HfO_2-based FTJs.

12.3.2 Memristive and Synaptic Emulations

Driven by the controllable polarization switching, memristive behaviors, and hence synaptic emulations have been reported in a number of HfO_2-based FTJs, such as

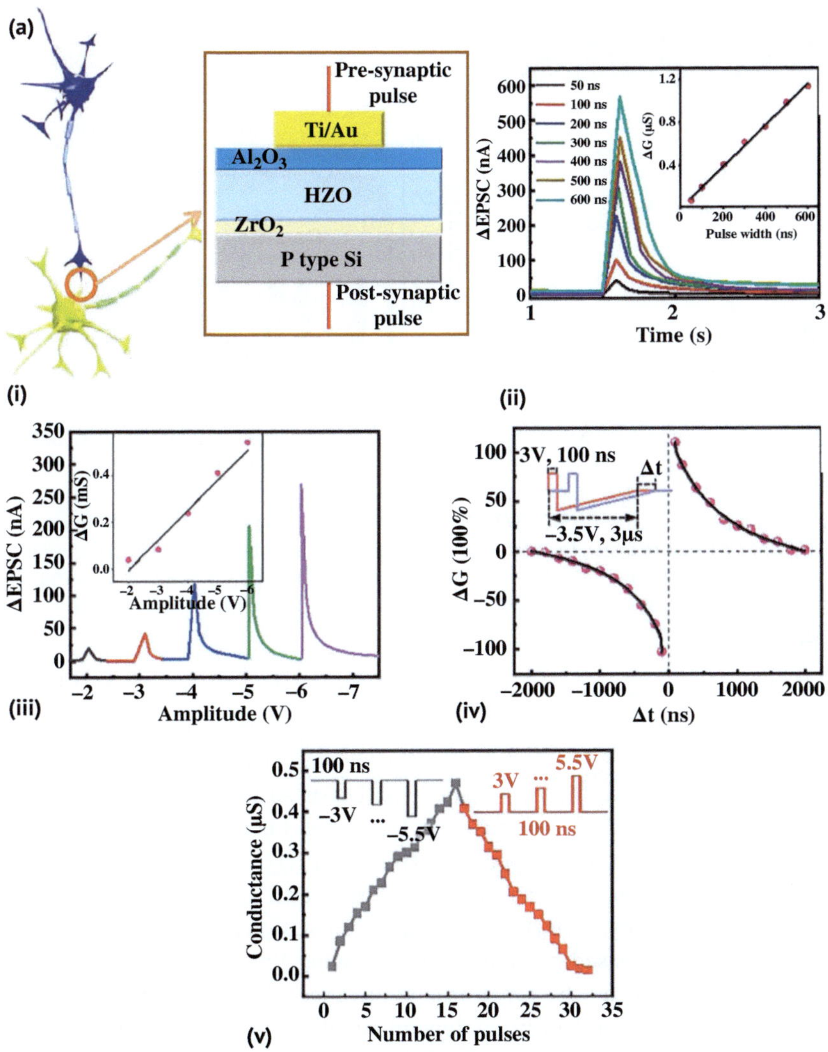

Figure 12.12 Memristive behaviors and synaptic emulations in HfO_2-based FTJs. (a) The Ti/Au/Al_2O_3/HZO/ZrO_2/*p*-Si devices: (i) sketches for the FTJ as a biological synapse, the modulation of EPSCs by controlling, (ii) pulse width and (iii) amplitude, (iv) the STDP behaviors, as well as (v) the long-term LTP/LTD of the artificial synapse. *Source*: [35] / John Wiley & Sons / CC BY 4.0.

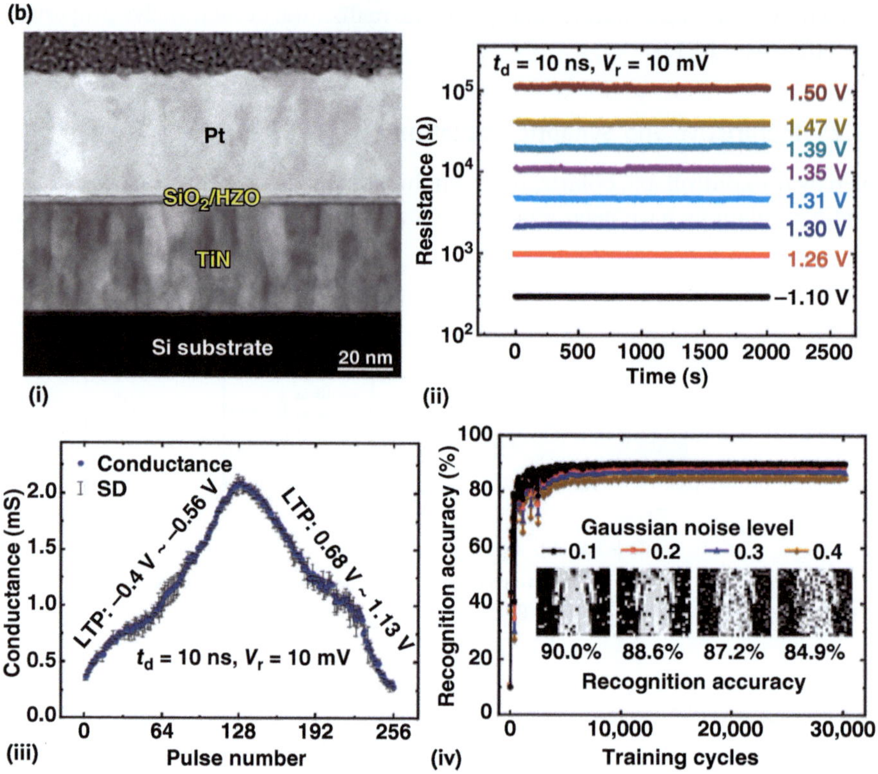

Figure 12.12 (Cont'd) (b) the $Pt/SiO_2/HZO/TiN$ devices: (i) cross-sectional HAADF image for the FTJ, (ii) eight distinct resistance states by varying pulse voltage, (iii) LTP/LTD curves showing linear and stable manipulations among 128 states, and (iv) CNN online training results on F-MNIST images with Gaussian noises. *Source*: [31] / with permission of John Wiley & Sons.

Pt/HZO/LSMO,[16] Pt/HZO/Nb:STO,[17] Ni/HZO/TiN,[55] $Pt/SiO_2/HZO/TiN$,[31] TiN/ $HfO_x/SiO_x/Si$,[32] $Ti/Au/Al_2O_3/HZO/p$-Si,[56] $Ti/Au/Al_2O_3/HZO/ZrO_2/p$-Si,[35] W/ HZO/TiN,[66] $Au/Ti/HfAlO_x/Pt$,[64] TiN/HZO/W,[63] and TiN/HZO/*p*-Si devices.[62] As shown in Figure 12.12a, the two-terminated $Ti/Au/Al_2O_3/HZO/ZrO_2/p$-Si FTJ can be regarded as an artificial synapse with the top and bottom electrodes denoting the pre- and post-neurons, respectively (the inset (i)).[35] By applying electric stimuli on the pre-neuron with varying pulse width and amplitude, typical synaptic behavior of excitatory postsynaptic currents (EPSCs) is obtained as indicative of the continuously increased current spikes, showing linear change of the synaptic weight (G, denoted by the junction conductance, the insets (ii and iii)).[35] The inset (iv) shows the spike-timing-dependent plasticity (STDP), an essential feature of Hebbian's rule of learning,[89] of the $Ti/Au/Al_2O_3/HZO/ZrO_2/p$-Si synapse by modulating the timing difference in the electric stimulus between the pre- and post-neurons $\left(\Delta t = t_{pre} - t_{post}\right)$. The long-term potentiation/depression (LTP/ LTD) of the junction conductance is shown in the inset (v).[35] Furthermore, in

Figure 12.12b, Wang et al. have demonstrated low-voltage and ultrahigh-speed artificial synapses based on SiO_2/HZO composite barriers, which exhibit gradually tunable conductance states up to 128 nonvolatile states with high linearity (< 1.0) obtained by 10 ns-wide pulses less than 1.2 V.[31] The authors also shown a convolutional neural network (CNN) built by the Pt/SiO_2/HZO/TiN synapses, yielding a high accuracy of 91.8% in recognizing fashion product images (F-MNIST database) associated with Gaussian noise (the inset (iv)).[31]

12.4 Conclusion

Undoubtedly, the employing of HfO_2-based barriers would boost industrialization of the FTJ devices owing to the good compatibility of HfO_2 with the CMOS process. Currently, some achievements, such as various device structures fabricated on Si wafers, giant TER ratios for data-storage margin, long-term retention, and switching endurance up to 10^8, as well as memristive properties for synaptic emulations, have been reported in a number of HfO_2-based FTJs (Table 12.1). However, more works are still required to further exploit the TER effects and improve related performance parameters. For example, the epitaxy of HfO_2-based barriers with pure *o* phase is important for facilitating the atomic-scale observation and analysis on dipole switching, charge aggregate, interfacial electronic structures, and ionic defects. With these microstructure diagrams clarified, polarization switching, especially the electric displacement of oxygen atoms, and its modulations on transport behaviors could be more deeply understood. In addition, device reliability, especially the switching endurance, hasn't been adequately investigated, which may limit broad applications. Comparative studies between HfO_2-based and perovskite-based FTJs, associated with the fatigue mechanism of these ferroelectric oxides, will give informative diagrams of the reliability issue. Besides, the exploitation of HfO_2-based barriers for electromagnetic multi-state memory may be intriguing because of their feature of oxygen hybridization and displacement on the ferroelectric/metal interfaces. Nevertheless, the integration on CMOS-compatible substrates, controllable multi-resistive states, as well as rich physics in phase transition and polarization switching, demonstrated great potential and viable pathways of the HfO_2-based FTJs toward high-density/high-speed data storage and brain-inspired computing for the future in-memory AI architectures.

Acknowledgments

The author acknowledges support from the Natural Science Foundation of China (Grant No. 52372113) and the Taishan Scholar Program of Shandong Province (Grant No. tstp20240511).

References

1 Garcia, V., Fusil, S., Bouzehouane, K. et al. (2009). Giant tunnel electroresistance for non-destructive readout of ferroelectric states. *Nature* 460: 81–84. https://doi.org/10.1038/nature08128

2 Gruverman, A., Wu, D., Lu, H. et al. (2009). Tunneling electroresistance effect in ferroelectric tunnel junctions at the nanoscale. *Nano Letters* 9: 3539–3543. https://doi.org/10.1021/nl901754t

3 Chanthbouala, A., Crassous, A., Garcia, V. et al. (2012). Solid-state memories based on ferroelectric tunnel junctions. *Nature Nanotechnology* 7: 101–104. https://doi.org/10.1038/NNANO.2011.213

4 Boscke, T. S., Muller, J., Brauhaus, D. et al. (2011). Ferroelectricity in hafnium oxide thin films. *Applied Physics Letters* 99 (10): 102903. https://doi.org/10.1063/1.3634052

5 Wen, Z. and Wu, D. (2020). Ferroelectric tunnel junction: Modulations on the potential barrier. *Advanced Materials* 32 (27): 1904123. https://doi.org/10.1002/adma.201904123

6 Zhuravlev, M. Y., Sabirianov, R. F., Jaswal, S. S. et al. (2005). Giant electroresistance in ferroelectric tunnel junctions. *Physical Review Letters* 94 (24): 246802. https://doi.org/10.1103/PhysRevLett.94.246802

7 Wen, Z., Li, C., Wu, D. et al. (2013). Ferroelectric-field-effect-enhanced electroresistance in metal/ferroelectric/semiconductor tunnel junctions. *Nature Materials* 12 (7): 617–621. https://doi.org/10.1038/NMAT3649

8 Wen, Z., Wu, D., Li, C. et al. (2017). Giant tunnelling electroresistance in metal/ferroelectric/semiconductor tunnel junctions by engineering the Schottky barrier. *Nature Communications* 8: 15217. https://doi.org/10.1038/ncomms15217

9 Pantel, D. and Alexe, M. (2010). Electroresistance effects in ferroelectric tunnel barriers. *Physical Review B—Condensed Matter and Materials Physics* 82 (13): 134105. https://doi.org/10.1103/PhysRevB.82.134105

10 Fowler, R. H. and Nordheim, L. (1928). Electron emission in intense electric fields. *Proceedings of the Royal Society of London. Series A, Containing Papers of a Mathematical and Physical Character* 119 (781): 173–181. http://doi.org/10.1098/rspa.1928.0091

11 Sze, S. M. (2007). *Physics of Semiconductor Devices* (3rd ed.). Hoboken: Wiley-Interscience.

12 Cao, X., Wu, T., Guo, J. et al. (2018). Tunneling current in HfO_2 and $Hf_{0.5}Zr_{0.5}O_2$-based ferroelectric tunnel junction. *Journal of Applied Physics* 123 (9): 094501. https://doi.org/10.1063/1.5016823

13 Prasad, B., Thakare, V., Kalitsov, A. et al. (2021). Large tunnel electroresistance with ultrathin $Hf_{0.5}Zr_{0.5}O_2$ ferroelectric tunnel barriers. *Advanced Electronic Materials* 7 (6): 2001074. https://doi.org/10.1002/aelm.202001074

14 Wei, Y. F., Matzen, S., Noheda, B. et al. (2019). Magneto-ionic control of spin polarization in multiferroic tunnel junctions. *npj Quantum Materials* 4: 62. https://doi.org/10.1038/s41535-019-0201-0

15 Jiao, P. J., Wu, Y., Wu, D. et al. (2021). Electroresistance in metal/ferroelectric/semiconductor tunnel junctions based on a $Hf_{0.5}Zr_{0.5}O_2$ barrier. *Applied Physics Letters* 118 (25): 252901. https://doi.org/10.1063/5.0053959

16 Du, X. Z., Yin, Y. W., Li, X. G. et al. (2022). High-speed switching and giant electroresistance in an epitaxial $Hf_{0.5}Zr_{0.5}O_2$-based ferroelectric tunnel junction memristor. *ACS Applied Materials & Interfaces* 14 (1): 1355–1361. https://doi.org/10.1021/acsami.1c18165

17 Gao, Z. M., Cheng, Y., Ge, C. et al. (2023). Giant electroresistance in hafnia-based ferroelectric tunnel junctions via enhanced polarization. *Device* 1 (1): 2666–9994. https://doi.org/10.1016/j.device.2023.100004

18 Chernikova, A., Kozodaev, M., Gruverman, A. et al. (2016). Ultrathin $Hf_{0.5}Zr_{0.5}O_2$ ferroelectric films on Si. *ACS Applied Materials & Interfaces* 8 (11): 7232–7237. https://doi.org/10.1021/acsami.5b11653

19 Goh, Y. and Jeon, S. (2018). The effect of the bottom electrode on ferroelectric tunnel junctions based on CMOS-compatible HfO_2. *Nanotechnology* 29 (33): 335201. https://doi.org/10.1088/1361-6528/aac6b3

20 Li, Z., Meng, J. L., Wang, T. Y. et al. (2023). Stabilizing the ferroelectric phase in HfAlO ferroelectric tunnel junction with different bottom electrodes. *IEEE Electron Device Letters* 44 (6): 947–950. https://doi.org/10.1109/LED.2023.3269070

21 Ambriz-Vargas, F., Ruediger, A., Kolhatkar, G. et al. (2017). A complementary metal oxide semiconductor process-compatible ferroelectric tunnel junction. *ACS Applied Materials & Interfaces* 9 (15): 13262–13268. https://doi.org/10.1021/acsami.6b16173

22 Athle, R., Troian, A., Borg, M. et al. (2022). Top electrode engineering for freedom in design and implementation of ferroelectric tunnel junctions based on $Hf_{1-x}Zr_xO_2$. *ACS Applied Electronic Materials* 4 (3): 1002–1009. https://doi.org/10.1021/acsaelm.1c01181

23 Ashcroft, N. W. and Mermin, N. D. (1976). *Solid State Physics*. New York: Saunders College Publishing.

24 Seo, J. and Shin, M. (2020). Atomistic Asymmetric effect on the performance of HfO_2-based ferroelectric tunnel junctions. *Physical Review Applied* 14 (5): 054018. https://doi.org/10.1103/PhysRevApplied.14.054018

25 Yamada, H., Dhongade, S., Sawa, A. et al. (2024). Electroresistance in all-oxide $RuO_2/Hf_{0.5}Zr_{0.5}O_2$/ITO ferroelectric tunnel junctions. *ACS Applied Electronic Materials* 6 (5): 3395–3402. https://doi.org/10.1021/acsaelm.4c00215

26 Katayama, K., Uchida, H., Funakubo, H. et al. (2016). Growth of (111)-oriented epitaxial and textured ferroelectric Y-doped HfO_2 films for downscaled devices. *Applied Physics Letters* 109 (11): 112901. https://doi.org/10.1063/1.4962431

27 Wang, L. F., Cho, M. R., Das, S. et al. (2016). Overcoming the fundamental barrier thickness limits of ferroelectric tunnel junctions through $BaTiO_3/SrTiO_3$ composite barriers. *Nano Letters* 16 (6): 3911–3918. https://doi.org/10.1021/acs.nanolett.6b01418

28 D. Wu, D., P. Wang, P., Li, A. D. et al. (2015). Li et al., "Improved memory functions in multiferroic tunnel junctions with a dielectric/ferroelectric composite barrier.," *Applied Physics Letters* Appl. Phys. Lett., 107 (, no. 23):, 232902. https://, 2015, doi.org/: 10.1063/1.4937390.

29 Lee, K., Byun, J., Chae, S. C. et al. (2022). Giant tunneling electroresistance in epitaxial ferroelectric ultrathin films directly integrated on Si. *Applied Materials Today* 26: 101308. https://doi.org/10.1016/j.apmt.2021.101308

30 Cheema, S. S., Shanker, N., Salahuddin, S. et al. (2022). One nanometer HfO_2-based ferroelectric tunnel junctions on silicon. *Advanced Electronic Materials* 8 (6): 2100499. https://doi.org/10.1002/aelm.202100499

31 Wang, H., Yin, Y. W., Li, X. G. et al. (2024). Silicon-compatible ferroelectric tunnel junctions with a $SiO_2/Hf_{0.5}Zr_{0.5}O_2$ composite barrier as low-voltage and ultra-high-speed memristors. *Advanced Materials* 36 (15): 2211305. https://doi.org/10.1002/adma.202211305

32 Min, K. K., Yu, J., and Kim, Y. (2022). Ferroelectricity of pure HfOx in metal-ferroelectric-insulator-semiconductor stacks and its memory application. *Applied Surface Science* 573: 151566. https://doi.org/10.1016/j.apsusc.2021.151566

33 Lee, Y., Song, S., Ham, W. et al. (2022). Si-doped HfO_2-based ferroelectric tunnel junctions with a composite energy barrier for non-volatile memory applications. *Materials* 15 (6): 2251. https://doi.org/10.3390/ma15062251

34 Hsiang, K. Y., Liao, C. Y., Chang, S. H. et al. (2021). Bilayer-based antiferroelectric $HfZrO_2$ tunneling junction with high tunneling electroresistance and multilevel nonvolatile memory. *IEEE Electron Device Letters* 42 (10): 1464–1467. https://doi.org/10.1109/LED.2021.3107940

35 Cao, Y. Y., Zhu, H., Chen, L. et al. (2023). Structural engineering of $Hf_{0.5}Zr_{0.5}O_2$-based ferroelectric tunneling junction for fast-speed and low-power artificial synapses. *Advanced Electronic Materials* 9 (5): 2201247. https://doi.org/10.1002/aelm.202201247

36 Shekhawat, A., Walters, G., Moghaddam, S. et al. (2020). Data retention and low voltage operation of $Al_2O_3/Hf_{0.5}Zr_{0.5}O_2$ based ferroelectric tunnel junctions. *Nanotechnology* 31 (39): 39LT01. https://doi.org/10.1088/1361-6528/ab9cf7

37 Hur, J., Park, C., Choe, G. et al. (2022). Characterizing HfO_2-based ferroelectric tunnel junction in Cryogenic temperature. *IEEE Transactions on Electron Devices* 69 (10): 5948–5951. https://doi.org/10.1109/TED.2022.3200919

38 Tsu, R. and Esaki, L. (1973). Tunneling in a finite superlattice. *Applied Physics Letters* 22 (11): 562–564. https://doi.org/10.1063/1.1654509

39 Tornos, J., Varela, M., Santamaria, J. et al. (2017). Resonant electron tunnelling assisted by charged domain walls in multiferroic tunnel junctions. *Nature Nanotechnology* 12 (7): 655–662. https://doi.org/10.1038/NNANO.2017.51

40 Du, R. F., Li, A. D., Wu, D. et al. (2014). Ferroelectric modulation on resonant tunneling through perovskite double-barriers. *Applied Physics Letters* 104 (14): 142907. https://doi.org/10.1063/1.4871277

41 Choi, W. S., Lee, S., Lee, H. N. et al. (2015). Resonant tunnelling in a quantum oxide superlattice. *Nature Communications* 6: 7424. https://doi.org/10.1038/ncomms8424

42 Wu, Z. J., Jiang, J., Yang, Q. et al. (2024). Ferroelectric tunnel junctions based on a HfO_2/dielectric composite barrier. *Applied Physics Letters* 125 (11): 112903 https://doi.org/10.1063/5.0216890

43 Chang, P. Y., Li, J., Guo, Y. R. et al. (2023). Bilayer-barrier ferroelectric-HfO_2 tunnel junction with high on current and giant on/off ratio arising from resonant tunneling. In: *2023 International Conference on Simulation of Semiconductor Processes and Devices (SISPAD)*. pp. 209–212. https://doi.org/10.23919/SISPAD57422.2023.10319602

44 Slesazeck, S., Hoffmann, M., Schroeder, U. et al. (2019). Ferroelectric $Hf_{1-x}Zr_xO_2$ memories: device reliability and depolarization fields. In: *2019 19th Non-Volatile Memory Technology Symposium (NVMTS)*. pp. 1–8. https://doi.org/10.1109/nvmts47818.2019.9043368

45 Max, B., Mikolajick, T., Hoffmann, M. et al. (2019). Retention characteristics of $Hf_{0.5}Zr_{0.5}O_2$-based ferroelectric tunnel junctions. In: *2019 IEEE 11th International Memory Workshop (IMW)*. pp. 40–43. https://doi.org/10.1109/imw.2019.8739765

46 Raffel, Y., Duhan, P., Revello, R. et al. (2022). Impact of temperature on reliability of MFIS HZO-based ferroelectric tunnel junctions. In: *2022 IEEE International Reliability Physics Symposium (IRPS)*. pp. P11-1–P11-5. https://doi.org/10.1109/IRPS48227.2022.9764585

47 Albe, K., Glaum, J., Hoffmann, M. et al. (2015). Mechanisms of aging and fatigue in ferroelectrics. *Materials Science and Engineering: B* 192: 52–82. https://doi.org/10.1016/j.mseb.2014.10.003

48 Lou, X. J. (2009). Polarization fatigue in ferroelectric thin films and related materials. *Journal of Applied Physics* 105 (2): 024101. https://doi.org/10.1063/1.3056603

49 Zeng, B. J., Xie, S. C., Zhang, S. R. et al. (2024). Polarization fatigue mechanism of laminated hafnium zirconium oxide ferroelectric thin films. *Acta Materialia* 272: 119920. https://doi.org/10.1016/j.actamat.2024.119920

50 Li, X. F., Li, C., Xu, Z. Y. et al. (2021). Ferroelectric properties and polarization fatigue of La:HfO_2 thin-film capacitors. *Physica Status Solidi RRL* 15 (4): 2000481. https://doi.org/10.1002/pssr.202000481

51 Pesic, M., Fengler, F. P. G., Larcher, L. L. et al. (2016). Physical Mechanisms behind the field-cycling behavior of HfO_2-based ferroelectric capacitors. *Advanced Functional Materials* 26 (25): 4601–4612. https://doi.org/10.1002/adfm.201600590

52 Huang, F., Chen, X., Liang, X. et al. (2017). Fatigue mechanism of yttrium-doped hafnium oxide ferroelectric thin films fabricated by pulsed laser deposition. *Physical Chemistry Chemical Physics* 19 (5): 3486–3497. https://doi.org/10.1039/c6cp07501k

53 Jiao, P. J., Cheng, H., Li, J. Y. et al. (2023). Flexoelectricity-stabilized ferroelectric phase with enhanced reliability in ultrathin La:HfO_2 films. *Applied Physics Review* 10 (3): 031417. https://doi.org/10.1063/5.0144958

54 Zheng, Y. Z., Zheng, Y. H., Gao, Z. M. et al. (2021). Atomic-scale characterization of defects generation during fatigue in ferroelectric $Hf_{0.5}Zr_{0.5}O_2$ films: Vacancy generation and lattice dislocation. In: *2021 IEEE International*

Electron Devices Meeting (IEDM). pp. 33.5.1–33.5.4. https://doi.org/10.1109/IEDM19574.2021.9720565

55 Wu, T. Y., Huang, H. H., Chu, Y. H. et al. (2019). Sub-nA low-current HZO ferroelectric tunnel junction for high-performance and accurate deep learning acceleration. In: *IEEE Internation Electron Devices Meeting (IEDM)*. https://doi.org/10.1109/iedm19573.2019.8993565

56 Ryu, H., Rao, F., Zhu, W. J. et al. (2019). Ferroelectric tunneling junctions based on aluminum oxide/zirconium-doped hafnium oxide for neuromorphic computing. *Scientific Reports* 9: 20383. https://doi.org/10.1038/s41598-019-56816-x

57 Wu, D., Wen, Z., Yang, Y. H. et al. (2021). Atomic-scale fatigue mechanism of ferroelectric tunnel junctions. *Science Advances* 7 (48): eabh2716. https://doi.org/10.1126/sciadv.abh2716

58 Zheng, N. C., Nie, Y. F., Shen, C. et al. (2025). Ferroelectric tunnel junctions integrated on semiconductors with enhanced fatigue resistance. *Science Advances* 11 (15): eads0724. https://doi.org/10.1126/sciadv.ads0724

59 Estandia, S., Gazquez, J., Sanchez, F. et al. (2021). Critical effect of the bottom electrode on the ferroelectricity of epitaxial $Hf_{0.5}Zr_{0.5}O_2$ thin films. *Journal of Materials Chemistry C* 9 (10): 3486–3492. https://doi.org/10.1039/d0tc05853j

60 Tan, H., Sanchez, F., Fontcuberta, J. et al. (2021). Polarization and resistive switching in epitaxial 2 nm $Hf_{0.5}Zr_{0.5}O_2$ tunnel junctions. *ACS Applied Electronic Materials* 3 (8): 3657–3666. https://doi.org/10.1021/acsaelm.1c00604

61 Sanchez, F., Fina, I., Fontcuberta, J. et al. (2020). Blocking of conducting channels widens window for ferroelectric resistive switching in interface-engineered $Hf_{0.5}Zr_{0.5}O_2$ tunnel devices. *Advanced Functional Materials* 30 (32): 2002638. https://doi.org/10.1002/adfm.202002638

62 Mikheev, V., Zarubin, S., Negrov, D. et al. (2020). Memristor with a ferroelectric HfO2 layer: in which case it is a ferroelectric tunnel junction. *Nanotechnology* 31 (21): 215205. https://doi.org/10.1088/1361-6528/ab746d

63 Goh, Y., Kim, M., Jeon, S. et al. (2020). Ultra-thin $Hf_{0.5}Zr_{0.5}O_2$ thin-film-based ferroelectric tunnel junction via stress induced crystallization. *Applied Physics Letters* 117 (24): 242901. https://doi.org/10.1063/5.0029516

64 Li, Z. H., Wang, T. Y., and Liu, P. (2023). CMOS compatible low power consumption ferroelectric synapse for neuromorphic computing. *IEEE Electron Device Letters* 44 (3): 532–535. https://doi.org/10.1109/LED.2023.3234690

65 Kobayashi, M., Tagawa, Y., Hiramoto, T. et al. (2019). Ferroelectric HfO2 tunnel junction memory with high TER and multi-level operation featuring metal replacement process. *IEEE Journal of the Electron Devices Society* 7 (1): 134–139. https://doi.org/10.1109/JEDS.2018.2885932

66 Athle, R. and Borg, M. (2023). Impact of temperature-induced oxide defects on $Hf_xZr_{1-x}O_2$ ferroelectric tunnel junction memristor performance. *IEEE Transactions on Electron Devices* 70 (3): 1412–1416. https://doi.org/10.1109/TED.2023.3240399

67 Yu, J. J., Wang, T. Y., and Xu, K. L. (2023). Improved ferroelectricity and tunneling electro resistance in Zr-rich $Hf_xZr_{1-x}O_2$ ferroelectric tunnel junction. *IEEE Electron Device Letters* 44 (2): 245–248. https://doi.org/10.1109/LED.2022.3228589

68 Thomas, R., Kolhatkar, G., Ruediger, A. et al. (2017). Tunneling electroresistance effect in a Pt/$Hf_{0.5}Zr_{0.5}O_2$/Pt structure. *Applied Physics Letters* 110 (9): 093106. https://doi.org/10.1063/1.4977028

69 Goh, Y. and Jeon, S. (2018). Enhanced tunneling electroresistance effects in HfZrO-based ferroelectric tunnel junctions by high-pressure nitrogen annealing. *Applied Physics Letters* 113 (5): 052905. https://doi.org/10.1063/1.5040031

70 Liu, Y. L., Zhu, H., Chen, L. et al. (2021). HfZrOx-based ferroelectric tunnel junction with crested symmetric band structure engineering. *IEEE Electron Device Letters* 42 (9): 1311–1314. https://doi.org/10.1109/LED.2021.3102226

71 Kwon, O., Kim, D., and Kim, S. (2023). Impact of annealing temperature on the remanent polarization and tunneling electro-resistance of ferroelectric Al-doped HfOx tunnel junction memory. *Physical Chemistry Chemical Physics* 25 (6): 4588–4597. https://doi.org/10.1039/d2cp05729h

72 Chen, L., Zhu, H., Xu, K. L. et al. (2024). Improved ferroelectricity and tunneling electroresistance by inducing the ZrO_2 intercalation layer in La:HfO_2 thin films. *ACS Applied Electronic Materials* 6 (2): 1055–1062. https://doi.org/10.1021/acsaelm.3c01496

73 Chen, L., Zhou, P., Zhu, H. et al. (2018). Ultra-low power $Hf_{0.5}Zr_{0.5}O_2$ based ferroelectric tunnel junction synapses for hardware neural network applications. *Nanoscale* 10 (33): 15826–15833. https://doi.org/10.1039/c8nr04734k

74 Goh, Y., Kim, M., jeon, S. et al. (2021). Selector- less ferroelectric tunnel junctions by stress engineering and an imprinting effect for high-density cross-point synapse arrays. *ACS Applied Materials & Interfaces* 13 (49): 59422–59430. https://doi.org/10.1021/acsami.1c14952

75 Hwang, J., Goh, Y., Jeon, S. et al. (2021). Effect of insertion of dielectric layer on the performance of hafnia ferroelectric devices. *IEEE Transactions on Electron Devices* 68 (2): 841–845. https://doi.org/10.1109/TED.2020.3043728

76 Sunbul, A., Ali, T., Mertens, K. et al. (2022). Optimizing ferroelectric and interface layers in HZO-based FTJs for neuromorphic applications. *IEEE Transactions on Electron Devices* 69 (2): 808–815. https://doi.org/10.1109/TED.2021.3131971

77 Shekhawat, A., Moghaddam, S., Lee, Y. et al. (2021). Effect of ferroelectric and interface films on the tunneling electroresistance of the Al_2O_3/$Hf_{0.5}Zr_{0.5}O_2$ based ferroelectric tunnel junctions. *Nanotechnology* 32 (48): 485204. https://doi.org/10.1088/1361-6528/ac1ebe

78 Chaudhary, P., Zarubin, S., Gruverman, A. et al. (2021). Electroresistance effect in MoS_2-$Hf_{0.5}Zr_{0.5}O_2$ heterojunctions. *Applied Physics Letters* 118 (8): 083106. https://doi.org/10.1063/5.0035306

79 Hoffmann, M., Schroeder, U., Mikolajick, T. et al. (2022). Atomic layer etching of ferroelectric hafnium zirconium oxide thin films enables giant tunneling electroresistance. *Applied Physics Letters* 120 (12): 122901. https://doi.org/10.1063/5.0084636

80 Chua, L. (1971). Memristor-the missing circuit element. *IEEE Transactions on Circuit Theory* 18 (5): 507–519. https://doi.org/10.1109/TCT.1971.1083337

81 Lee, D. H., Yu, G. T., Park, J. Y. et al. (2023). Effect of residual impurities on polarization switching kinetics in atomic-layer-deposited ferroelectric $Hf_{0.5}Zr_{0.5}O_2$

thin films. *Acta Materialia* 247: 118732. https://doi.org/10.1016/j.actamat.2023.118732
82 Li, Y. X., Li, J. Z., Liang, R. R. et al. (2019). Switching dynamics of ferroelectric HfO_2-ZrO_2 with various ZrO_2 contents. *Applied Physics Letters* 114 (14): 142902. https://doi.org/10.1063/1.5093793
83 Ma, L. Y. and Liu, S. (2023). Structural polymorphism kinetics promoted by charged oxygen vacancies in HfO_2. *Physics Review Letters* 130 (9): 096801. https://doi.org/10.1103/PhysRevLett.130.096801
84 Buragohain, P., Erickson, A., Mimura, T. et al. (2022). Effect of film microstructure on domain nucleation and intrinsic switching in ferroelectric Y:HfO_2 thin film capacitors. *Advanced Functional Materials* 32 (9): 2108876. https://doi.org/10.1002/adfm.202108876
85 Zhou, C., Liu, S., Chen, Z. H. et al. (2024). Enhanced polarization switching characteristics of HfO_2 ultrathin films via acceptor-donor co-doping. *Nature Communication* 15 (1): 2893. https://doi.org/10.1038/s41467-024-47194-8
86 Noh, T. W., Han, H. S., Jo, J. Y. et al. (2007). Domain switching kinetics in disordered ferroelectric thin films. *Physics Review Letters* 99 (26): 267602. https://doi.org/10.1103/PhysRevLett.99.267602
87 Chanthbouala, A., Garcia, V., Grollier, J. et al. (2012). A ferroelectric memristor. *Nature Materials* 11 (10): 860–864. https://doi.org/10.1038/NMAT3415
88 Ryu, T. H., Min, D. H., and Yoon, S. M. (2020). Comparative studies on ferroelectric switching kinetics of sputtered $Hf_{0.5}Zr_{0.5}O_2$ thin films with variations in film thickness and crystallinity. *Journal of Applied Physics* 128 (7): 074102. https://doi.org/10.1063/5.0013487
89 Bi, G. Q. and Poo, M. M. (2001). Synaptic modification by correlated activity: Hebb's postulate revisited. *Annual Review of Neuroscience* 24: 139–166. https://doi.org/10.1146/annurev.neuro.24.1.139

13

HfO_2-based Ferroelectric Materials for Energy Storage Applications

*Wentao Shuai and Jiyan Dai**

Department of Applied Physics, The Hong Kong Polytechnic University, Hong Kong, SAR, China

*Corresponding author: jiyan.dai@polyu.edu.hk

With the development of the modern semiconductor industry, it is highly desired for high energy storage density of energy storage components in pulsed power devices. Compared to other energy storage devices, capacitors have the advantages of high discharge power, high utilization efficiency, and large room for improvement in energy storage density. They are gradually becoming energy storage components in pulsed power devices and are widely used. However, existing pulse capacitors have disadvantages such as low energy storage density, low discharge current, and short discharge lifespan, making it difficult to meet the needs of further development of new technologies. Recently, the performance of ferroelectric and antiferroelectric HfO_2 thin films in energy-related and memory applications is attracting great deal of attention. Compared to conventional materials, HfO_2-based materials have many advantages, such as compatibility with silicon-based semiconductor technology, ultrathin thickness (nanoscale), and a simple composition, enabling them to be integrated into three-dimensional nanostructures. Due to the extraordinary ferroelectric properties of HfO_2, it has significant advantages in energy storage devices, especially in high-energy-density capacitors and supercapacitors. In this chapter, we comprehensively discussed the basic properties of HfO_2-based ferroelectric materials in energy storage systems, including the excellent energy storage performance, and the optimization of material properties and structural design.

13.1 Introduction: Understanding Dielectric Energy Storage Capacitors

Dielectric energy storage capacitors (DECs) are essential components in modern electronics, offering a path for efficient energy storage and quick release. These capacitors operate on the principle of storing electrical energy in an electric field within a dielectric material. The energy storage mechanism involves the polarization of the dielectric material when an electric field is applied, and the energy is stored as potential energy in the dipole moments of the material.[1]

In the rapid development of pulsed power technology, high-performance dielectric capacitors have become indispensable key components as their applications are gradually popularized in various fields such as consumer electronics, new energy automobiles, Internet of Things, artificial intelligence, cloud computing, and 5G communications. As a key component for energy supply in electronic circuit systems, energy storage units are often expected to have characteristics such as ultrahigh integration, high energy storage density (ESD), high-power density (PD), exceptional fatigue durability, and good temperature stability. Currently, the mainstream energy storage methods include dielectric capacitors, electrochemical capacitors, batteries, and solid oxide fuel cells (SOFCs); a comparison of them can be seen in Figure 13.1.

Due to the different operating principles, there is a clear difference between the PD and energy density of several energy storage systems. Batteries, a key energy storage technology, dominate the market with their high energy density (up to 180 Wh/kg), providing a continuous and stable source of energy for modern power and electronic devices.[2] Despite the relatively high ESD of batteries, fatigue durability has become a key limiting factor in their development, with capacity gradually decaying with use. In addition, lithium-ion batteries can experience thermal runaway in high-temperature environments, leading to combustion and explosion accidents. Restricted charging and discharging rates, large internal resistance, as well as uneven distribution of lithium resources and large fluctuations in market prices have constrained the further development of lithium-ion batteries. Electrochemical capacitors, also known as supercapacitors or supercells, are devices for rapid energy storage and release.

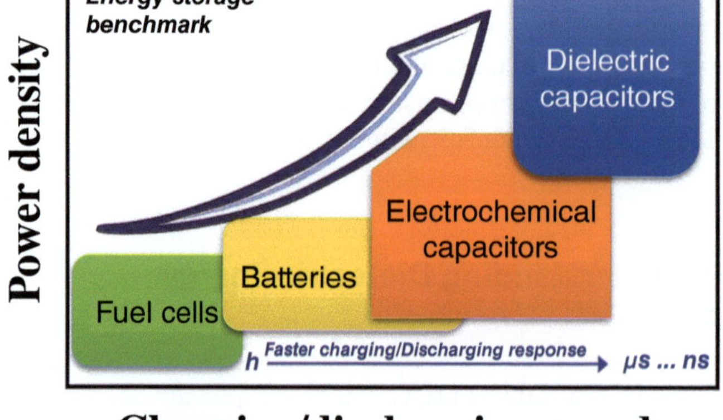

Figure 13.1 Comparisons analysis of PD and charging/discharging dynamics for different energy storage systems.

Despite their relatively low energy density (about 5 Wh/kg), electrochemical capacitors have a high-PD (10 kW/kg) and fast charging and discharging speeds (microseconds or even nanoseconds). Thanks to these advantages, supercapacitors are more suitable for applications requiring fast charge/discharge and instantaneous high-current output, such as acceleration of electric vehicles, braking energy recovery, uninterruptible power supply (UPS), and other fields with high application value. In recent years, electrochemical capacitors have received a lot of attention in the academic and commercial fields and are considered to be as important as batteries. However, electrochemical capacitors also have some drawbacks, such as high internal resistance, low operating voltage, and narrow operating temperature range, which limit their applications in high-frequency, high-voltage, and high-temperature fields. SOFC, as an efficient and environmentally friendly energy storage technology, has advantages in high fuel conversion efficiency, fuel flexibility, and environmental friendliness. However, the high operating temperature of SOFC limits its application range and challenges the material and system stability. In addition, the relatively low PD of SOFCs limits them in certain application scenarios with high instantaneous power demand. Also, cost issues and long startup times are problems that SOFCs need to overcome.

DECs have significant advantages in some key performance indicators such as PD, charge/discharge speed, and cycle life. Primarily, as shown in Figure 13.2, the PD and response speed are extremely high, which gives them an obvious advantage in scenarios that require fast charging and discharging. Secondly, dielectric capacitors have a long cycle life and can work stably in high-temperature environments, reducing maintenance costs, and the risk of failure. In addition, dielectric capacitors have a low dependence on raw materials, contributing to lower manufacturing costs and increased sustainability.

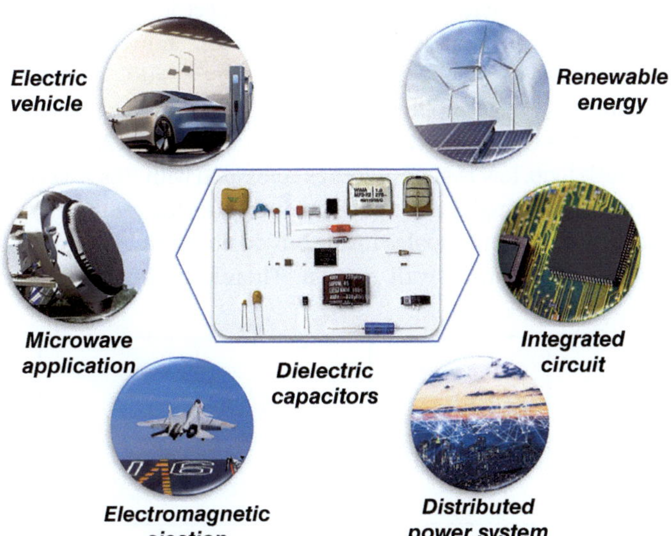

Figure 13.2 The Applications for novel dielectric film capacitors.

Ferroelectric (FE) and antiferroelectric (AFE) materials have broad prospects in energy-related applications such as energy harvesting, energy storage, infrared detection, and cooling. Historically, these properties have been studied in a variety of materials, including polymers, ceramics, single crystals, and glasses. However, due to their relatively large layer thicknesses (micrometers or more), these materials are not suitable for future nanoscale devices.[3] Recently, there has been a focus on HfO_2-based FE thin films for energy-related applications. Compared to traditional materials, HfO_2-based materials offer numerous advantages, such as compatibility with silicon-based semiconductor technology, ultrathin thickness (nanometers), simple composition, and suitability for integration into three-dimensional (3D) nanostructures. HfO_2-based materials have great potential in energy-related applications, including energy storage, thermoelectric energy harvesting, infrared sensing, and solid-state cooling. This chapter introduces some fundamental knowledge about these energy-related properties.

13.1.1 Working Principles for DECs

Dielectric capacitors are energy storage systems that charge and discharge energy through the polarization and depolarization processes of the dielectric under an electric field. Based on the relationship between their dielectric constant and the electric field strength, they can be classified into linear and nonlinear dielectrics. The dielectric constant of linear dielectrics does not change with the electric field strength, which is commonly found in ceramic and polymer materials.[4] Due to the absence of permanent electric dipoles, the dielectric constant of linear dielectrics is a constant independent of the electric field strength.

Nonlinear dielectric materials are typically affected by internal dipole reorientation, domain switching, and domain wall motion, showing a nonlinear relationship between the polarization intensity and the electric field strength under an applied electric field. Commonly used nonlinear dielectric materials in the field of dielectric energy storage include: paraelectric dielectrics (PE), FE materials, AFE materials, and relaxor ferroelectrics (RFEs).[4,5] Several classes of the dielectrics (linear, PE, FE, AFE, and RFE) have been studied for the energy-storage applications. Typical polarization behavior of all linear and nonlinear dielectrics is schematically presented in Figure 13.3. AFE materials exhibit a double hysteresis loop, low hysteresis loss as compared with FEs, and zero remanent polarization, resulting in a high ESD, can be attained from these materials.

PE materials have internal permanent dipoles inside but do not form FE domains, with less domain switching and domain wall motion. Therefore, their dielectric

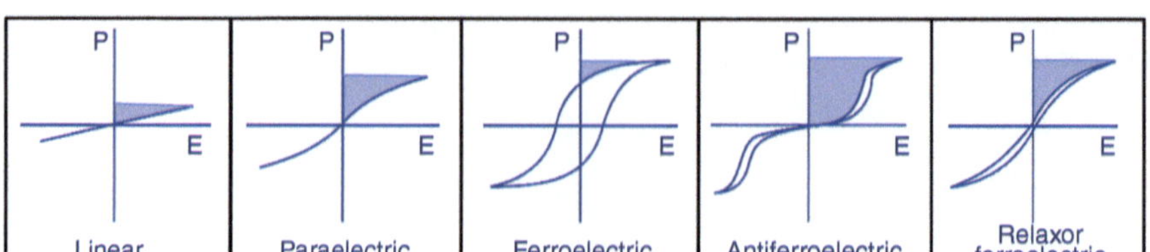

Figure 13.3 Schematic diagrams of five typical types of dielectrics, where blue shaded area is W_{rec}.

constant is typically higher than that of linear dielectrics but lower than that of FE materials, while having lower dielectric loss and higher dielectric breakdown strength. PE materials have low hysteresis loss; $SrTiO_3$ (STO) and $Ba_{1-x}Sr_xTiO_3$ (BST, $x > 0.35$) exhibit PE characteristics at room temperature and are typical PE materials. Under the action of an external electric field, polarization intensity is generated, and upon removal of the electric field, it returns to a non-polarized state. FE materials are a class of materials with spontaneous polarization, and their polarization intensity can change significantly under the action of an external electric field. The polarization intensity of FE materials usually exhibit nonlinear characteristics under the action of field strength, and due to the presence of remnant polarization (P_r), they usually have larger dielectric losses, affecting the recoverable energy density (W_{rec}) and efficiency (η) that can be released. Common FE energy storage materials include: FE ceramic materials, such as barium titanate ($BaTiO_3$), zirconium titanate ($ZrTiO_3$), lead zirconate titanate ($PbZr_xTi_{1-x}O_3$, PZT), etc.; FE polymer materials, such as PVDF, P(VDF-HFP), etc. AFE materials are a class of materials in which the internal electric dipoles are arranged in an antiparallel manner in the absence of an external electric field; in the initial state, the dipoles cancel each other out and do not show polarity externally. However, under the action of an appropriate external electric field, the dipoles realign in parallel, and the AFE materials undergo a reversible phase transition from the AFE phase to the FE phase, exhibiting a high dielectric constant under high electric field strength. The AFE ceramic materials that have been widely studied include: $PbZrO_3$-based AFEs, $(Na_{0.5}Bi_{0.5})TiO_3$-based AFEs, HfO_2-based AFEs, etc. RFE materials are a special type of FE material whose dielectric properties change significantly with temperature and frequency, first discovered by Smolensky et al. in 1954. Compared to ordinary FE materials, RFE materials possess a high dielectric constant, but a wider phase transition temperature range, a weaker FE domain structure, and stronger polarization relaxation phenomena; this allows them to achieve almost zero P_r and higher saturation polarization (P_s). In RFEs, FE domains are usually nano-sized and can flip more freely under the action of an electric field, showing lower P_r and dielectric loss than FEs. However, traditional RFEs, such as $Pb(Mg_{1/3}Nb_{2/3})O_3$-$PbTiO_3$, are limited by their lower coercive field (E_c), and the W_{rec} in dielectric energy storage applications is not satisfactory.[6] Therefore, achieving lead-free RFE thin film with polarization and breakdown strength similar to FEs has become a hot topic in the research field of dielectric energy storage in recent years.[7]

13.2 Electrical Performance Metrics and Characterizations

13.2.1 Energy Storage Density

Dielectric energy storage thin film-based capacitors are primarily evaluated based on several key performance metrics, of which the W_{rec} is one of the most critical. In the context of modern electronic circuitry, where there is an increasing demand for micro- and nanoscale components and higher levels of integration, it is essential for DECs to achieve and improve their W_{rec}. This capability ensures that these capacitors can effectively store and release energy within the constraints of increasingly compact and complex electronic systems. Despite years of research in the field of dielectric energy storage, compared to commercial BOPP (biaxially oriented polypropylene) capacitors, the W_{rec} of dielectric film capacitors has seen significant improvement. However, when compared to other devices such as batteries, its W_{rec} still lags behind. Thus, how to further improve the ESD of film capacitors remains a focal point of current application research in the field of energy storage.[8]

$$W_{rec} = \int_{P_r}^{P_{max}} E\, dP \tag{13.1}$$

As depicted in Figure 13.4, the gray area represents energy loss due to hysteresis loss. The breakdown strength (E_b) is also a critical factor. An ideal dielectric capacitor should exhibit a large ratio of maximum polarization (P_{max}) to remanent polarization (P_r), a high breakdown field strength (BDS), and minimal polarization switching hysteresis. The W_{rec} can be calculated using formula (13.1), represented by the integration area, where P_{max} and P_r denote the maximum polarization and the remaining polarization, respectively.

13.2.2 Energy Storage Efficiency

Loss energy density (W_{loss}) and energy storage efficiency are both key metrics for evaluating the performance of energy storage film capacitors. In practical applications, minimizing the W_{loss} is essential to enhance energy conversion and transfer efficiency. The W_{loss} is calculated by integrating the area enclosed by the *PE* hysteresis

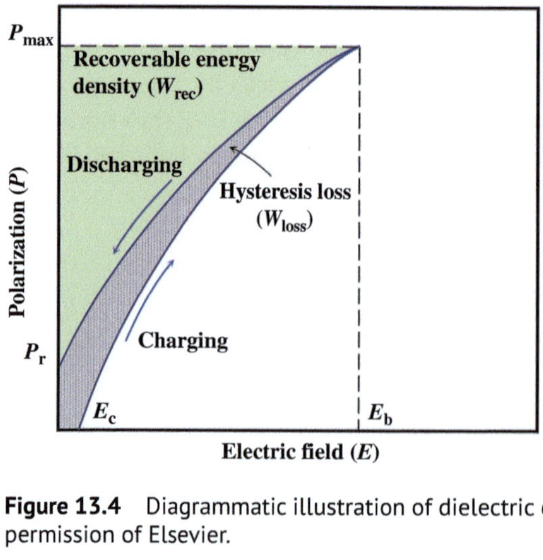

Figure 13.4 Diagrammatic illustration of dielectric energy storage. *Source:* [9] / with permission of Elsevier.

loop, as indicated in the gray area of Figure 13.4. Formulas for total energy (W_{total}), W_{loss}, and efficiency are as follows:

$$W_{total} = \int_{0}^{P_{max}} E\, dP \tag{13.2}$$

$$W_{loss} = W_{total} - W_{rec} \tag{13.3}$$

$$\eta = \frac{W_{rec}}{W_{total}} = \frac{W_{rec}}{W_{rec} + W_{loss}} \times 100\% \tag{13.4}$$

The energy release efficiency of a capacitor indicates the energy conversion efficiency during charging and discharging, which is the ratio of the energy discharged from the input. A higher efficiency reduces heat generation during operation, mitigates potential performance degradation, slows down the aging process, and extends the capacitor's lifespan. Consequently, it is imperative for a high-quality energy storage capacitor to narrow the *P-E* loop, thereby enhancing efficiency and practical performance.

13.2.3 Charging-discharging Characteristics

The charging-discharging speed is one of the key parameters for measuring the performance of energy storage thin-film capacitors, and it directly affects the performance of capacitors in practical applications. Fast charging and discharging can meet the demands of high-speed, high-power devices, such as pulse power systems, electric vehicles, etc. In these application scenarios, capacitors need to release a large amount of energy in a short period and be quickly recharged. Therefore, improving the charge and discharge speed of energy storage thin-film capacitors is of great significance for achieving high-performance electronic devices. The energy storage capacitors not only need to exhibit high W_{rec} and a high efficiency (η), but also need to have a fast charge-discharge process. The principles of charging-discharging testing are close to those of actual power pulse circuits. Therefore, an appropriate charge-discharge test method is crucial to accurately testing the actual discharge performance of the energy storage performance.[10]

13.2.3.1 Charging and Discharging Speed Test

The time constant (τ) is a circuit parameter r with the unit of time. The time constant determines the rate at which a capacitor charges or discharges through a resistor (Figure 13.5a). If the value of RC is much less than 1, the capacitor will quickly reach its final charge value; if the value of RC is much higher than 1, the capacitor will reach its final charge value very slowly. In this RC circuit, the charging and discharging processes of a capacitor can be described by the following formulas. During the charging process, the voltage across the capacitor, $V_C(t)$, as a function of time t is given by the formula:

$$V_C(t) = V_s\left(1 - e^{-\frac{t}{RC}}\right) \tag{13.5}$$

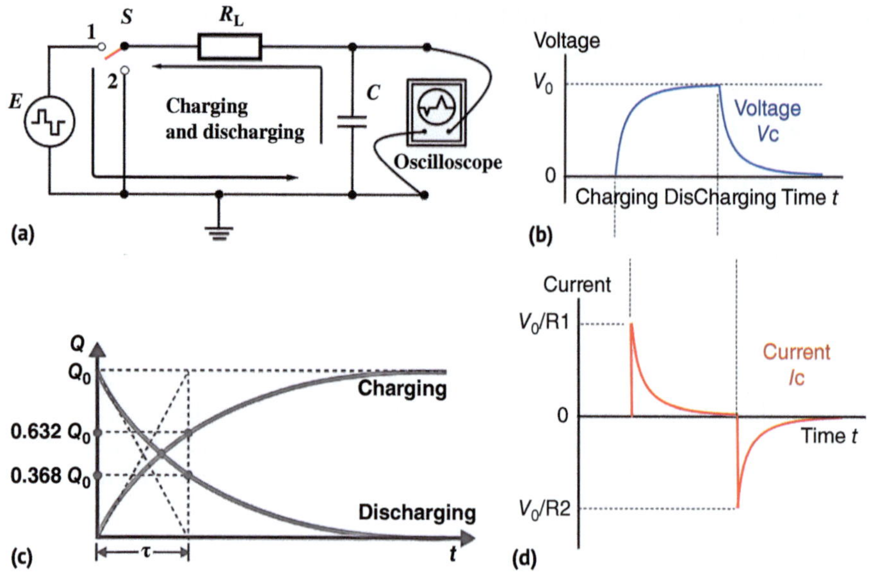

Figure 13.5 Schematic diagram of the charging-discharging test circuit. (a) RC circuit with a square wave. (b, d) The charging-discharging behavior for the period of the square T_0 wave larger than 2τ and (c) the changes of Q_0 from 0 to maximum.

As shown in Figure 13.5b,d, where V_s is the supply voltage, R is the resistance, C is the capacitance, and e is the base of the natural logarithm. And, the current through the resistor, $I(t)$, as a function of time t is given by the formula:

$$I(t) = \frac{V_s}{R} e^{-\frac{t}{RC}} \tag{13.6}$$

In these formulas, R is the resistance value, C is the capacitance value, and RC is the time constant, which determines the speed at which the capacitor charges or discharges. The time constant $\tau = RC$ represents the time it takes for the capacitor to charge to approximately 63.2% of V_s or discharge to approximately 36.8% of V_0. As illustrated in Figure 13.5c, it can be observed that during the charging process of a capacitor, the voltage across the capacitor increases rapidly, while the current decreases rapidly in accordance with an exponential decay. When time t is equal to the time constant τ (i.e., CR), the charge Q on the capacitor will reach 63.2% of its maximum value Q_0. The charge on the capacitor, $Q(t)$, as a function of time t is given by the formula:

$$Q(t) = CV_C(t) = CV_s\left(1 - e^{-\frac{t}{RC}}\right) \tag{13.7}$$

And the discharge is the opposite process to charging. Theoretically, it takes an infinite amount of time (typically more than 5τ) for the charge on the capacitor to reach its maximum Q_0. At this point, the exponential function $e^{-t/CR}$ converges to 0.

13.2.3.2 Dynamic ESD Test

In practical applications, the DEC is affected by factors such as circuit losses and parasitic capacitance. Therefore, to precisely assess the discharge performance of energy storage capacitors, it is essential to employ suitable testing methods. In addition to the static method (Section 13.2.1), which calculates energy density by determining the area under the *PE* hysteresis loop, another approach involves measuring the discharge current of a sample within a resistance–inductance–capacitance (RLC) circuit.[11]

This alternative technique is known as the dynamic method, and its testing principle is illustrated in Figure 13.6, where Figure 13.6a shows the RLC circuit diagram for charge and discharge tests, and Figure 13.6b shows the schematic diagram of underdamping, overdamping, and critical damping. The RLC charging and discharging test circuit employs a test sample that functions as a capacitive element. When the switch is connected to contact 1, the capacitor is connected to a pulsed voltage source and charged. Conversely, when the switch is connected to contact 2, the capacitor is discharged via the circuit on the right. The circuit incorporates a series inductor (L) and a resistor (R), and the measurement current is recorded by the oscilloscope. According to Kirchhoff's voltage law, the discharge circuit can be expressed by Equation (13.8):

$$Ri(t) + L\frac{di(t)}{dt} + \frac{1}{C}\int i(t)dt = 0 \tag{13.8}$$

Let $a = \frac{R}{2L}$, $\omega = \frac{1}{\sqrt{LC}}$, then we have:

$$\frac{d^2i}{dt^2} + 2a\frac{di}{dt} + \omega^2 i = 0 \tag{13.9}$$

1. When $a = \omega$, i.e., $R > 2\sqrt{\frac{L}{C}}$, the circuit is in an over-damped state, and the current waveform is shown in Figure 13.6b. The current function can be expressed as:

$$i(t) = \frac{U_d e^{-at}}{2L\sqrt{a^2 - \omega^2}}\left(e^{-\left(\sqrt{a^2-\omega^2}\right)t} - e^{\left(\sqrt{a^2-\omega^2}\right)t}\right) \tag{13.10}$$

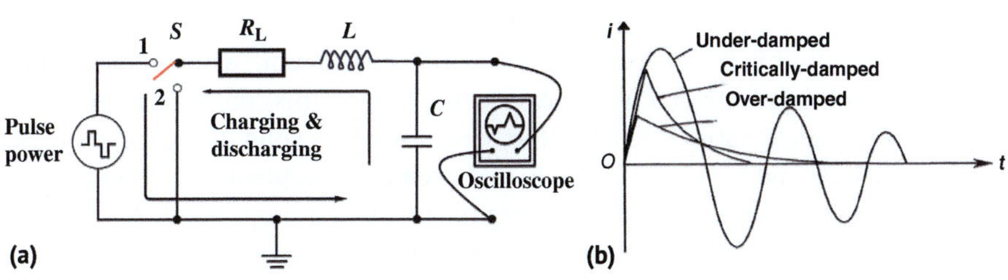

Figure 13.6 Schematic diagram of the dynamic method test. (a) RLC circuit with a square wave and (b) the damped behavior during this period.

2. For over-damped states, the load resistance should be much greater than the equivalent series resistance of the capacitor, so that most of the stored energy is dissipated in the load resistance. The measured energy density is almost the same as the stored energy density. In the critically damped state, when $a = \omega$, i.e., $R = 2\sqrt{\frac{L}{C}}$, the circuit is in a critically damped state, and the current decays to 0 in the shortest time. The current function at this time is as formula (13.11):

$$i(t) = \frac{U_{te} - w}{L} = \frac{U_{te}}{L} \cdot \frac{L}{L} \tag{13.11}$$

3. When $a < \omega$, i.e., $R < 2\sqrt{\frac{L}{C}}$, the circuit is in an under-damped state, and the current waveform exhibits oscillatory characteristics, also known as damped oscillation. The current waveform is shown in Figure 13.6b, and it is a sinusoidal decay oscillation. Let $a_3 = \sqrt{\omega^2 - a^2}$, then the under-damped discharge current waveform function is:

$$i(t) = \frac{U_0}{\sqrt{\frac{L}{C} - \frac{R^2}{4}}} \exp\left(-\frac{Rt}{2L}\right) \sin\left(\sqrt{\frac{1}{LC} - \frac{R^2}{4L^2}}\right) t \tag{13.12}$$

where $\alpha = R/2L$ is known as the "damping coefficient," indicating that the under-damped discharge current decays exponentially. The oscillation period T of the current is expressed as:

$$T = \frac{2\pi}{\omega} = \frac{2\pi}{\sqrt{\frac{1}{LC} - \frac{R^2}{4L^2}}} \tag{13.13}$$

Therefore, for circuits require a fast and stable output, a shorter oscillation period can facilitate the attainment of a steady state. Conversely, an extended oscillation period may result in a circuit charging-discharging process responds more gradually.

13.2.4 Breakdown Strength

When the applied electric field exceeds a certain value, solid dielectrics lose their insulating properties due to the effects of heat, chemistry, and force, forming conductive pathways within the dielectric that lead to its destruction, which is called the breakdown of solid dielectrics. The critical electric field in this context is referred to as the BDS, or dielectric strength. After the breakdown of a solid dielectric, there is typically a burnt or melted channel crack, and even if the applied electric field disappears, the dielectric cannot recover its original insulating properties.[12]

The BDS of HfO_2 thin film materials is affected by a variety of factors, such as growth rate, crystallinity, and internal defects in the material. The BDS is closely related to the uniformity of the electric field and is generally independent of the thickness. According to the ESD formula, the ESD is proportional to the square of the BDS. This means that increasing the BDS of a material can greatly increase its ESD.

At the same time, increasing the BDS means that the DEC can operate stably at a higher voltage, that is, it has a higher rated working voltage and a wider operating frequency range, thereby improving the safety and reliability of the capacitor. In summary, the BDS is critical for the energy storage performance and long-term operational stability of DECs.

13.2.5 Dielectric Loss and Leakage Current

In the context of HfO_2-based materials for energy storage applications, understanding and managing dielectric loss and leakage current are critical aspects that influence the performance and reliability of devices. Dielectric loss refers to the energy lost as heat when an electric field is applied across a dielectric material, while leakage current pertains to the unintended flow of electrical current through the dielectric under the influence of an electric field. For HfO_2-based thin films used in energy storage, it is essential that these materials maintain the minimal fluctuations in the dielectric properties and polarization performance over a wide temperature range.

Specifically, the dielectric constant should not vary more than ±15% from −55 °C to 125 °C, with some applications requiring this stability even up to 150~300 °C for high-temperature environments.[11,13,14] This stringent requirement highlights the necessity for HfO_2-based materials to possess robust thermal stability.

The dielectric loss in HfO_2-based thin films can be categorized into several types: intrinsic losses due to the material's inherent properties, extrinsic losses caused by defects or impurities within the material, and interface losses occurring at the boundaries between different layers or phases. Intrinsic losses are closely related to the material's crystal structure and phase composition, whereas extrinsic losses can often be mitigated by improving the purity of the starting materials and refining the fabrication process. Interface losses are particularly significant in multilayer structures, where the quality of interfaces plays a crucial role in determining device performance.

Leakage current is another important factor that impacts the efficiency and longevity of energy storage systems incorporating HfO_2-based thin films. Leakage currents can arise from various mechanisms, including direct tunneling, Fowler–Nordheim tunneling, and trap-assisted conduction. Direct tunneling becomes increasingly problematic as the thickness of the dielectric layer decreases, making it especially relevant for ultrathin HfO_2 films. Fowler–Nordheim tunneling occurs at higher electric fields and can lead to accelerated degradation of the dielectric. Trap-assisted conduction involves charge carriers moving through localized states within the bandgap of the material, contributing to increased leakage current.[8,15]

Reducing leakage current can be achieved by composition adjustment (e.g., doping) and optimizing the microstructure of the HfO_2 film. Moreover, the leakage current characteristics of HfO_2-based thin films can also be influenced by the presence of interfacial layers or passivation treatments.[16] For instance, growing a thin Al_2O_3 or SiO_2 layer at the HfO_2/Si interface has been shown to significantly reduce leakage current without compromising the dielectric constant.[17] Similarly, employing atomic layer deposition (ALD) techniques allows for precise control over HfO_2-based film thickness and composition, thereby minimizing leakage paths and enhancing device reliability. Achieving low dielectric loss and leakage current in HfO_2-based materials for energy storage applications involves a multifaceted approach that includes material selection, process optimization, and innovative design strategies.

13.2.6 Reliability Test

The reliability of operation is a pivotal factor influencing the performance of dielectric materials and pulse capacitors, regardless of the specific material or component in question. In general, operational reliability necessitates that the electrical properties of materials and equipment remain unchanging when subjected to external stimulation, including thermal, electrical, mechanical, and magnetic fields. In particular, fatigue endurance is a very important issue affecting the working capability of capacitors. In 2009, Lou [18] reviewed different fatigue mechanisms in ceramics, films, and single crystals, which helps us better understand the fatigue behavior of materials. Additionally, mechanical fields and magnetic fields can both influence the polarization behavior of materials, the importance of which may play a key role in future multifunctional coupling requirements. Therefore, the assessment criteria for work reliability should be a complex engineering task.

The Weibull distribution is widely used as a statistical tool in evaluating the operational reliability and life analysis of dielectric materials.[19] The Weibull distribution can well describe the probability distribution characteristics of the BDS of the material, and plays a vital role in understanding and predicting the failure probability of the material under different electric field conditions. According to the Weibull distribution, the cumulative probability $F(E)$ (formula 13.14) indicates the cumulative probability of a material breakdown at a given breakdown field E:

$$F(t) = 1 - e^{-\left(\frac{E}{E_0}\right)^m} \tag{13.14}$$

where E is the measured or applied BDS, which is the electric field strength that causes dielectric materials to fail. E_0 is the characteristic of BDS, which corresponds to the BDS when the cumulative failure probability is 63.2%. And m is the Weibull modulus, also known as the shape parameter, which reflects the degree of dispersion of the data; a higher m value means a more consistent data set, while a lower m value indicates greater variability. In experiments, a logarithmic transformation is often used to convert the above nonlinear relationship into a linear relationship for the purpose of analyzing the experimental data. Specifically, a linear equation is obtained by taking the natural logarithm twice:

$$\ln(-\ln(1 - F(E))) = m \cdot \ln(E) - m \cdot \ln(E_0) \tag{13.15}$$

or simplified as:

$$\ln(-\ln(1 - F(E))) = m \cdot \ln(E) + C \tag{13.16}$$

where $C = -m \cdot \ln(E_0)$ is a constant term. By plotting $\ln(-\ln(1 - F(E)))$ against $\ln(E)$, a straight line with a slope of m can be obtained, and the intercept can be used to solve for E_0. For practical works, these formulas help us understand the failure probability of materials under various electric fields to evaluate the reliability of devices.

13.3 HfO_2 in Dielectric Energy Storage: Advances, Status, and Challenges

HfO_2-based thin film materials have shown tremendous potential in the energy sector, particularly in the field of energy storage.[20,21] Over the past decade, doped and undoped $Hf_xZr_{1-x}O_2$-based FE thin films have emerged as dielectric materials suitable for energy storage in dielectric capacitors. Zirconium (Zr) is considered an attractive doping element because its atomic radius is very close to that of hafnium (Hf), which allows ZrO_2 and HfO_2 to have lattice parameters that are very close, thus enabling them to mix in any proportion to form a single-phase solid solution. The transition from FE to AFE behavior by adjusting the Zr doping ratio makes $Hf_xZr_{1-x}O_2$ stand out in the field of energy storage. As shown in Figure13.7a, Park et al. first reported the energy storage characteristics of $Hf_xZr_{1-x}O_2$ thin films in 2014.[9] It can be observed that as the Zr doping ratio increases (x: 0.4→0.7), $Hf_xZr_{1-x}O_2$ thin films transition from AFE behavior to enhanced AFE polarization behavior. Among films of different compositions, the $Hf_{0.3}Zr_{0.7}O_2$ film exhibits the best energy storage performance, showing the highest W_{rec} of about 42J/cm^3 and efficiency of about 46% at a BDS of 4.5MV/cm (Figure 13.7b). Therefore, compared to HfO_2-based FE materials with large residual polarization, $Hf_xZr_{1-x}O_2$ films with AFE behavior, due to their lower P_r (close to 0), coercive field, and greater P_s, are more suitable for energy storage in static capacitors, showing significant application potential.[22]

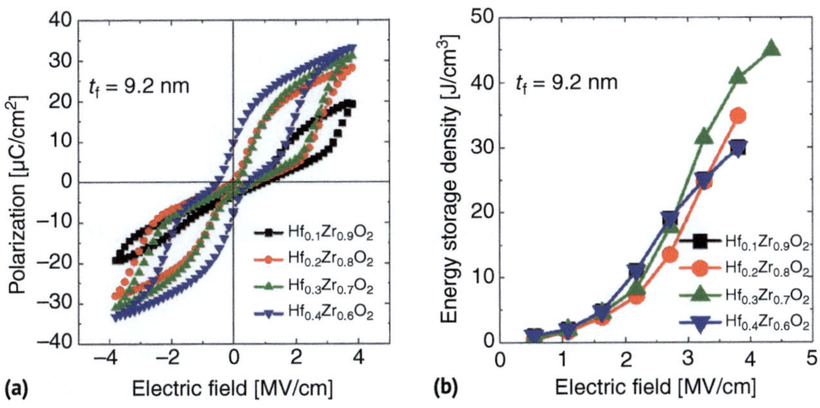

Figure 13.7 (a) *PE* loops and (b) ESD property of $Hf_xZr_{1-x}O_2$ thin films with different Zr doping. *Source*: [9] / with permission of EBSCO Industries.

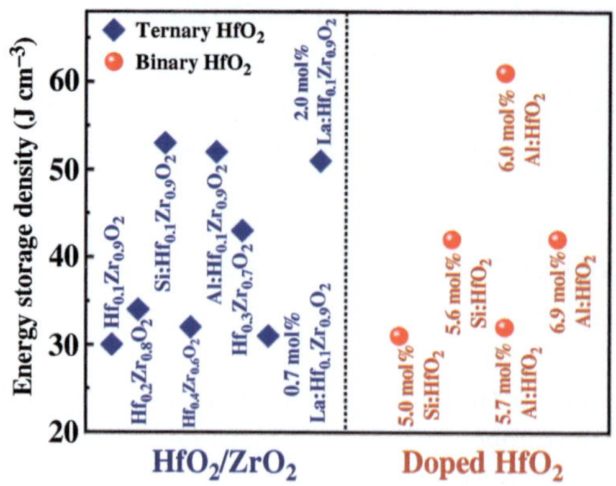

Figure 13.8 Room temperature ESD of doped $Hf_xZr_{1-x}O_2$ thin films and doped HfO_2 thin films.

Subsequently, research on the energy storage characteristics of HfO_2-based thin films with AFE characteristics has continued to emerge, with representative studies including element doping strategies (the doping strategies for HfO_2 can be divided into metal doping and nonmetal impurity doping). As shown in Figure 13.8, the energy storage characteristics of representative doped HfO_2-based AFE thin films are compared. These studies mainly introduce different types of elements (such as Zr, Al, La, and Si) into HfO_2 materials, using their AFE characteristics to increase ESD and enhance energy efficiency. The energy storage characteristics of Si:HfO_2 thin films were first reported by Hoffmann et al. in 2015.[23]

After the addition of Si doping, HfO_2-based thin films can exhibit both FE and AFE characteristics. For example, at a lower Si concentration (such as 4.4mol%), Si:HfO_2 thin films exhibit FE behavior; while with an increase in Si doping ratio (5.6mol%), Si:HfO_2 thin films exhibit AFE behavior, also showing a larger W_{rec} of about 40J/cm^3, with an efficiency (η) as high as 80%. Similarly, for a higher Al doping concentration (6.9mol%), HfO_2-based thin films can also exhibit AFE characteristics; at this point, the film's W_{rec} and η values are about 42J/cm^3 and about 55%, respectively. HfO_2-based AFE thin film capacitors achieved through element doping have significantly improved in energy density compared to $Hf_{0.5}Zr_{0.5}O_2$ (HZO) FE thin films.[22,24–27]

13.4 Novel HfO_2-based Materials and Structural Design

13.4.1 AFE-based Period

Despite the fact that AFE HfO_2-based materials exhibit unique advantages in the field of energy storage, HfO_2-based AFE thin films generally exhibit decreasing W_{rec} and η values with increasing number of charge-discharge cycles due to field-induced phase transitions between the AFE and FE states, and the wake-up effect, which is mainly determined by the composition and applied field strength. Moreover, the ESD is still relatively small (about 50J/cm^3) compared with that of traditional energy storage medium capacitors, and these issues need to be addressed urgently.

13.4.2 From AFE to New Era

Although AFE-HZO thin films exhibit high P_s and low P_r, the broadening of the *PE* hysteresis loop after field-induced phase transition results in certain losses. Moreover, the unavoidable W_{loss} (energy loss during switching) in AFE materials further limits the improvement of their energy storage efficiency. In contrast, RFE materials with slender hysteresis loops typically have narrower loops and low remanent polarization, making them very suitable for fabricating dielectric capacitors with high recoverable energy and high energy storage efficiency. The electrical properties of $Hf_xZr_{1-x}O_2$ thin films are related to their crystal structure and sensitively depend on their composition and processing conditions. According to the formula for calculating ESD, a large ratio of maximum polarization to remanent polarization (P_{max}/P_r) and a slender hysteresis loop are crucial for improving energy storage performance. Therefore, some recent reports have indicated that the merits of both can be synergistically integrated by incorporating a relaxed state within the AFE-HZO film. The utilization of a high-entropy strategy has been identified as a viable approach to achieving ultrahigh ESD in high-entropy lead-free RFEs, through the augmentation of configurational entropy. Moreover, the incorporation of field-driven negative capacitance effects can substantially augment the ESD of AFE HfO_2 thin films. This augmentation, in turn, increases the device applications and ESD of HfO_2 thin films. The abovementioned methods offer novel ideas and possibilities in the field of energy storage performance for device applications.

As Figure 13.9 shows, this approach significantly reduces W_{loss} while maintaining a large P_{max}, providing new insights into developing HfO_2-based thin film energy storage materials with superior comprehensive performance. In essence, this strategy leverages the unique characteristics of RFEs, where the material exhibits a diffuse phase transition without long-range order, characterized by polar nanoregions (PNRs) – to mitigate the inherent limitations of AFE materials. The result is an optimized material system that not only achieves higher ESD but also enhances the efficiency of energy recovery, thereby addressing critical challenges in the development of advanced energy storage devices. This innovative approach underscores the importance of tailoring the microstructure and phase composition of HfO_2-based thin films through precise compositional control and process optimization, paving the way for next-generation high-performance energy storage solutions.[28] Table 13.1 shows the energy storage performance of HfO_2-based dielectric capacitors since first reported.

13.4.2.1 High Entropy Strategy

In recent work, we introduced Al-doping into HZO thin films to enhance the disorder of the crystalline lattice and form nanodomains. By precisely controlling the

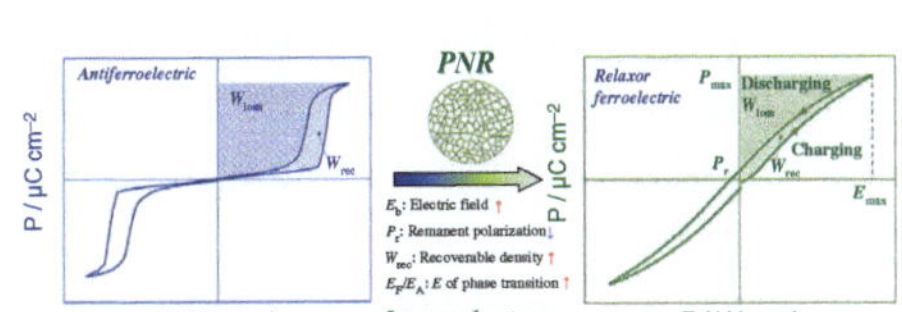

Figure 13.9 High entropy strategy from AFE to RFEs.

Table 13.1 Comparisons of W_{rec} and E_b between recently reported HfO_2-based energy storage thin films.

Dielectric films	Thickness	W_{rec} (J cm^{-3})	η (%)	E_b (MV cm^{-1})	References
$Hf_xZr_{1-x}O_2$	9.2nm	46	~50	4.5	[22]
Si-HfO_2	9nm	~40	~80	~3.3	[23]
Si-HfO_2	10nm	61.2	65	4.5	[27]
Al-HfO_2	10nm	40.6	50	4	[29]
Si-$Hf_{0.5}Zr_{0.5}O_2$	10nm	40–53	~80	4.75	[30]
La-$Hf_{0.5}Zr_{0.5}O_2$	10nm	49	70	~4	[24]
Al-$Hf_{0.5}Zr_{0.5}O_2$	10nm	52	80	5.0	[26]
Al_2O_3:DIL-$Hf_{0.5}Zr_{0.5}O_2$	2–8nm	54.3	51.3	4–5.5	[31]
$Hf_{0.67}Zr_{0.33}O_2/Al_2O_3/Hf_{0.67}Zr_{0.33}O_2$	10nm	70	50	6	[32]
$Hf_{0.5}Zr_{0.5}O_2/Hf_{0.25}Zr_{0.75}O_2$	25nm	72	57.8	7.2	[33]
ZrO_2	450nm	75.4	88	5.8	[34]
$TiO_2/ZrO_2/TiO_2$	8.7~40nm	80~94	80	5.5	[35]
HfO_2/ZrO_2	8.8nm	49.9	61.9	4.0	[36]
Al-HfO_2	50nm	18.2	51.8	3	[37]
Al_2O_3/BST-Ce/ZrO_2	–	71.5	86.1	7.9	[38]
$TiO_2/ZrO_2/TiO_2$ (Unipolar)	8~10	114.5	76	5.4	[39]
RAFE-$Hf_{0.5}Zr_{0.5}O_2$	~15nm	101.5	~80	~6.17	[9]
NC-HZO-Al_2O_3	~100nm	80–115	80~90		[40]
Ba-Hf-O system	~100nm	155	87	12	[41]

Al-doping concentration, HZO can achieve a high breakdown electric field while still possessing a comparatively high polarization value.

When discussing the electrical properties of aluminum-doped hafnium zirconium oxide (Al:HZO) films, we observe a transition from FE to AFE, and then to relaxor antiferroelectric (RAFE) states. As the aluminum content x increases, *PE* loops at an electric field of 4MV·cm^{-1} illustrate these transitions (Figure 13.10a,b).[9]

Comparing the energy storage performance metrics – recoverable energy density (W_r), efficiency (η), and maximum PD relative to remanent polarization (P_{max}/P_r) of films with varying aluminum content at their breakdown fields reveals the impact of aluminum doping on these parameters (Figure 13.10c). Additionally, the dependence of compensation current versus electric field (*I–E*) sweeps for FE/AFE/RAFE Al:HZO capacitors with three distinct levels of aluminum doping is depicted in Figure 13.10d.[9]

Additionally, under high electric field conditions, Al-doped HZO films does not exhibit a significant hysteresis window, which enables a high energy conversion

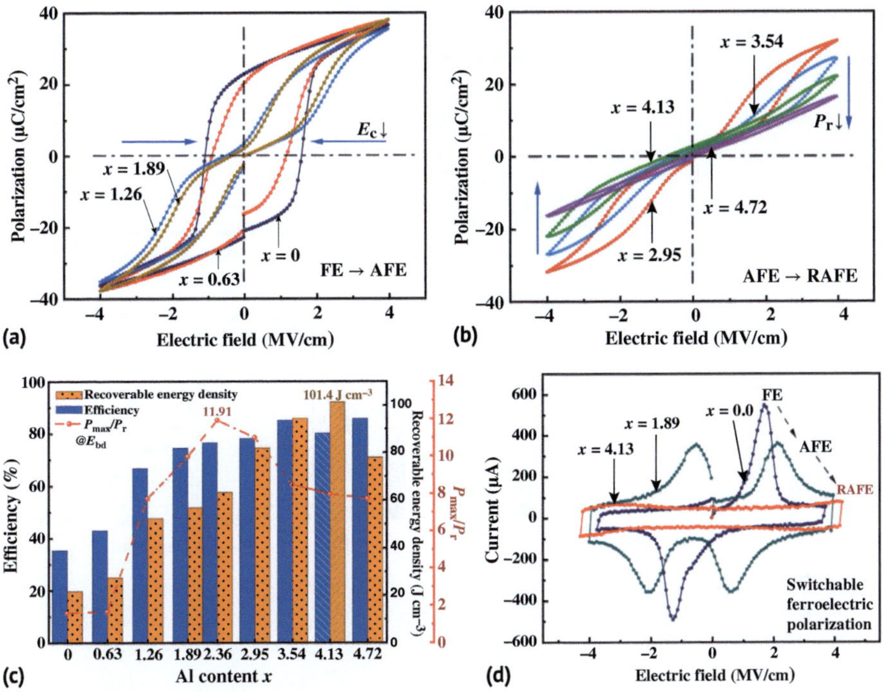

Figure 13.10 Evolution of *PE* loops, *I*–*E* sweeps and energy storage performance increased Al content. (a) *P-E* loops from FE to AFE and (b) from AFE to RAFE at an electric field of 4 MV cm^{-1} with increased Al content *x*. (c) Comparison of the energy storage performance W_r, η, and P_{max}/P_r of the films with different Al content at their breakdown fields. (d) Dependence of compensation current versus electric field (*I*–*E*) sweeps of FE/AFE/RAFE Al:HZO capacitors with three different Al doping levels. *Source*: [9] / with permission of Elsevier.

efficiency. For the first time, we experimentally demonstrate that a superparaelectric-like RAFE characteristic can be achieved in HfO_2-based thin films by doping Al into the AFE HZO. In a 15 nm-thick Al:HZO film, a corresponding W_{rec} over 100 J/cm^3 is achieved with an efficiency higher than 80% as well as a much higher breakdown strength > 6 MV/cm at ultrafast charging/discharging speed. Compared to those perovskite oxide-based thin film capacitors, HfO_2 possesses promising features of high dielectric breakdown strength as well as ultralow leakage current.[42–44] Furthermore, the low-temperature ALD and annealing broadens its applications to flexible and wearable electronic devices.

Figure 13.11 provides the PFM phase images for the FE pure HZO thin films under the same voltage condition. The corresponding phase plots for both RAFE and FE conditions are shown in Figure 13.11c,f, respectively. As Figure 13.11h shows, we attribute the RFE characteristics to the formation of nanodomains due to the doping of AlO_x to the HZO film. By comparing the PFM images of the Al:HZO film with $x = 4.13$ (Figure 13.11a,b) to the pure HZO film (Figures 13.11d,e), one can see that, the domain size of the Al-doped film is much smaller (a few nanometers) than

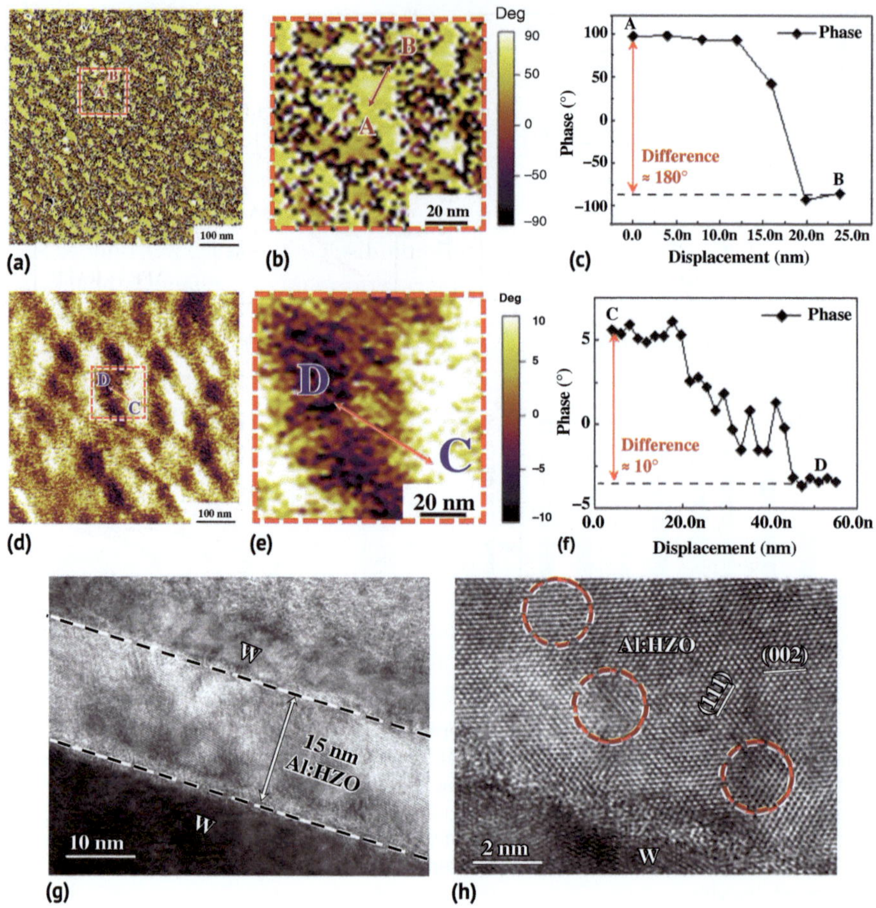

Figure 13.11 Disordered nanostructures and nanodomains in RAFE HZO film. PFM phase images of (a, b) the RAFE ($x = 4.13$) and (d, e) the FE pure HZO thin films with applying a DC bias write voltage (V_{dc}) of 3.0V as well as (c, f) the corresponding phase plots. (g, h) Cross-section TEM images of $x = 4.13$ mol% Al-doped HZO films. *Source*: [9] / with permission of Elsevier.

that of the pure HZO film, which is a few tens of nanometers. It is also worthy to note that even at 3V, the $x = 4.13$ Al:HZO film realizes polarization switching manifested by the nearly 180° phase angle change (Figure 13.11c) due to polarization switching (see the area indicated by red line); while the same voltage only induces a small phase change as shown in Figure 13.11f, implying very limited polarization switching in HZO film. The different phase change behaviors further verify the existence of nanodomains in the Al:HZO ($x = 4.13$) samples. Al doping into the HZO film interrupts the long-range periodic structure of HZO and introduces disordered nanostructures, as can be suggested by the cross-sectional HRTEM images shown in Figure 13.11g,h, in which the disordered structure can be clearly seen in the circled areas.

This disordered structure of the Al:HZO phase contributes not only to the formation of nanodomains but also to facilitate the switching of nanodomains and reduce energy dissipation, that is, a characteristic of RFE. The RAFE characteristics greatly enhance the W_{rec} and efficiency, and the weak interactions between nanodomains may also give rise to faster charging/discharging speed of the capacitor. This high-entropy strategy demonstrated a new superparaelectric approach to realize superior energy storage performance (W_r over 100 J/cm^3 and efficiency higher than 80%) along with ultrahigh charging/discharging speed in Al-doped RFE HZO films. Besides superior energy storage performance, excellent fatigue properties (>10^{10} cycles) and thermal stability up to ~100 °C. Based on the advantages of ALD preparation and compatibility with standard CMOS manufacturing, it shows great advantages for integrated energy storage capacitor applications that require miniaturization and integration. These favorable features indicate that HfO_2-based superparaelectric-like dielectric thin films could exhibit great potential for high-energy-density applications in microdevices.

13.4.2.2 Amorphous HfO_2-based Oxide

As discussed above, the high entropy strategy approach has been demonstrated to further improve the energy performance of superparaelectric RFE thin films of complex perovskite oxides. By introducing local disorder, the E_b values of BFSTO and $Ba(Zr_{0.2}Ti_{0.8})O_3$ films can be increased up to 4.5 MV/cm and 5.7 MV/cm, accordingly, excellent energy storage densities have been achieved. The high energy storage performance was also observed in high-entropy stabilized $Bi_2Ti_2O_7$-based dielectric films.[28] Hence, it is also reasonable to expect the existence of nanodomains and their superior energy storage properties by combining the advantages of RFEs and AFE materials. Recently, Wen et al. reported the use of a new structural strategy, extensively doping HfO_2 with divalent ions (A_{2+}) to induce a structural transformation from fluorite (HfO_2) to perovskite ($AHfO_3$), shown in Figure 13.12a,b. The two structures have similar metal ion frameworks but different lattice sites for oxygen ions. In the transition region of the structural evolution, the destabilization of the oxygen ions leads to the destruction of the long-range periodicity, while the maintenance of the metal-ion framework helps to suppress the formation of heterogeneous phases under strong nonstoichiometric conditions, thus forming a noncrystalline structure with only short-range order (Figure 13.12c,d).[41] This amorphous hafnium-based oxide formed at high temperature retains the lattice strain of the divalent doped ions and has a higher degree of disorder and density compared to the HfO_2-based amorphous/crystalline thin films reported in the literature. Its dielectric breakdown strength is as high as 12 MV/cm (Figure 13.13a,b), and it retains the high-k relative dielectric constant of HfO_2 (>18 @10^6 Hz), breaking the negative correlation limit of $\varepsilon_r - E_b$ commonly found in dielectric materials (Figure 13.13c,d).

The amorphous A-Hf-O capacitors exhibit a W_{rec} of up to 155J/cm^3, which is the highest record to date for binary high-k materials represented by HfO_2 shown in Figure 13.13e. The amorphous nature of this material is attributed to the instability

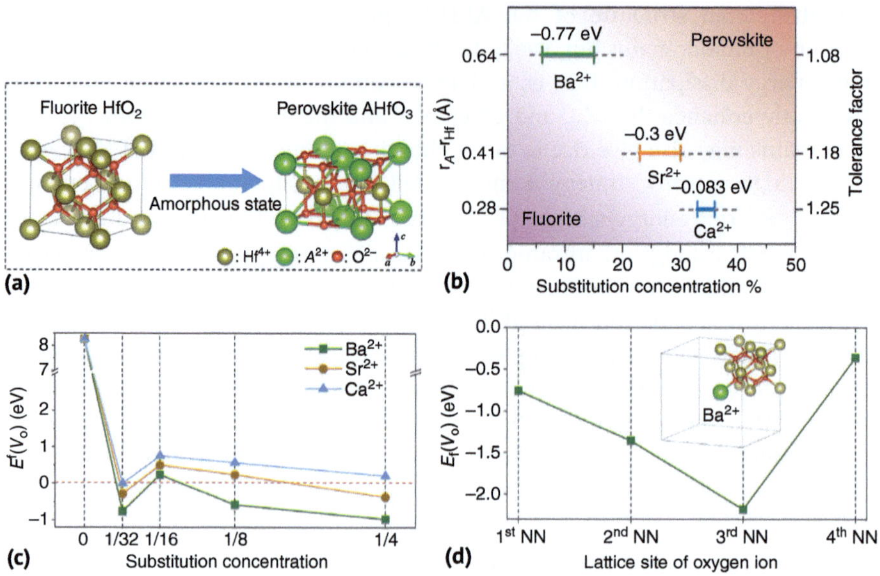

Figure 13.12 Amorphization of hafnium-based oxides. (a) Schematic drawing for the structure evolution from fluorite HfO_2 to perovskite $AHfO_3$, where the HfO_2 is drawn in normal coordinates of <100> (*a* axis), <010> (*b* axis), and <001> (*c* axis) while the $AHfO_3$ is drawn in the coordinates of <110> (*a* axis), <11¯0> (*b* axis), and <001> (*c* axis). (b) Amorphous regions of the Ba-Hf-O, Sr-Hf-O, and Ca-Hf-O systems, respectively, as functions of the difference in ionic radii between A^{2+} and Hf^{4+} (r_A-r_{Hf}) and the tolerance factor of $AHfO_3$. (c) Formation energy of oxygen vacancy $[E_f(V_O)]$ at the first nearest-neighbor (NN) site as a function of the substitution concentration in t-phase HfO_2. (d) $E_f(V_O)$ at different nearest-neighbor sites for the Ba concentration of 1/32. The inset depicts the lattice structure. *Source:* [41] / Springer Nature/CC BY 4.0.

of oxygen, which leads to long-range periodic collapse and short-range coexistence of multiple symmetries, thereby suppressing carrier avalanche effects and improving fracture toughness. In addition, research has shown that the microstructure and properties of the material can be effectively controlled by regulating the type and concentration of substitute metal ions (such as Ba, Sr, or Ca), providing a new methodology for designing high-performance dielectric energy storage systems. In the amorphous structures, oxygen instability results in strong disordering in both short- and long-ranges and hence yields ultrahigh breakdown strengths, accompanied with high permittivity; this will give rise to giant energy storage densities. In addition to the E_b = 12 MV/cm in the Ba-Hf-O system, the amorphous Sr-Hf-O and Ca-Hf-O also show very high E_b of ~10 and ~8.3 MV/cm, yielding large W_{rec} of ~117 and ~72 J/cm^3, respectively. These findings suggest highly controllable and widely applicable amorphous state with a variety of fluorite and perovskite structures, which provide a new platform for designing high-performance dielectric energy storage.

13.4.2.3 Negative Capacitance Systems

In many single-crystal perovskite systems, it has been shown that the capacitance of the FE superlattice can be enhanced by a negative capacitance.[45] Recently,

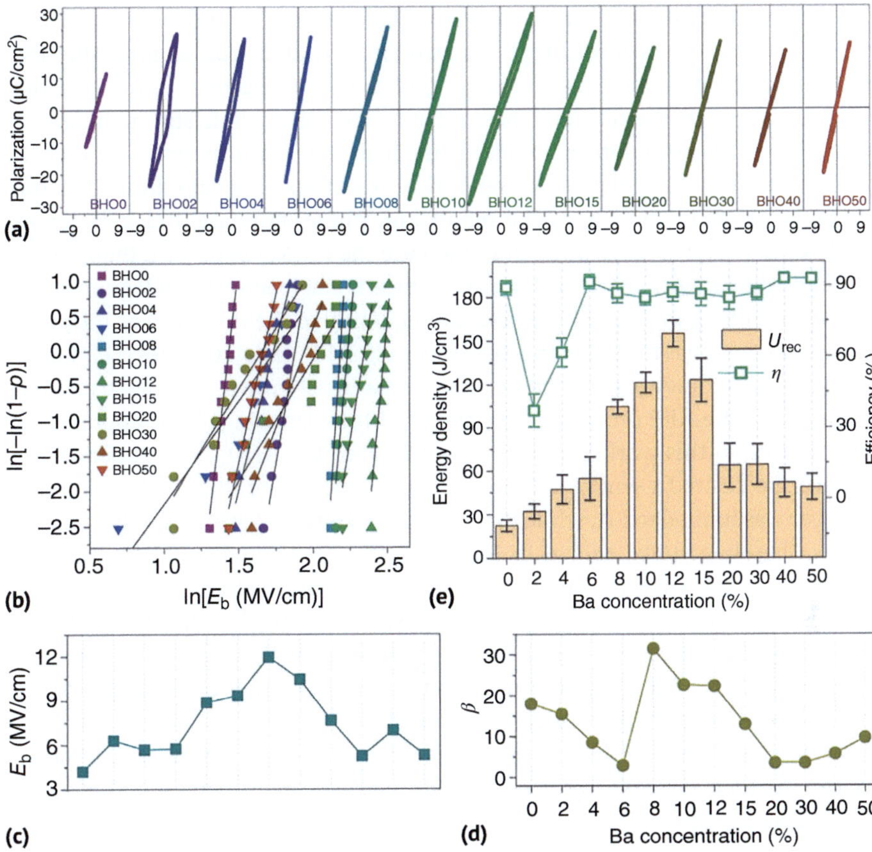

Figure 13.13 Dielectric energy storage of BHO thin-film capacitors. (a) *P-E* hysteresis loops of Pt/BHO/LSMO capacitors measured at 10 kHz. (b) Two-parameter Weibull distribution analysis of breakdown strengths over 12 capacitors for each Ba concentration. (c) Statistical E_b and (d) Weibull modulus β extracted from (b) plotted as a function of Ba concentration. (e) Energy storage performance of the BHO capacitors calculated from (a), and the error bars are averaged over 12 capacitors. *Source*: [41] / Springer Nature / CC BY 4.0.

Suraj S. Cheema et al. reported the electrostatic ESD and PD of HfO_2- ZrO_2-based thin-film micro-capacitors integrated on a silicon wafer. The authors considered the HfO_2-ZrO_2 (HZO) dielectric phase space of a 9 nm-thick HZO film grown by ALD on TiN-buffered silicon. Capacitance-voltage (*C-V*) measurements across the composition phase space illustrate the expected evolution from a FE orthorhombic phase (o-phase: *Pca*2$_1$) at intermediate Zr content to an AFE tetragonal phase (t-phase: P42/nmc) at high Zr content; the parent nonpolar t-phase transforms to the polar o-phase under the influence of an electric field. During this field-driven FE phase transformation, there is a superlinear charge response (called "second phase"), which improves energy storage capacity. Energy density as a function of composition shows that the volumetric energy storage (115 J/cm^3) peaks at 80% Zr content, corresponding to the compressed AFE state of the *C-V* loop. The authors in this work considered using a different NC mechanism, field-driven NC, to generate enhanced charges to enhance energy storage, shown in Figure 13.14a.[40] Although the 9 nm-thick HZO film shows a record high recoverable ESD after iron engineering, the overall stored energy is still small from an application perspective. To increase the overall stored energy, the film thickness must be increased while still maintaining the field-driven NC behaviors that underlie the high ESD performance. This is challenging for FE and AFE phases, as the critical thickness is usually limited to the 10 nm range before the bulk nonpolar monoclinic phase (m-phase, P21/c) becomes favored. Moreover, it is more difficult to increase the thickness of a FE than it is to increase the thickness of a PE: in HZO, the t-phase is stable at very small grain sizes, and then transitions to the o-phase (and subsequently to the m-phase) as the grain size increases, which is related to the film thickness. ALD nanolaminates (i.e., superlattices) can overcome this barrier, because in the ALD-grown fluorite-structured nanolaminates, the AFE properties persist up to 50 nm. Drawing on the approach of nanolaminates, in this work, the authors synthesized the Al_2O_3-HZO superlattices to achieve the desired t-symmetry. For the ALD superlattice, the thickness of the Al_2O_3 layer was limited to 5 Å, while the HZO layer contained 80% Zr-content, based on a 9 nm baseline study, shown in Figure 13.14b. Compared to the continuous HZO approach, the superlattice approach maintains the desirable energy storage properties even after increasing the thickness, i.e., (i) field-induced NC charge up, (ii) enhanced dielectric constant during the AFE-to-FE phase transition, and (iii) high breakdown field. The continuous ~100 nm HZO film is consistent with the FE o-phase, consistent with the electrical behavior.[46]

In addition to the superlattice thickness scaling, the geometric scaling strategies can further increase the planar ESD without increasing the thickness. Metal-insulator-metal (MIM) structures, which require the deposition of conformal

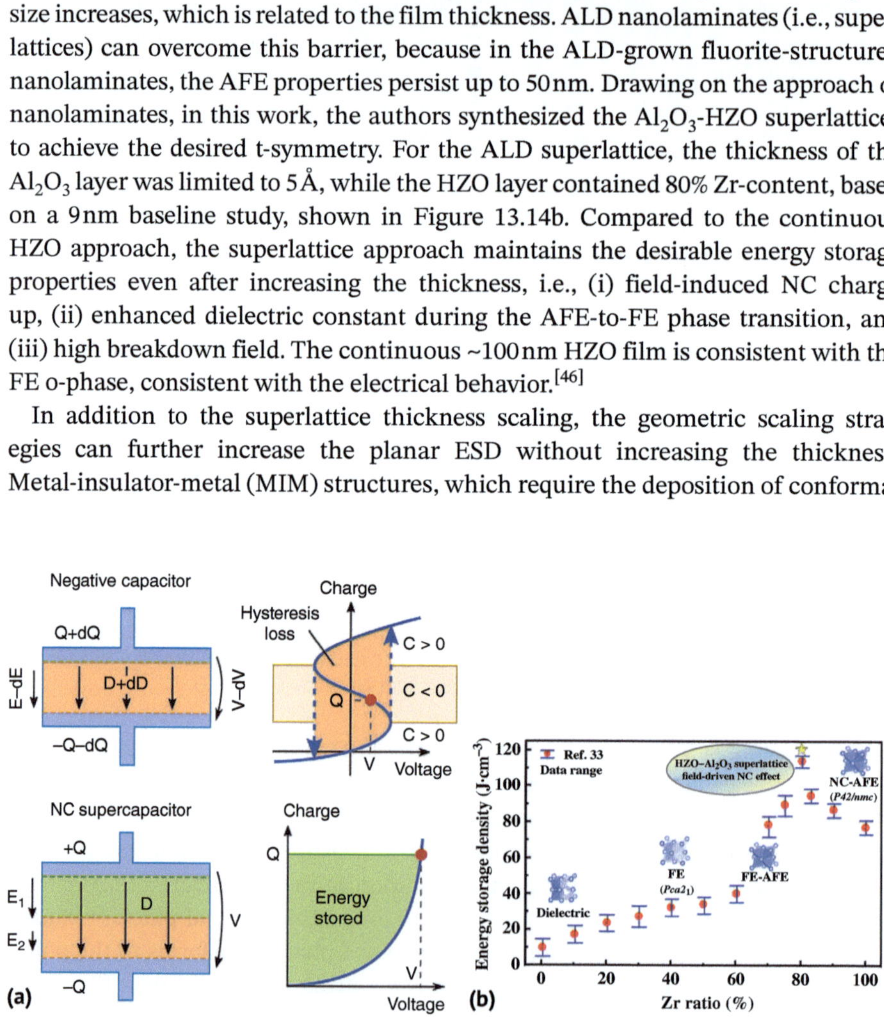

Figure 13.14 (a) Schematic of negative capacitance to enhance energy storage and (b) the extended ultrahigh energy storage via HZO-Al_2O_3 NC superlattice. *Source:* [46] / with permission of John Wiley & Sons.

metals and insulators within the porous nanostructures, have successfully increased the capacitance density, thereby increasing the energy storage per unit planar area (Figure 13.15a,b). Due to the increased material volume per unit area of the 3D-stack capacitors, the surface capacitance is greatly amplified compared to the 2D-planar capacitors. With the AFE behavior, the NC behavior is also transferred to the 3D capacitor (Figure 13.15c). The 3D-stack capacitor has an area ESD of more than 100 times at 4 MV/cm, and the recoverable area ESD of the trench capacitor is 80 mJ/cm^2. The integration of NC superlattices into 3D-stack capacitors achieves record values of 170 times the ESD (80 mJ/cm^2) of the electrostatic capacitors.

With this method, which has both ultrahigh energy density and PD, overcomes the capacity-speed trade-off in the traditional electrostatic-electrochemical energy storage hierarchy. In addition, the integration of ultrahigh-density and ultrafast-charging thin films in a BEOL-compatible process enables the monolithic integration of micro-capacitors on a chip, thereby providing electronic microsystems with tremendous energy storage and power transmission performance.

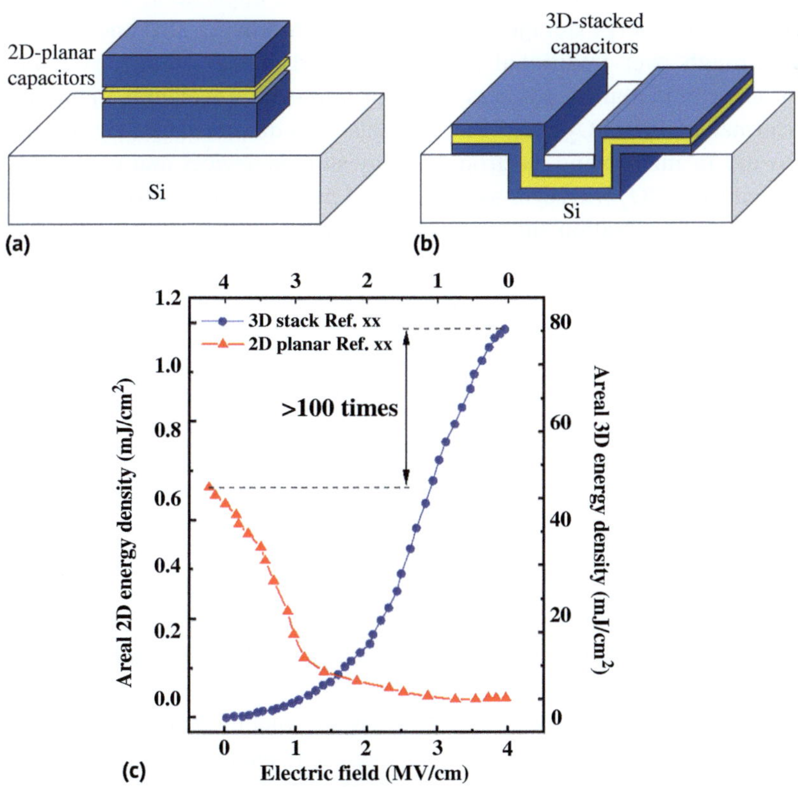

Figure 13.15 (a, b) Schematic of 2D-planar and 3D stacked AFE NC micro capacitors on Si substrate and (c) areal recoverable ESD as the function of electric field for 2D planar and 3D-integrated capacitors.

13.5 Summary and Perspectives

In this chapter, we reviewed the theory and development of HfO_2-based FE materials for energy storage applications and highlighted HfO_2-based materials' unique position in the realm of energy storage. We explored the electrical performance metrics that define the efficiency and reliability of these capacitors, such as energy density and efficiency, and characterization method. The advances in HfO_2 for dielectric energy storage have been significant, with the material showing promise due to its compatibility with CMOS technology, maturity of ALD technology, and excellent polarization performance at ultrathin thicknesses. However, challenges remain, particularly in enhancing the dielectric breakdown field while maintaining high polarization values with minimal hysteresis under high electric fields, and further research on energy storage system integration. The material properties and structure design of HfO_2-based FE materials are crucial for optimizing energy storage performance. Recent research has also shown that high breakdown fields and high polarization can be achieved through nanodomain engineering, high entropy strategies for dopant introduction, and the application of negative capacitance effects in energy storage systems, thereby improving ESD and achieving a breakthrough over traditional AFE HfO_2-based energy storage systems. Due to their environmental friendliness and continuous improvements in energy storage performance, the HfO_2-based FE materials have demonstrated significant potential for use in micro-nano electronic energy storage devices and circuits. It can be expected that HfO_2 will serve as an essential material and contribute to the development of next-generation energy storage solutions.

Acknowledgment

This work is supported by the Hong Kong Research Grants Council (Project No. 15302023) and the Guangdong Basic and Applied Basic Research Foundation (Grant No. 2023B1515130003).

References

1 Puli, V. S., Jayakrishnan, A. R., Pradhan, D. K. et al. (2022). Review on energy storage in lead-free ferroelectric films. *Energy Storage* 5: e359. https://doi.org/10.1002/est2.359

2 Chen, D., Zhu, X., Yang, X. et al. (2023). A review on structure–property relationships in dielectric ceramics using high-entropy compositional strategies. *Journal of the American Ceramic Society* 106: 6602–6616. https://doi.org/10.1111/jace.19349

3 Dai, J. Y., Lee, P. F., Wong, K. H. et al. (2003). Epitaxial growth of yttrium-stabilized HfO_2 high-k gate dielectric thin films on Si. *Journal of Applied Physics* 94: 912–915. https://doi.org/10.1063/1.1585116

4 Su, R., Tseng, J. K., Lu, M. S. et al. (2012). Ferroelectric behavior in the high temperature paraelectric phase in a poly(vinylidene fluoride-co-trifluoroethylene) random copolymer. *Polymer* 53: 728–739. https://doi.org/10.1016/j.polymer.2012.01.001

5 Ali, F., Zhou, D., Sun, N. et al. (2020). Fluorite-structured ferroelectric-/antiferroelectric-based electrostatic nanocapacitors for energy storage applications. *ACS Applied Energy Materials* 3: 6036–6055. https://doi.org/10.1021/acsaem.0c00987

6 Nishigaki, S., Murano, K., and Ohkoshi, A. (1982). Dielectric properties of ceramics in the system ($Sr_{0.5}0Pb_{0.25}Ca_{0.25}$)$TiO_3$-$Bi_2O_3$·$3TiO_2$ and their applications in a high-voltage capacitor. *Journal of the American Ceramic Society* 65: 554–560. https://doi.org/10.1111/j.1151-2916.1982.tb10781.x

7 Kursumovic, A., Li, W. W., Cho, S. et al. (2020). Lead-free relaxor thin films with huge energy density and low loss for high temperature applications. *Nano Energy* 71. https://doi.org/10.1016/j.nanoen.2020.104536

8 Park, M. H., Kim, H. J., Kim, Y. J. et al. (2015). Toward a multifunctional monolithic device based on pyroelectricity and the electrocaloric effect of thin antiferroelectric $Hf_xZr_{1-x}O_2$ films. *Nano Energy* 12: 131–140. https://doi.org/10.1016/j.nanoen.2014.09.025

9 Park, M. H., Lee, Y. H., Kim, H. J. et al. (2015). Ferroelectricity and antiferroelectricity of doped thin HfO_2-based films. *Advanced Materials* 27: 1811–1831. https://doi.org/10.1002/adma.201404531

10 Yang, L., Kong, X., Li, F. et al. (2019). Perovskite lead-free dielectrics for energy storage applications. *Progress in Materials Science* 102: 72–108. https://doi.org/10.1016/j.pmatsci.2018.12.005

11 Luo, C., Zhang, Y., Shuai, W. et al. (2023). Ferroelectric $Hf_{0.5}Zr_{0.5}O_2$-gated synaptic transistors with large conductance dynamic range and multilevel states. *Science China Materials* 66: 2372–2382. https://doi.org/10.1007/s40843-022-2359-6

12 Park, M. H., Kim, H. J., Kim, Y. J. et al. (2016). Giant negative electrocaloric effects of $Hf_{0.5}Zr_{0.5}O_2$ thin films. *Advanced Materials* 28: 7956–7961. https://doi.org/10.1002/adma.201602787

13 Mueller, S., Muller, J., Schroeder, U. et al. (2013). Reliability characteristics of ferroelectric Si:HfO_2 thin films for memory applications. *IEEE Transactions on Device and Materials Reliability* 13: 93–97. https://doi.org/10.1109/TDMR.2012.2216269

14 Wang, D., Zhang, Y., Wang, J. et al. (2022). Enhanced ferroelectric polarization with less wake-up effect and improved endurance of $Hf_{0.5}Zr_{0.5}O_2$ thin films by implementing W electrode. *Journal of Materials Science & Technology* 104: 1–7. https://doi.org/10.1016/j.jmst.2021.07.016

15 Park, M. H., Schenk, T., Hoffmann, M. et al. (2017). Effect of acceptor doping on phase transitions of HfO_2 thin films for energy-related applications. *Nano Energy* 36: 381–389. https://doi.org/10.1016/j.nanoen.2017.04.052

16 Zhang, Y., Wang, D., Luo, C. et al. (2022). Controllable coercive field of ferroelectric HfO_2 films via UV-ozone surface modification. *IEEE Transactions on Electron Devices* 69: 3094–3099. https://doi.org/10.1109/TED.2022.3164856

17 Wang, D., Zhang, Y., Guo, Y. et al. (2025). Improved leakage and fatigue properties of W/$Hf_{0.5}Zr_{0.5}O_2$/W capacitor through the insertion of Pt metallic layer. *Materials Science in Semiconductor Processing* 188: 109165. https://doi.org/10.1016/j.mssp.2024.109165

18 Lou, X. J. (2009). Polarization fatigue in ferroelectric thin films and related materials. *Journal of Applied Physics* 105 (2): 024101. https://doi.org/10.1063/1.3056603

19 Pan, H., Lan, S., Xu, S. et al. (2021). Ultrahigh energy storage in superparaelectric relaxor ferroelectrics. *Science* 374: 100–104. https://doi.org/10.1126/science.abi7687

20 Shuai, W., Dai, J. Y., Xu, Z. et al. (2023). Superior and ultrafast energy storage performance of relaxorantiferroelectric HfO_2-based supercapacitors. *Energy Storage Materials* 62: 102931. https://doi.org/10.1016/j.ensm.2023.102931

21 Shao, M.-H., Zhao, R. T., Liu, H. et al. (2024). Challenges and recent advances in HfO_2-based ferroelectric films for non-volatile memory applications. *Chip* 3: 100101. https://doi.org/10.1016/j.chip.2024.100101

22 Park, M. H., Kim, H. J., Kim, Y. J. et al. (2014). Thin $Hf_xZr_{1-x}O_2$ films: A new lead-free system for electrostatic supercapacitors with large energy storage density and robust thermal stability. *Advanced Energy Materials* 4: 1400610. https://doi.org/10.1002/aenm.201400610

23 Hoffmann, M., Schroeder, U., Künneth, C. et al. (2015). Ferroelectric phase transitions in nanoscale HfO_2 films enable giant pyroelectric energy conversion and highly efficient supercapacitors. *Nano Energy* 18: 154–164. https://doi.org/10.1016/j.nanoen.2015.10.005

24 Kozodaev, M. G., Chernikova, A. G., Khakimov, R. R. et al. (2018). La-doped $Hf_{0.5}Zr_{0.5}O_2$ thin films for high-efficiency electrostatic supercapacitors. *Applied Physics Letters* 113: 123902. https://doi.org/10.1063/1.5045288

25 Kim, K. D., Lee, Y. H., Gwon, T. et al. (2017). Scale-up and optimization of HfO_2-ZrO_2 solid solution thin films for the electrostatic supercapacitors. *Nano Energy* 39: 390–399. https://doi.org/10.1016/j.nanoen.2017.07.017

26 Lomenzo, P. D., Chung, C.-C., Zhou, C. et al. (2017). Doped $Hf_{0.5}Zr_{0.5}O_2$ for high efficiency integrated supercapacitors. *Applied Physics Letters* 110: 232904. https://doi.org/10.1063/1.4985297

27 Ali, F., Liu, X., Zhou, D. et al. (2017). Silicon-doped hafnium oxide anti-ferroelectric thin films for energy storage. *Journal of Applied Physics* 122: 144105. https://doi.org/10.1063/1.4989908

28 Yang, B., Zhang, Y., Pan, H. et al. (2022). High-entropy enhanced capacitive energy storage. *Nature Materials* 21: 1074–1080. https://doi.org/10.1038/s41563-022-01274-6

29 Zhang, W. L., Mao, Y. H., Cui, L. et al. (2020). Impact of the radiation effect on the energy storage density and wake-up behaviors of antiferroelectric-like Al-doped $HfO(_2)$ thin films. *Physical Chemistry Chemical Physics* 22: 21893–21899. https://doi.org/10.1039/d0cp04196c

30 Sun, Z., Ma, C., Liu, M. et al. (2017). Ultrahigh energy storage performance of lead-free oxide multilayer film capacitors via interface engineering. *Advanced Materials* 29. https://doi.org/10.1002/adma.201604427

31 Silva, J. P. B., Silva, J. M. B., Sekhar, K. C. et al. (2020). Energy storage performance of ferroelectric ZrO_2 film capacitors: Effect of HfO_2:Al_2O_3 dielectric insert layer. *Journal of Materials Chemistry A* 8: 14171–14177. https://doi.org/10.1039/d0ta04984k

32 Das, D., Gaddam, V., and Jeon, S. (2021). Insertion of dielectric interlayer: A new approach to enhance energy storage in $Hf_xZr_{1-x}O_2$ capacitors. *IEEE Electron Device Letters* 42: 331–334. https://doi.org/10.1109/led.2021.3055140

33 He, Y., Zheng, G., Wu, X. et al. (2022). Superhigh energy storage density on-chip capacitors with ferroelectric Hf($_{0.5}$)Zr($_{0.5}$)O($_2$)/antiferroelectric Hf($_{0.25}$)Zr($_{0.75}$) O($_2$) bilayer nanofilms fabricated by plasma-enhanced atomic layer deposition. *Nanoscale Advances* 4: 4648–4657. https://doi.org/10.1039/d2na00427e

34 Wang, Y., Wang, Y., Zeng, H. et al. (2022). Ultra-high energy storage density of transparent capacitors based on linear dielectric ZrO_2 thin films with the thickness scaled up to hundreds nanometers. *Applied Physics Letters* 120: 023904. https://doi.org/10.1063/5.0076929

35 Yi, S.-H., Lin, H.-C., and Chen, M.-J. (2021). Ultra-high energy storage density and scale-up of antiferroelectric TiO_2/ZrO_2/TiO_2 stacks for supercapacitors. *Journal of Materials Chemistry A* 9: 9081–9091. https://doi.org/10.1039/d0ta11991a

36 Yang, K., Lee, E. B., Lee, D. H. et al. (2022). Energy conversion and storage using artificially induced antiferroelectricity in HfO_2/ZrO_2 nanolaminates. *Composites Part B: Engineering* 236: 109824. https://doi.org/10.1016/j.compositesb.2022.109824

37 Payne, A., Alex Hsain, H., Lee, Y. et al. (2022). Thermal stability of antiferroelectric-like Al:HfO_2 thin films with TiN or Pt electrodes. *Applied Physics Letters* 120: 232901. https://doi.org/10.1063/5.0083656

38 Ge, S., Liu, S., Feng, W. et al. (2023). Achieving high energy storage performance and thermal stability concurrently in the cost-cutting Al_2O_3/$Ba_{0.6}Sr_{0.4}Ti_{0.95}Ce_{0.05}O_3$/$ZrO_2$ composite films for energy storage applications. *Ceramics International* 49: 9155–9164. https://doi.org/10.1016/j.ceramint.2022.11.075

39 Yi, S.-H., Chan, Y.-C., Mo, C.-L. et al. (2022). Enhancement of energy storage for electrostatic supercapacitors through built-in electric field engineering. *Nano Energy* 99: 107342. https://doi.org/10.1016/j.nanoen.2022.107342

40 Cheema, S. S., Shanker, N., Hsu, S. L. et al. (2024). Giant energy storage and power density negative capacitance superlattices. *Nature* 629: 803–809. https://doi.org/10.1038/s41586-024-07365-5

41 Yu, Y., Zhang, Q., Xu, Z. et al. (2023). Structure-evolution-designed amorphous oxides for dielectric energy storage. *Nature Communications* 14: 3031. https://doi.org/10.1038/s41467-023-38847-1

42 Böscke, T. S., Müller, J., Bräuhaus, D. et al. (2011). Ferroelectricity in hafnium oxide thin films. *Applied Physics Letters* 99: 102903. https://doi.org/10.1063/1.3634052

43 Muller, J., Boscke, T. S., Schroder, U. et al. (2012). Ferroelectricity in simple binary ZrO_2 and HfO_2. *Nano Letters* 12: 4318–4323. https://doi.org/10.1021/nl302049k

44 Mueller, S., Mueller, J., Singh, A. et al. (2012). Incipient ferroelectricity in Al-doped HfO_2 thin films. *Advanced Functional Materials* 22: 2412–2417. https://doi.org/10.1002/adfm.201103119

45 Schroeder, U., Park, M. H., Mikolajick, T. et al. (2022). The fundamentals and applications of ferroelectric HfO_2. *Nature Reviews Materials* 7: 653–669. https://doi.org/10.1038/s41578-022-00431-2

46 Hoffmann, M., Fengler, F. P. G., Max, B. et al. (2019). Negative capacitance for electrostatic supercapacitors. *Advanced Energy Materials* 9: 1901154. https://doi.org/10.1002/aenm.201901154

14

HfO_2-based Ferroelectric Materials for Piezoelectric Applications

Guoliang Yuan, Guodong Zhang, and Xiao Xu*

School of Materials Science and Engineering, Nanjing University of Science and Technology, Nanjing, Jiangsu, China

*Corresponding author: yuanguoliang@njust.edu.cn

14.1 Background of $Hf_{1-x}M_xO_2$ Piezoelectric Films

The piezoelectric effect has catalyzed numerous significant studies and applications. The presence of piezoelectric properties in a crystal is determined by its structural symmetry. Among the 32 crystal classes, conventionally categorized into seven crystal systems, only 20 classes exhibit piezoelectric properties. Piezoelectricity is exclusively observed in crystal classes lacking a center of symmetry. The asymmetry responsible for piezoelectric effects originates from dipoles formed by atomic groups within the crystal lattice. Ferroelectric materials represent a specific subclass of pyroelectrics, which themselves are a subset of piezoelectrics. All these categories fall under the broader classification of dielectric materials (Figure 14.1).[1,2] Consequently, ferroelectrics inherently exhibit both piezoelectric and pyroelectric properties. Ferroelectric compounds must adopt symmetry-breaking space groups of the 10 polar point groups, including: hexagonal 6 mm ($C6v$) and 6 ($C6$), trigonal 3 m ($C3v$) and ($C3$), tetragonal 4 ($C4$) and 4 mm ($C4v$), orthorhombic $mm2$ ($C2v$), monoclinic m (Cs), and 2 ($C2$), as well as triclinic 1 ($C1$). In polycrystalline $Hf_{1-x}M_xO_2$ (A = Zr, La, Y, Si, Al, Ta, Sr, and others) films that comprise various orthorhombic, tetragonal phases, et al., the orthorhombic $mm2$ phase shows high ferroelectric polarization and strong piezoelectric response.

As illustrated in Figure 14.2, piezoelectric materials exhibit the ability to generate surface charges on electrodes when subjected to deformation, a phenomenon comes from the direct piezoelectric effect. For example, most piezoelectric solids expand along the applied tensile forces. This effect is triggered when either tensile

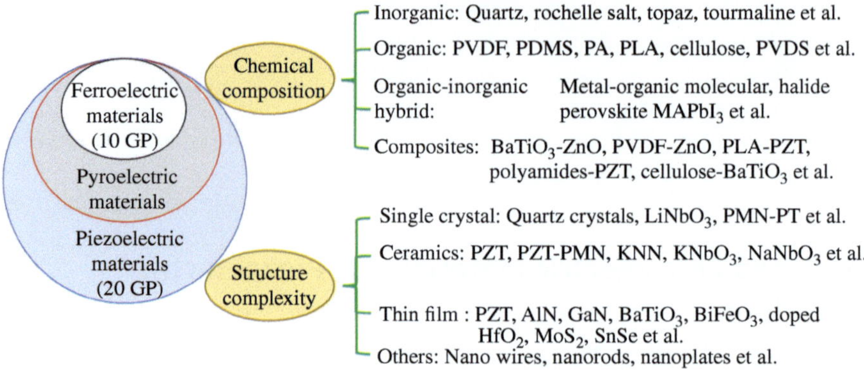

Figure 14.1 The relationship among ferroelectric, piezoelectric, and pyroelectric materials, as well as the classification of piezoelectric materials.

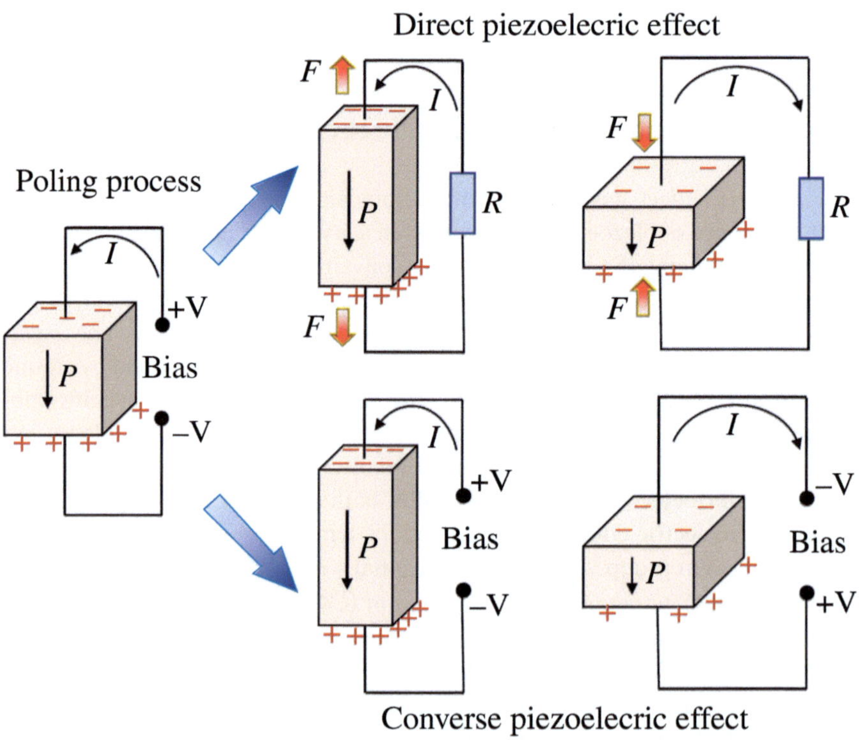

Figure 14.2 Sketches of direct piezoelectric effect and converse piezoelectric effect, where + and− mean positive and negative binding charge of ferroelectric polarization (P), and F and R mean force and resistance.

or compressive stress vibrations disrupt the established dipole equilibrium within the crystal structure. The orientation of polarization (P) within the solid is indicated by the negative and positive signs of binding polarization charges on the piezoelectric material. It should be noted that the generated screening charges on the covered electrodes are opposite in sign to the polarization. The resulting disturbance leads to an unbalanced charge distribution, contributing to the surface charge change that can be captured by electrodes or external circuits.[3] Through this process, mechanical energy converts to electrical energy. When high-frequency vibrations exist in a piezoelectric body, a corresponding high-frequency current can be induced according to the direct piezoelectric effect.

On the contrary, an electric field can induce mechanical strain in piezoelectric materials through the converse piezoelectric effect, which explains the conversion of electrical energy into mechanical energy. The piezoelectric body under a high-frequency electrical signal can generate high-frequency mechanical vibrations and result in ultrasonic signals. These phenomena illustrate the reciprocal relationship between the converse and direct piezoelectric effects. It is important to note that solid materials exhibiting the converse piezoelectric effect invariably possess the direct piezoelectric effect, and vice versa. Consequently, piezoelectric materials possess the capability to convert electrical energy into mechanical energy and vice versa. The collective term "piezoelectric effect" encompasses both the converse and direct piezoelectric effects.

The intrinsic equations and four piezoelectric coefficients $\left(d_{ij}, e_{ij}, g_{ij}, h_{ij}\right)$ can be expressed as follows:[1,4,5]

$$D = dT + \varepsilon E \tag{14.1}$$

$$S = sT + dE \tag{14.2}$$

$$d_{ij} = \left(\frac{\partial D_i}{\partial T_j}\right)^E = \left(\frac{\partial S_j}{\partial E_i}\right)^T \tag{14.3}$$

$$e_{ij} = \left(\frac{\partial D_i}{\partial S_j}\right)^E = -\left(\frac{\partial T_j}{\partial E_i}\right)^T \tag{14.4}$$

$$g_{ij} = -\left(\frac{\partial E_i}{\partial T_j}\right)^D = \left(\frac{\partial S_j}{\partial D_i}\right)^T \tag{14.5}$$

$$h_{ij} = -\left(\frac{\partial E_i}{\partial S_j}\right)^D = -\left(\frac{\partial T_j}{\partial D_i}\right)^S \tag{14.6}$$

In these equations, E is electric field, D denotes electric displacement, S indicates strain, T represents stress, s stands for mechanical flexibility, ε represents the dielectric constant, and d signifies the piezoelectric constant of the material. The two terms after the equal sign in Equation (14.3) represent, respectively, the direct and inverse piezoelectric effects.

The direct piezoelectric effect facilitates the conversion of external forces into electrical signals, which is crucial for various sensors and energy harvesters (Figure 14.3). When subjected to dynamic forces resulting from acceleration, deceleration, contact, bending, vibration, or ultrasonic waves, piezoelectric sensors, and energy harvesters generate new charges in accordance with the direct piezoelectric effect. Sensors utilize these generated charges to assess the status of target objects, while energy harvesters collect the free charges for electrical energy applications.[3,4]

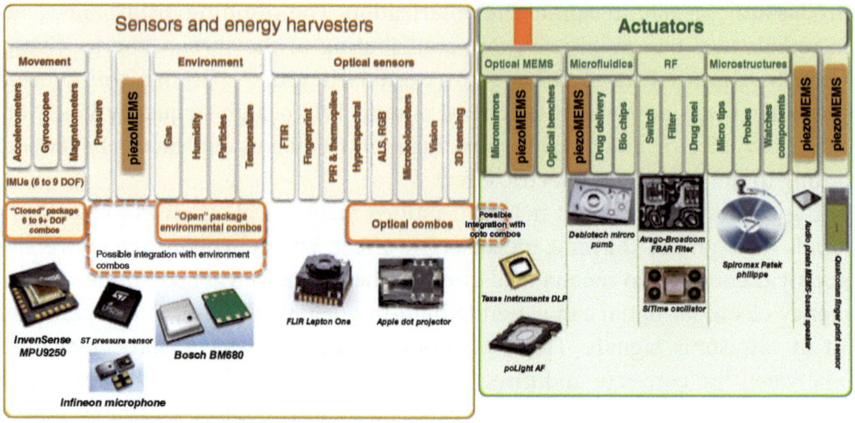

Figure 14.3 Application of piezoelectric materials in actuators, sensors, and energy harvesters. *Source*: [7] / IEEE / CC BY 4.0.

Conversely, the converse piezoelectric effect enables the transformation of electrical signals into mechanical deformation, which is essential for diverse piezoelectric actuators. A piezoelectric actuator functions as a transducer, converting electrical signals into precisely controlled physical displacements or strokes. This conversion is achieved through the converse piezoelectric effect, which relies on electromechanical coupling rather than electromagnetic induction. The precise movement control offered by these actuators is primarily utilized in high-precision applications such as machine tools, lithography equipment, mirrors, lenses, microscopes, telescopes, and other instruments requiring fine adjustments.

In certain complex applications, both direct and converse piezoelectric effects are employed simultaneously. This dual utilization is evident in medical ultrasound, ultrasonic flaw detectors, and sonar systems.[6,7] Sonar technology employs sound waves to map or locate objects in the surrounding environment. Initially, a piezoelectric transducer, subjected to an AC voltage bias, vibrates based on the converse piezoelectric effect, emitting a cluster of sound waves towards the target object. While some waves are absorbed by the object, others are reflected to the transducer, which then generates an electrical signal through the direct piezoelectric effect. These electrical signals are subsequently analyzed to produce a visualized image on computer. In contrast, surface acoustic wave (SAW) devices, such as resonators and filters, find widespread application in radio frequency (RF) communication.[8] A typical SAW device incorporates two interdigital transducers at each end (input and output). When RF AC voltage signals are applied to the input transducer, the piezoelectric layer undergoes continuous compression and expansion on the chip, corresponding to the converse piezoelectric effect. The mechanical vibrations can be converted into electrical signals with specific RFs by the output transducer, in accordance with the direct piezoelectric effect.

Numerous renowned piezoelectric materials have found extensive applications across a variety of fields. These applications encompass a wide range of piezoelectric microelectromechanical systems (MEMS), various sensors for monitoring motion and environmental conditions, resonators, piezoelectric motors, piezoelectric energy harvesters, piezoelectric speakers, SAW devices for radio communication, ultrasonic medical devices, and so on.[6] Many well-known piezoelectric materials, including $PbZr_{1-x}Ti_xO_3$ (PZT), $LiNbO_3$, $LiTaO_3$, ZnO, AlN, GaN, and polyvinylidene fluoride (PVDF), have been extensively employed in industrial applications due to their outstanding and reliable piezoelectric properties, straightforward preparation methods, and high cost-effectiveness (Table 14.1). However, there are inherent trade-offs among material selection, fabrication processes, and performance. Although the current market share of large-scale piezoelectric ceramics, single crystals, and micrometer-thick films remains significant, challenges such as the need for

Table 14.1 Piezoelectric coefficients of $Hf_{1-x}M_xO_2$ piezoelectric films and some well-known thin films.

Piezoelectric film	Growth method	d_{33} (pC/N)	d_{33}^* (pm/V)	d_{11}^* (pm/V)	d_{31}^* (pm/V)	g_{33} (m V·m/N)	References
$Hf_{0.5}Zr_{0.5}O_2$	PLD	37.8				120	[19]
$Hf_{0.5}Zr_{0.5}O_2$	ALD		−2 to 3				[23]
$Hf_{0.5}Zr_{0.5}O_2$	ALD		−23.9		−3.1		[24]
$Hf_{0.5}Zr_{0.5}O_2$	ALD		−0.3				[25]
$Hf_{0.962}Si_{0.038}O_2$	ALD				−11.5		[15]
$Hf_{0.973}Si_{0.027}O_2$	ALD		17.8				[26]
$Hf_{0.96}Si_{0.04}O_2$	ALD		20				[27]
$Hf_{0.93}Y_{0.07}O_2$	PLD		2.5				[28]
$Hf_{0.95}Y_{0.05}O_2$	Sputtering		5				[29]
$Hf_{0.9}Ce_{0.1}O_2$	Sputtering		7.2				[30]
AlN	MBE		5.1			30	[31]
AlN	Sputtering		4.25		−2.7	30	[32]
AlN	Sputtering	6					[33]
GaN	MBE		3.1			70	[31]
GaN	MBE		2.13				[34]
GaN	CVD		2.8		−1.4		[35]
ZnO	Solgel		44.3				[36]
PbO	Exfoliation		29.6	3.9			[37]
$LiNbO_3$	Sputtering		12.5				[38]

(Continued)

Table 14.1 (Continued)

Piezoelectric film	Growth method	d_{33} (pC/N)	d_{33}^* (pm/V)	d_{11}^* (pm/V)	d_{31}^* (pm/V)	g_{33} (m V·m/N)	References
$LiTaO_3$	Solgel		24			50	[39]
$BaTiO_3$	Sputtering	12				13	[40]
$BaTiO_3$	Sputtering				−1.8		[41]
PZT	PLD	200	180			20–40	[42]
PVDF	Fusion	7.3				280	[43]
MoS_2	CVD		1.4				[44]
WSe_2	CVD			3.3			[45]
SnS_2	Exfoliation		~2.2	~1			[46]
In_2Se_3	Exfoliation		0.3				[47]

reduced power consumption and miniaturization continue to arise in response to the growing demand for nanoelectromechanical systems (NEMS).

The ultrathin $Hf_{1-x}M_xO_2$ ferroelectric films, such as $Hf_{0.5}Zr_{0.5}O_2$ (HZO) have recently garnered significant attention due to their elevated piezoelectric coefficient, high Curie temperature, and compatibility with the advanced CMOS technology in integrated circuits (ICs). Atomic layer deposition (ALD) technology is capable of producing large-sized, high-quality $Hf_{1-x}M_xO_2$ films at temperatures below 500°C. Through ongoing research into modified ALD processes, various nanomaterials such as thin films, monolayers, core-shell structures, nanoparticles, and single atoms have been developed (Figure 14.4).[9] ALD has emerged as a highly effective technique for thin film fabrication, owing to its capability to deposit materials with angstrom-level precision in thickness control and uniformity across wafer-scale dimensions. Nowadays, ALD is preferred over other thin-film fabrication methods when nanoscale material design is required. Ultrathin $Hf_{1-x}M_xO_2$ films commonly show multiple crystal phases and metastable ferroelectric phases.[10,11] The unique ferroelectric properties arise from the metastable orthorhombic phase with non-centrosymmetric *mm*2 symmetry.[12,13] The amorphous $Hf_{1-x}M_xO_2$ films are deposited by ALD at 260–300°C. Subsequently, these films undergo annealing in a rapid thermal anneal (RTA) furnace at temperatures between 380 and 550°C to achieve a polycrystalline structure. Utilizing RTA in conjunction with either tungsten (W) or titanium nitride (TiN) electrodes can improve crystallinity and increase the ratio of the ferroelectric orthorhombic phase. For most $Hf_{1-x}M_xO_2$ ultrathin films, there is a direct correlation between ferroelectric polarization and the direct piezoelectric coefficient d_{33}; specifically, an increase in ferroelectric polarization typically results in a larger d_{33} value.

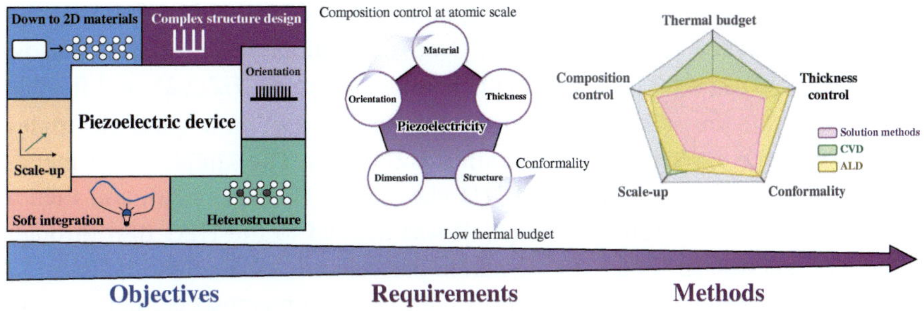

Figure 14.4 Interrelationship among objectives, requirements, and methods in the piezoelectric device field. *Source*: [9] / with permission of Elsevier.

Although ALD is the most prevalent method for growing $Hf_{1-x}M_xO_2$ polycrystalline films with mixed phases, PLD is frequently employed for $Hf_{1-x}M_xO_2$ film growth due to at least two factors. Firstly, single-crystalline $Hf_{1-x}M_xO_2$ ferroelectric films can be prepared on specific single-crystal substrates using PLD at 700–800°C. Secondly, the orthorhombic phase is inherently stable in $Hf_{1-x}M_xO_2$ films grown by PLD, and its stability is not significantly dependent on the clamping strain from either TiN or W electrodes. Consequently, gold (Au) and platinum (Pt) films with high conductance can be utilized as electrodes for $Hf_{1-x}M_xO_2$ ferroelectric films in many cases.

14.2 Direct Piezoelectric Effect of $Hf_{1-x}M_xO_2$ Films

The direct piezoelectric coefficient d_{33} (pC/N) is typically measured using commercial d_{33} meters, while the converse piezoelectric coefficient d_{33}^* (pm/V) is commonly assessed with Doppler instruments for both piezoelectric ceramics and most micrometer-thick films. The rapid advancement of industrial technology has necessitated the miniaturization of sensors and actuators to the nanometer scale to ensure ultrahigh resolution of integrated NEMS devices.[14] The ultrathin $Hf_{1-x}M_xO_2$ films offer an outstanding solution for the fabrication of high-quality NEMS utilizing current Si-based CMOS technology.[15,16] Consequently, precise characterization of the piezoelectric coefficient of nanometer-scale thick $Hf_{1-x}M_xO_2$ films is crucial for the design of high-quality NEMS devices.[17,18]

Accurately measuring the d_{33} value of large-size, nanometer-thick piezoelectric films using commercial d_{33} meters has been historically challenging. This difficulty arises from two primary factors. Firstly, the induction of macroscopic piezoelectricity requires polarization of the large-sized ferroelectric film with a high electric field. However, ferroelectric films with millimeter-sized electrodes and nanometer-scale thickness typically exhibit substantial leakage current at high electric fields, often resulting in electrical breakdown during the ferroelectric poling process. Secondly, high inhomogeneous strains can easily damage the large-sized and ultrathin piezoelectric films. Therefore, homogeneous strain within the nanometer-level amplitude is necessary to induce piezoelectric current in nanometer-thick ferroelectric films without causing physical damage.

Recently, three types of HZO films with 30 nm thickness were fabricated using PLD at 10 Pa oxygen pressure and 720°C on three different substrates, that is, hard $SrRuO_3/SrTiO_3$ (HZO-1), hard $SrRuO_3$/Mica (0.2 mm) (HZO-2), and flexible $SrRuO_3$/Mica (10 μm) (HZO-3).[19] Subsequently, the 1×1 mm^2 gold electrodes were grown at the surface of these ultrathin HZO films. The samples were meticulously positioned between two rigid ITO glass plates (Figure 14.5a), and then a time (t)-dependent alternating force ($F(t)$) was applied to the HZO film through the two ITO glasses. In theory, the dynamic piezoelectric current can be induced by the alternating force in the HZO films with nonzero macroscopic polarization (Figure 14.5b). Because the original HZO-1 film has a small macroscopic polarization,

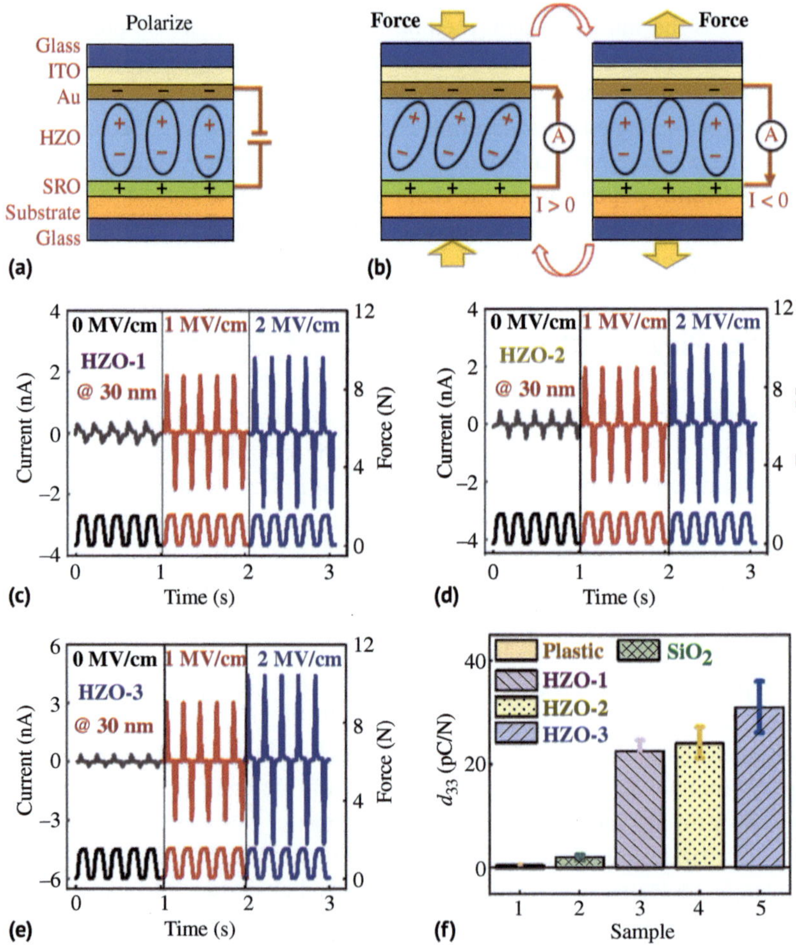

Figure 14.5 (a, b) Schematic illustrations depicting the polarization of HZO films and the generation of piezoelectric current induced by applied dynamic forces. Time-dependent current and force curves for the 30 nm thick (c) HZO-1, (d) HZO-2, and (e) HZO-3 films. (f) Comparative analysis of piezoelectric d_{33} coefficients for different samples. *Source*: [19] / with permission of Elsevier.

it shows a minimal short-circuit piezoelectric current (I_{sc}). After the macroscopic polarization was induced by a DC electric field (E_p), significant piezoelectric currents can be generated by the same F(t) (Figure 14.5c). The piezoelectric current gradually increased with E_p increasing. For instance, the HZO films polarized at 2 MV/cm exhibited a greater piezoelectric current compared to those polarized at 1 MV/cm. Furthermore, the principles are equally applicable to the HZO-2 film (Figure 14.5d) and the HZO-3 film (Figure 14.5e), too. This observation indicates that the alternating force induces a piezoelectric current rather than the triboelectric current in HZO films, as the triboelectric current is theoretically independent of ferroelectric polarization.[18]

In fact, the piezoelectric coefficient d_{33} can be determined using the following equation:[19,20]

$$d_{33} = \Delta Q(\mathrm{t})/\Delta F(\mathrm{t}) = I_{sc}\,\Delta \mathrm{t}/\Delta F(\mathrm{t}) \tag{14.7}$$

Here, Q(t) represents the piezoelectric charge. The HZO-1 (30 nm), HZO-2 (30 nm), and HZO-3 (30 nm) exhibit the d_{33} values of 23 pC/N, 24 pC/N, and 31 pC/N, respectively (Figure 14.5f). It is noteworthy that the flexible HZO-3 shows the highest d_{33} value among all HZO films because its ferroelectric domains exhibit greater mobility under dynamic force. For reference, the d_{33} values of 0.2 mm quartz and thin plastic wrap were characterized by the above method, yielding d_{33} values of 2 pC/N and 0, respectively, which are consistent with measurements obtained using a commercial d_{33} meter.

Following polarization with $E_p = 2$ MV/cm, the maximum d_{33} values were 20 pC/N, 31 pC/N, and 38 pC/N for HZO-3 (200 nm), HZO-3 (30 nm), and HZO-3 (10 nm) films, respectively (Table 14.1). Notably, the HZO-3 (10 nm) film demonstrates the highest d_{33} coefficient among all HZO films tested. This d_{33} value increases as the film thickness of HZO-3 reduces from 200 to 10 nm, mainly because the proportion of the ferroelectric phase rises. Furthermore, the d_{33} values of these three HZO films decreased by less than 10% after 21 days at room temperature. Therefore, these ultrathin piezoelectric films hold significant potential for applications in many innovative NEMS/MEMS devices.[14,21]

Figure 14.6a illustrates the flexible HZO (60 nm) film integrated with interdigitated electrodes (HZO-4) on a flexible mica substrate with 10 μm in thickness. Initially, the in-plane macroscopic polarization was achieved by applying a high E_p between digitated electrodes on the HZO film surface. Open-circuit voltage (V_{oc}) and I_{sc} were induced by the dynamic bending strain on the flexible HZO-4 film. Both V_{oc} and I_{sc} reached maximum values during periodic bending and unbending of the HZO-4 film. Different dynamic strains were achieved by varying the bending radius of HZO-4 sample. When the bending strain changed from 0.09% to 0.156% at a frequency of 1 Hz, the V_{oc} improved from 0.6 to 0.9 V (Figure 14.6b) and the I_{sc} rose from 3.6 to 6.5 nA (Figure 14.6c). The dependence of V_{oc} and I_{sc} on bending frequency was characterized at a constant strain of 0.102%. V_{oc} increased from 0.7 to 1.1 V (Figure 14.6d) and I_{sc} increased from 4.7 to 7.6 nA (Figure 14.6e) as

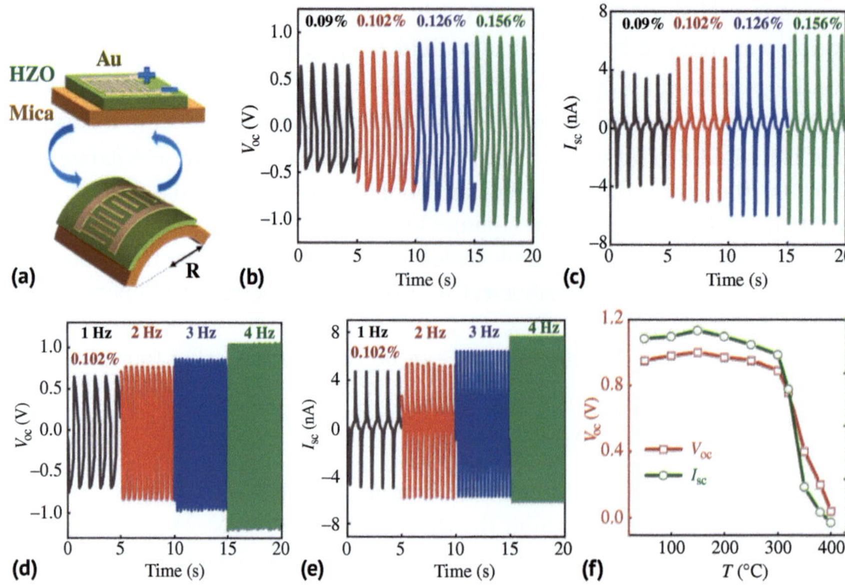

Figure 14.6 (a) Schematic representation of the flat and bent states of the HZO-4 film with the thickness of 60 nm and inter-digitated electrodes. (b) Open-circuit voltage V_{oc} and (c) I_{sc} generated by various strains during bending-flattening cycles. (d) V_{oc} and (e) I_{sc} produced by bending-flattening cycles at frequencies of 1–4 Hz. (f) V_{oc} and I_{sc} of the flexible HZO-4 film after annealing at 25–400°C. *Source*: [19] / with permission of Elsevier.

the frequency increased from 1 to 4 Hz. These results demonstrate that HZO films serve as outstanding piezoelectric functional layers for flexible, lightweight, and lead-free piezoelectric devices.[22] To evaluate the high-temperature stability of piezoelectric properties, the relationship between room-temperature V_{oc} and I_{sc} and post-annealing temperature is presented in Figure 14.6f. Both V_{oc} and I_{sc} exhibit exceptional thermal stability, with only a ~10% decrease after the HZO-4 sample was post-annealed at 300°C. This finding indicates that HZO films show great promise for application in high-temperature piezoelectric devices in high-temperature environments.

Moreover, the piezoelectric voltage coefficient g_{33} is of paramount importance for all piezoelectric sensors, since it seriously influences the signal acceptance capabilities. A higher sensitivity of piezoelectric sensors requires a larger g_{33} value. The HZO film achieves a g_{33} value as high as 20 mV·m/N because of a moderate d_{33} and a low relative permittivity. As presented in Table 14.1, the g_{33} value of the HZO film is higher than most traditional piezoelectric films, including AlN, ZnO, PZT, $BaTiO_3$, and $LiTaO_3$; however, it is slightly lower than that of the polymer PVDF. The comparison highlights the merits of $Hf_{1-x}M_xO_2$ ultrathin films for high-performance piezoelectric sensors.

14.3 Converse Piezoelectric Effect of $Hf_{1-x}M_xO_2$ Films

Beyond the positive piezoelectric effect, the converse piezoelectric effect is also critically important for piezoelectric material. The thickness change and corresponding converse piezoelectric coefficient d_{33}^* (pm/V) of ceramic and μm-scale thick films can be easily measured by Doppler instruments, laser interferometers, and other techniques. However, it is challenging to fully polarize large-size ferroelectric films with nm-scale thickness and mm-scale electrodes as mentioned above. Moreover, the thickness change resulting from the converse piezoelectric effect is typically around 10 pm for a 10 nm-thick piezoelectric film, making it difficult for Doppler and laser interferometers to accurately measure this minute thickness change. In summary, there is currently no mature method to precisely determine the d_{33}^* coefficient in nm-thick piezoelectric films.

Fortunately, piezoelectric force microscopy (PFM) is a well-established technique for characterizing ferroelectric domains and estimating d_{33}^*. The nanoscale conductive tip, when subjected to a high DC bias, can effectively polarize a small region of an ultrathin ferroelectric film. Additionally, the tip is capable of applying a low AC voltage bias to the nanoscale contact region of the piezoelectric film. Subsequently, a laser can be utilized to measure both the vibrating amplitude of the tip head and that of this contact region. This amplitude may considerably surpass the thickness variation induced by the piezoelectric effect. Consequently, the proportion of vibrating amplitude to AC voltage bias – amplitude/bias (pm/V) – serves as a valid parameter for approximately evaluating d_{33}^* in the nanoscale region of ferroelectric films with thicknesses on the order of several nanometers. When different films exhibit identical thicknesses and substrates, possess comparable mechanical properties, and are subjected to the same measurement conditions, the amplitude/bias is directly proportional to the d_{33}^* value in piezoelectric ultrathin films. However, the vibrating amplitude measured by PFM is much higher than the pure thickness variation of the piezoelectric film due to the converse piezoelectric effect. The amplitude/bias strongly depends on the piezoelectric strain, mechanical properties, film and substrate thickness, AC bias frequency, PFM tip mass and radius, and other factors. Consequently, it remains a challenge to accurately estimate the d_{33}^* of ultrathin piezoelectric films using the PFM technique. Scientists utilize PFM to measure the target piezoelectric film and subsequently calibrate its d_{33}^* value against that of standard film with identified d_{33}^* values, such as PZT and quartz. As a result, PFM has become one of the most prevalent and convenient methods to observe ferroelectric domains and estimate d_{33}^* value of the nm-scale region of ultrathin ferroelectric films.

Epitaxial HZO (6.5nm)/$La_{0.7}Sr_{0.3}MnO_3$ (LSMO, 25nm) films were prepared on (001) $SrTiO_3$ (STO) substrate, (110) STO substrate, and (111) STO substrate by

PLD method at 700–800°C, respectively.[48] The remnant polarization (P_r) values were 18 μC/cm² for Pt/HZO/LSMO/STO (001) capacitors, 32 μC/cm² for Pt/HZO/LSMO/STO (110) capacitors, and 12μC/cm² for Pt/HZO/LSMO/STO (111) capacitors, respectively.[49] In this case, the ferroelectric polarization and piezoelectricity of these epitaxial HZO films grown by PLD do not significantly depend on the top electrode and the corresponding clamping strain. Figure 14.7a–c present archetypal PFM phase images with a 180° phase contrast of two inner regions poled with −8 V bias (P_{up} orienting to the top electrode) and +8 V bias (P_{down} orienting the bottom electrode), respectively. The outer region with the as-grown state shows P_{down} in all cases. The corresponding amplitude/bias (pm/V) images (Figure 14.7d–f) exhibit small contrast between the P_{down} and P_{up} regions.[49,50]

To better evaluate the piezoelectric response, a 25 nm-thick LSMO film and a 9 nm-thick $Hf_{1-x}M_xO_2$ film were grown on (001) STO substrate by PLD at temperatures ranging from 700 to 800°C under an oxygen pressure of 13.3 Pa. The piezoelectric properties of these ferroelectric films were measured by PFM using a conductive tip with 1 V AC voltage.[51] The applied AC voltage induces vibration in these films through the converse piezoelectric effect, and the vibration amplitude is dependent on factors such as the tip bias and frequency. Figure 14.8a,b present the amplitude versus frequency curves for La-doped and Zr-doped HfO_2 films, respectively. The $Hf_{0.95}La_{0.05}O_2$ and $Hf_{0.98}La_{0.02}O_2$ films exhibit similar resonance peak amplitudes, which are larger than that of the HfO_2 film without piezoelectricity (Figure 14.8a). In the case of $Hf_{1-x}Zr_xO_2$ films, the resonance peak amplitude rises with increasing Zr content (Figure 14.8b). Figure 14.8c,d show the evolution of the amplitude of

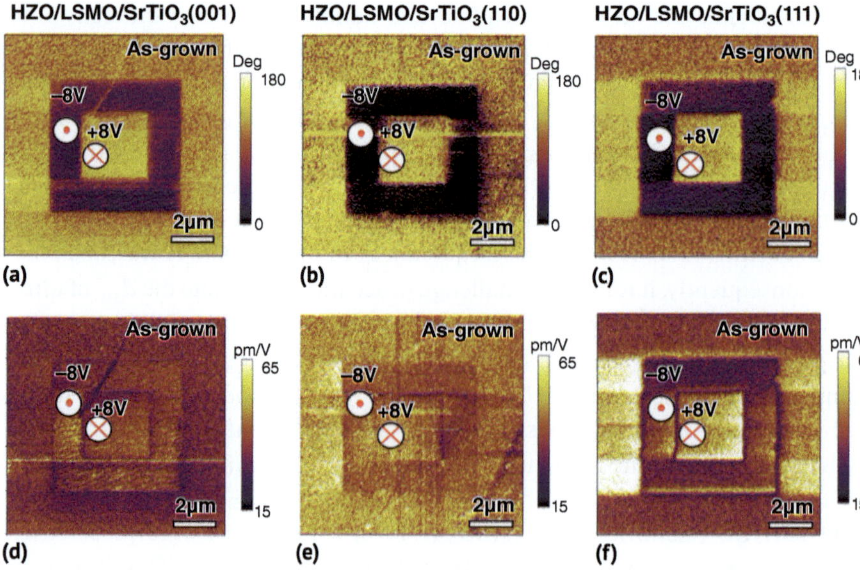

Figure 14.7 PFM out-of-plane (a–c) phase image and (d–f) amplitude/bias images of three single-crystal HZO films on LSMO/$SrTiO_3$(001), LSMO/$SrTiO_3$(110), and LSMO/$SrTiO_3$(111), respectively. *Source*: [48] / with permission of The Royal Society of Chemistry.

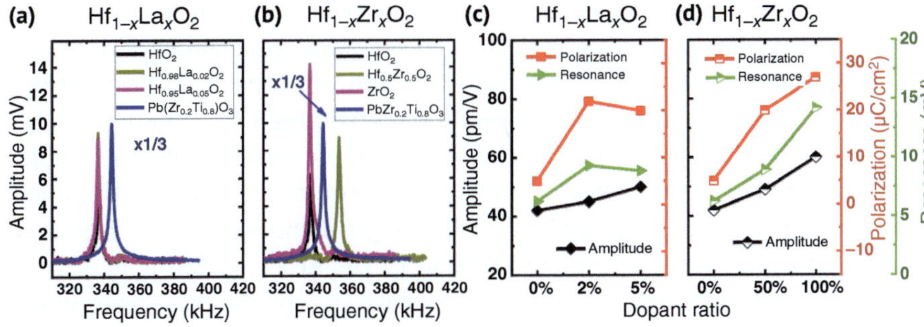

Figure 14.8 PFM amplitude versus frequency of AC signal of conductive tip upon the as-grown region of (a) $Hf_{1-x}La_xO_2$ and (b) $Hf_{1-x}Zr_xO_2$ films. Amplitude/bias (black), remnant polarization (red), and resonance amplitude (green) as a function of the doping concentration of (c) $Hf_{1-x}La_xO_2$ and (d) $Hf_{1-x}Zr_xO_2$ films. *Source*: [51] / with permission of American Chemical Society.

the resonance peak and the average piezoelectric amplitude as a function of La and Zr doping, respectively.[52] In both instances, the amplitude exhibits an increase with doping, with the exception of the $Hf_{0.95}La_{0.05}O_2$ film. This observation is consistent with the slightly lower polarization exhibited by the $Hf_{0.95}La_{0.05}O_2$ film,[53] which agrees with its lower piezoelectric amplitude and, consequently, the lower proportion of the orthorhombic phase.

As presented in Table 14.1, different d_{33}^* values have been reported for various $Hf_{1-x}M_xO_2$ thin films.[15,27] The limitations of PFM technology and the electrode effect are two factors that hinder the accurate measurement of d_{33}^* in $Hf_{1-x}M_xO_2$ films at present. At first, the significant variation in the reported d_{33}^* values by different groups can be attributed to the difficulty in accurately estimating d_{33}^* from the amplitude/bias value obtained by PFM. Furthermore, $Hf_{1-x}M_xO_2$ films grown by ALD are typically a mixture of phases, and the ratio of the ferroelectric orthorhombic phase is strongly influenced by the strain imposed by the top TiN or W electrodes. However, most PFM images and curves reported in the literature were measured with the PFM tip directly on the $Hf_{1-x}M_xO_2$ film without a top electrode, resulting in a significantly lower ratio of the ferroelectric phase compared to the electrode/$Hf_{1-x}M_xO_2$/electrode structure. It is anticipated that with advancements in instrumentation and the development of high-quality, ultrathin HZO films with large-size electrodes fully polarized by high electric fields, precise measurement of d_{33}^* will become feasible soon.

14.4 Positive and Negative Longitudinal Piezoelectric Coefficients

As a novel member of the ferroelectric family discovered in the twenty-first century, $Hf_{1-x}M_xO_2$ ferroelectric films exhibit intriguing physical phenomena, such as the negative longitudinal piezoelectric effect. The decrease/increase in polarization (P)

with the application of tensile strain (σ) along the polar direction represents the negative/positive longitudinal piezoelectric effects. Most piezoelectric materials demonstrate positive longitudinal piezoelectric effect, resulting in positive piezoelectric coefficients d_{33} and d_{33}^*.[54] Conversely, there have been few materials exhibiting negative longitudinal piezoelectricity to date. Ferroelectric PVDF is a notable exception. Its polarization decreases under tensile strain in the polar direction, leading to negative values for d_{33} and d_{33}^*.[55] The origin of the negative d_{33}^* has been ascribed to the intermixed crystalline phase in PVDF. Recently, You et al. experimentally demonstrated the occurrence of negative d_{33}^* in the two-dimensional material $CuInP_2S_6$. This phenomenon was attributed to its van der Waals layered structure and the reduced dimensionality of its lattice.[56]

Dependence of PFM phase/amplitude on low-frequency tip voltage can be characterized when the low-frequency tip voltage is superimposed on the high-frequency AC tip voltage during PFM measurement. These loops are widely employed to investigate the piezoelectric properties of ultrathin ferroelectric films. For the 200 nm-thick PZT polycrystalline films, the observed clockwise rotation of the PFM phase versus tip bias loop indicates a positive d_{33}^* value (Figure 14.9a).[57] The piezoelectric amplitude is lowest near the coercive voltage for ferroelectric materials. In contrast, the PVDF film demonstrates a negative d_{33}^* value, as evidenced by the

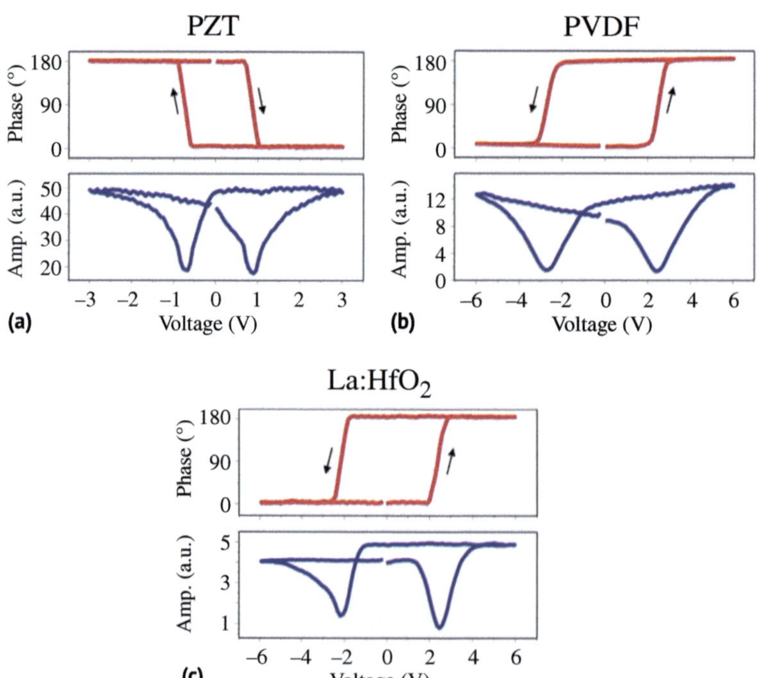

Figure 14.9 PFM phase (top) and amplitude (bottom) versus tip voltage loops of the (a) IrO_2/PZT/Pt capacitor, (b) PVDF film, and (c) Pt/TiN/$Hf_{0.88}La_{0.12}O_2$/TiN capacitor. *Source*: [57] / Springer Nature / CC BY 4.0.

anticlockwise rotation observed within the PFM phase loops (Figure 14.9b).[55] The 20 nm-thick 12 at% La:HfO_2 (i.e., $Hf_{0.88}La_{0.12}O_2$) film had been deposited on TiN bottom electrodes by ALD, and then it was capped with a TiN top electrode. The TiN/$Hf_{0.88}La_{0.12}O_2$/TiN stack was subsequently annealed at 800°C in a N_2 atmosphere for 20 seconds. Under identical conditions, as depicted in Figure 14.9c,[57] the PFM phase loop of TiN/$Hf_{0.88}La_{0.12}O_2$/TiN capacitors exhibit an anticlockwise rotation. This behavior unequivocally indicates the negative d_{33}^* in the $Hf_{0.88}La_{0.12}O_2$ film. Given that the polycrystalline $Hf_{0.88}La_{0.12}O_2$ film exhibits a negative longitudinal piezoelectric effect, it bears resemblance to the behavior of PVDF organic compounds, which contrasts with traditional ferroelectrics such as PZT.

Besides, the piezoelectric properties of HZO films exhibit sample-history dependence, as illustrated in Figure 14.10.[23] It is crucial to identify the intrinsic and extrinsic factors governing the piezoelectric response, including its sign.[58] Simulations have demonstrated that the negative d_{33}^* mainly originates from the active oxygen atoms and their special chemical coordination. The sign of the piezoelectric d_{33}^* in $Hf_{1-x}M_xO_2$ films can be modulated by controlling active oxygens that

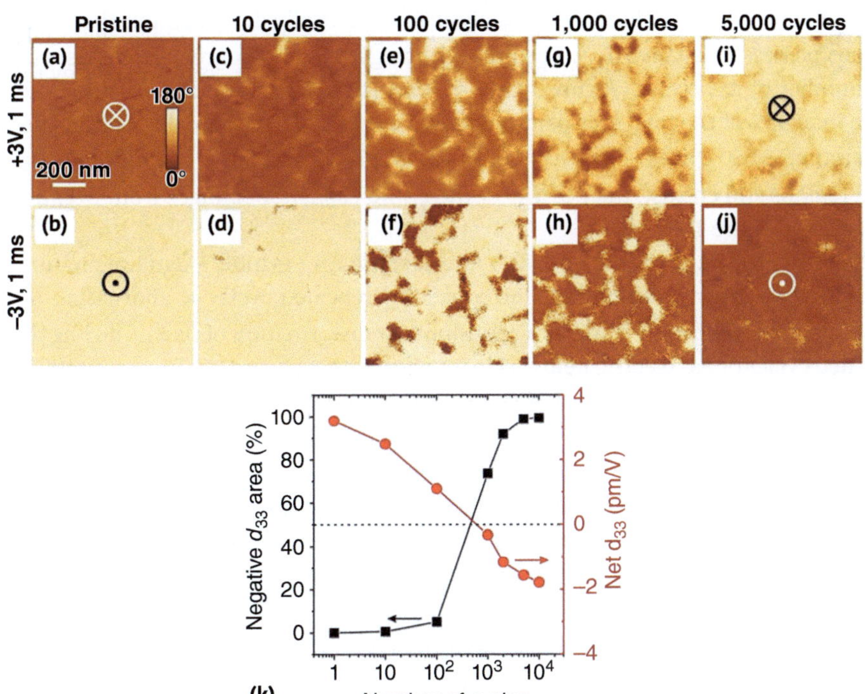

Figure 14.10 PFM phase images of the W/HZO/W capacitor polarized by positive (top row) and negative (bottom row) biases. The images were collected at the same location for (a, b) 1 cycle, (c, d) 10 cycles, (e, f) 100 cycles, (g, h) 1000 cycles, and (i, j) 5,000 cycles, utilizing a pulse amplitude of ±3 V with a duration of 1 ms per cycle. Panel (k) illustrates the percentage of the area exhibiting negative d_{33}^* values alongside the average d_{33}^* value across the entire scanned area as a function of AC cycles. *Source*: [23] / with permission of Springer Nature.

depend on the sample history. Here W (30 nm)/HZO (10 nm)/W (25 nm) capacitors were fabricated on p^+-type Si substrates using ALD for HZO and magnetron sputtering for W films. Many PFM phase images were acquired from the same location during successive cycles of the AC field applied to the W/HZO/W capacitor. For the pristine capacitor, the PFM images exhibit uniform contrast following poling (Figure 14.10a,b), indicating that the polarization is fully switchable. This observation aligns with the large P_r observed in the macroscopic polarization versus electric field loop of the pristine state. The HZO film poled by 3 V and −3 V biases show the downward (dark contrast in Figure 14.10a) and upward (bright contrast in Figure 14.10b) polarization states, respectively.[23] These results indicate a positive d_{33}^* in the pristine HZO.

After a series of ferroelectric polarization switching events triggered by multiple AC field cycles, these switching events lead to the formation of regions with negative d_{33} value. These regions display out-of-phase PFM signals (bright contrast) after applying 3 V bias and in-phase signals (dark contrast) after applying −3 V bias (Figure 14.10c,d).[23] Although this inverted PFM response may seem indicative of polarization switching contrary to the applied field – referred to as "anomalous" switching – PFM phase signal analysis unequivocally demonstrates that these regions possess a negative d_{33}^*. The area with negative d_{33}^* progressively rises with the number of AC cycles (Figure 14.10e–j). Compared to its original pristine state, more than 95% of the total area demonstrates a reversed PFM phase distinction after undergoing 5,000 cycles (Figure 14.10i,j). Although both negative and positive d_{33}^* local regions still coexist, the net d_{33}^* of the entire capacitor changes from negative to positive (Figure 14.10k).

The piezoelectric properties in HZO ferroelectric film exhibit a dynamic nature, susceptible to intrinsic alterations introduced by repeated cycles of alternating AC electrical fields. The HZO film has a continuous transformation of its d_{33}^* during the repeated cycles of AC electric fields. This transformation progresses from an initially positive d_{33}^* state to a negative d_{33}^* state, accompanied by a monotonic increase in polarization. The manipulation of d_{33}^* extends beyond mere sign inversion across the entire capacitor volume. The net longitudinal piezoelectric response can be nullified through appropriate AC training, while maintaining full switchability of polarization. The theoretical calculations of density functional theory (DFT) prove that the sign change of d_{33}^* is attributed to a phase transition between two distinct polar orthorhombic phases.

Epitaxial strain has been demonstrated to modulate the sign of d_{33}^* in HZO, as illustrated in Figure 14.11. Here, the epitaxial $La_{0.67}Sr_{0.33}MnO_3$ (LSMO, 20 nm) and HZO (10 nm) films were sequentially grown on (110)-orientated STO, $LaAlO_3$ (LAO), and $(LaAlO_3)_{0.33}$-$(Sr_2AlTaO_6)_{0.67}$ (LSAT) single crystal substrates by the PLD method at 750°C, respectively.[59] Figure 14.11a shows the XRD patterns of HZO/LSMO films on the three substrates. The diffraction angle of the (110) crystal planes in the LSMO film exhibits an increase as the substrate exhibits a larger lattice constant, indicating a compressive strain in the LSMO lattice in the direction perpendicular to the plane. The HZO films consistently exhibit a primary peak at

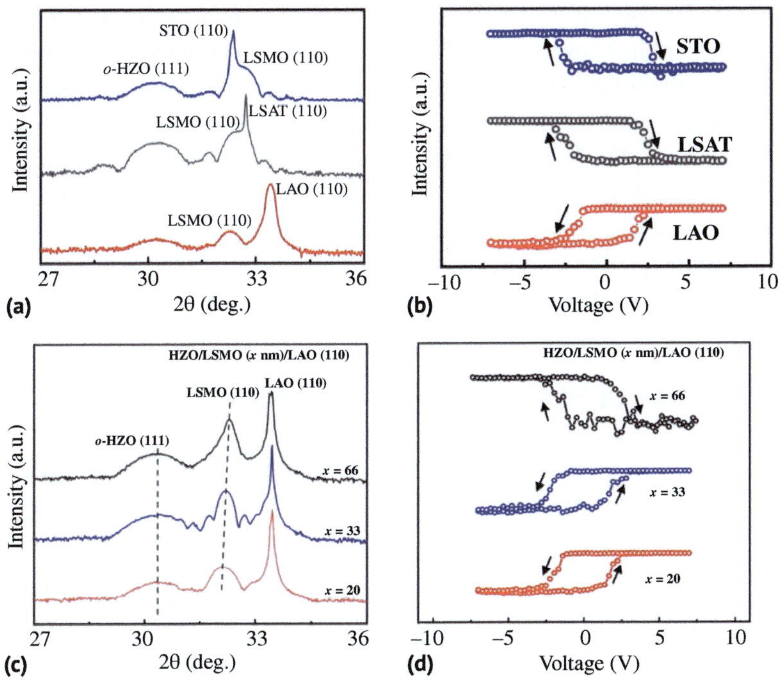

Figure 14.11 (a) XRD patterns of HZO/LSMO bilayer films grown on (110)-oriented STO, LAO, and LSAT substrates and (b) PFM phase versus tip voltage loops of these HZO films. (c) XRD patterns of HZO films grown on (110)-oriented LAO with LSMO buffer layers of 66, 33, and 20 nm thickness, and (d) PFM phase versus tip voltage loops of these HZO films. *Source*: [59] / Springer Nature / CC BY 4.0.

approximately 30.2°, corresponding to the (111) crystal plane of the orthorhombic ferroelectric phase. The PFM phase versus tip bias loop directly correlates with the sign of piezoelectricity, a phenomenon previously confirmed in PVDF, a well-established negative piezoelectric material.[55,60] As shown in Figure 14.11b, the clockwise rotation of the phase versus tip bias loops proves positive d_{33}^* values in HZO films grown on LSAT and STO substrates.[59] Conversely, the HZO film deposited on the LAO substrate demonstrates an anticlockwise phase versus tip voltage loop, indicative of a negative d_{33}^*.[57]

As illustrated in Figure 14.11c,d, the buffer LSMO films, with thicknesses ranging from 20 to 66 nm, impart varying epitaxial strains to the HZO film, thereby influencing the positive and negative signs of d_{33}^*.[59] A series of HZO (10 nm)/LSMO epitaxial films were synthesized on LAO substrates by the PLD method. Since the thickness of the LSMO films ranges from 20 to 66 nm, these films exhibit varying degrees of strain relaxation. XRD patterns of these samples are presented in Figure 14.11c.[59] The angle change of the (110) diffraction peak with increasing thickness suggests gradual strain relaxation in these LSMO buffer layers. This approach facilitates the continuous adjustment of epitaxial strain within the HZO film, ranging from compressive to tensile values. Figure 14.11d illustrates the PFM phase versus tip bias loops for HZO films deposited on LSMO (20 nm), LSMO (33 nm), and LSMO (66 nm), respectively.[59] The HZO films on both LSMO (33 nm) and LSMO (20 nm) exhibit negative d_{33}^* values. In contrast, for the 66 nm LSMO buffer, d_{33}^* becomes positive as the strains in LSMO are fully relaxed. According to the DFT calculations, the remarkable piezoelectric properties – both positive and negative d_{33}^* – arise from the competitive interactions among the four types of oxygen atoms in response to epitaxial strain in these single-crystal HZO films. The positive and negative d_{33}^* of three single-crystal HZO films on different substrates align well with theoretical calculations.

14.5 Applications of $Hf_{1-x}M_xO_2$ Piezoelectric Films

$Hf_{1-x}M_xO_2$ films, as novel ferroelectric and piezoelectric materials, hold great potential for applications in piezoelectric resonators, actuators, sensors, and energy harvesters. The piezoelectric coefficient d_{33} of $Hf_{1-x}M_xO_2$ films significantly surpasses that of AlN and GaN ultrathin films, which are widely employed in various actuators and sensors for MEMS and NEMS. Consequently, actuators incorporating $Hf_{1-x}M_xO_2$ films are expected to exhibit superior performance characteristics.

The ultrathin piezoelectric film serves as the most essential functional layer in a wide array of MEMS and NEMS applications,[61] owing to the piezoelectric effect's inherent electromechanical nature, which induces robust electromechanical behavior. Equation (14.8) delineates the relationship between spontaneous polarization $\left(P_{\rm j}(\rm E)\right)$ and polarization-induced strain $(\sigma_{\rm i})$ as:[14]

$$\sigma_{\rm i} = Q_{\rm ij}\left(P_{\rm j}(\rm E)\right)^2 \tag{14.8}$$

Where $P_{\rm j}(\rm E)$ represents the ferroelectric polarization and $Q_{\rm ij}$ denotes the electrostrictive coefficients of ferroelectric films. The resultant $\sigma_{\rm i}$ gives rise to the electromechanical phenomenon, and the $d_{\rm ij}$ of a ferroelectric can be more comprehensively expressed through the parameters $Q_{\rm ij}$ and $P_{\rm j}$:[14]

$$d_{\rm ij} = 2Q_{\rm ij}P_{\rm j}\left(\varepsilon_{\rm ij} + \varepsilon_{\rm ij}{}'E_{\rm j} + \varepsilon_{\rm ij}{}''E_{\rm j}^2 + \ldots\right) \tag{14.9}$$

$$\text{Linear } d_{\rm ij} \approx 2Q_{\rm ij}P_{\rm j}\varepsilon_{\rm ij} \tag{14.10}$$

Here, $\varepsilon_{\rm ij}$ denotes the dielectric permittivity, $d_{\rm ij}$ is direct proportional to $P_{\rm j}$ that can be modulated by $E_{\rm j}$, while $\varepsilon_{\rm ij}{}'$ and $\varepsilon_{\rm ij}{}''$ represent the first-order derivative and second-order derivative of $\varepsilon_{\rm ij}$ with respect to $E_{\rm j}$.

MEMS/NEMS piezoelectric resonators belong to electromechanical devices that facilitate reciprocal conversion between mechanical and electrical energies, especially at a resonance frequency ($f_{\rm re}$). This conversion is achieved through the bidirectional coupling of the electric field and stress within a piezoelectric functional layer. Furthermore, the mechanical energy of the resonator and the electrical energy supplied through metallic electrodes oscillate between potential and kinetic forms during each half-cycle of vibration. Consequently, the comprehensive properties of a resonant system are determined by energy conversion mechanisms. The quality factor (Q_M) is generally defined as $Q_M = 2\pi B/A$ to estimate the energy loss in a resonator, where A represents energy dissipated per cycle and B denotes peak energy stored.[61] Integrated NEMS enables the miniaturization of sensors and actuators, resulting in nanoscale devices with ultrahigh resolution. These NEMS devices potentially exhibit high Q_M and f_{re}, providing wideband spectral processors, stable frequency references, and high-resolution resonant sensors in the centimeter-wave and millimeter-wave regimes.[62] The sizes of mechanical resonators need to be reduced to the nanoscale to operate effectively at these high-frequency regimes. This represents a reduction of nearly two orders of magnitude compared to the currently available MEMS devices.

Ultrasound wave generators and resonators represent the most traditional MEMS devices that were initially employed in various industries. The primary thin film materials under investigation included zinc oxide (ZnO) and aluminum nitride (AlN) (Table 14.1) during the 1970s and 1980s. PZT was introduced in the 1990s as a stronger piezoelectric functional layer for actuators. For higher-frequency applications, such as passband filters for GHz telecommunication, AlN and ZnO with wurtzite structures remained superior due to their significantly higher mechanical quality factors compared to PZT. Additionally, AlN and ZnO films can be prepared more readily than $LiNbO_3$ film.[61] Currently, the ultrathin $Hf_{1-x}M_xO_2$ ferroelectric film is regarded as one of the most suitable materials for NEMS devices.[63] This preference is attributed to three main factors. Firstly, the ALD growth conditions of $Hf_{1-x}M_xO_2$ super films are compatible with current CMOS techniques. Secondly, large-size $Hf_{1-x}M_xO_2$ films with thicknesses of several nanometers exhibit high piezoelectric coefficients (d_{33} and d_{33}^*) that significantly surpass those of AlN, GaN, and ZnO films. Furthermore, the Young's modulus of free-standing ferroelectric HZO films is measured using nondestructive optical transduction, and it is as high as ~320 GPa.[64]

As shown in Figure 14.12, two NEMS resonators have been developed through the integration of HZO films onto Si substrates and AlN-on-silicon membranes.[14] The fabrication process involves several steps. Initially, TiN/HZO/TiN films are grown using the ALD technique and then post-annealed at 500°C for 20 seconds, with the capping TiN electrode promoting the ferroelectric orthorhombic phase. Subsequently, a ~20 nm-thick Pt film is sputtered onto the capping electrode to serve as electrodes for the two-port resonators. Finally, the free-standing devices and electrical pads are set up for both excitation and sensing signals. Figure 14.12a shows the NEMS resonator with the TiN (10 nm)/HZO (10 nm)/TiN (30 nm)/Si (70 nm)/SiO_2 structure on Si substrate. The HZO film, fully polarized by 3.5 V, can produce linear electromechanical transduction under a low AC electric field. This HZO film is excited by high-frequency electric fields, and the resonance frequency of 339 kHz is confirmed by the use of a holographic optical probe. When the resonators work at room temperature in an air environment, the Q_M attains a significant value of 250.

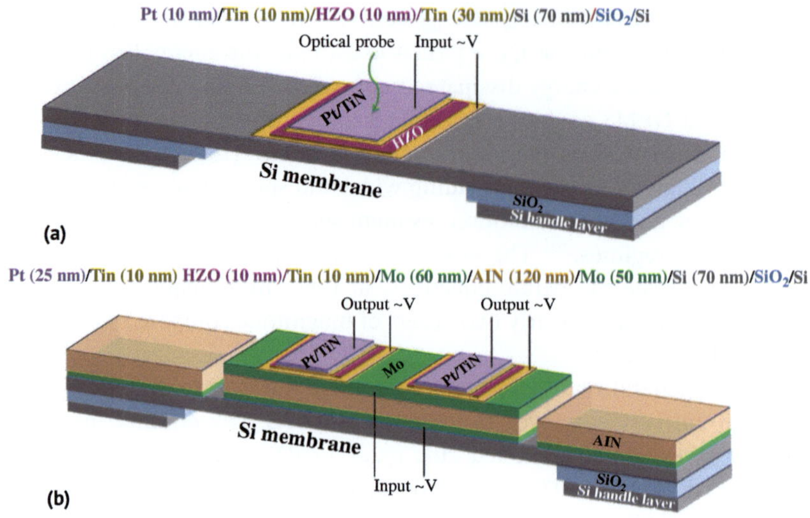

Figure 14.12 Sketches of two NEMS resonators with HZO films integrated on (a) Si and (b) AlN-on-Si membranes. HZO and AlN are piezoelectric films, while TiN, Pt/TiN, and Mo serve as electrodes. *Source*: Adapted from [14].

Figure 14.12b depicts a NEMS resonator featuring multilayer structure consisting of the following layers: 10 nm Pt, 10 nm TiN, 10 nm HZO, 10 nm TiN, 60 nm Mo, 120 nm AlN, 50 nm Mo, 70 nm Si, all on a SiO_2/Si substrate.[14] In this structure, the HZO and AlN films serve as the piezoelectric functional layers for the Pt/TiN/HZO/TiN capacitor and the Mo/AlN/Mo capacitor, respectively. The combination of RTA with the thermal strain induced by TiN effectively stabilized the HZO ferroelectric phase. Furthermore, high-voltage pulses were employed to activate the HZO film through the wake-up effect, thereby enhancing its ferroelectric and piezoelectric properties. During this resonators operating in centimeter-wave regimes, alternating current (AC) electric fields applied to the AlN film induce dynamic mechanical strain through the converse piezoelectric effect. Meanwhile, the HZO piezoelectric film generates dynamic AC electric signals under mechanical strain through the direct piezoelectric effect, enabling electrical detection. When the resonator operates in lateral-extensional modes in air, it exhibits a high Q_M of 530 and a high resonance frequency f_{re} of 413 MHz. When operating in thickness-extensional modes in air, the f_{re} is determined by its thickness, achieving an impressive 12.8 GHz with a quality factor of 310. These results yield a quality factor product (f_{re} Q_M) as high as 4×10^{12}. These resonators can be used in high-resolution resonant sensors, wideband spectral processors, and stable frequency references.

Furthermore, another HZO-based NEMS resonator interlayer achieves a super high Q_M of 171,200. Figure 14.13a,b illustrates the sketch of this nano-resonator and its measurement apparatus.[21] There are 9 nm-thick HZO layers intercalated with 1 nm-thick Al_2O_3 layers, which stabilize the metastable ferroelectric phase of HZO through strain. These multilayer films were fabricated using the ALD method, and

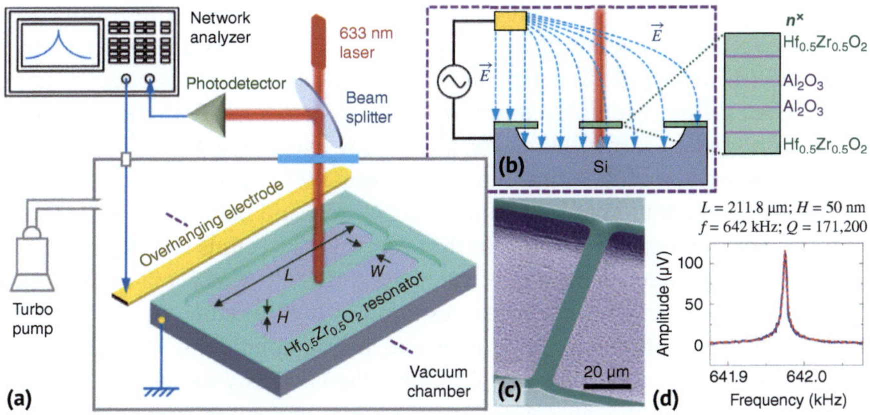

Figure 14.13 (a) Sketch and (b) cross-sectional view of HZO resonators with noncontact electrode and optical detection. (c) SEM image of a doubly clamped HZO resonator. (d) A typical measured resonance spectrum of the HZO resonator. *Source*: [21] / with permission of American Chemical Society.

they underwent post-annealing in nitrogen at 500°C for a duration of 20 seconds. Notably, the multilayers exhibit significantly higher ferroelectric polarization and enhanced piezoelectric properties compared to pure HZO films of equivalent thickness. Figure 14.13c exhibits the image of the HZO film from a nano-resonator, captured using scanning electron microscopy.

The resonator was characterized through a combination of optical interferometry detection and electromechanical motion excitation to achieve a high Q_M without external damping effect. The non-contacting electrode is hundreds of micrometers from the HZO film. The upper metal electrode, strategically positioned away from the HZO film, effectively prevents any obstruction of the detection laser. This arrangement minimizes contact between the electrode and the HZO film. An AC field applied to the HZO film through electrodes induces an oscillating mechanical strain, thereby facilitating the lightweight device motion in accordance with the converse piezoelectric effect. Furthermore, the high-frequency vibrations of the resonator were detected using interferometric techniques with a 633 nm laser. The vacuum environment reduces the damping effects during the vibration process and enhances the Q_M. The resonator exhibited excellent comprehensive performance, including a high resonance frequency f_{re} of 642 kHz (Figure 14.13d), a Q_M as high as 171,200, and a super-high $f_{re}\,Q_M$ of 10^{11} Hz. These excellent parameters can be attributed to three key factors. Firstly, a uniform driving force was applied across the lightweight device without additional loading and lossy electrodes. Secondly, the high vacuum environment effectively mitigated air damping effects. Thirdly, laser interferometry was employed to achieve ultrasensitive motion detection.

Because ultrathin $Hf_{1-x}M_xO_2$ films demonstrate exceptional electromechanical coupling and energy density characteristics, they are essential materials for the fabrication of miniaturized energy transducers. In fact, significant advancements have been made in the development of many energy conversion devices that utilize ultrathin $Hf_{1-x}M_xO_2$ piezoelectric films in the past decade.

Table 14.1 presents a compilation of representative piezoelectric thin films in scientific research and all kinds of device applications. PZT films, characterized by large ferroelectric polarization, high pyroelectric coefficients, and high piezoelectric coefficients, are primarily employed in micron-thick form for MEMS devices such as acceleration sensors, infrared imagers, and energy harvesting devices. However, the growth temperature of PZT films typically exceeds 600°C, presenting challenges in the industrial preparation of high-quality, large-area, and nano-thick PZT films. In contrast, large-area uniform and ultrathin AlN and GaN films can be prepared using CMOS-compatible technologies such as ALD.[6,61] These materials lack ferroelectric and pyroelectric properties, resulting in highly stable d_{33} values that are not significantly influenced by growth and poling conditions. While widely used in various MEMS and NEMS applications, the small d_{33} values of AlN and GaN films severely limit the piezoelectric performance of devices. Recent scientific research has uncovered numerous two-dimensional piezoelectric films, including MoS, $MoSe_2$, SnSe, and GeSe.[1,65] Some of these two-dimensional materials exhibit excellent piezoelectric and semiconductor properties simultaneously, offering immense potential for application in various optoelectronic devices. While one or more performance parameters of these new materials are outstanding, further validation is needed for the overall performance of the materials and devices.

$Hf_{1-x}M_xO_2$ piezoelectric films offer several distinct advantages compared to the piezoelectric films. Firstly, the ALD process for preparing ultrathin and large-area $Hf_{1-x}M_xO_2$ piezoelectric films is fully compatible with existing CMOS fabrication techniques. In fact, both HfO_2 and $Hf_{1-x}M_xO_2$ films are employed in many commercial chips. Secondly, the d_{33} value of $Hf_{1-x}M_xO_2$ piezoelectric films significantly surpasses that of GaN and AlN materials commonly used in MEMS and NEMS applications. Thirdly, research on $Hf_{1-x}M_xO_2$ piezoelectric properties continues to advance rapidly, and substantial improvements in piezoelectric properties are anticipated soon. Lastly, ultrathin $Hf_{1-x}M_xO_2$ films have been intensively studied as the piezoelectric functional layer for many NEMS/MEMS devices, resulting from their high energy density and electromechanical coupling properties.

14.6 Summary and Perspective

The crystal structures, piezoelectric properties, and typical applications of ultrathin $Hf_{1-x}M_xO_2$ films are summarized in this chapter. The $Hf_{1-x}M_xO_2$ thin films include several different crystal structures, though there is only a straightforward molecular formula. The monoclinic/orthorhombic structures are the ground/metastable state. Additionally, the polycrystalline $Hf_{1-x}M_xO_2$ films on silicon substrate exhibit a multiphase structure. To achieve the metastable orthorhombic structure and enhance both ferroelectricity and piezoelectricity in $Hf_{1-x}M_xO_2$ films, various strategies are thoughtfully employed. These include metal element doping, the use of specialized electrodes, the application of strain, and the implementation of multilayer structures. Suitable metal doping elements – including Zr, La, Si, Y, Al, Ta, and Sr – are incorporated into the $Hf_{1-x}M_xO_2$ matrix. The thickness is often reduced to nanometer scales to amplify interfacial effects of these $Hf_{1-x}M_xO_2$ films; this reduction can adjust interfacial strain from the electrode or the buffered layer of ferroelectric capacitors.

The d_{33} (pC/N) of these $Hf_{1-x}M_xO_2$ piezoelectric films is commonly characterized using charge integration methods and the ultrathin HZO films achieve d_{33} values as high as 37.8 pC/N. Despite it is challengeable to accurately measure d_{33}^* of ultrathin $Hf_{1-x}M_xO_2$ film, PFM provides an approximate estimation of d_{33}^* in local regions at the nanometer scale. Most conventional piezoelectric films, such as PZT, typically exhibit positive values for both d_{33} and d_{33}^*. However, certain ultrathin HZO ferroelectric films exhibit a distinct transition from positive d_{33}^* to negative d_{33}^* when subjected to epitaxial strain or during repeated polarization switching cycles. This novel d_{33}^* evolution stems from subtle transitions in the polar orthorhombic structure, which are influenced by mechanical strain, vacancies, and electric fields. Such characteristics introduce unprecedented flexibility for innovative designs in piezoelectric devices.

The ultrathin $Hf_{1-x}M_xO_2$ films serve as piezoelectric functional layers in various NEMS and MEMS devices. Especially, resonators incorporating HZO piezoelectric films on silicon wafers have been successfully prepared. The resonance frequencies of the resonator range from 100 kHz to 10 GHz, all while preserving a high Q_M. This merit is beneficial for the development of faster communication technologies in centimeter-wave and millimeter-wave ranges. The investigation of $Hf_{1-x}M_xO_2$ ferroelectric films has been conducted for only a few decades, and it is expected that four significant advancements in their applications will emerge in the near future.

At first, various novel physical phenomena of $Hf_{1-x}M_xO_2$ ultrathin films may be uncovered when the thickness approaches that of typical two-dimensional semiconductor materials. Furthermore, the interplay between negative piezoelectric coefficients and negative capacitance may give rise to novel physical behaviors.

Secondly, the existence of multiple structural phases may facilitate the design of innovative piezoelectric morphotropic phase boundaries (MPBs) in $Hf_{1-x}M_xO_2$ ultrathin films. MPBs are possible to largely improve d_{33} and d_{33}^*. These films show promise for advancing transducers, resonators, filters, microphones, switches, accelerometers, and other novel energy harvesters.

Thirdly, the g_{33} coefficient of $Hf_{1-x}M_xO_2$ piezoelectric films exhibited by these films surpasses that found in most conventional piezoelectric thin films. Consequently, they have the potential to serve as the piezoelectric functional layer for high-performance MEMS/NEMS sensors.

Fourthly, NEMS and MEMS with $Hf_{1-x}M_xO_2$ as a piezoelectric functional layer can be integrated with ICs. The integration of NEMS/MEMS with logic and memory chips enhances cost efficiency, enables more compact designs, and promotes faster speeds along with higher performance. Furthermore, these MEMS/NEMS can be fabricated in conjunction with electronic circuits on the same silicon substrate. This integration facilitates the development of system on chip (SoC).

Acknowledgment

The author thanks National Key Research and Development Program of China (2024YFA1208602) and the National Natural Science Foundation of China (62374090) for their support.

References

1 Li, B., Xie, Z., Liu, H. et al. (2023). A review of ultrathin piezoelectric films. *Materials* 16: 3107.
2 Gao, W., Brennan, R., Hu, Y. et al. (2018). Energy transduction ferroic materials. *Materials Today* 21: 771–784.
3 Mishra, S., Unnikrishnan, L., Nayak, S. K. et al. (2019). Advances in piezoelectric polymer composites for energy harvesting applications: A systematic review. *Macromolecular Materials and Engineering* 304: 1800463.
4 Covaci, C. and Gontean, A. (2020). Piezoelectric energy harvesting solutions: A review. *Sensors* 20: 3512.
5 Meitzler, A. H., Berlincourt, D., Welsh, F. S. et al. (1988). *IEEE Standard on Piezoelectricity*. IEEE.
6 Haider, S. T., Shah, M. A., Lee, D.-G. et al. (2023). A review of the recent applications of aluminum nitride-based piezoelectric devices. *IEEE Access* 11: 58779–58795.
7 Manbachi, A. and Cobbold, R. S. (2011). Development and application of piezoelectric materials for ultrasound generation and detection. *Ultrasound* 19: 187–196.
8 Xu, Z. and Yuan, Y. J. (2018). Implementation of guiding layers of surface acoustic wave devices: A review. *Biosensors and Bioelectronics* 99: 500–512.
9 Li, Y., Goei, R., Ong, A. J. et al. (2024). Atomic layer deposition of piezoelectric materials: A timely review. *Materials Today Energy* 39: 101457.
10 Böscke, T., Teichert, S., Bräuhaus, D. et al. (2011). Phase transitions in ferroelectric silicon doped hafnium oxide. *Applied Physics Letters* 99: 112904.
11 Hyuk Park, M., Joon Kim, H., Jin Kim, Y. et al. (2013). Evolution of phases and ferroelectric properties of thin $Hf_{0.5}Zr_{0.5}O_2$ films according to the thickness and annealing temperature. *Applied Physics Letters* 102: 242905.
12 Park, M. H., Lee, Y. H., Kim, H. J. et al. (2015). Ferroelectricity and antiferroelectricity of doped thin HfO_2-based films. *Advanced Materials* 27: 1811–1831.
13 Park, M. H., Kim, H. J., Kim, Y. J. et al. (2015). Toward a multifunctional monolithic device based on pyroelectricity and the electrocaloric effect of thin antiferroelectric $Hf_xZr_{1-x}O_2$ films. *Nano Energy* 12: 131–140.
14 Ghatge, M., Walters, G., Nishida, T. et al. (2019). An ultrathin integrated nanoelectromechanical transducer based on hafnium zirconium oxide. *Nature Electronics* 2: 506–512.

15 Mart, C., Kämpfe, T., Hoffmann, R. et al. (2020). Piezoelectric response of polycrystalline silicon-doped hafnium oxide thin films determined by rapid temperature cycles. *Advanced Electronic Materials* 6: 1901015.

16 Cheema, S. S., Kwon, D., Shanker, N. et al. (2020). Enhanced ferroelectricity in ultrathin films grown directly on silicon. *Nature* 580: 478–482.

17 Narayan, B., Malhotra, J. S., Pandey, R. et al. (2018). Electrostrain in excess of 1% in polycrystalline piezoelectrics. *Nature Materials* 17: 427–431.

18 Hajra, S., Vivekananthan, V., Sahu, M. et al. (2021). Triboelectric nanogenerator using multiferroic materials: An approach for energy harvesting and self-powered magnetic field detection. *Nano Energy* 85: 105964.

19 Liu, N., Zhang, X., Ding, Y. et al. (2023). Strong piezoelectricity of the nm-thick flexible $Hf_{0.5}Zr_{0.5}O_2$ ferroelectric film. *Journal of Alloys and Compounds* 968: 172083.

20 Yong, H., Tang, W., Xu, X. et al. (2025). Structure, properties and application of HfO_2-based piezoelectric films. *Nanoscale* 17: 19613–19624.

21 Zheng, X.-Q, Tharpe, T., Enamul Hoque Yousuf, S. et al. (2022). High quality factors in superlattice ferroelectric $Hf_{0.5}Zr_{0.5}O_2$ nanoelectromechanical resonators. *ACS Applied Materials & Interfaces* 14: 36807–36814.

22 Chen, H., Liu, Y., Yan, M. et al. (2022). Enhanced energy density in sandwich-structured P (VDF-HFP) nanocomposites containing $Hf_{0.5}Zr_{0.5}O_2$ nanofibers. *Chemical Engineering Journal* 436: 131123.

23 Lu, H., Kim, D.-J., Aramberri, H. et al. (2024). Electrically induced cancellation and inversion of piezoelectricity in ferroelectric $Hf_{0.5}Zr_{0.5}O_2$. *Nature Communications* 15: 860.

24 Guberna, E., Margolin, I., Kalika, E. et al. (2023). Giant electromechanical effect in piezoelectric nanomembranes based on $Hf_{0.5}Zr_{0.5}O_2$. *ACS Applied Materials & Interfaces* 16: 975–984.

25 Chouprik, A., Kirtaev, R., Korostylev, E. et al. (2022). Nanoscale doping and its impact on the ferroelectric and piezoelectric properties of $Hf_{0.5}Zr_{0.5}O_2$. *Nanomaterials* 12: 1483.

26 Kirbach, S., Lederer, M., Eßlinger, S. et al. (2021). Doping concentration dependent piezoelectric behavior of Si: HfO_2 thin-films. *Applied Physics Letters* 118: 012904.

27 Falkowski, M. and Kersch, A. (2020). Optimizing the piezoelectric strain in ZrO_2- and HfO_2-based incipient ferroelectrics for thin-film applications: An ab initio dopant screening study. *ACS Applied Materials & Interfaces* 12: 32915–32924.

28 Shimura, R., Mimura, T., Tateyama, A. et al. (2021). Preparation of 1 μm thick Y-doped HfO_2 ferroelectric films on (111) $Pt/TiO_x/SiO_2$/(001) Si substrates by a sputtering method and their ferroelectric and piezoelectric properties. *Japanese Journal of Applied Physics* 60: 031009.

29 Shimura, R., Mimura, T., Tateyama, A. et al. (2022). No-heating deposition of 1-μm-thick Y-doped HfO_2 ferroelectric films with good ferroelectric and piezoelectric properties by radio frequency magnetron sputtering method. *Physica Status Solidi (RRL) – Rapid Research Letters* 16: 2100574.

30 Shiraishi, T., Konno, T. J., and Funakubo, H. (2022). Ferroelectric and piezoelectric properties of 100 nm-thick CeO_2-HfO_2 epitaxial films. *Applied Physics Letters* 120: 132901.

31 Lueng, C., Chan, H. L., Surya, C. et al. (2000). Piezoelectric coefficient of aluminum nitride and gallium nitride. *Journal of Applied Physics* 88: 5360–5363.

32 Stoeckel, C., Kaufmann, C., Hahn, R. et al. (2014). Pulsed dc magnetron sputtered piezoelectric thin film aluminum nitride–technology and piezoelectric properties. *Journal of Applied Physics* 116: 034102.

33 Kao, K.-S., Chung, C.-J., Chen, Y.-C. et al. (2004). The influence of varied sputtering condition on piezoelectric coefficients of AlN thin films. In: *14th IEEE International Symposium on Applications of Ferroelectrics*. ISAF-04 2004. IEEE. pp. 181–184.

34 Lueng, C., Chan, H., Surya, C. et al. (1999). Piezoelectric coefficient of GaN measured by laser interferometry. *Journal of Non-Crystalline Solids* 254 (1–3): 123–127.

35 Guy, I., Muensit, S., and Goldys, E. (1999). Extensional piezoelectric coefficients of gallium nitride and aluminum nitride. *Applied Physics Letters* 75 (26): 4133–4135.

36 Ghosh, M. and Rao, M. G. (2013). Growth mechanism of ZnO nanostructures for ultra-high piezoelectric d_{33} coefficient. *Materials Express* 3 (4): 319–327.

37 Ghasemian, M. B., Zavabeti, A., Abbasi, R. et al. (2020). Ultra-thin lead oxide piezoelectric layers for reduced environmental contamination using a liquid metal-based process. *Journal of Materials Chemistry A* 8 (37): 19434–19443.

38 Baklanova, K. D., Solnyshkin, A. V., Kislova, I. L. et al. (2018). Pyroelectric properties and local piezoelectric response of lithium niobate thin films. *Physica Status Solidi (A)* 215 (5): 1700690.

39 Verma, A., Panayanthatta, N., Ichangi, A. et al. (2021). Interdependence of piezoelectric coefficient and film thickness in $LiTaO_3$ cantilevers. *Journal of the American Ceramic Society* 104 (5): 1966–1977.

40 Czekaj, D., Surowiak, Z., Bakirov, A. et al. (2001). Piezoelectric effect in RF sputtered ferroelectric thin films. *Journal of the European Ceramic Society* 21 (10–11): 1609–1613.

41 Drezner, Y., Nitzani, M., and Berger, S. (2005). Piezoelectric ultrathin $BaTiO_3$ films. *Applied Physics Letters* 86 (4): 042906.

42 Zhang, R., Ding, Y., Liu, N. et al. (2024). High macroscopic piezoelectric d_{33} of the nm-thick flexible PZT ferroelectric film. *Journal of Materials Science: Materials in Electronics* 35 (4): 298.

43 Tao, R., Shi, J., Rafiee, M. et al. (2022). Fused filament fabrication of PVDF films for piezoelectric sensing and energy harvesting applications. *Materials Advances* 3 (12): 4851–4860.

44 da Cunha Rodrigues, G., Zelenovskiy, P., Romanyuk, K. et al. (2015). Strong piezoelectricity in single-layer graphene deposited on SiO_2 grating substrates. *Nature Communications* 6 (1): 7572.

45 Lee, J. H., Park, J. Y., Cho, E. B. et al. (2017). Reliable piezoelectricity in bilayer WSe_2 for piezoelectric nanogenerators. *Advanced Materials* 29 (29): 1606667.

46 Wang, Y., Vu, L.-M., Lu, T. et al. (2020). Piezoelectric responses of mechanically exfoliated two-dimensional SnS_2 nanosheets. *ACS Applied Materials & Interfaces* 12 (46): 51662–51668.

47 Xue, F., Zhang, J., Hu, W. et al. (2018). Multidirection piezoelectricity in mono- and multilayered hexagonal α-In2Se_3. *ACS Nano* 12 (5): 4976–4983.

48 Tan, H., Song, T., Dix, N. et al. (2023). Vector piezoelectric response and ferroelectric domain formation in $Hf_{0.5}Zr_{0.5}O_2$ films. *Journal of Materials Chemistry C* 11 (22): 7219–7226.

49 Tan, H., Lyu, J., Sheng, Y. et al. (2023). A transversal approach to predict surface charge compensation in piezoelectric force microscopy. *Applied Surface Science* 607: 154991.

50 Lyu, J., Fina, I., Solanas, R. et al. (2018). Robust ferroelectricity in epitaxial $Hf_{1/2}Zr_{1/2}O_2$ thin films. *Applied Physics Letters* 113 (8): 082902.

51 Tan, H., Estandía, S., Sánchez, F. et al. (2023). Effects of doping, stress, and thickness on the piezoelectric response and its relation with polarization in ferroelectric HfO_2. *ACS Applied Electronic Materials* 5 (12): 6630–6639.

52 Celano, U., Gomez, A., Piedimonte, P. et al. (2020). Ferroelectricity in Si-doped hafnia: probing challenges in absence of screening charges. *Nanomaterials* 10 (8): 1576.

53 Song, T., Tan, H., Bachelet, R. et al. (2021). Impact of La concentration on ferroelectricity of La-doped HfO_2 epitaxial thin films. *ACS Applied Electronic Materials* 3 (11): 4809–4816.

54 Jing, H., Guan, C., and Zhu, H. (2024). Understanding and tuning negative longitudinal piezoelectricity in hafnia. *npj Computational Materials* 10 (1): 160.

55 Katsouras, I., Asadi, K., Li, M. et al. (2016). The negative piezoelectric effect of the ferroelectric polymer poly (vinylidene fluoride). *Nature Materials* 15 (1): 78–84.

56 You, L., Zhang, Y., Zhou, S. et al. (2019). Origin of giant negative piezoelectricity in a layered van der Waals ferroelectric. *Science Advances* 5 (4): eaav3780.

57 Dutta, S., Buragohain, P., Glinsek, S. et al. (2021). Piezoelectricity in hafnia. *Nature Communications* 12 (1): 7301.

58 Buragohain, P., Lu, H., Richter, C. et al. (2022). Quantification of the electromechanical measurements by piezoresponse force microscopy. *Advanced Materials* 34 (47): 2206237.

59 Cheng, H., Jiao, P., Wang, J. et al. (2024). Tunable and parabolic piezoelectricity in hafnia under epitaxial strain. *Nature Communications* 15 (1): 394.

60 Liu, J., Liu, S., Yang, J.-Y. et al. (2020). Electric auxetic effect in piezoelectrics. *Physical Review Letters* 125 (19): 197601.

61 Bhugra, H. and Piazza, G. (2017). *Piezoelectric MEMS Resonators*. Springer, Switzerland.

62 Ramezany, A. and Pourkamali, S. (2018). Ultrahigh frequency nanomechanical piezoresistive amplifiers for direct channel-selective receiver front-ends. *Nano Letters* 18 (4): 2551–2556.

63 Ghatge, M., Walters, G., Nishida, T. et al. (2020). A 30-nm thick integrated hafnium zirconium oxide nano-electro-mechanical membrane resonator. *Applied Physics Letters* 116 (4): 043501.

64 Tharpe, T., Zheng, X.-Q., Feng, P. X.-L. et al. (2021). Resolving mechanical properties and morphology evolution of free-standing ferroelectric $Hf_{0.5}Zr_{0.5}O_2$. *Advanced Engineering Materials* 23 (12): 2101221.

65 Zhang, Q., Zuo, S., Chen, P. et al. (2021). Piezotronics in two-dimensional materials. *InfoMat* 3 (9): 987–1007.

Index

Note: Page numbers in *italics* and **bold** refers to figures and tables respectively.